CPSS Power Electronics Series

This series comprises advanced textbooks, research monographs, professional books, and reference works covering different aspects of power electronics, such as Variable Frequency Power Supply, DC Power Supply, Magnetic Technology, New Energy Power Conversion, Electromagnetic Compatibility as well as Wireless Power Transfer Technology and Equipment. The series features leading Chinese scholars and researchers and publishes authored books as well as edited compilations. It aims to provide critical reviews of important subjects in the field, publish new discoveries and significant progress that has been made in development of applications and the advancement of principles, theories and designs, and report cutting-edge research and relevant technologies. The CPSS Power Electronics series has an editorial board with members from the China Power Supply Society and a consulting editor from Springer.

Readership: Research scientists in universities, research institutions and the industry, graduate students, and senior undergraduates.

More information about this series at http://www.springer.com/series/15422

Dong Jiang · Zewei Shen · Qiao Li · Jianan Chen ·
Zicheng Liu

Advanced Pulse-Width-Modulation: With Freedom to Optimize Power Electronics Converters

Springer

Dong Jiang
School of Electrical and Electronic Engineering
Huazhong University of Science and Technology
Wuhan, Hubei, China

Qiao Li
College of Electrical and Information Engineering
Hunan University
Changsha, Hunan, China

Zicheng Liu
School of Electrical and Electronic Engineering
Huazhong University of Science and Technology
Wuhan, Hubei, China

Zewei Shen
University of Electronic Science and Technology of China
Shenzhen, Guangdong, China

Jianan Chen
School of Electrical and Electronic Engineering
Huazhong University of Science and Technology
Wuhan, Hubei, China

ISSN 2520-8853 ISSN 2520-8861 (electronic)
CPSS Power Electronics Series
ISBN 978-981-33-4384-9 ISBN 978-981-33-4385-6 (eBook)
https://doi.org/10.1007/978-981-33-4385-6

This Springer imprint is published by the registered company Springer Nature Singapore Pte Ltd.
The registered company address is: 152 Beach Road, #21-01/04 Gateway East, Singapore 189721, Singapore

Foreword

The value of a technical book should lie in two kinds of aspects: one is to summarize the existing technologies in the field, and the other is to illustrate the recent technologies from the authors. Then, the readers can benefit from the background of the related area as well as the specific knowledge from the authors.

The book *Advanced Pulse-Width-Modulation: With Freedom to Optimize Power Electronics Converters* is trying to cover both of the two aspects. Pulse width modulation (PWM) for power electronics conversion has been developed for decades, and the fundamental knowledge and analysis tools are summarized in this book. This part can help the readers to review the classic PWM technologies.

But this book is much more than that. The core idea of this book is to understand the time-domain model of PWM and its impact on the system, and to optimize the performance by utilizing the control freedoms in the PWM. This part is based on the original contribution of the authors in the past 10 years.

Previous work on PWM optimization is mainly based on frequency-domain analysis, which is difficult for real-time implementation. For understanding the time-domain model of PWM, the optimization can be actuated together with the pulse generation. The power losses, switching ripple and Electromagnetic Interference (EMI) of the power electronics converters can be controlled and improved. Especially, there will be more freedom to optimize the impact of PWM on the common-mode voltage. This book has done a deep study on this topic.

The first author, Prof. Dong Jiang, was my Ph.D. student from 2007 to 2011. I was working together with him for the beginning study of the topics in this book. I understand the origin of this book well. He and his students have continued to work hard on this topic for more than 10 years. Their work has been recognized by scholars and engineers all over the world, with 50+ journals and conference papers and 10+ granted patents. Professor Jiang has presented this topic in four different IEEE conferences as tutorial. This book is the integration of the best and most important parts of their work.

I am sure that the scholars, students and engineers of power electronics will all enjoy this book, and by following the approach of this book, they can learn and benefit from it.

Knoxville, TN, USA Fred Wang

Preface

Pulse Width Modulation (PWM) is a concept together with the history of power electronics. When power electronics devices began to be used in the power electronics converter to obtain the controllable output voltage/current, PWM technology was utilized for output actuation.

Semiconductor devices can work in two kinds of modes: linear mode and saturated mode. In linear mode, output voltage/current can be linearly amplified, but the device will tolerate high voltage and current together and cause high power losses. In saturated mode, or switching mode, the output voltage/current is in on/off style. The power losses on the semiconductor devices can be significantly reduced, but the binary output characteristics might be different from the amplification target of continuous waveform. Power electronics devices are with the second mode: switching mode. Therefore, the capability of working with high power is together with the challenge of binary output characteristics.

Then, PWM technology was developed to deal with this challenge. Based on "electronics average" principle, power electronics converter with PWM can obtain high-frequency pulse series in output voltage or current, to "approach" the reference continuous voltage or current. The main purpose of PWM is to make the average effect of the output pulse voltage/current to be the same as the continuous reference voltage/current.

In the past few decades, two major groups of PWM technologies have been developed and applied in multiphase AC/DC and DC/AC power conversion. One is carrier-based PWM and the other is space vector PWM. Both of these two methods are based on average balance principle, and have been proved to be equivalent.

Frequent switching pulses characteristics of PWM bring many side effects for power electronics converters. There are mainly three side effects of PWM. First, the switching losses of power semiconductor devices will accumulate together with PWM. Second, the switching actions of PWM will generate voltage/current ripple in the converter. Third, the high-frequency component of PWM will introduce Electromagnetic Interference (EMI) in the converter, especially the Common-Mode (CM) EMI. These three side effects are very critical for the system performance and should be limited, which are discussed in Chap. 3 of this book.

To actively improve the side effects of PWM, the direct way is through the PWM itself. Conventional PWM is mainly aiming at the basic function: to obtain the reference, but without optimization or control of the side effects. Some freedoms have not been utilized, including switching frequency and pulse location. One exception is the "Random PWM" and its relatives, which are using random variation of PWM frequency to mitigate EMI. In statistical principle, they can reduce the EMI peak. However, they are lacking control of switching losses and voltage/current ripple because they are based on random effects and without understanding the model of side effects. To do real-time control of PWM freedoms, analysis and understanding of these side effects should be included.

Mathematical analysis tools have been developed in the past few decades for PWM, especially in frequency domain. Fourier analysis can derive the spectrum of voltage and current with PWM. To analyze more deeply, double Fourier analysis can identify the PWM's effect on switching harmonics and their side-bands. Frequency-domain analysis for PWM has become a very mature technology since the beginning of the twenty-first century.

However, there is an obvious problem for frequency-domain analysis for PWM: it can only analyze the effect of PWM but cannot help to do the real-time control of PWM, since it is a post-process tool. To do real-time control of PWM, there must be a real-time analysis tool. Among the three major side effects, voltage/current ripple is synchronous with PWM. Time-domain prediction of ripple is possible in power electronics converters. Then, with this real-time prediction tool, real-time control of PWM is possible. With this idea, the control freedoms of PWM, especially the switching frequency can be actively updated to control the voltage/current ripple, and optimize the other side effects as well. Therefore, time-domain analysis and prediction tool are the basis of the "advanced PWM" in this book. The concept of real-time prediction model is introduced in Chap. 4 of this book.

A typical example of this idea is the model predictive variable switching frequency PWM (VSFPWM) in this book. The basis of VSFPWM is the real-time current/voltage ripple prediction model introduced in Chap. 4 of this book. The control freedom is the switching cycle, or the carrier cycle, which is proportional to the ripple amplitude. Based on the prediction model, switching cycle can be controlled to adapt the ripple requirement in every cycle. Then the switching frequency is varied according to the prediction model; thus it not only realizes control of ripple, but also improves the switching losses and EMI. This concept is introduced in Chap. 5, with regular two-level voltage source converter as an example.

With the development of power electronics converter topology, the crucial switching ripple will be different and the freedoms for PWM will also be more than those of the regular two-level voltage source converter. For example, multilevel converter is with concern of capacitor voltage ripple which is caused by PWM current charging/discharging process. With the real-time prediction model for different kinds of ripple in different kinds of topologies, VSFPWM can be developed to control the corresponding ripple and optimize other parameters including power losses and EMI. In Chap. 6, many kinds of converter topologies are analyzed for VSFPWM, as the deeper study of Chap. 5.

For the influence of PWM on EMI, CM EMI is a special issue. The generation of CM voltage is the switching actions of all the phases. Also, the CM EMI conduction path is highly associated with the non-ideal phenomenon of the system, especially the parasitic capacitance to the ground. Reduction of CM EMI through advanced PWM technologies is discussed in Chap. 7. The evolution of converter topologies gives more freedom to optimize CM EMI; this book is mainly introducing the novel technologies of "zero CM PWM" for paralleled converters, and its application with different kinds of load.

Finally, the implementation of advanced PWM technologies in simulation and experiments is a very important issue, and is introduced in Chap. 8. MATLAB/SIMULINK software and Texas Instruments' DSP are used as target of application.

The authors have worked in this area for more than 10 years, with the purpose of controlling power electronics converters with better performance for industry motor drive and grid-tied power conversion. We have published more than 50 technical papers in this area, with several projects delivered to sponsor successfully. In the year 2016, Springer Press contacted us to publish a book in the Springer-CPSS Power Electronics series. We began to work on this book then. In the year of 2018, we published the first book on this topic in China Machine Press, with language in Chinese. But this book in Springer Press is more than an English version of the Chinese book. From 2018 to 2020, many new contributions have been made with this topic, and contained in this book. A clear roadmap has been sketched for the advanced PWM to optimize the performance of power electronics converter. We hope this book can be helpful for the scholars and engineers all around the world, and new technologies can be stimulated in industry applications.

The authors want to thank all the collaborators in the past 10 years for the technologies in this book. As the first author, I would like to thank Professor Fred Wang from University of Tennessee most. He was my Ph.D. advisor, giving me the topic of studying how to improve "Random PWM" more than 10 years ago. The original works of VSFPWM and zero-CM PWM are under his supervision. He wrote the foreword of this book with his continuous enthusiasm for this topic. Dr. Ruxi Wang and other colleagues in my Ph.D. period also gave me a lot of help for this topic. I would like to thank Dr. Vladimir Blasko and Dr. Brian McCabe also. When we were working together in United Technologies Research Center, they supported my study of advanced PWM. Since I joined Huazhong University of Science and Technology (HUST) in 2015, Prof. Ronghai Qu, Dr. Dawei Li, Dr. Tianjie Zou and other colleagues have given us much help especially for electrical machine support. The management groups of School of Electrical and Electronics Engineering (SEEE) and Department of Applied Electronics Engineering in HUST have given us much help to realize the related technologies in lab. I also want to thank Miss Hongyan Qu, for her help in the final checking of this book.

Thanks to all my students in the past 5 years for their contribution to this book. Three of them have made significant contributions for this book and their names are in the author list. Dr. Zewei Shen and my colleague Dr. Zicheng Liu are the main contributors on the zero-CM PWM technologies in Chap. 7. Dr. Qiao Li and

Mr. Jianan Chen contributed mainly on the VSFPWM technologies in Chaps. 4–6. Many other students have also contributed to this book, without their names in the author list. But I want to show special thanks to them, especially to Dr. Xun Han, Mr. Xuan Zhao, Mr. An Li and Mr. Ruodong Wang.

Just in the period of writing the preface of this book, I heard the news that Professor Thomas A. Lipo had just passed away. The pioneer monograph "Pulse Width Modulation for Power Converters: Principles and Practice" authored by Profs. Holmes and Lipo is a guideline and bible for us to pursue the knowledge in our book. I always thank the colleagues around the world for their contribution and influence on this book.

The technologies in this book have been with many research projects' funding. We want to thank the sponsors for these projects. Especially, thanks to the National Natural Science Foundation of China Project under grant U1866211.

I would like to thank my wife most for her love and support during the writing of this book.

During the last stage of revising this book, we suffered from the COVID-19 pandemic. All the authors, including me, were quarantined at home, especially in the city of Wuhan. The revising of this book is done together with the pandemic situation. This period of working at home gives us a chance to think deeply about this book. The purpose of technology is to make a better world for the mankind. We hope the researchers of medical technologies can find new ways to optimize the control of virus soon. When the people in the world can be united to fight against virus, the hope is always there. This book contains our best wishes to the people in the world who have suffered in the COVID-19 pandemic.

Wuhan, China Dong Jiang
May 2020

Contents

Abbreviations

ADC	Analog-to-digital converter
AFE	Active-front-end
ANPC	Active neutral point clamped
ASPMSM	Asymmetrical six-phase PMSM
AZSPWM	Active zero-state PWM
BJT	Bipolar junction transistor
CBPWM	Carrier-based PWM
CCVS	Current control voltage source
CI	Coupled inductor
CM	Common mode
CMC	Common-mode current
CMI	Common Mode Inductor
CMV	Common-mode voltage
CRP	Current ripple prediction
CSC	Current source converter
CSFPWM	Constant switching frequency PWM
CSI	Current source inverter
DM	Differential mode
DMCC	Differential-mode circulating current
DPWM	Discontinuous PWM
DPWM_Max	Maximum clamped DPWM
DPWM_Min	Minimum clamped DPWM
DPWM1	Minimum-loss DPWM
DSP	Digital signal processor
DTC	Dead time compensation
DTPM	Dual three-phase machines
EMC	Electromagnetic compatibility
EMF	Electromotive force
EMI	Electromagnetic interference
ETO	Emitter turn-off thyristor
FC	Flying capacitor
FC-MMC	Flying-capacitor MMC

FEA	Finite-element analysis
FET	Field-effect transistor
FPGA	Field programmable gate array
GaN	Gallium-nitrite
GTO	Gate turn-off thyristor
HIPWM	Harmonics injection PWM
HVDC	High-voltage direct current
IC	Integrated circuit
ICSF	Inductor current state-feedback
IGBT	Insulated-gate bipolar transistor
IGCT	Integrated gate-commutated thyristor
JFET	Junction field-effect transistor
LISN	Line impedance stabilization network
MMC	Modular multilevel converter
MOSFET	Metal–oxide–semiconductor field-effect transistor
MPC	Model predictive control
MPP	Model predicted PWM
NDPWM	Normal distributed PWM
NP	Neutral point
NPC	Neutral point clamped
NSPWM	Near-state PWM
OW	Open-winding
PC	personal computer
PD_PWM	Phase disposition PWM
PI	Proportional integral
PLL	Phase locking loop
PMSM	Permanent magnet synchronous machine
PS_SPWM	Phase-shift SPWM
PV	Photovoltaic
PWM	Pulse width modulation
RMS	Root mean square
RPWM	Random PWM
SCR	Silicon-controlled rectifier
SHE	Selective harmonic elimination
SiC	Silicon-carbide
SM	Submodule
SPWM	Sinusoidal PWM
SVPWM	Space vector PWM
THD	Total harmonics distortion
UDPWM	Unified distributed PWM
VSC	Voltage source converter
VSD	Variable speed drive
VSFPWM	Variable switching frequency PWM
VSFPWM1	VSFPWM for current ripple peak value control
VSFPWM2	VSFPWM for current ripple RMS value control

VSFPWM-NP	VSFPWM based on neutral point voltage balance
VSI	Voltage source inverter
WTHD	Weight-THD
ZCC	Zero-current clamping
ZCMV	Zero Common Mode Voltage
zero-CM VSFPWM	Zero-CM PWM with variable switching frequency
ZSC	Zero-sequence current
ZSCC	Zero-sequence circulating current
ZSV	Zero-sequence voltage

Chapter 1
Brief Introduction of Power Electronics and Pulse Width Modulation

In the past decades, microelectronics and power electronics technologies have been with significant progress. Microelectronics has brought the modern world to the information age. In comparison, the impact of power electronics on human civilization is still with great potential. Realization of electrical power conversion through switching of power semiconductor devices is the essence of power electronics, and the major approach of switching is through pulse width modulation (PWM). As the first chapter of this book, the background of power electronics and the progress of PWM technology will be introduced, as well as the opportunities and challenges for power electronics.

1.1 Power Electronics Devices and Power Electronics Converters

Human's effective utilization of electrical power began in the nineteenth century. After the finding of electromagnetic phenomenon by Faraday and the establishment of electromagnetic theory by Maxwell in the nineteenth century, the basis of modern electromagnetics and electrical engineering has been established. In the next century, two branches of industry have been developed based on electromagnetic theory: electrical power industry with electrical machines and power transmission/distribution facilities as representatives and the telecommunication industry with telephone and radio as representatives. These two branches of industry have significantly changed the world in the first half of twentieth century.

Since the mid-twentieth century, semiconductor technologies have been developing with rapid progress. The continuous and linear electrical theory was extended to the discrete and nonlinear area. Significant change has been made for both electrical power industry and telecommunication industry. In telecommunication area, a lot of

D. Jiang et al., *Advanced Pulse-Width-Modulation: With Freedom to Optimize Power Electronics Converters*, CPSS Power Electronics Series,
https://doi.org/10.1007/978-981-33-4385-6_1

electronics devices have been applied. Computer, integrated circuit (IC) and artificial intelligence have appeared. It can be concluded that the modern world has been pushed to the "zero and one" age by semiconductor technologies. The massive driving force of semiconductor technologies in telecommunication industry is because of the strong capability of signal processing. For electrical power industry, the object for processing is electrical power, not signal. It is the reason why the progress of semiconductor in electrical power is with a few years' delay than in telecommunication. However, in the recent few decades, important progress of semiconductor in electrical power industry has been obtained, that is power electronics.

Development of electrical power industry has been relying on the facilities, including electrical machine and transformer in the past few decades, to realize energy conversion between electrical energy and mechanical energy or electrical energy and electrical energy. This kind of conversion is continuous and linear. Application of semiconductor in power conversion begins from the application of power diode. The physical principle of power diode is similar to the regular diode with P-N junction. It can conduct current in positive direction and block current in the negative direction. Power diode can be used to transfer AC power to DC power as rectifier. However, with only power diode the circuit is not controllable and can only achieve simple function. In 1957, thyristor was invented by General Electric Company and the conventional power conversion style has been changed. Thyristor is a solid-state switch based on silicon (Si) material. It is with four-layer structure of P-N-P-N. It can achieve small conducting resistance and high impeding resistance, which can be controlled with gate. Then thyristor can be used for controllable rectifier to achieve AC-to-DC power conversion. It is the reason why thyristor is also called semiconductor-controlled rectifier (SCR). Then, thyristor has been used in DC-AC converter (inverter), AC-AC converter (varying frequency and voltage) and DC-DC converter (chopper). Because of the gate control capability, it can be with more flexible application in power converter and active control. The area of power electronics began to appear in that period, too.

Thyristor had ruled the area of power electronics for a few decades. But it has an obvious problem: it can be turned on by the gate signal, but it cannot be turned off by external gate signal. Turn-off of thyristor should be with reversed voltage in main circuit to reduce the current to zero. So, thyristor is a half-controlled power electronics device. To effectively turn-off the thyristor, external circuit is needed to assist. In early period of power electronics, assistant circuit for thyristor turn-off is a popular research area. Then, gate turn-off thyristor (GTO) was invented based on regular thyristor. It can be turned on and turned off by gate trigger, which is a fully controlled switch. GTO is a high-voltage, high-current power electronics device with active gate control capability. It has been widely used in railway traction and industrial motor drives in the late twentieth century.

The main problem for GTO is that it is with gate current trigger, which is not easy and the switching speed is also limited. Based on GTO structure, integrated gate driver has been developed in the past two decades and emitter turn-off thyristor (ETO) and integrated gate-commutated thyristor (IGCT) have been developed. The core of these kinds of devices is still thyristor, but with low-inductance integrated gate driver circuit. For the users, their gate drivers are easier and simpler than GTO. Switching speed is also increased. In recent years, IGCTs have been applied in many high-voltage high-power conversion areas and show their superior performance.

All these power electronics devices (thyristor, GTO, ETO, IGCT) are with the core of P-N-P-N structure. In the same period, another power semiconductor structure is the transistor structure. Transistor structure contains bipolar junction transistor (BJT) and field-effect transistor (FET). BJT is based on two kinds of PN junctions, including PNP structure and NPN structure, which can be controlled by the external circuit through injection of base current. FET is controlled by external electrical field to change its conducting mode. The most popular FET device is the metal–oxide–semiconductor field-effect transistor (MOSFET). Also, because of the fabrication process, junction field-effect transistor (JFET) is also a popular device in wide-band-gap semiconductors.

BJT used in power conversion is a fully controlled power device, which is controlled by gate current. It has strong voltage and current capability. The major problems are the high-power requirement for gate driver and slow switching speed. Power MOSFET device is also a fully controlled power device, controlled through gate voltage. It needs low gate driver power and can be with fast switching speed. However, MOSFET's major problem is bigger voltage drop in conducting mode and it is difficult to be used for high-power application. Combining the advantages of the two kinds of devices, insulated-gate bipolar transistor (IGBT) was invented in 1980s and became the major fully controlled power semiconductor device which is driven by gate voltage. IGBT is equivalent as BJT, which is driven by a MOSFET in the gate. It contains the fast switching characteristics of MOSFET and the high-power capability of BJT. In the recent few decades, IGBT has been widely used in many different power conversions including motor control, renewable energy conversion, grid connected converter and power system.

At the same time, materials for power electronics devices have been with new development in the past two decades. Novel wide-band-gap semiconductor materials, including silicon-carbide (SiC), gallium-nitrite (GaN) and diamond materials, have been applied in power semiconductor devices. Wide-band-gap semiconductor materials are with higher energy level difference and can be with better performance in conducting resistance, voltage blocking capability, switching speed and high temperature capability than conventional Si devices. Since the twenty-first century, major semiconductor companies have devoted to the research and development of SiC power electronics devices. 1200 V level SiC devices have appeared in market in 2010s. Materials' progress is an important propulsion for power electronics technologies.

The core idea of this book is studying the power electronics converter's pulse width modulation (PWM) technologies based on fully controlled fast switching devices. Power electronics device is the physics basis of the study. It is because of the power electronics device's own characteristics that PWM has its research meaning. In this book, IGBT and power diode are the major object for study. In fact, this book's content is also suitable for other fully-controlled power electronics devices.

Power electronics devices together with other assisting components and passive components can form power electronics converters to achieve controllable power conversion.

Based on the voltage blocking and current conducting capability of the power electronics devices, the switches in power electronics devices can be divided into two major combinations, shown in Fig. 1.1. Figure 1.1a shows the combination of one active switch with an anti-paralleled diode. In this structure, positive direction (A-B) current can be controlled through gate signal, and negative direction (B-A) current can conduct through the diode. When the gate is turned off, this switch can block the positive direction (A-B) voltage, but cannot block negative direction (B-A) voltage. So, in the voltage/current plane, this switch's effective area is shown in Fig. 1.2a. Figure 1.1b shows the structure of voltage bidirectional switch, which is an active switch with a series-connected diode. In this switch, positive direction (A-B) voltage can be blocked by turning off of active switch, and negative direction (B-A) voltage can be blocked by diode. Then, it is a voltage bidirectional switch. But because the diode is connected in series, it can only conduct positive direction (A-B) current. Its effective area is shown in Fig. 1.2b.

Besides the switch, another kind of important component in power electronics converter is passive component. In classic circuit theory, passive components include resistors, inductors and capacitors. Resistor is the simplest passive component, which satisfies the voltage/current linear relationship instantaneously. Capacitor and inductor are with integral and differential relationship in voltage/current curve, respectively. Figure 1.3 shows the symbols of inductor and capacitor. In inductor, energy is stored in magnetic field. The current slope is proportional to the voltage

Fig. 1.1 Two kinds of switches in power electronics converters: **a** current bidirectional switch, **b** voltage bidirectional switch

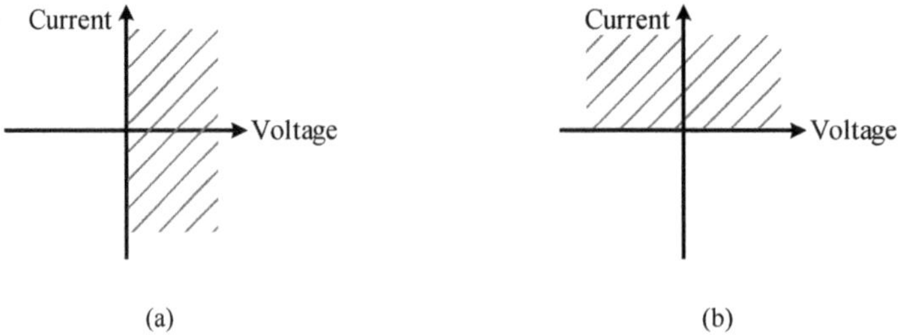

Fig. 1.2 Voltage/current plane of the two kinds of switches: **a** current bidirectional switch, **b** voltage bidirectional switch

Fig. 1.3 Two typical passive components' symbols: **a** inductor, **b** capacitor

Fig. 1.4 Two typical DC/DC converter topologies: **a** Buck converter, **b** Boost converter

on the inductor. Current in inductor cannot be changed suddenly and can be treated as a state variable in the system. In capacitor, energy is stored in electrical field. The voltage slope is proportional to the current flowing into the capacitor. Voltage in capacitor cannot be changed suddenly and can be treated as another state variable in the system.

With the two kinds of switches in Fig. 1.1 and passive components, most of the power electronics converters can be built up. Taking the simple DC/DC converter for example, two typical topologies are shown in Fig. 1.4. Figure 1.4a is the buck converter, with the load voltage in the capacitor as a state variable. Switch (S) is continuously switching to change the voltage on the inductor and the current in *L* is a stable value with ripple. Because of the on and off status in S, load voltage is reduced from the source voltage. Figure 1.4b shows the boost converter, also with load voltage in the capacitor as a state variable. Continuous switching of S changes the inductor voltage and makes the inductor current in two modes: charging the load and freewheeling. Load voltage will be increased from the source voltage.

DC/DC converter topologies in Fig. 1.4 are the basic topologies. Most of the power electronics converters can be derived from these basic topologies. In DC/DC converter, the control targets are DC values and they do not need the bidirectional capabilities for voltage or current in many cases. However, with application to the converter with AC variable (voltage or current), voltage or current will be bidirectional. Then, the switch in Fig. 1.1 which is with voltage or current bidirectional capability should be used. Figure 1.5 is the simple single-phase DC-AC converter, or inverter. Input part is DC voltage source, so it is also called voltage source inverter (VSI). Load inductor makes the load with current source mode: alternating current in continuous mode. Because the current flowing through the switch is with bidirectional characteristics, current bidirectional switch in Fig. 1.1a should be used. Then, with varying (sinusoidal) duty cycles in the switch, output side can be with AC characteristics. This topology can be extended to three-phase, shown in Fig. 1.6,

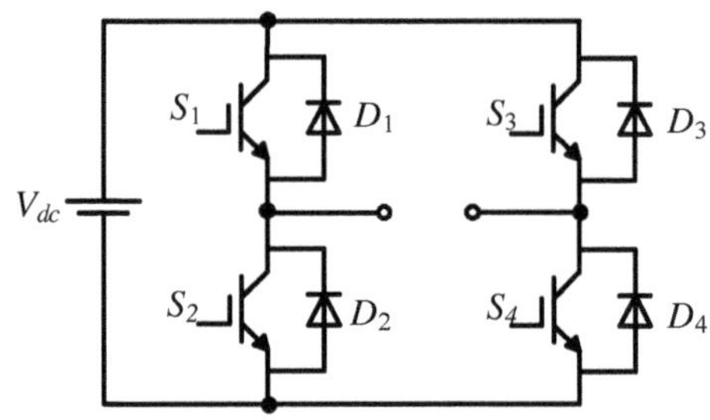

Fig. 1.5 Single-phase inverter structure

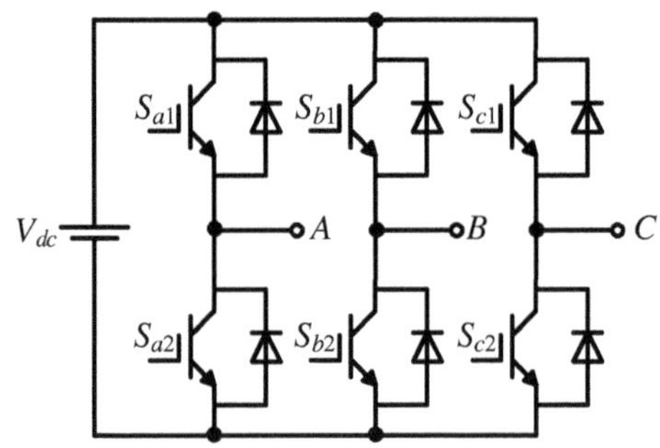

Fig. 1.6 Three-phase inverter structure

which is the most widely used topology in high-power application. This is also the main topology for this book.

In comparison with voltage source inverter, another typical topology is the current source inverter (CSI). In inverter, if the DC input is with current source characteristics, it is called CSI. Figure 1.7 is a three-phase CSI structure. The DC-link inductor makes the DC input side with current source characteristics and the DC side is with stable current input. In the meantime, the load side needs voltage source characteristics, so the AC side is with capacitors parallel to the load terminals. Because the DC input is a unidirectional current and the load is with alternating voltage, the device in the CSI should be with voltage bidirectional characteristics like Fig. 1.1b: one active switch

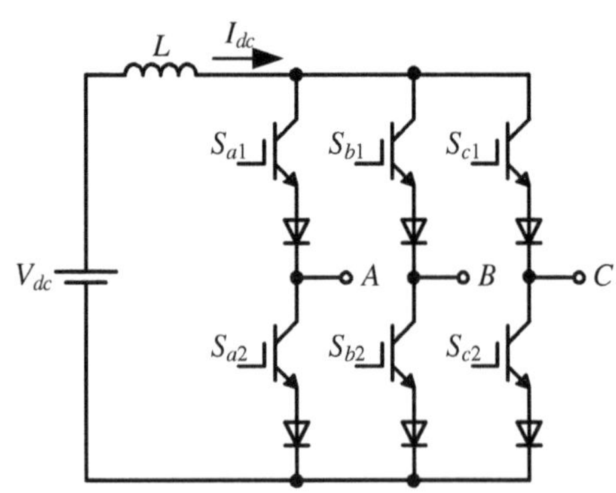

Fig. 1.7 Three-phase current source inverter structure

in series with a diode. In three-phase CSI, there are three phase-legs, each of which is with two voltage bidirectional switches.

Figures 1.6 and 1.7 are the basic three-phase inverter structures. In fact, when load and source are reversed, the same topology of Fig. 1.6 can be used as AC-DC converter, or rectifier. For the topology in Fig. 1.7, the reversing of source and load should be together with reversing of switch directions and to form current source rectifier. In the major three-phase power electronics topologies, including inverter and rectifier, the topologies are from Figs. 1.6 and 1.7. In real three-phase applications, voltage source converter (inverter or rectifier) is the most widely used topology and is also the major object of this book. In Fig. 1.6, AC side voltage can be back-electromotive force (EMF) of electrical machine, or grid voltage. This topology can represent the major applications of three-phase power conversion. In some applications, current source converter has some advantages. It is also introduced in this book as supplementary.

1.2 Introduction of Pulse Width Modulation

After the discussion of power electronics devices and converter topologies, the next step is to understand the control techniques for the converters. Figure 1.8 is a typical control system diagram. It is with the following process: the error between the reference value and the feedback value from sensor is sent to the controller, obtaining the reference actuation value through the controller. In this case, an actuator is needed to transfer the reference actuation value to physical actuation value to add on the control object for output. Without actuator, control system is only mathematical model and not with the real control capability. In power electronics converter, pulse width modulation is the actuator to transfer the reference actuation value to physical actuation value.

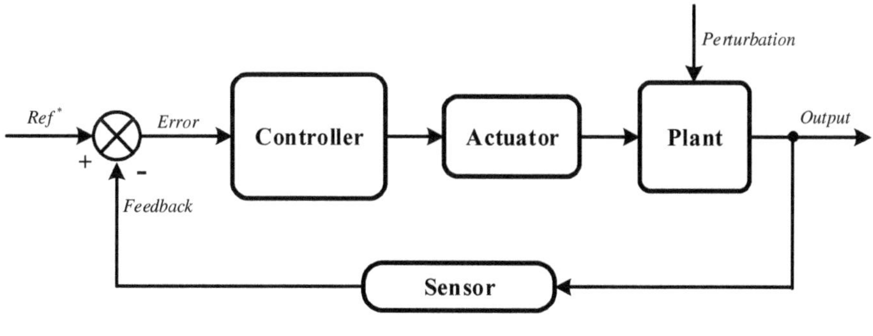

Fig. 1.8 A typical control system diagram

References [1, 2] have provided the medium and high-power converter system multilayer control structure which is published by IEEE as a standard, shown in Fig. 1.9. In power electronics converter system, control is divided into five layers.

1. The first layer is the system-level control. This layer is associated with top-level application, which is usually with time scale bigger than 10 ms, for example, electrical vehicle's running mode.
2. Under the system-level control is the application control, which is the actuation used to satisfy the system-level control, usually with time scale between 1 and 10 ms, for example, electrical motor speed control which is used to satisfy the vehicle running mode in the first layer.
3. The next layer is the power electronics converter control, which is the actuation of power converter to satisfy the application control target, usually with time scale between 10 μs and 1 ms, for example, motor controller's current control to satisfy the speed control requirement in the second layer.

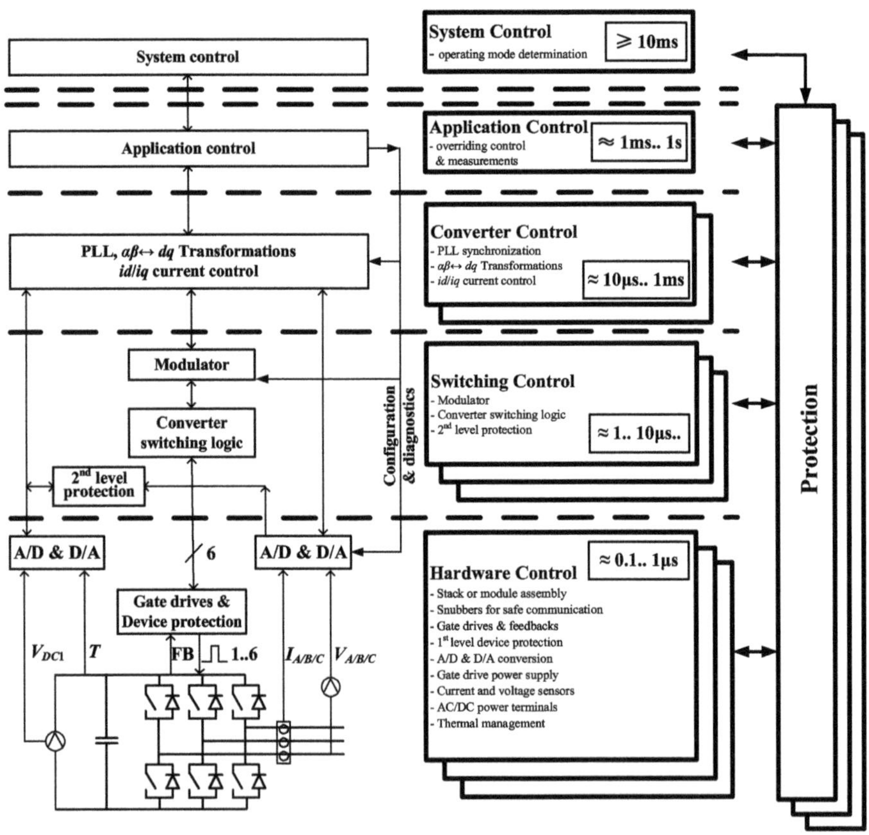

Fig. 1.9 Multilayer control structure of power electronics system [1, 2]

4. The next layer is the switch control. It can meet the converter control target of the third layer through active selection of switch combinations, with PWM as the representative and the time scale is usually between 1 and 10 μs.
5. The bottom layer is the hardware control. Under switch control, there is a process to transfer the digital switch logic to power electronics switching, for example, the gate driver for power electronics devices. It is the bottom layer control for the whole system, with time scale less than 1 μs.

From Fig. 1.9 it can be found that the power electronics converter's control is from top to bottom level with different functions and time scales. This book is focusing on the fourth layer: switch control. The essence of switch control layer is to arrange the switch combination and dwelling time, then pulse series is generated to approach the third-layer's control reference and is transmitted to the hardware in the fifth layer. Taking VSC as an example: in order to control its output current, the current controller is generating reference voltage. This reference voltage is a continuous value, but VSC cannot obtain continuous voltage output, it can only generate equivalent pulse series voltage. Then, the process of using pulse series voltage to approach the reference voltage is the switch control in the fourth layer. PWM is the most typical method for switch control.

PWM is the acronym for pulse width modulation. From the name, it is clear that the method is achieved through modulation of pulse width. It initially appeared in telecommunication area as one modulation method for radio signal, and was introduced to power electronics area later.

Figure 1.10 shows the basic principle of PWM with a single phase-leg in VSC. For VSC, the DC-link voltage is V_{dc}. With the DC bus midpoint as reference, the phase-leg output voltage V_x is switching between positive and negative half DC-link voltage, which can be realized by comparing the reference voltage V_{ref} and carrier in each switching cycle T_s. When the reference voltage is bigger than the carrier (triangle wave), the upper switch in the phase-leg is turned on and the bottom switch is turned off. Based on the geometrical relationship, the time T_0 for negative voltage ($-V_{dc}/2$) satisfies $T_0/T_s = (V_{dc}/2 - V_{ref})/V_{dc}$. Then, the average value of V_x in one switching cycle T_s is calculated shown in (1.1). It proves that the phase-leg output average voltage in one switching cycle is equal to the reference voltage. Based on this relationship, PWM can achieve the equivalent reference voltage.

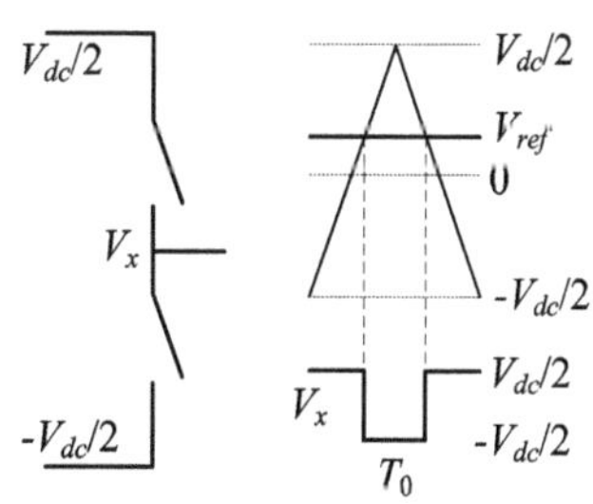

Fig. 1.10 PWM to achieve reference voltage in the phase-leg in VSC

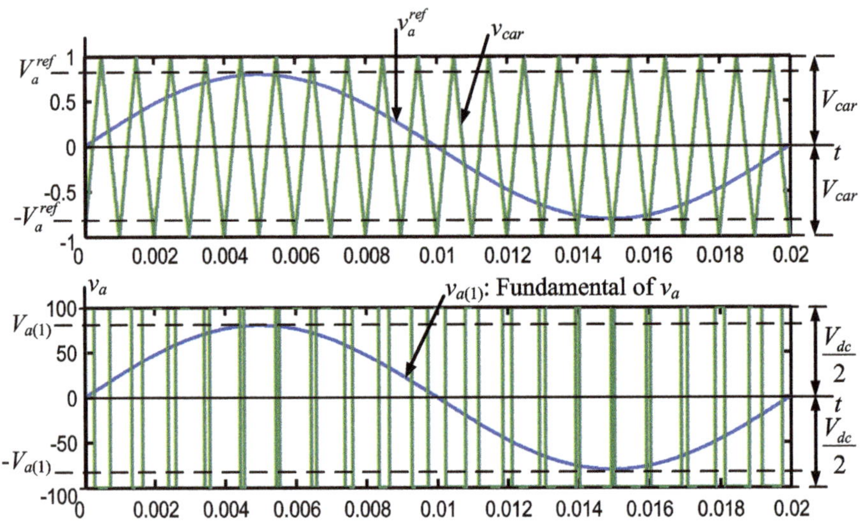

Fig. 1.11 Carrier-based PWM for AC reference voltage

$$V_x = \frac{1}{T_s} \times \left[(T_s - T_0) \times \frac{V_{dc}}{2} + T_0 \times (-\frac{V_{dc}}{2}) \right] = V_{ref} \tag{1.1}$$

For constant reference voltage, this carrier comparison method can approach the value. If the reference voltage is alternating, the duty cycle will also vary with the variable reference voltage. Figure 1.11 shows the AC reference voltage v_a^{ref} to compare with triangle carrier with amplitude of V_{car} to generate PWM. The generated pulse series v_a is with the same value of continuous reference voltage in each switching cycle. If the carrier frequency is high enough, the filtered result of v_a is equivalent to v_a^{ref}.

Besides the carrier comparison to approach the reference, another PWM method is through the space vector combination. Space vector PWM (SVPWM) is not approaching one voltage reference, but the reference voltage vector formed by several phase voltages. Taking three-phase VSC as an example, labeling "1" for positive voltage and "0" for negative voltage for each phase-leg, then there are $2^3 = 8$ combinations, defined as eight switch voltage vectors. There are two zero voltage vectors (000 and 111) in all eight voltage vectors which will not generate effective output voltage. Other six voltage vectors form a regular hexagon with the length of each vector being V_c. The space voltage plane is divided into six zones, shown in Fig. 1.12. The reference voltage V_{ref} can be combined by the two adjacent switch voltage vectors in the located zone. In each switching cycle T_s, two switch voltage vector dwelling time t_1 and t_2 can be calculated by triangle function, shown in (1.2). Besides t_1 and t_2, the rest of the time t_0 in the switching cycle will be compensated by zero voltage vectors. Then, the volt-seconds balance is obtained in the whole switching cycle. For three-phase AC reference voltage, it will be a rotating voltage vector in the

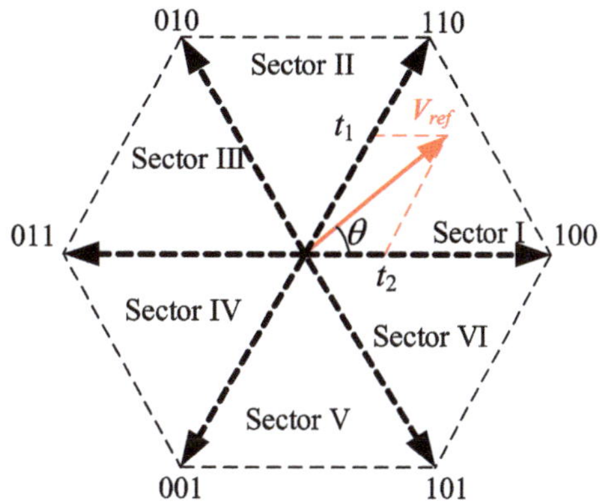

Fig. 1.12 Space vector PWM in three-phase VSC

space vector plane. By voltage combination and time arrangement in each switching cycle, reference voltage can be achieved by SVPWM in three-phase VSC.

$$\begin{cases} \dfrac{V_c \cdot t_1}{\sin\theta} = \dfrac{V_c \cdot t_2}{\sin(\frac{\pi}{3} - \theta)} = \dfrac{V_{ref} \cdot T_s}{\sin(\frac{2\pi}{3})} \\ t_0 = T_s - t_1 - t_2 \end{cases} \quad (1.2)$$

For current source converter (CSC), the combination target is reference current (or current vector), and the combination effect will be different. The carrier-based PWM method and SVPWM method for CSC will also be different.

The fundamental PWM method in three-phase converter is only briefly introduced in this chapter. More details will be discussed in the next chapter.

1.3 Development and Challenges for Power Electronics

Power electronics have been developed since the mid-twentieth century. Universities began to open related courses in the late twentieth century. Power electronics technology is a mixed area of power technology, electronics technology and control technology. Its essence is to realize controllable power conversion through power semiconductors. In the past few decades, power electronics have made significant impact on industry.

(1) Motor drives in industry and home appliance: more than half of the generated power is consumed by motor. Conventional constant frequency supplied motors are with the problem of low efficiency. Power electronics have pushed the development of motor with variable speed drive (VFD) and saved a lot of energy. VFD has been applied not only in pump and fan but also in home appliances.

(2) Lighting area: novel power electronics ballasts have been applied for fluorescent lights. Power suppliers have been applied to novel LED light. They have improved efficiency for lighting.
(3) Power system: fast switching power electronics facilities have been applied in conventional electric power system, which effectively improved system stability, harmonics and power quality. Power electronics technology has directly developed high-voltage DC (HVDC) power transmission and flexible AC transmission system (FACTS).
(4) Transportation electrification area: power electronics technology has been applied in high-speed railway traction, electrical and hybrid electrical vehicle, more electrical aircraft and electrical propulsion of ship that significantly changed the transportation area.
(5) Renewable energy area: solar power, wind power and other renewable energy are with the characteristics of unstable operation and uneven distribution. Power electronics technology can make them to be stable and reliably transferred to the electrical power which can be utilized.
(6) Information technology (IT) area: in modern IT area, power suppliers of cellphones, computers to data centers are based on power electronics technology.

There are six basic functions in power electronics [3], which are:

(1) Switching function of power electronics converters: to control the electromagnetic energy flow;
(2) Conductive function: to guide the electromagnetic energy flow in the converter;
(3) Electromagnetic energy storage function: to keep the energy persistent with the interruption from switching function;
(4) Information function: to work as the "brain" and keep the previous three functions to work together well;
(5) Thermal function: to keep the thermal balance in the power electronics converters;
(6) Mechanical function: to guarantee the structure stability of the converter.

These six basic functions determine that the power electronics converter system is a multiphysics operation system. They can be mapped to the next nine technical areas:

(1) Power electronics device technology: including device, gate driver, snubber circuit, protection and so on;
(2) Power electronics switch network technology: including switch control and converter topologies;
(3) Passive components technology: including magnetic components, capacitors and conductive components;
(4) Package technology: including material science and technology, contacting technology and distribution technology;
(5) Electromagnetic environment technology: including harmonics, electromagnetic interference (EMI) and electromagnetic compatibility (EMC);

(6) Physical environment technology: including acoustic interference and environment pollution;
(7) Cooling technology: including fluid cooling, circulating, thermal conductivity and exchange;
(8) Manufacture technology;
(9) Sensor and control technology.

All these technologies are tied together closely and impact the design, development and operation of power electronics converter system. Among them, PWM technology is included in power electronics switching functions, and is also the key part of power electronics switch network technology. The physics behind it and the scientific problems are the core of this book.

Power electronics technologies are with two major driving forces: the first is the power electronics itself, especially the development of devices; the second is the requirement of applications.

Power electronics device is a major driving force for power electronics technologies. In the early time when thyristor was dominating, the first generation of power electronics technology appeared based on the thyristor rectifier and assistant circuit, with the same value for fundamental frequency and switching frequency. Then, with the development of fully controlled high-power devices of BJT and GTO, second generation of power electronics technologies appeared, with medium voltage drive and AC motor control as representatives and with switching frequency from hundreds to thousands Hertz. Also, in low-power area, because of the development of power MOSFET, switching power suppliers, resonant converters and power factor corrections appeared, with switching frequency up to hundreds of kHz. Then, since the birth of IGBT in 1980s, power electronics technologies have been with revolutionary impact. Because IGBT is with the advantages of high power of BJT and fast switching of MOSFET, high switching frequency in high-power applications becomes possible. Based on voltage source converter, HVDC and microgrid technologies, high performance AC motor control technologies, renewable energy application and complex topologies have been developed since then and brought the power electronics to the golden age.

In the recent decades, a new generation of power electronics devices is bringing a new revolution in power electronics technologies, that is in wide-band-gap power electronics devices. Compared with conventional Si power devices, wide-band-gap devices are with better voltage-blocking capability, less-conducting resistance and switching losses and better high-temperature performance. With the progress of fabrication and package technologies, products of wide band gap power electronics devices have appeared in the market with SiC devices as representative. The cost is also approaching to that of Si devices. Facing with the development of novel devices, many important problems need solving.

The other driving force for the development of power electronics technologies is the requirement of application. The earliest power electronics technologies are from the requirement of high level applications including military and aerospace areas, for example, the aerospace power supplier technology. These kinds of applications are considering little cost, and the technologies are not suitable for regular customers. From 1980s to 1990s, the requirements from information technologies pushed the development of switching power suppliers and power electronics technologies entering into regular civil usage. Also, the requirement of industrial automation pushed the development of variable speed drive. Since the turn of century, the requirements of transportation electrification and flexible power system have pushed the development of high-performance motor control and high-power conversion technologies.

In recent years, transportation electrification and power system have been with strong demand for power electronics. In the meantime, the requirement of renewable energy utilization and wireless power transmission are continuously opening new directions and becoming the driving force for power electronics technologies in the new century.

Because of the two major driving forces, while the conventional power electronics technologies are becoming mature and into products, the next generation of power electronics technologies are coming into being now. The main characteristics of next-generation power electronics include:

(1) More integrated: power electronics devices, microelectronics devices and package to be integrated, power circuit and control circuit to be integrated, source-converter-load to be integrated. The level of integration and modularity and power density are enhanced;
(2) With higher switching frequency: switching time scale is pushed from hundreds of nano-seconds to tens of nano-seconds, switching frequency is also pushed to MHz level. Passive components become smaller;
(3) More intelligent: smart control theory is applied and system's steady and dynamic performance is improved. Power electronics devices and converters themselves become sensors of the network.

All these novel characteristics bring not only opportunities but also challenges. More integrated power electronics bring stricter demands for thermal and EMC in smaller areas; higher switching speed also pushes the power conversion into a wider range in frequency domain; intelligent power electronics require more demands from information technologies.

1.4 Summary

As the background of this book, this chapter has introduced the development history, current status and future perspective of power electronics technologies. Power electronics devices are the basis for power electronics technologies. Power electronics devices have been with the progress from power diode to the half-controlled thyristor, then to fully-controlled BJT, MOSFET and IGBT. It is because of fast-switching IGBT's development that PWM technology found its application basis. Also, it was because of the power electronics devices' nonideal characteristics, PWM technology found the space to improve the system performance. Power electronics converters are built based on power electronics devices to achieve power conversion. Two kinds of power electronics converters: voltage source converter and current source converter are introduced. With the most widely used voltage source converter, the principle of PWM has been briefly introduced in this chapter. Two major driving forces for power electronics development have been illustrated in this chapter: the development of power electronics devices and the demand of applications. With the two driving forces, the next-generation power electronics technologies are emerging and bringing new opportunities and challenges [4–6].

References

1. Hingorani N, Ginn H, Sullivan J (2011) Control/protection architecture for power electronic converters. In: Proceedings of the IEEE electric ship technologies symposium, Alexandria, VA, pp 472–477
2. IEEE B E. P1676 (2010) IEEE draft guide for control architecture for high power electronics (1 MW and Greater) used in electric power transmission and distribution systems. IEEE Xplore, pp. 1–8
3. van Wyk JD, Lee FC (2013) On a future for power electronics. IEEE J Emerg Sel Topics Power Electron 1(2):59–72
4. Erickson RW, Maksimovic D (2001) Fundamental of power electronics. Springer Press
5. Mohan N, Undeland TM, Robbins WP (2002) Power electronics: converters, applications, and design, 3rd ed. Wiley
6. Holmes DG, Lipo TA (2003) Pulse width modulation for power converters: principles and practice. Wiley

Chapter 2
Principle of Pulse Width Modulation

In Chap. 1, the basic concept of pulse width modulation has been introduced briefly. In this chapter, the principle of pulse width modulation will be further discussed. Based on fundamental principle, space vector PWM (SVPWM) and carrier-based PWM (CBPWM) will be presented, respectively, together with the relationship between these two approaches. With the brief introduction of frequency-domain analysis method for PWM, the opportunities and challenges for PWM technologies will be summarized.

2.1 Space Vector PWM (SVPWM)

The basic principle of pulse width modulation has been introduced in the first chapter: to achieve the equivalence between pulse voltage and continuous voltage based on the volt-seconds balancing principle. To achieve this equivalence, there are two major ways: space vector combination and carrier comparison. Space vector combination is building the concept of space vector in 2-D plane of voltage or current, with the combination of standard vectors in the converter to obtain the reference vector. Carrier-based method is based on the average-equivalence between every line/phase voltage (or current) of pulse and reference in each switching cycle. In this part, the principle of SVPWM is introduced at first with the most widely used three-phase voltage source converters (VSC). In the last part, a brief introduction of SVPWM for current source converter (CSC) will be made, too [7].

Figure 2.1 shows a typical equivalent circuit for VSC. This circuit is a general structure for either three-phase inverter or three-phase rectifier. AC side is with current source characteristics and DC side is with voltage source characteristics. Switches in three phases can switch the corresponding phase to either positive or negative DC bus. The DC bus current is combined by the three-phase currents with the corresponding phases. In each phase leg, the switch combination is only with two choices. The two

D. Jiang et al., *Advanced Pulse-Width-Modulation: With Freedom to Optimize Power Electronics Converters*, CPSS Power Electronics Series,
https://doi.org/10.1007/978-981-33-4385-6_2

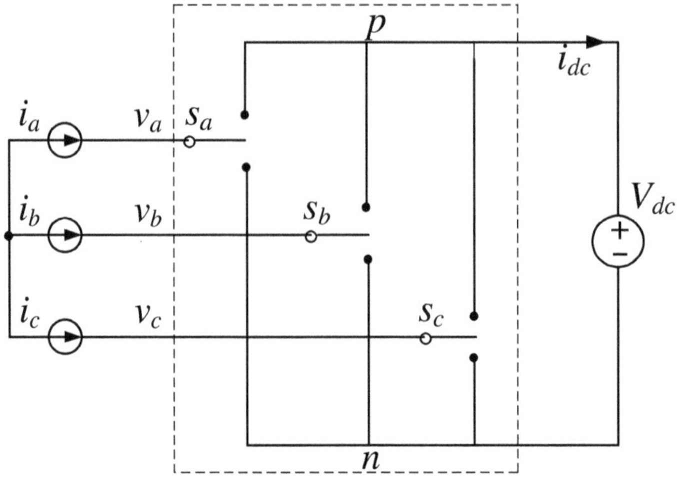

Fig. 2.1 The topology of three-phase VSC

switches cannot be turned on or turned off simultaneously, otherwise the DC bus will be with shoot-through (both on) or the AC side current source will be open circuit (both off). So, a three-phase converter is with eight different switch combinations. With the DC-bus midpoint as the reference, the phase voltage and DC bus current in eight combinations are shown in Table 2.1.

The switching function $S_x = 1$ ($x = a, b, c$) means the corresponding phase is connected to the positive DC bus, the corresponding phase voltage is $V_{dc}/2$ and the corresponding phase current is flowing into the positive DC bus. The switching function $S_x = 0$ ($x = a, b, c$) means the corresponding phase is connected to the negative DC bus, the corresponding phase voltage is $-V_{dc}/2$ and the corresponding phase current is flowing into the negative DC bus.

Then, eight different output phase voltages can be obtained using the eight different switch combinations through Table 2.1. In order to transfer three-phase

Table 2.1 The switching table of three-phase VSC: with phase voltage as an example

S_a	S_b	S_c	Switching state	i_{dc}	v_a	v_b	v_c
0	0	0	nnn	0	$-V_{dc}/2$	$-V_{dc}/2$	$-V_{dc}/2$
0	0	1	nnp	i_c	$-V_{dc}/2$	$-V_{dc}/2$	$V_{dc}/2$
0	1	0	npn	i_b	$-V_{dc}/2$	$V_{dc}/2$	$-V_{dc}/2$
0	1	1	npp	i_b+i_c	$-V_{dc}/2$	$V_{dc}/2$	$V_{dc}/2$
1	0	0	pnn	i_a	$V_{dc}/2$	$-V_{dc}/2$	$-V_{dc}/2$
1	0	1	pnp	i_a+i_c	$V_{dc}/2$	$-V_{dc}/2$	$V_{dc}/2$
1	1	0	ppn	i_a+i_b	$V_{dc}/2$	$V_{dc}/2$	$-V_{dc}/2$
1	1	1	ppp	$i_a+i_b+i_c$	$V_{dc}/2$	$V_{dc}/2$	$V_{dc}/2$

voltage to the 2-D plane, coordinate transformation is used, as shown in (2.1).

$$T_{abc/\alpha\beta} = \frac{2}{3}\begin{bmatrix} 1, & -\frac{1}{2}, & -\frac{1}{2} \\ 0, & \frac{\sqrt{3}}{2}, & -\frac{\sqrt{3}}{2} \end{bmatrix} \tag{2.1}$$

Through the coordinate transformation in (2.1), three-phase voltage can be transformed to α-β coordinate, shown in (2.2).

$$\begin{bmatrix} v_\alpha \\ v_\beta \end{bmatrix} = T_{abc/\alpha\beta}\begin{bmatrix} v_a \\ v_b \\ v_c \end{bmatrix} \tag{2.2}$$

Among all the eight vectors in three-phase VSC, there are six vectors (100, 110, 010, 011, 001, 101) whose output voltage vectors $[v_a, v_b, v_c]^T$ are mapped to six standard voltage vectors in α-β coordinate, shown in Fig. 2.2. There is 60° difference between each of the two adjacent vectors. The length of each vector is $2/3V_{dc}$.

Besides these six standard vectors, there are two other combinations: 000 and 111. With (2.2), the output voltage will be zero for either of these two vectors, so they are called zero vectors. Physically, these two vectors are mapped with the combination of all these phase-legs connected to positive bus or negative bus. There is no connection between the DC source and the AC load and there is no contribution to the AC side voltage from the DC bus.

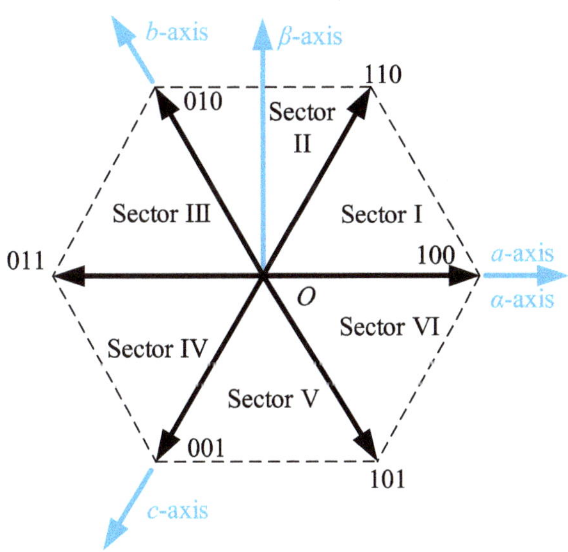

Fig. 2.2 Standard voltage vectors in α-β coordinate

In three-phase system, any voltage vector $[v_a, v_b, v_c]^T$ could be transferred to α-β plane with (2.2). Different from six standard vectors, three-phase voltage vector can be an arbitrary value. For symmetric three-phase sinusoidal voltage, it will be a rotational vector with angular frequency of ω_0 in α-β plane. However, three-phase VSC can only provide six standard vectors and two zero vectors. So, generation of arbitrary three-phase voltage with VSC in α-β plane is through the combination of standard vectors and zero vectors to approach the arbitrary vector. This is the basic principle of space vector PWM (SVPWM).

As discussed in Chap. 1, the 2-D plane can be divided into six sectors in Fig. 2.2. Any voltage vector in α-β plane will be in one sector. The simplest way is to combine the reference voltage vector with the two adjacent standard voltage vectors in the corresponding zone. With this approach, the error between the reference vector and standard vector will be minimized. In view of the converter performance, the voltage on the three-phase inductor will be minimized, and the current ripple will be minimized, too. Actually, the reference vector combination is with more than one method, and other standard vectors can also be used for combination, to pursue other performance targets, which will be introduced in the following chapters. The method using the two adjacent vectors is the most widely used SVPWM approach.

The first step for SVPWM is to determine the sector of the reference vector and the angle θ of the vector in this sector. Then, the action time for the two adjacent standard vectors can be calculated, shown in Fig. 2.3. V_{ref} can be combined by standard vector V_1 with time t_1 and standard vector V_2 with time t_2 in the switching cycle T_s. Based on the relationship of similar triangles, the corresponding volt-seconds can meet (2.3). V_c is the standard vector with length of $2/3V_{dc}$. Then, the action time of t_1 and t_2 can be derived in (2.4). However, the switching cycle T_s is not exactly the sum of t_1 and t_2. The rest of the cycle can be compensated by zero vector, which is t_{zero}, shown in (2.5).

$$\frac{V_c \cdot t_1}{\sin(\frac{\pi}{3}-\theta)} = \frac{V_c \cdot t_2}{\sin\theta} = \frac{V_{ref} \cdot T_s}{\sin(\frac{2\pi}{3})} \tag{2.3}$$

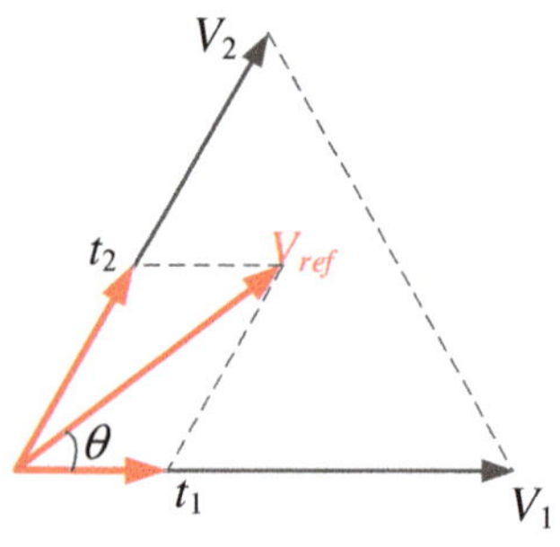

Fig. 2.3 The vector combination in SVPWM in a typical sector

$$\begin{cases} t_1 = \dfrac{V_{ref} \cdot T_s \sin(\frac{\pi}{3} - \theta)}{V_c \sin(\frac{2\pi}{3})} \\ t_2 = \dfrac{V_{ref} \cdot T_s \sin\theta}{V_c \sin(\frac{2\pi}{3})} \end{cases} \tag{2.4}$$

$$t_{zero} = T_s - t_1 - t_2 \tag{2.5}$$

One thing needs to be emphasized: in (2.5), it is not guaranteed that zero vector action time t_{zero} is positive. In fact, when the reference line voltage V_{ref} is with amplitude bigger than $\sqrt{3}V_{dc}/3$, the result in (2.5) will be negative. This case is called over-modulation, which means the DC-link voltage is not big enough to provide the voltage with this amplitude for V_{ref}. If defining the modulation index m as the reference phase voltage amplitude over half of the DC-link voltage, it should satisfy (2.6). It means that using SVPWM, the maximum output phase voltage amplitude is 1.1547 times of half DC voltage.

$$m \leq (\sqrt{3}V_{dc}/3)/(V_{dc}/2) = 1.1547 \tag{2.6}$$

After calculating action times for each standard vector with (2.4) and (2.5), the average output is determined. However, the sequence of vectors can impact the output effect, too. Reasonable sequence should be obtained to make sure each phase is only switching twice in one switching cycle and the output current ripple can be minimized. This is so-called seven-segment SVPWM, shown in Fig. 2.4. With this kind of SVPWM, pulses are symmetrical in each switching cycle. In two sides, there are two zones with vector 000 and the length is $t_{zero}/4$ for each. The second and sixth zones are with vector V_1 and action time of $t_1/2$ for each. The third and fifth zones are with vector V_2 and action time of $t_2/2$ for each. The central zone is with vector 111 and action time of $t_{zero}/2$. Then, the switching cycle is divided into these seven

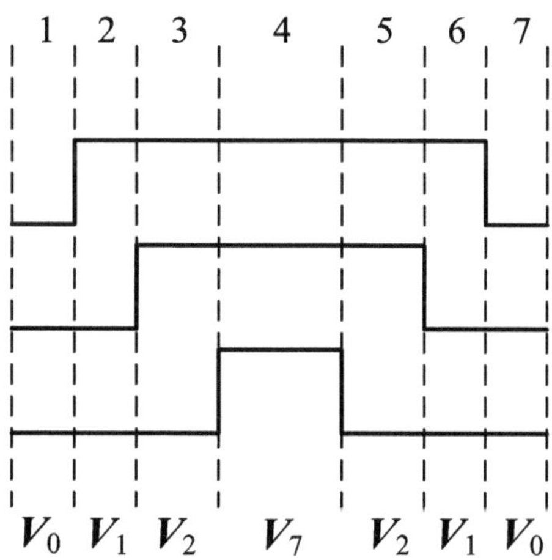

Fig. 2.4 Seven-segment SVPWM

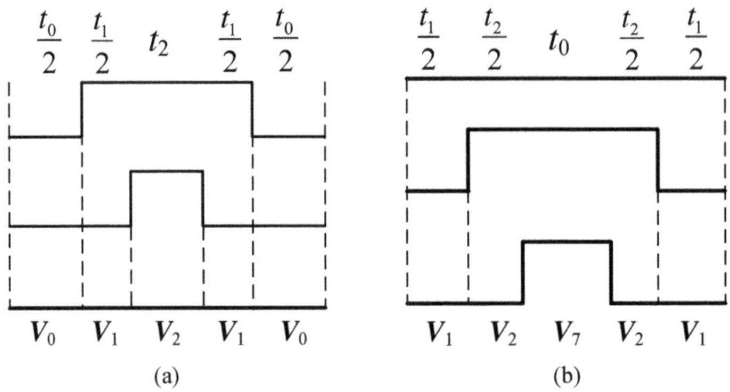

Fig. 2.5 Five-zone SVPWM (DPWM): **a** DPWM_Min, **b** DPWM_Max

zones and each of them is with a certain vector. The action times for all the vectors have been satisfied.

In the seven-segment SVPWM in Fig. 2.4, the switching actions for each phase are exactly twice in each switching cycle. Zero vector is equally distributed to 000 and 111. They are located in two sides and the center with $t_0 = t_7 = t_{zero}/2$. Adjusting of zero vector distribution will not impact on the vector combination effect. Then, the zero-vector action time can be fully given to 000 or 111. That is the so-called five-zone SVPWM, shown in Fig. 2.5. Figure 2.5a is the case with all zero vector action time given to 000, with action time of $t_{zero}/2$ in two sides; Fig. 2.5b is the case with all zero vector action time given to 111, with action time of t_{zero} in the center. Compared with seven-zone SVPWM, five-zone SVPWM is with five zones in each switching cycle, instead of seven zones. Also, in either Fig. 2.5a or Fig. 2.5b, there is one phase keeping the same level without switching actions. In Fig. 2.5a, Phase-C is keeping low level. In Fig. 2.5b, Phase-A is keeping high level. Because of the discontinuous switching there is another name given to five-zone SVPWM: Discontinuous PWM (DPWM). In Fig. 2.5a, the minimum duty cycle is kept to be zero and this kind of DPWM is called minimum clamped DPWM (DPWM_Min). In Fig. 2.5b, the maximum duty cycle is kept to be one and this kind of DPWM is called maximum clamped DPWM (DPWM_Max). Compared with seven-zone SVPWM, DPWM is with less switching actions and less switching losses, but the current ripple will increase. This issue will be mentioned in Chap. 3.

Two kinds of DPWM methods can keep one phase without switching actions in each switching cycle and reduce the full switching losses. However, choosing which phase to be clamped is also a freedom for optimization. DPWM_Min in Fig. 2.5a chooses the minimum duty cycle to be clamped to negative DC bus; DPWM_Max in Fig. 2.5b chooses the maximum duty cycle to be clamped to positive DC bus. In fact, the switching losses in power electronics devices are approximately proportional to the product of device voltage and current. For VSC, power electronics device voltage stress equals with the DC bus voltage, so the phase with maximum switching losses is the phase with maximum current (absolute value). Then, combining DPWM_Min and DPWM_Max, a switching loss optimized by DPWM method can be derived: the minimum loss DPWM. By comparing the phase current absolute value, the phase with bigger value is selected to be clamped. Then, the total switching losses can be further reduced and the converter is working between the two modes of DPWM_Max and DPWM_Min.

Earlier analysis of SVPWM is based on phase voltage. When using line voltage to be the object in three-phase system, SVPWM will be different and will be briefly introduced.

When dealing with line voltage, the switching table is shown in Table 2.2. Using the coordinate transformation in (2.2), the line voltage vectors in α-β plane are shown in Fig. 2.6. Compared with Fig. 2.2, the standard line voltage vector is rotated at 30° and the length is $2V_{dc}/\sqrt{3}$. Two zero vectors are still with zero length.

For line voltage combination, the approach is similar to SVPWM with phase voltage: dividing the plane into six sectors and determining the sector for the reference line voltage vector, then using the two adjacent standard line voltage vectors and zero

Table 2.2 Switching status of three-phase VSC: with line voltage vector

S_a	S_b	S_c	Status	i_{dc}	v_{ab}	v_{bc}	v_{ca}
0	0	0	nnn	0	0	0	0
0	0	1	nnp	i_c	0	$-V_{dc}$	V_{dc}
0	1	0	npn	i_b	$-V_{dc}$	V_{dc}	0
0	1	1	npp	i_b+i_c	$-V_{dc}$	0	V_{dc}
1	0	0	pnn	i_a	V_{dc}	0	$-V_{dc}$
1	0	1	pnp	i_a+i_c	V_{dc}	$-V_{dc}$	0
1	1	0	ppn	i_a+i_b	0	V_{dc}	$-V_{dc}$
1	1	1	ppp	i_a+i_b+i_c	0	0	0

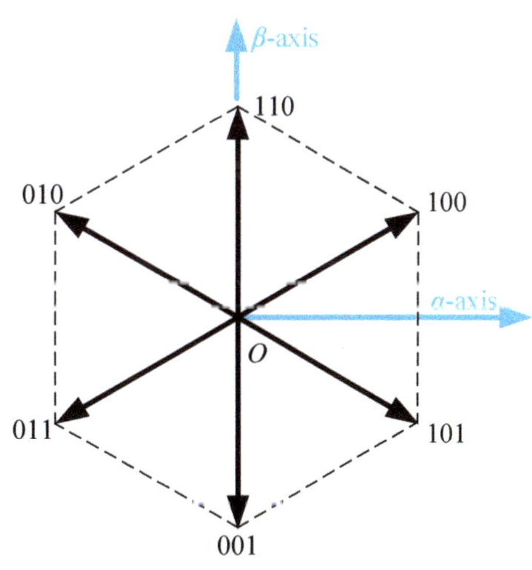

Fig. 2.6 Standard line voltage vectors in α-β plane

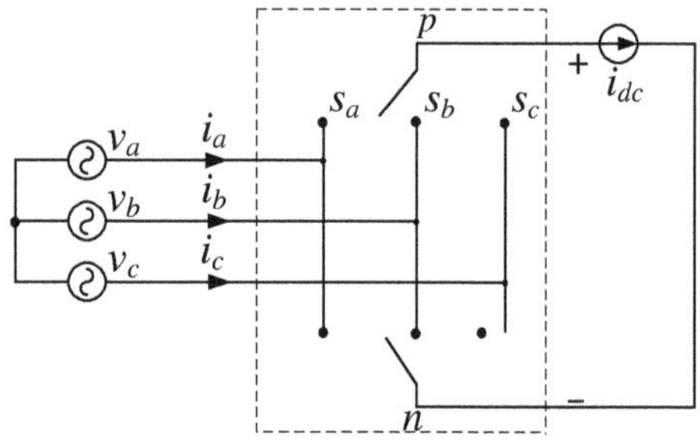

Fig. 2.7 Topology for three-phase CSC

vector to combine the reference line voltage vector and obtain the action time t_1, t_2 and the rest of zero vector action time t_{zero}. Based on standard vector action time, the sequence of vectors can be determined, including five-zone and seven-zone SVPWM.

In the standard line voltage vectors in Fig. 2.6, every standard vector is with length of $2V_{dc}/\sqrt{3}$. In order to avoid over-modulation, the reference line voltage vector should be inside the inscribed circle. When using symmetrical three-phase voltage, the line voltage vector should be with $2V_{dc}/\sqrt{3} \times \sqrt{3}/2 = V_{dc}$. Defining the modulation index m to be the ratio between line voltage peak value and the full DC bus voltage, m should satisfy $m \leq V_{dc}/V_{dc} = 1$. Compared with (2.6), the definition of modulation index with line voltage is reduced from 1.1547 to 1, which means the output line voltage peak value cannot exceed the full DC bus voltage.

Besides the voltage source converter, current source converter (CSC) is another kind of typical topologies for three-phase power conversion, shown in Fig. 2.7. For CSC, the DC side is with current source characteristics and the AC side is with voltage source characteristics, which is opposite from VSC. Also, the power electronics switch is different from VSC. In CSC, positive or negative DC bus is selected to be connected to the corresponding phase. Since the DC bus is not allowed to be open-circuit, p and n should be connected to one phase. Also, the upper and lower switches in the same phase-leg is allowed to be turned on simultaneously. So, there are totally $3 \times 3 = 9$ switch combinations and also 9 standard current vectors. Three standard vectors are zero vectors which have no contribution to the output. Six standard vectors are active current vectors. Space vector combination is based on those six active current vectors and three zero current vectors. In this book, the most widely used two-level VSC is treated as standard topology to study PWM and CSC is treated as one of the advanced topologies. SVPWM for three-phase CSC will be introduced in Chap. 6.

2.2 Carrier-Based PWM (CBPWM)

Space vector PWM is using a virtual 2-D plane for three-phase system to realize the approximation of pulse voltage to continuous voltage through vector combination.

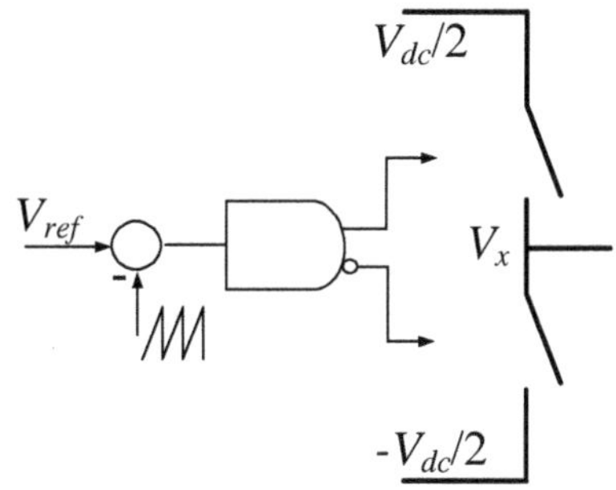

Fig. 2.8 Carrier-based PWM structure in single phase-leg

Another approximation method is directly through the volt-seconds equivalence, that is carrier-based PWM method. It has been briefly introduced in Chap. 1 also.

Figure 2.8 shows the basic structure of carrier-based PWM in single phase-leg. The positive and negative bus voltages are $V_{dc}/2$ and $-V_{dc}/2$ with reference to the DC neutral point. The output terminal voltage V_x needs to approach the reference voltage. Reference voltage is compared with the triangle carrier and determines the gate signals for the upper and lower switches, then determines the output voltage V_x to be switched between positive and negative DC bus voltage.

The reference voltage in Fig. 2.8 is arbitrary and continuous voltage. It can be sinusoidal or nonsinusoidal, but the frequency should be lower than the carrier. Otherwise, it cannot be synthetized by PWM. The carrier can be with different triangle series. It will not impact the output voltage average value, but will impact the pulse location. Figure 2.9 shows the four cases of carrier with the output voltage. The upper and lower limits for triangle wave are $V_{dc}/2$ and $-V_{dc}/2$. After comparing with V_{ref}, the output voltage V_x is pulse voltage switching between $V_{dc}/2$ and $-V_{dc}/2$. Figure 2.9a, b are two extreme cases: the triangles are right-angled and the vertical side is either in the left (Fig. 2.9a) or in the right (Fig. 2.9b). Then the carrier is sawtooth and the output pulses are either aligned with right or left of the switching cycle. Figure 2.9c is a general case: the summit of the triangle is arbitrarily distributed in the switching cycle and the slope of the two sides can be arbitrary, too. In this case, the rising and falling edges of the output pulse can be arbitrarily distributed in the switching cycle, but the distance between them is fixed by duty cycle. Figure 2.9d is the case with symmetrical triangle wave and the summit of the triangle is in the center of the switching cycle. Then, the output pulse voltage is also central-aligned symmetrical. Based on the principle of similar triangles, the low-level time T_0 satisfies $T_0/T_s = (V_{dc}/2 - V_{ref})/V_{dc}$. The reference voltage determines the duty cycle of the output pulse and the triangle wave curve determines the pulse location in the switching cycle. In application, the symmetrical triangle in Fig. 2.9d is mostly used and is also the carrier for five-zone SVPWM and seven-zone SVPWM. However, the carrier can be a variable to adjust the location of the pulses and improve the system performance. It will be introduced in the following chapters.

The discussion in Fig. 2.9 is for the carrier's curve. The curve for the reference is also an important factor. There are two kinds of reference for comparison with triangle

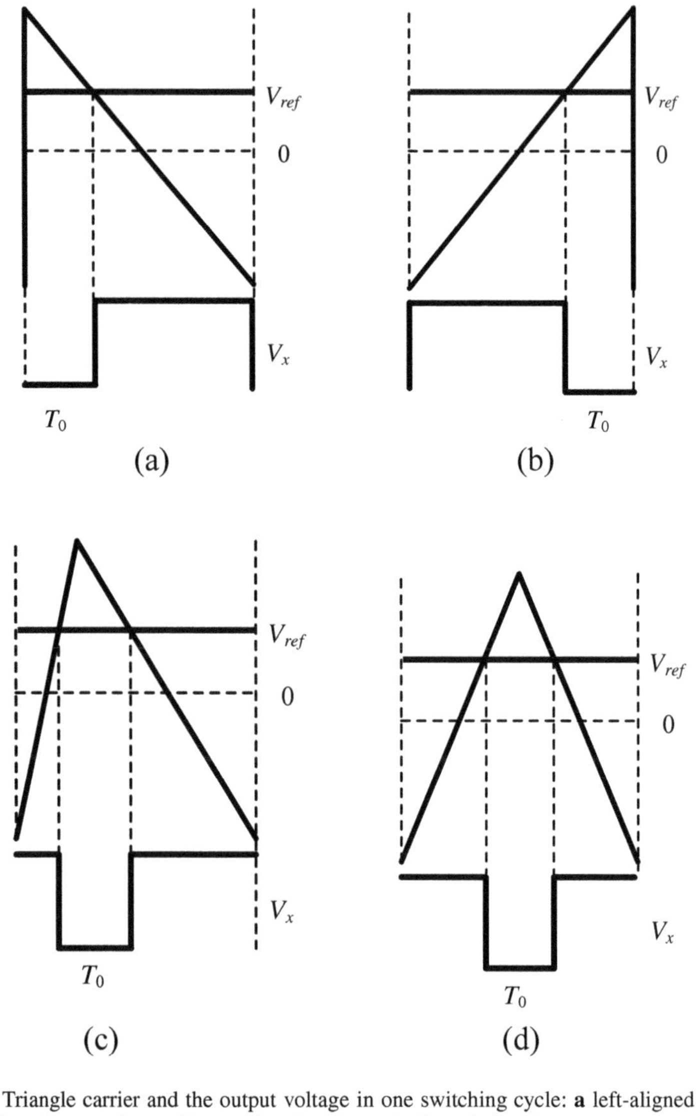

Fig. 2.9 Triangle carrier and the output voltage in one switching cycle: **a** left-aligned sawtooth, **b** right-aligned sawtooth, **c** arbitrary triangle, **d** central-aligned triangle

carriers: natural sampling method and regular sampling method (also called uniform sampling method), shown in Fig. 2.10. Figure 2.10a shows the natural sampled carrier-based PWM, which is using continuous reference voltage V_{ref} to compare with triangle carriers. The crossing point between V_{ref} and carrier determines the rising and falling edges of the output pulse voltage. This method can approach the reference voltage precisely since the comparison is directly between continuous reference and the carrier. However, because of the continuous variation of V_{ref}, the

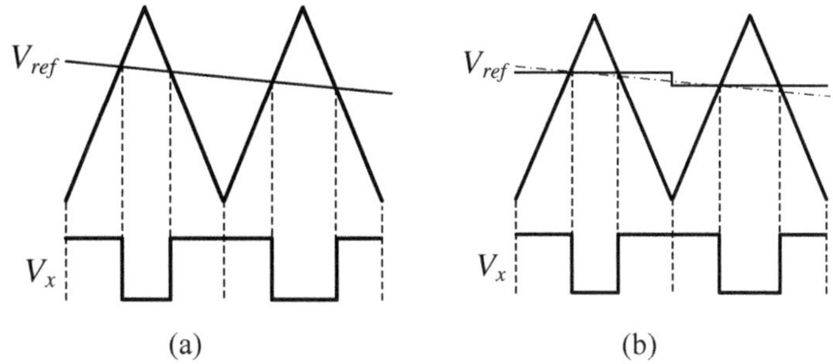

Fig. 2.10 Two kinds of carrier-based PWM: **a** natural sampled PWM, **b** regular sampled PWM

crossing points with the triangle carrier will have small variation, which makes the pulses relatively asymmetrical. Natural sampling is usually used in PWM generation with analog circuit since the reference voltage is continuous in analog system and can be used for comparison directly. However, in widely used digital system, the reference will be held to be constant in the sampling cycle. It comes up with the regular sampling method in Fig. 2.10b. With regular sampled PWM, the reference will be kept constant in each switching cycle. In Fig. 2.10b, V_{ref} is with digital sampling and changes to steps to compare with triangle carrier. Once it is compared with symmetrical triangle carrier, the rising and falling edges will be symmetrical in the switching cycle, which makes the generated pulse to be symmetrical, too. When carrier frequency is much higher than the reference frequency, the regular sampled step voltage is very close to the continuous reference. Then, the pulse generated by regular sampled PWM can approach the reference, too. Regular sampled PWM has been widely used in digital system like the digital signal processor (DSP). The sampled time can be in the center of the switching cycle and can be in other places also. Also, if the reference can be sampled more than once in the switching cycle, it can be closer to the continuous reference voltage.

Symmetrical triangle carrier with regular sampling is the most popular carrier-based PWM method. The carrier-based PWM method introduced in this book is mainly with this approach.

For carrier-based PWM in three-phase system, reference voltage can be different for different applications. The most popular application is the sinusoidal PWM (SPWM) and its modifications. SPWM means that the three-phase reference voltage follows three-phase symmetrical sinusoidal function to compare with carrier and generate PWM. If normalized to the range between −1 and 1, the three-phase reference function is shown in (2.7).

$$\begin{cases} m_a = m\cos\omega_0 t \\ m_b = m\cos(\omega_0 t - 2\pi/3) \\ m_c = m\cos(\omega_0 t + 2\pi/3) \end{cases} \tag{2.7}$$

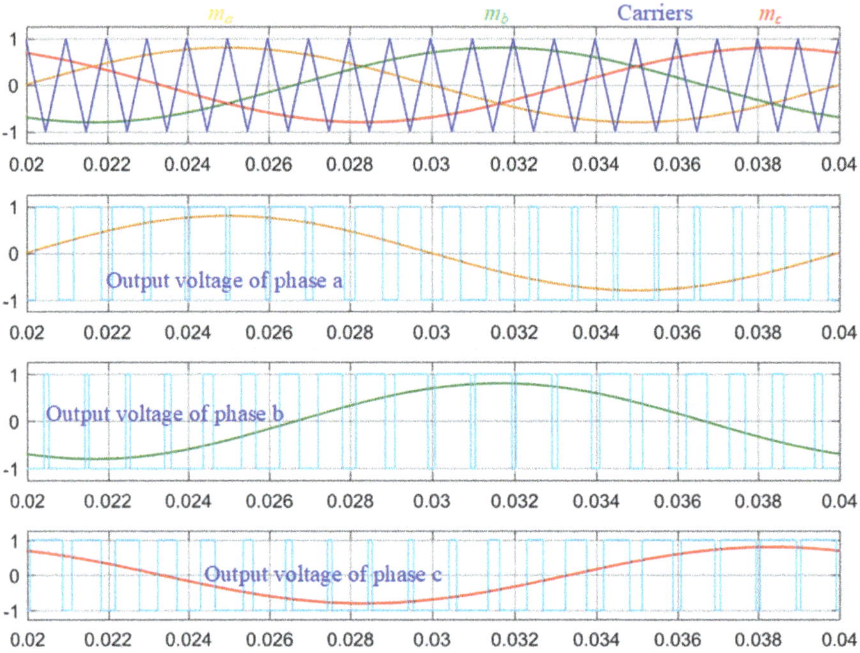

Fig. 2.11 Reference, carrier and output voltage with three-phase SPWM: natural sampling

In (2.7), m is the amplitude of reference which is normalized to 0–1, named modulation index, which means the ratio between the output phase voltage amplitude and the half DC-bus voltage. The three-phase reference is compared with triangle carrier which is with amplitude of 1. The PWM generation effect is shown in Fig. 2.11: when the reference is bigger than triangle carrier, the corresponding phase-leg is turned to 1 (positive DC bus voltage of $V_{dc}/2$); when the reference is smaller than triangle carrier, the corresponding phase-leg is turned to -1 (negative DC bus voltage of $-V_{dc}/2$). Then the output three-phase pulse series are approaching the reference in each switching cycle with average value. The output voltage's fundamental component is also approaching the sinusoidal reference value.

Figure 2.11 shows the method of natural sampling, so the three-phase reference voltages m_a, m_b and m_c are continuous waveforms. The crossing points with the triangle carrier in each switching cycle are asymmetrical, too. In digital control system, sample-holder is used to make the three-phase voltage discrete and realize regular sampling which is shown in Fig. 2.12. The three-phase reference voltages are discontinuous steps and keep constant in each switching cycle. So, the output voltage pulse is symmetrical in each switching cycle.

The difference between output phase voltages is the line voltage. Figure 2.13 shows the line voltage in three-phase SPWM. The line voltage is with three levels and the amplitude is two. It means the instantaneous line voltage can be full DC-link voltage V_{dc} or negative DC-link voltage $-V_{dc}$.

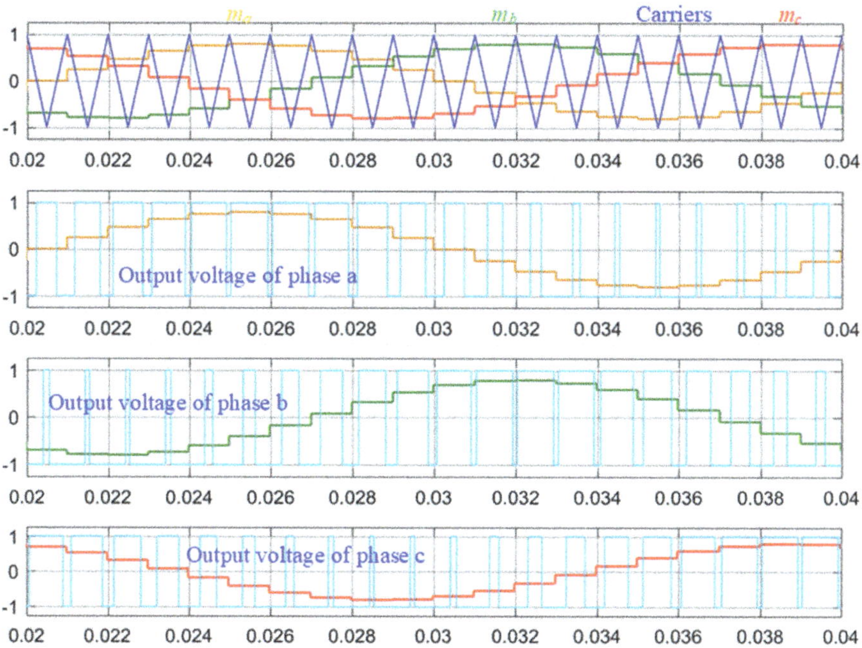

Fig. 2.12 Reference, carrier and output voltage with three-phase SPWM: regular sampling

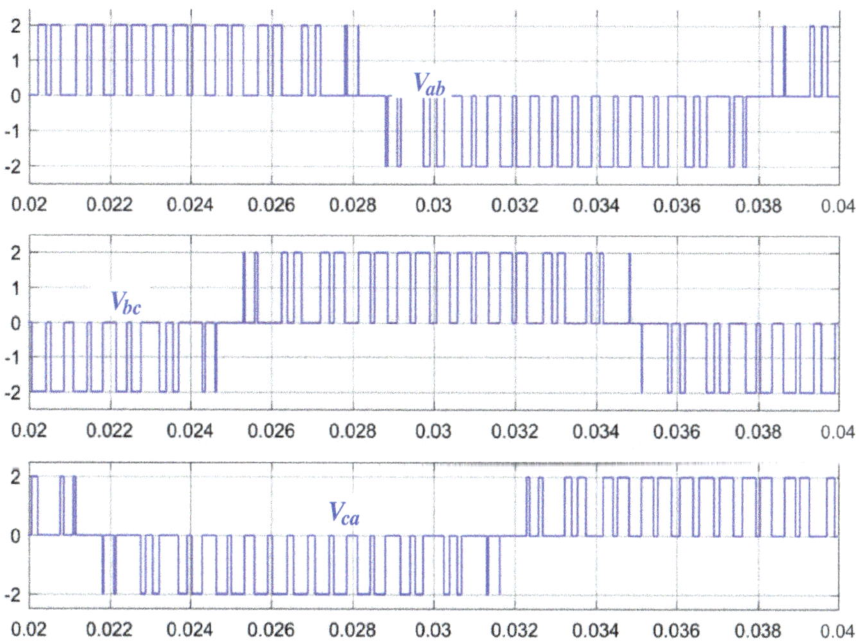

Fig. 2.13 Line voltage with three-phase SPWM

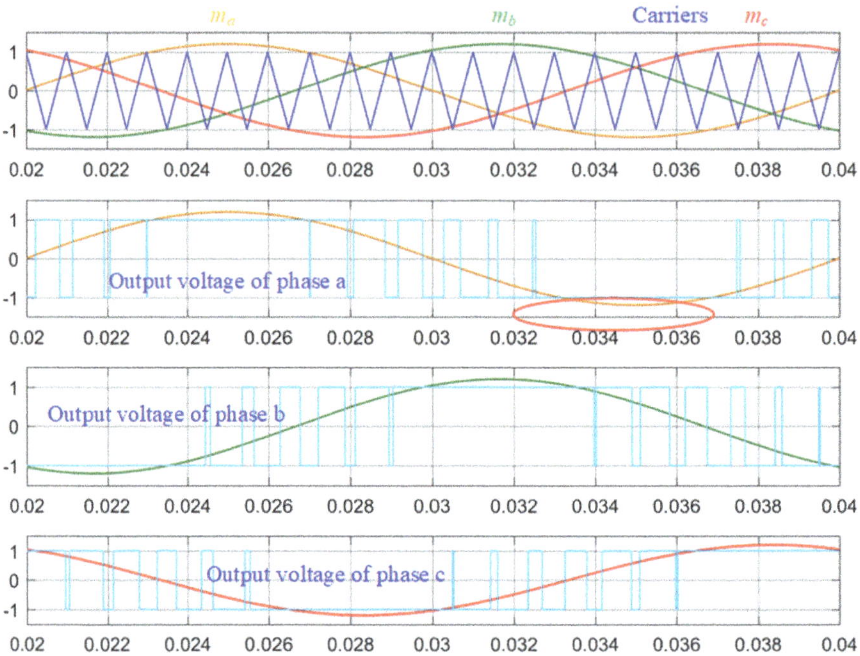

Fig. 2.14 Three-phase SPWM with over-modulation

However, there is an obvious problem for SPWM: modulation index in (2.7) cannot exceed 1. Otherwise the amplitude of the reference will exceed the amplitude of the triangle carrier and cause over-modulation problem. Figure 2.14 shows the example of over-modulation with m to be 1.1. In this case, the reference will keep higher or lower than the triangle carrier in some regions and cannot make effective PWM. The output voltage cannot approach the reference.

Because of the limit of modulation index, SPWM can generate maximum output phase voltage to be half of the DC-link voltage. Compared with the SVPWM which can generate maximum output phase voltage to be 1.1547 times of half DC-link voltage in (2.6), the modulation index is reduced to be 0.866 times of SVPWM by SPWM. In order to improve the modulation index, third-order harmonics injection PWM has been developed.

Because the voltage added on the load is line voltage, adding the same component on the three-phase voltage will not affect the output line voltage since the difference between each two-phase voltages will be the same as before. So, it is possible to inject suitable "common-mode" voltage based on sinusoidal reference to reduce its peak value. One typical method is third-order harmonics injection, shown in (2.8), which is also called harmonics injection PWM (HIPWM). In HIPWM, three-phase reference voltages are added with the same third-order harmonic $\mu\cos(3\omega_0 t)$ based on SPWM. By solving the value of this function (2.8), it is clear that (2.8) will be with its minimum value when $\mu = -1/6$. In that case, m can be as high as 1.1547 to make sure m_a, m_b and m_c not exceed 1. Then the modulation index of HIPWM will be as high as that of SVPWM.

$$\begin{cases} m_a = m(\cos \omega_0 t + \mu \cos(3\omega_0 t)) \\ m_b = m(\cos(\omega_0 t - 2\pi / 3) + \mu \cos(3\omega_0 t)) \\ m_c = m(\cos(\omega_0 t + 2\pi / 3) + \mu \cos(3\omega_0 t)) \end{cases} \tag{2.8}$$

Figure 2.15 shows the three-phase reference voltage, carrier and output voltage for HIPWM with $m = 1.1$. Compared with Fig. 2.14, the reference voltage of HIPWM can be kept between -1 and 1 and no over-modulation effect appears even though m exceeds 1. It is because the injected third-order harmonics effectively reduce the original sinusoidal reference amplitude. Since the injected components for three phases are the same, the output line voltage is the same as the original sinusoidal case. This figure proves that HIPWM can effectively increase the modulation index of SPWM and the DC voltage utilization.

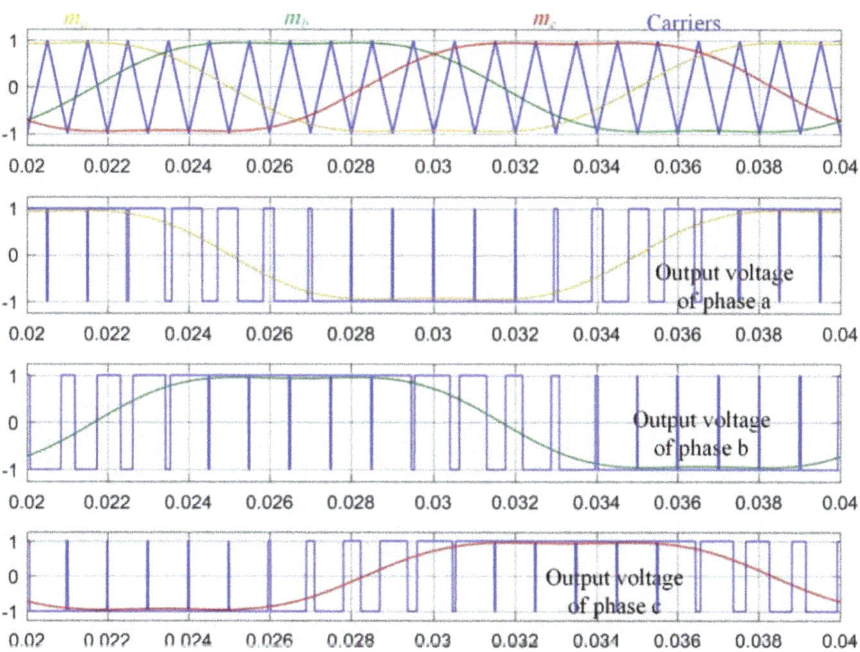

Fig. 2.15 Reference, carrier and output phase voltage of HIPWM

2.3 Relationship Between SVPWM and CBPWM

This chapter has introduced two kinds of PWM realization methods: SVPWM and CBPWM. In fact, for converter itself, there are only unique and certain gate driver signals to physically achieve PWM, but no difference between space vector combination and carrier comparison. Therefore, SVPWM and CBPWM should be equivalent physically. This section is to build up the relationship between two kinds of PWM [1, 8].

First, since the reference voltage is kept the same in each switching cycle for SVPWM, then it is possible to use the two active voltage vectors to combine it and use zero vectors to compensate the rest of time, and SVPWM should be equivalent to regular sampled CBPWM, which is keeping the reference voltage to be constant in each switching cycle.

Figure 2.4 shows the seven-segment SVPWM and the corresponding vectors in the seven zones: taking the first sector as an example, seven vectors are V_0, V_1, V_2, V_7, V_2, V_1 and V_0. The output voltage in each phase is a symmetrical single pulse. Then, it is possible to use a symmetrical triangle wave (with amplitude of 1) to be the carrier, and to compare it with the three references m_a, m_b, m_c, shown in Fig. 2.16. Based on the area equivalent principle, the three references can be calculated by (2.9). Three-phase voltage can be expressed by (2.10), with DC-link voltage of V_{dc}.

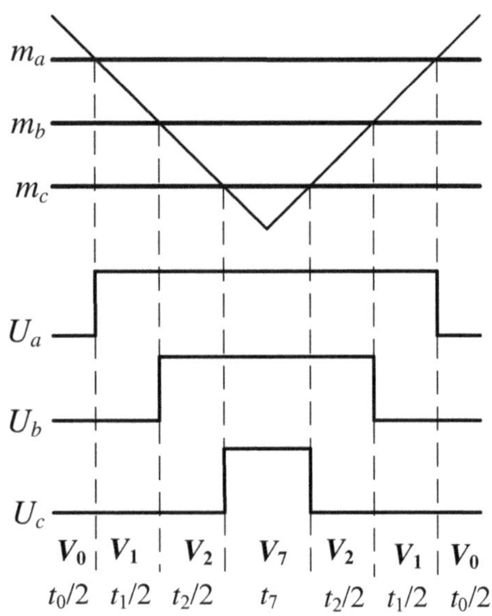

Fig. 2.16 SVPWM's equivalency with CBPWM (taking the first sector as an example)

$$\begin{cases} m_a = \dfrac{(t_1 + t_2 + t_7 - t_0)}{T_s} \\ m_b = \dfrac{(t_2 + t_7 - t_0 - t_1)}{T_s} \\ m_c = \dfrac{(t_7 - t_0 - t_1 - t_2)}{T_s} \end{cases} \tag{2.9}$$

$$\begin{cases} U_a = \dfrac{(t_1 + t_2 + t_7 - t_0)}{T_s} \cdot \dfrac{V_{dc}}{2} \\ U_b = \dfrac{(t_2 + t_7 - t_0 - t_1)}{T_s} \cdot \dfrac{V_{dc}}{2} \\ U_c = \dfrac{(t_7 - t_0 - t_1 - t_2)}{T_s} \cdot \dfrac{V_{dc}}{2} \end{cases} \tag{2.10}$$

By adding the three modulation functions together in (2.9), the common-mode modulation function can be expressed in (2.11). It has been explained that because the line voltage is the difference between phase voltages, the common-mode modulation function will be cancelled out in line voltage. The line voltage expressions are shown in (2.12). It can be found that the line voltage is not related to zero vector. It means that the common-mode modulation function is not influencing the output line voltage, and is the major difference for different PWM methods.

$$e = \frac{1}{3}(m_a + m_b + m_c) = \frac{3t_7 + t_2 - t_1 - 3t_0}{3T_s} \tag{2.11}$$

$$\begin{cases} U_{ab} = U_a - U_b = V_{dc} \cdot t_1/T \\ U_{bc} = U_b - U_c = V_{dc} \cdot t_2/T \\ U_{ca} = U_c - U_a = -V_{dc} \cdot (t_1 + t_2)/T \end{cases} \tag{2.12}$$

In (2.9), the actuation time (t_1 and t_2) for vector V_1 and V_2 are calculated through (2.4). It means that when the reference voltage vector is fixed, the actuation time is also fixed. However, the output effects of zero vectors 000 and 111 are the same. Then, how to distribute the zero vector actuation time t_{zero} to t_0 and t_7 is an extra freedom to control. It can be defined with a coefficient k ($0 \leq k \leq 1$) to adjust the ratio of 000 and 111 in total zero vector actuation time, shown in (2.13). The bigger k means the ratio of 111 is higher, the smaller k means the ratio of 000 is higher.

$$\begin{cases} t_0 = (1 \quad k) \cdot t_{zero} \\ t_7 = k \cdot t_{zero} \end{cases} \tag{2.13}$$

The analysis above is based on the case in the first sector. Similar analysis can be done for six different sectors, and the results are shown in Table 2.3.

In Table 2.3, the nonzero vector's action time can be calculated by (2.4) with the two adjacent standard voltage vectors. The unified function of common-mode

Table 2.3 Space vector action time arrangement and common-mode modulation function [1]

Sector	Action time arrangement	Common-mode modulation function e
1	$T_s = t_1 + t_2 + t_0 + t_7$	$e = (3t_7 + t_2 - t_1 - 3t_0)/(3T_s)$
2	$T_s = t_2 + t_3 + t_0 + t_7$	$e = (3t_7 + t_2 - t_3 - 3t_0)/(3T_s)$
3	$T_s = t_3 + t_4 + t_0 + t_7$	$e = (3t_7 + t_4 - t_3 - 3t_0)/(3T_s)$
4	$T_s = t_4 + t_5 + t_0 + t_7$	$e = (3t_7 + t_4 - t_5 - 3t_0)/(3T_s)$
5	$T_s = t_5 + t_6 + t_0 + t_7$	$e = (3t_7 + t_6 - t_5 - 3t_0)/(3T_s)$
6	$T_s = t_6 + t_1 + t_0 + t_7$	$e = (3t_7 + t_6 - t_1 - 3t_0)/(3T_s)$

modulation function is shown in (2.14). Since k can be arbitrarily settled, every common-mode modulation function e of CBPWM can be calculated with (2.14) to obtain the corresponding k. Then, all kinds of SVPWM can be equivalently achieved by injecting corresponding e for CBPWM.

$$e = k(1 - m_{\max}) + (1 - k)(-1 - m_{\min}) \tag{2.14}$$

SPWM and HIPWM are the two examples to study the corresponding SVPWM of CBPWM.

For the most typical three-phase SPWM, the three-phase reference voltages are the standard sinusoidal waveforms with 120° phase-shift, shown in (2.7), and the common-mode modulation function is $e = 0$. Solving the equation of (2.14), the expression of k is shown in (2.15). Then, after calculating t_1 and t_2 for SVPWM, the action time for 000 and 111 can be derived through (2.13) and (2.15).

$$k = \frac{1 + m_{\min}}{2 - m_{\max} + m_{\min}} \tag{2.15}$$

For HIPWM expressed in (2.8), the common-mode modulation function is $e = \mu m\cos(3\omega_0 t)$. With (2.14), the expression for k can be shown in (2.16). Then, the action time for 000 and 111 can be arranged through (2.16). The corresponding SVPWM is obtained.

$$k = \frac{1 + m_{\min} + \mu m \cos(3\omega_0 t)}{2 - m_{\max} + m_{\min}} \tag{2.16}$$

Figure 2.17 shows the ratio of 000 and 111 in the full zero vector, varying in one line-cycle with modulation index of 0.9.

Besides the method to obtain equivalent SVPWM from the CBPWM, the reversed method is also feasible, which is achieving equivalent CBPWM from SVPWM through (2.14) with the common-mode modulation function e and the three-phase references m_a, m_b and m_c. Then, the corresponding CBPWM is with the same output effect of the original SVPWM.

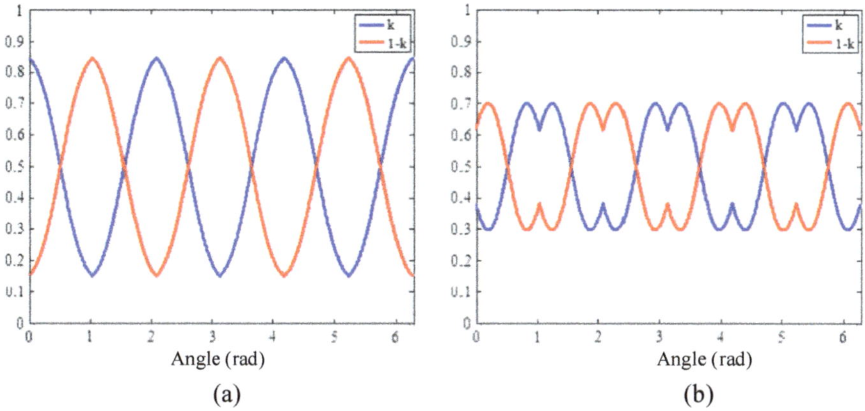

Fig. 2.17 Distribution of 000 and 111 action time in **a** SPWM, **b** HIPWM

With the typical seven-zone SVPWM discussed in Sect. 2.2, k is fixed to be 0.5, which means the action times of 000 and 111 are the same in each switching cycle: $t_0 = t_7$. Then, the common-mode modulation function is $e = 0.5(1 - m_{max}) + 0.5(-1 - m_{min})$.

For five-zone DPWM, only one of 000 and 111 is utilized: DPWM_Min is with $k = 0$ and DPWM_Max is with $k = 1$. Then minimum-loss DPWM1 is switching between the case of $k = 0$ and $k = 1$ with the current direction. The common-mode modulation functions for these three DPWM methods are expressed in (2.17).

$$\begin{cases} e = \begin{cases} 1 - m_{\max} & \text{DPWMMax} \\ -1 - m_{\min} & \text{DPWMMin} \end{cases} \\ e = \begin{cases} 1 - m_{\max} & (|i_{\max}| \geq |i_{\min}|) \\ -1 - m_{\min} & (|i_{\max}| < |i_{\min}|) \end{cases} \quad \text{DPWM1} \end{cases} \tag{2.17}$$

Then, the reference modulation function and the common-mode modulation function for seven-zone SVPWM, DPWM_Min, DPWM_Max, DPWM1 can be derived, as shown in Fig. 2.18. In Fig. 2.18d, DPWM1 is in the case of unity power factor, which means the modulation function is in phase with the current. Modulation index is selected to be 0.9 in Fig. 2.18.

Since SVPWM and DPWM can be realized by adding common-mode modulation function e on the basis of SPWM, it is possible to achieve space vector combination through carrier comparison in PWM. Figure 2.19 shows the diagram of achieving SVPWM/DPWM through carrier comparison. On the basis of SPWM, common-mode modulation function e is calculated with the duty cycles. Adding the sinusoidal modulation function with e, the modulation function for SVPWM/DPWM can be obtained. To compare with triangle carrier, the output pulses will be PWM signals for SVPWM/DPWM. This method is easier than space vector combination in implementation and suitable for digital controller.

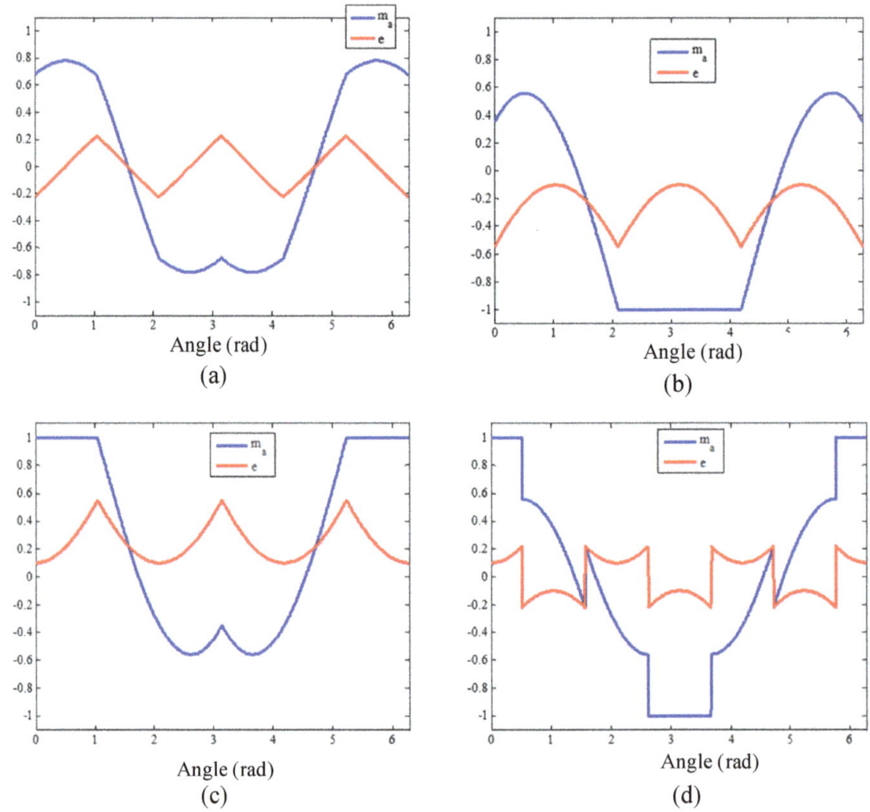

Fig. 2.18 Typical modulation function and common-mode modulation function: **a** seven-zone SVPWM, **b** DPWM_Min, **c** DPWM_Max, **d** DPWM1

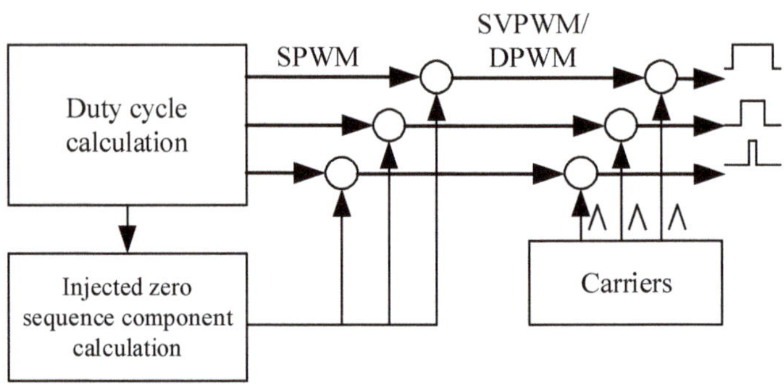

Fig. 2.19 SVPWM/DPWM realization through carrier comparison

2.4 Some Nonideal Factors in PWM Generation

The previous discussions for SVPWM and CBPWM are for ideal cases where digital pulses can be amplified to power pulses directly. In fact, there is an essential difference between digital pulses and power pulses. Because of the equivalent passive parameters (parasitic capacitance and inductance) embedded in the power electronics converters, power conversion should have transient behavior and the delay effects will be more obvious than digital pulses. Also, for safety consideration, there should be more margins in power pulses than digital pulses. Therefore, some nonideal factors should be considered in application.

Taking the most typical voltage source converter for example: Fig. 2.20a is a typical switching unit with controllable switch and anti-paralleled diode. Two of this kind of units will form a phase-leg in Fig. 2.20b. The digital signals of switching signals will be added to the gates of the two switches and achieve the output voltage pulse V_a finally. Because of the transient behavior of power electronics devices and the artificial safety margin, the output V_a is not exact the amplified signal pulse. That is the nonideal factors' impact.

Two groups of nonideal factors will be introduced in this section as well as the improving methods: the first group of nonideal factors is based on the physical characteristics of the converter and the second group of nonideal factors is based on artificial safety margins.

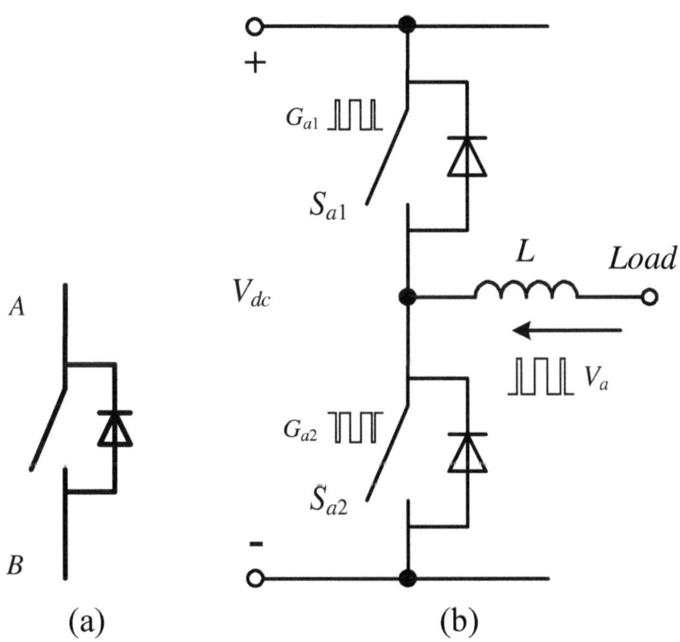

Fig. 2.20 The switch in voltage source converter: **a** switch unit, **b** phase-leg

For the first group of nonideal factors, power electronics devices' steady-state and dynamic characteristics are not ideal. They have both steady-state nonideal characteristics of voltage drop in on-state and leakage current in off-state and dynamic characteristics of switch delay and reverse recovery. Also, the parasitic parameters in propagation loop will make the impact on the power pulses to be worse.

Power electronics devices' steady-state characteristics are mainly the voltage drop in on-state and leakage current in off-state. In the switch unit in Fig. 2.20a, the current will mainly conduct through the switch in positive direction and through the anti-paralleled diode in the negative direction. For the switches like IGBT and MOSFET, the conduction characteristics are shown in Fig. 2.21. A V–I curve for IGBT in conducting mode is shown in Fig. 2.21a. With sufficient gate voltage, the on-mode characteristics of IGBT can be treated as a constant voltage with a series-connected resistor. The V–I curve of MOSFET in conducting mode is shown in Fig. 2.21b. MOSFET is conducting current through the channel and it can be equivalent with a resistor. When calculating the voltage drop in positive direction, the method can be different for IGBT and MOSFET. When considering the negative direction current, the characteristics of the diode is the determining factor. A typical diode conducting curve is shown in Fig. 2.22. It is equivalent with a constant voltage and a series-connected resistor, like IGBT. When current is conducting in negative direction, this model can be used to calculate the voltage drop on the diode.

Because of positive and negative voltage drop, the output voltage pulse is the positive and negative DC voltage together with the voltage drop on the switch units. This voltage drop is determined by the load current. Because of this voltage drop, output voltage is with small difference with the reference voltage and will cause errors for output voltage and current. Artificial voltage compensation is needed in the reference voltage to make the output voltage to be the same with original reference.

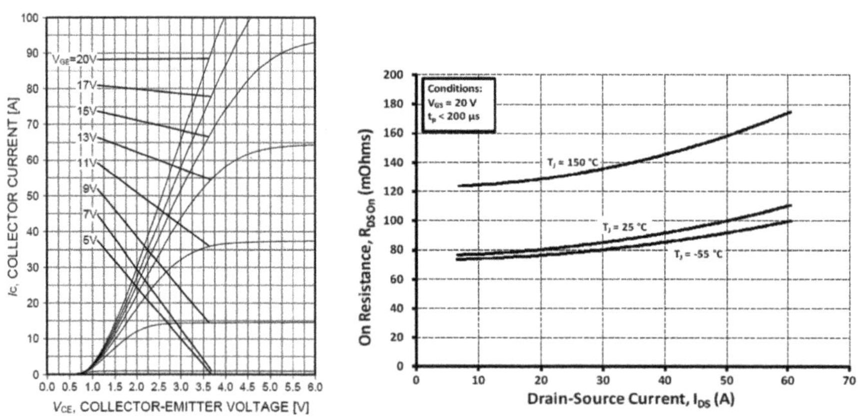

(a) (b)

Fig. 2.21 Typical conducting characteristics of **a** IGBT, **b** MOSFET [2, 3]

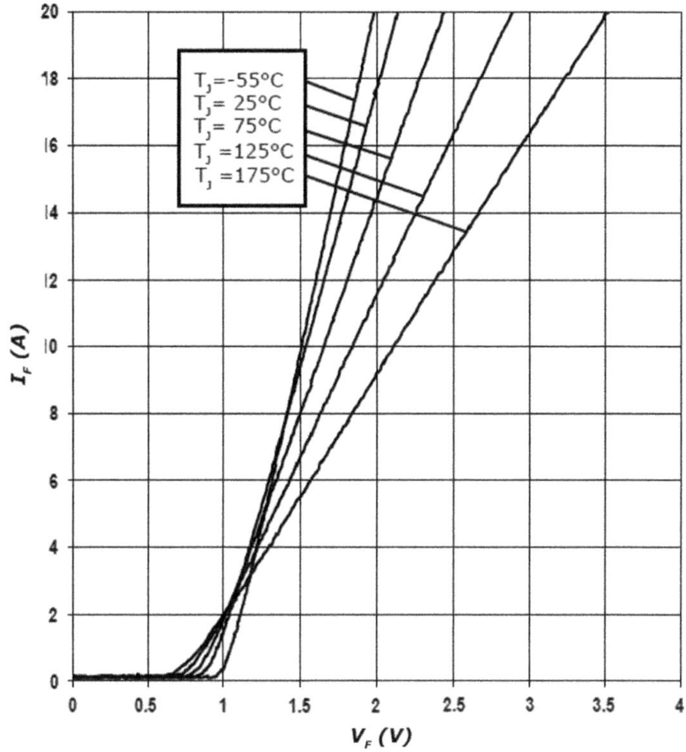

Fig. 2.22 Typical conducting characteristics of diode [4]

Besides the steady-state error, another nonideal factor is the dynamic behavior of the power electronics devices, which is the switching transient of the power electronics devices. Compared with the digital pulses which are close to rectangles, the switching transient of power electronics devices is slower and may be together with reverse recovery, which will make the voltage/current pulses to be different from digital pulses.

In power electronics devices, the first stage of delay appears in gate driver: the gate capacitance and gate driver circuit will bring the first stage delay for the pulses. After that, the switching behavior of power electronics devices will bring the second stage delay. Taking IGBT for example, its switching process is finished in a few hundred nanoseconds usually. The delay is also with this level of time. The output voltage will be more like a trapezoid series than a rectangle series. The slope in the trapezoid is determined by switching speed. SiC devices can significantly reduce the switching time and increase the slope in the trapezoid, making it closer to rectangle series.

The second group of impact is based on the artificial safety margin. Generally speaking, it includes the impact from deadtime and minimum pulse width.

For the phase-leg in Fig. 2.20b, in ideal case, the gate driver signal G_{ap} and G_{an} should be exactly reciprocal and achieve the output voltage which is determined by the reference duty cycle. However, because of the turn-on and turn-off transient, there will be a period of time which two switches are both in on-state, called shoot-through. So, deadtime is needed to guarantee that the gate driver signal of one switch is off but the other is not turned on. It can be compared to the traffic light: when one line's traffic signal is turned from green to red, the other line's signal should be kept red for a while to make sure safety. Generally speaking, deadtime is to make a delay time of T_d for the turn-on signal. For power electronics converter, deadtime is associated with switching speed. The deadtime should be longer than the device turn-on and turn-off time together.

Because of deadtime, power electronics converter's output power pulse will be different from signal pulse of PWM and it is also associated with current direction. Figure 2.23a shows the case when current is flowing out from the phase-leg. In ideal situation, the output voltage U_{AN} is synchronous with G_{ap}. With deadtime, the rising edges of G_{ap} and G_{an} are delayed with T_d. During deadtime period, the gate driver signals for two switches are both low and the two switches are both off. However, the output current cannot be with open-circuit and the current will flow through the anti-paralleled diode. Since the current is flowing out of the phase-leg, the current will flow through the anti-paralleled diode of the lower switch and the output voltage will be clamped to negative DC bus. Compared with gate signal G_{ap}, the output voltage pulse

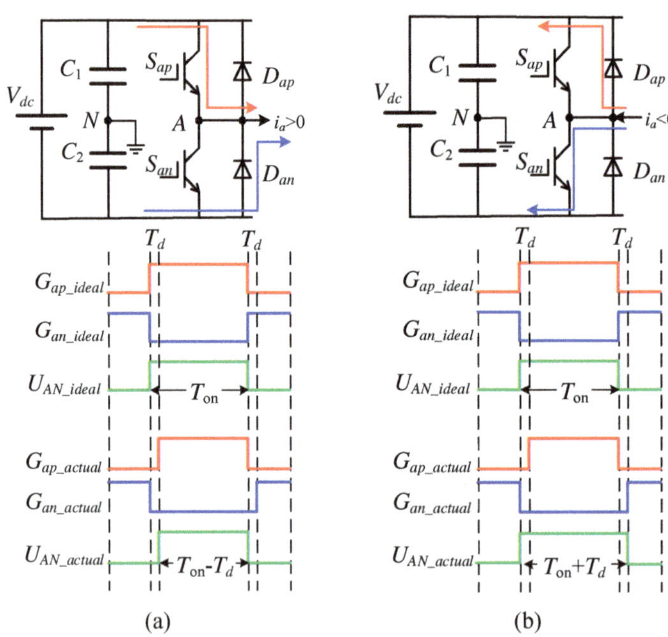

Fig. 2.23 Output voltage and gate driver signal with deadtime: **a** current flow out of phase-leg, **b** current flow toward the phase-leg

is losing duty cycle of T_d/T_s. When current is flowing toward the phase-leg, the case will be opposite, shown in Fig. 2.23b. Although the gate driver signals are the same with Fig. 2.23a, during deadtime period, current will flow through the anti-paralleled diode of the upper switch and the output voltage will be clamped to positive DC bus. Compared with the gate signal G_{ap}, the output voltage pulse is adding duty cycle of T_d/T_s. In these two cases, the output voltage will be different from the reference. But because of the current direction, one case will generate smaller voltage than the reference and the other will generate bigger voltage than the reference.

Based on this principle, Fig. 2.24 shows the reference between the ideal case and the case with deadtime with SPWM, in unit power factor case where current is in-phase with voltage. It can be found that in positive half line-cycle, the case with deadtime will have smaller reference than the ideal case and in the negative half line-cycle, the case with deadtime will have bigger reference than the ideal case. In the period near zero-crossing points, the voltage is clamped to zero. Doing FFT analysis for the two waveforms, the results are shown in Fig. 2.25. From the overview in Fig. 2.25a, because of the deadtime's impact, the fundamental component of the reference voltage will be reduced from the ideal case. But it is more than that: because of the deadtime's impact, the reference voltage will have low-order harmonics like the third-order harmonics, as shown in Fig. 2.25b. It means the impact of deadtime is serious and should be compensated. Otherwise the output voltage amplitude and quality will be influenced.

Another nonideal factor is the minimum pulse width. When the output duty cycle is small, the turn-on and turn-off time will reduce the pulse width. With deadtime, the pulse width will be further reduced. The pulse width can be negative in mathematics, but it cannot be realized physically. So, the minimum pulse width should be limited when the original pulse width is smaller than a threshold. This will also impact on the original reference and should be compensated.

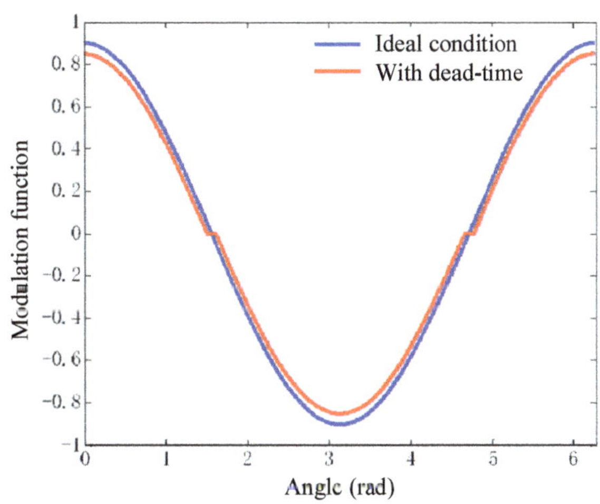

Fig. 2.24 Ideal reference voltage and the equivalent reference with deadtime

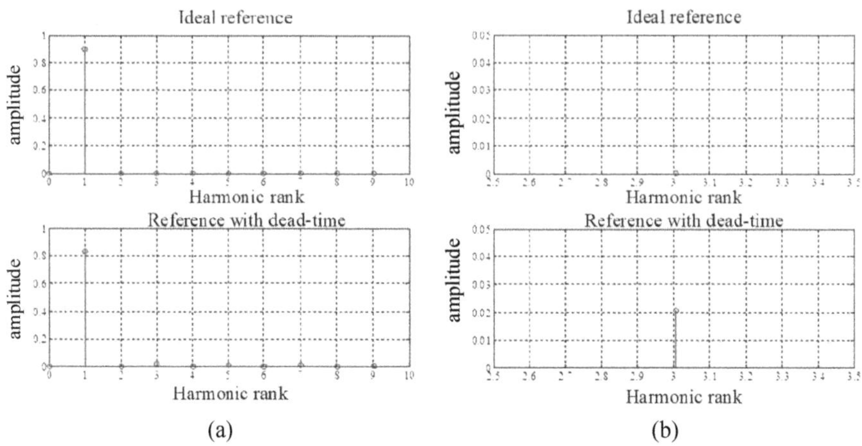

Fig. 2.25 Comparison of the spectrum of reference voltage between ideal case and the case with deadtime: **a** overview, **b** zoom-in details for third-order harmonics

2.5 Mathematical Analysis Method for PWM

PWM is approaching reference with pulse series, which must be with harmonics and distortion. In time domain, these distortions and harmonics represent as the errors from the continuous sinusoidal voltage/current, but are hard to quantify. Mathematical analysis tools are necessary to quantify these errors. Fourier transformation is the major tool to achieve time–frequency domain transformation [5].

With Fourier transformation for voltage or current waveform $x(t)$ which is with main cycle period of T, it can be shown in (2.18):

$$x(t) = X_0 + X_1 \cos(\omega_1 t + \varphi_1) + X_2 \cos(2\omega_1 t + \varphi_2) + X_3 \cos(3\omega_1 t + \varphi_3) + \cdots \tag{2.18}$$

In (2.18), ω_1 is the fundamental frequency, X_1 and ϕ_1 are the amplitude and phase of the fundamental component, respectively, which are the target for the PWM. Other components are unwanted harmonics, including the DC component X_0 and high-frequency harmonics of X_2, X_3 and so on.

Although the output voltage/current can be positive or negative, the load power losses mainly work as ohmic losses. For evaluation of harmonics, RMS value is an important parameter. To calculate the mean square root of two sides in (2.18) and consider the orthogonality of the different harmonics, the RMS value can be expressed in (2.19).

$$X_{rms} = \sqrt{\frac{1}{T} \int_0^T \sum_{n=0}^{\infty} X_n^2 \cos^2(n\omega_1 t + \varphi_n) dt}$$

$$= \sqrt{\frac{1}{T}\int_0^T \left[X_0^2 + \sum_{n=1}^{\infty} \frac{X_n^2}{2}(1 + \cos 2(n\omega_1 t + \varphi_n))\right] dt} \quad (2.19)$$

Because the integral of cosine function in the line-cycle T will be zero, (2.19) can be simplified as (2.20). It means the RMS value of time-domain waveform is equal to the square root of the sum of all harmonics' square value.

$$X_{rms} = \sqrt{\frac{1}{T}\int_0^T \left[X_0^2 + \sum_{n=1}^{\infty} \frac{X_n^2}{2}\right] dt} = \sqrt{X_0^2 + \sum_{n=1}^{\infty} \frac{X_n^2}{2}}$$
$$= \sqrt{X_{0,rms}^2 + \sum_{n=1}^{\infty} X_{n,rms}^2} \quad (2.20)$$

In most cases, the expected output is the fundamental component X_1 and other components are treated as harmonics. Equation (2.20) can be expressed as (2.21).

$$X_{rms} = X_{1,rms}\sqrt{1 + \left(\frac{X_{0,rms}}{X_{1,rms}}\right)^2 + \sum_{n=2}^{\infty}\left(\frac{X_{n,rms}}{X_{1,rms}}\right)^2} \quad (2.21)$$

Then, the total harmonics distortion (THD) can be expressed as (2.22). It is the square root of the sum of all harmonics' square value together with the DC component's square value, expressed as the relative value with the fundamental component. It can represent the distortion of all harmonics together.

$$THD = \sqrt{\left(\frac{X_{0,rms}}{X_{1,rms}}\right)^2 + \sum_{n=2}^{\infty}\left(\frac{X_{n,rms}}{X_{1,rms}}\right)^2} \quad (2.22)$$

With the defined THD, X can be current or voltage. If X is current, THD is with more clear physical meaning since the ohmic losses are proportional to square of different harmonic currents. THD's square is proportional to the sum of all harmonics' square and can represent the ohmic losses of all harmonics. However, when X is voltage, THD is not able to represent the ohmic losses. Because most of the typical loads are inductive, like electrical machine, the impedance is proportional to frequency. With different harmonics, the impedance is also proportional to the order of harmonics. In this case, weight-THD (WTHD) is defined, and shown in (2.23). Treating DC component to be zero, the calculation of the sum of square values of each order of harmonics is with a weight coefficient of $1/n$ for each. WTHD can express the fact that the impedance is proportional to the order of harmonics. Figure 2.26 shows the voltage and current waveforms of a PWM converter as well as the harmonics of voltage, weighted harmonics of voltage and current harmonics.

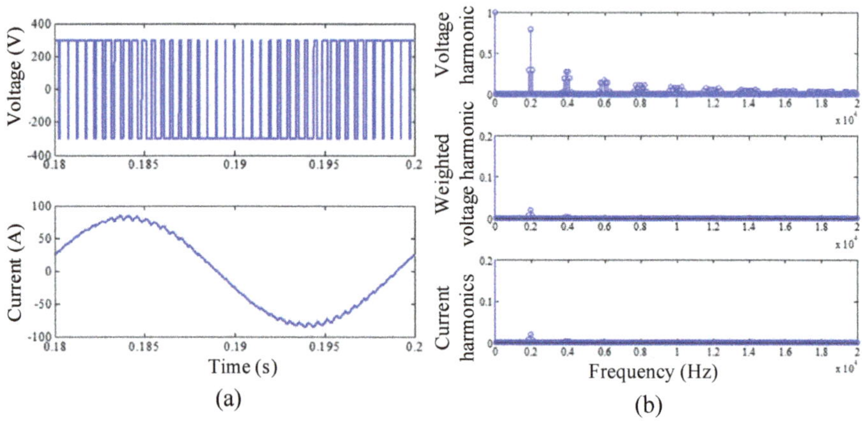

Fig. 2.26 Output voltage and current of a PWM inverter: **a** time domain, **b** harmonics spectrum

It is clear that harmonics of voltage is much higher than current harmonics. The weighted harmonics of voltage is close to the harmonics of current. Voltage THD is 1.21, voltage WTHD is 0.023 and the current THD is 0.025.

$$WTHD = \frac{\sqrt{\sum_{n=2}^{\infty}\left(\frac{X_{n,rms}}{n}\right)^2}}{X_{1,rms}} \tag{2.23}$$

Also, different load can be with different impact on harmonics impedance and losses. For example, in induction machine, rotor resistance, rotor leakage inductance and copper losses will impact the weight coefficient. With consideration of all these factors, more complex weight coefficient will be generated than (2.22). The evaluation of harmonics should be together with load [5].

The above-mentioned mathematical analysis is through Fourier transformation. Fourier transformation is called "the eyes for signal processing", which can analyze arbitrary waveform in time-domain and obtain the frequency-domain results. Since the invention of fast Fourier transformation (FFT), it has been widely implemented. But Fourier transformation has a conflict between precision and calculation for high-frequency PWM system and it is hard to express each component simply. Because of these reasons, double-Fourier analysis has been applied in PWM system which is with analytical expression. It has achieved better results than FFT in many cases [5].

One special characteristic for PWM system is that it contains two major frequency components: fundamental frequency component and carrier frequency component. These two kinds of frequency components are relatively independent and can be expressed in (2.24) as two variables in time domain.

$$\begin{cases} x(t) = \omega_c t + \theta_c \\ y(t) = \omega_0 t + \theta_0 \end{cases} \tag{2.24}$$

ω_c and θ_c are frequency and phase of carrier, ω_0 and θ_0 are frequency and phase of fundamental component, respectively.

For a time-domain signal of $f(t)$, it can be expressed as two-dimensional function of $f(x, y)$. Then, with similar analysis like Fourier transformation, it can be expressed in (2.25).

$$\begin{aligned} f(x, y) = {} & \frac{A_{00}}{2} + \sum_{n=1}^{\infty} (A_{0n} \cos ny + B_{0n} \sin ny) + \sum_{m=1}^{\infty} (A_{m0} \cos mx + B_{m0} \sin mx) \\ & + \sum_{m=1}^{\infty} \sum_{\substack{n=-\infty \\ n \neq 0}}^{\infty} \left[A_{mn} \cos(mx + ny) + B_{mn} \sin(mx + ny) \right] \end{aligned} \tag{2.25}$$

The coefficients for double-Fourier analysis are shown in (2.26).

$$\begin{cases} A_{mn} = \dfrac{1}{2\pi^2} \displaystyle\int_{-\pi}^{\pi} \int_{-\pi}^{\pi} f(x, y) \cos(mx + ny) dx dy \\ B_{mn} = \dfrac{1}{2\pi^2} \displaystyle\int_{-\pi}^{\pi} \int_{-\pi}^{\pi} f(x, y) \sin(mx + ny) dx dy \end{cases} \tag{2.26}$$

Combining (2.24) and (2.25), (2.27) is derived:

$$\begin{aligned} f(x, y) = {} & \frac{A_{00}}{2} + \sum_{n=1}^{\infty} (A_{0n} \cos n(\omega_0 t + \theta_0) + B_{0n} \sin n(\omega_0 t + \theta_0)) \\ & + \sum_{m=1}^{\infty} (A_{m0} \cos m(\omega_c t + \theta_c) + B_{m0} \sin m(\omega_c t + \theta_c)) \\ & + \sum_{m=1}^{\infty} \sum_{\substack{n=-\infty \\ n \neq 0}}^{\infty} [A_{mn} \cos(m(\omega_c t + \theta_c) + n(\omega_0 t + \theta_0)) \\ & + B_{mn} \sin(m(\omega_c t + \theta_c) + n(\omega_0 t + \theta_0))] \end{aligned} \tag{2.27}$$

The first item in the right side of (2.27) is DC bias, the second item is fundamental frequency component, and the third item is switching frequency component. The fourth item contains every order of switching frequency and the sidebands. It has been introduced in [5] and is not further introduced in this chapter.

Both Fourier transformation and double Fourier transformation are time-to-frequency transformation. They have their own shortages. First, lack of time and frequency location capability: for certain signal, Fourier-based methods cannot locate the frequency of a certain time period and cannot locate the time for a certain frequency component. This shortage limits the implementation for switching frequency variation. Second, Fourier-based methods are not able to deal with nonsteady-state signal. They usually assume the frequency is not varying with time and not able to deal with the transient process and so on. Third, Fourier-based methods are limited in resolution. These shortages limit the implementation for PWM analysis. In recent years, short-time Fourier transformation and wavelet analysis have been introduced in PWM analysis, too. But this chapter will not go to the details [6].

2.6 Summary

As the basis for this book, this chapter has introduced the basic principles and critical problems for PWM technologies. With the most typical three-phase voltage source converter as an example, space vector PWM and carrier-based PWM have been introduced as the two major methods. The introduction of SVPWM has been with the standard voltage vectors and the space vector combinations. The introduction of CBPWM has been with natural sampling and regular sampling. Based on SPWM and CBPWM, injection of third-order harmonics can help to increase the maximum modulation index, which is HIPWM. With mathematical analysis, the relationship between CBPWM and SVPWM has been built up: each SVPWM combination has an equivalent CBPWM; the only difference is the common-mode modulation function; each CBPWM is also has an equivalent SVPWM; and the only difference is the arrangement of zero vectors. Then, a few of the nonideal factors in PWM have been introduced in this chapter, including the natural factors of power electronics converter itself and the artificial factors for safety margins. These factors include power electronics devices' voltage drop, switch delay, deadtime and minimum pulse width and so on. Mathematical analysis methods have been introduced in this chapter also, including the methods based on Fourier transformation, THD and WTHD.

Continuous progress of PWM has been made in the past half century. With the book published in 2003 [5] as a milestone, PWM has been with a system. However, there are still many topics inside the system of PWM: power losses will be accumulated together with PWM. PWM will cause current ripple in the converter and will also bring electromagnetic interference (EMI) problem for the converter. These three problems are always existing. Previous studies of PWM are mainly with the PWM basic functions and the side effects have not been with serious consideration. This book is aiming at this issue.

References

1. Zhou K, Wang D (2002) Relationship between space-vector modulation and three-phase carrier-based PWM: a comprehensive analysis [three-phase inverters]. IEEE Trans Ind Electron 49(1):186–196
2. Datasheet, IGW25N120H3, Infineon
3. Datasheet, C2M0080120D, CREE
4. Datasheet, C4D10120E, CREE
5. Holmes DG, Lipo TA (2003) Pulse width modulation for power converters: principles and practice. Wiley
6. Qian S (2001) Introduction to time-frequency and wavelet transforms. Prentice Hall
7. Mohan N, Undeland TM, Robbins WP (2002) Power electronics: converters, applications, and design, 3rd ed. Wiley
8. Blasko V (1997) Analysis of a hybrid PWM based on modified space-vector and triangle-comparison methods. IEEE Trans Ind Appl 33(3):756–764

Chapter 3
Pulse Width Modulation's Impact

PWM technologies have multiple impacts on the power electronics converters. First, the switching losses of power electronics devices will be accumulated together with PWM and impact the converter power losses. Second, pulse voltage will cause ripple on the passive components in the output side. Also, high-frequency components in PWM will generate electromagnetic interference to the system. In the last section of this chapter, there is an overview on several typical spread spectrum PWM strategies, which are opening words for the following chapters.

3.1 Introduction of PWM on the System Performance

Power electronics converter realizes energy conversion through switching, which is not an ideal process. In power conversion process, power electronics devices and passive devices are with different stress. Also, since the switching action is in a short time scale, there is a wide-band excitation being added to the system and the influence on the system will be more complex in frequency domain. PWM as the actuation element in the power conversion has significant impact on the system in multiple aspects. Modification and optimization for PWM should be based on understanding of these impacts.

Taking the most typical two-level voltage source converter as an example, Fig. 3.1 shows the phase-leg architecture. With gate signals, midpoint of the phase-leg will generate PWM voltage series which is synchronous with the gate signal. Power electronics devices will bear the voltage stress in the switching process; also, current in the power electronics devices will have fast change in the switching process. Power electronics devices will have this process in every switching action and generate switching losses. Switching losses will continuously accumulate with PWM. PWM's first impact on the system performance is the accumulation of switching losses.

D. Jiang et al., *Advanced Pulse-Width-Modulation: With Freedom to Optimize Power Electronics Converters*, CPSS Power Electronics Series,
https://doi.org/10.1007/978-981-33-4385-6_3

Fig. 3.1 Typical phase-leg in voltage source converter

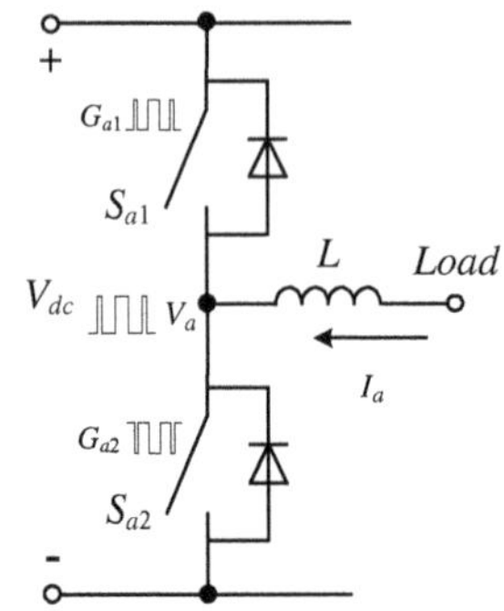

In Fig. 3.1, the midpoint voltage is a series voltage switching between positive and negative DC bus voltage. However, the load voltage is mainly with continuous characteristics. Taking motor drive as an example, load voltage is mainly the back-EMF of the motor. Taking grid-connected converter as an example, load voltage is mainly the sinusoidal grid voltage. Although as discussed in Chap. 2 that PWM can make the average voltage of the pulse voltage to be the same as the reference voltage, there is an obvious difference between the pulse voltage and the load continuous voltage in every switching cycle. Therefore, inductive components should be added between the converter and the load to decouple the switching voltage and continuous voltage. The error between the pulse voltage and the load voltage will be added on the inductors and it generates current ripple. Because of the switching characteristics of the power electronics converter, this kind of PWM current ripple is inevitable. PWM will determine the source of the current ripple: voltage errors. For current source converter, the output is pulse current and the ripple is mainly voltage ripple. PWM's second impact on converter is (voltage/current) ripple.

In Fig. 3.1, because the pulse voltage series in the midpoint of phase-leg has complex spectrum, the excitation on the system is not only with the fundamental frequency and its harmonics but also with wide-band characteristics. The high-frequency components in the pulse voltage will generate electromagnetic interference (EMI) noise and it will conduct in the system. Based on conductive path difference, EMI can be divided as differential-mode (DM) EMI and common-mode (CM) EMI. DM EMI conducts between positive and negative DC bus and between phases, which are the main electrical paths in the converter. CM EMI conducts in the same direction in the DC bus and AC phases, through the parasitic parameters to the ground which is the parasitic path of the converter. EMI is mainly determined by two major parts. One is the devices' switching characteristics (di/dt and dv/dt), which is usually with time scale below μs and EMI frequency range above MHz. This part of high-frequency EMI is usually determined by power electronics devices and gate drivers. The other is the combination of pulse series, which is with time scale determined by switching frequency and the EMI frequency range is below MHz usually. This part of low-frequency EMI is influenced by PWM. Especially, CM EMI's source is the combination of the switching actions of all phases and is more complex than DM EMI. Therefore, the third major impact of PWM on the converter is EMI.

In fact, PWM's three major impacts are the by-products of PWM and are the necessary issues for PWM technologies. Sections 3.2, 3.3 and 3.4 will introduce the details for these three impacts.

3.2 PWM and Switching Losses

Power semiconductor devices are not ideal switches. Because of the ohmic effect of the path of the carrier and junction voltage drop, there will be conduction losses in the devices. Also, during the transient period of switching, main voltage and current will exist in the device together and generate switching losses. Generally speaking, the conduction losses are more related to load. However, the switching losses are more related to modulation scheme. The power losses are not only influencing the efficiency of power conversion but also bringing thermal stress on the devices and influencing the design of thermal management components like heat-sinks.

The power losses in power electronics converters are mainly contributed by the conduction losses and switching losses of the power electronics devices. Switching losses will be accumulated together with PWM which will determine the accumulation locations and action times. PWM is the main factor to determine the switching losses in the converter. For conduction losses, PWM is also a factor to impact, because PWM will impact the current ripple and will indirectly influence the conduction losses caused by current ripple. However, in general case, the conduction losses are mainly determined by load current which is not associated with PWM directly. PWM's contribution to the full conduction losses is small. This chapter is mainly discussing the PWM's impact on switching losses.

Figure 3.2 shows a typical phase-leg in voltage source converter and the voltage and current waveforms of the two switches in turn-on and turn-off transient. Figure 3.2a defines the directions of the two switches (S_1 and S_2) and their voltages (V_1 and V_2) and currents (I_1 and I_2). The DC-link voltage is V_{dc} and the load

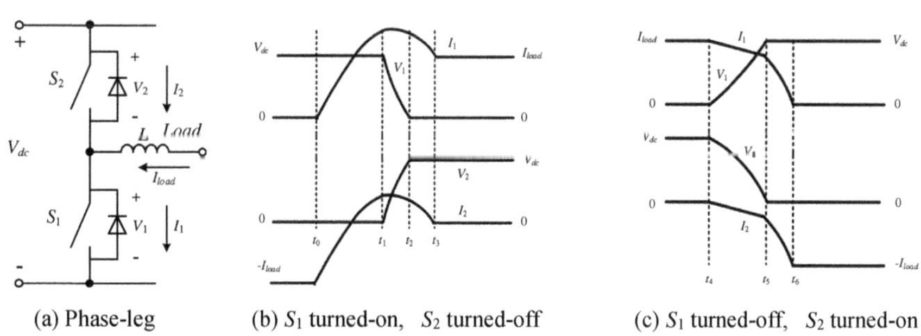

Fig. 3.2 Phase-leg in voltage source converter and the corresponding waveforms

current is I_{load}. Because of the DC-link capacitor and load inductor, V_{dc} and I_{load} will not change in the switching transient period. Then, based on Kirchhoff's law for voltage and current, the voltage and current waveforms of S_2 can be derived from those of S_1, shown in Fig. 3.2b, c. In Fig. 3.2b, c, the two switches are both with voltage/current overlap areas. Figure 3.2b shows the case when S_1 is turned on and S_2 is turned off: I_1 will rise at t_0 together with diode reverse-recovery, V_1 will drop from V_{dc} at the time of t_1 and finish at t_2. Reverse-recovery will be finished at t_3 and I_1 will be balanced with I_{load}. The turn-on transient is finished. The commutation transient of S_2 is synchronous with this process: based on $V_1 + V_2 = V_{dc}$ and $I_1 - I_2 = I_{load}$, the waveforms of V_2 and I_2 can be derived. Figure 3.2c shows the case when S_1 is turned off and S_2 is turned on: I_1 will drop slowly from I_{load} at t_4 and the voltage V_1 will begin to rise. When V_1 reaches to the stable value of V_{dc} at t_5, I_1 will drop quickly and reach zero at t_6 and the commutation is finished. Similar to Fig. 3.2b, the waveforms of S_2 can be derived from those of S_1 [1].

Based on Joule's law, the integral of the product of voltage and current on each switch in each switching transient period is the switching loss in this switching action. Therefore, the sum of the integral results of each switch in the phase-leg in one switching cycle is the total switching energy losses of the phase-leg in the switching cycle, as shown in (3.1).

$$E = E_1 + E_2 + E_3 + E_4 = \int_{t=t_0}^{t_2} V_1 I_1 dt + \int_{t=t_1}^{t_3} V_2 I_2 dt + \int_{t=t_4}^{t_6} V_1 I_1 dt + \int_{t=t_4}^{t_5} V_2 I_2 dt \tag{3.1}$$

In (3.1), there are three major factors to influence the switching loss: voltage, current and time. In voltage source converter, the DC-link voltage is constant; current will change with load; and switching time will be determined by the power electronics devices and the gate drivers. For the full converter switching loss calculation, sum of the losses in the whole line-cycle is needed. One line-cycle can be divided into N switching cycles. The total switching losses are the sum of the losses in N switching cycles. $E_{sw,x}(n)$ is the switching loss in the nth switching cycle ($n = 1, 2, 3, \ldots, N$) in the xth phase-leg ($x = a, b, c$). $E_{sw,x}(n)$ is proportional to the absolute value of the phase current $|i_x(n)|$. If defining the switching loss under unit current to be e_{unit}, the full switching losses in the three-phase converter in the whole line-cycle will be expressed in (3.2). Power losses can be expressed as $P_{sw} = E_{sw}f$, where f is the fundamental frequency.

$$E_{sw} = \sum_{x=a,b,c} \sum_{n=1}^{N} E_{sw,x}(n) = \sum_{x=a,b,c} \sum_{n=1}^{N} |i_x(n)| e_{unit} \tag{3.2}$$

From (3.2) it can be derived that there are a few methods to reduce switching losses in the converter. One method is to reduce the per-unit switching losses e_{unit}, which is mainly through the faster switching devices and gate drivers. The other method is to reduce the equivalent switching actions N, and the corresponding load current $|i_x(n)|$ at the switching point. These are impacted by PWM directly. Reduction of equivalent switching frequency can effectively reduce the switching losses. Switching at proper time can further reduce the switching losses. However, the switching frequency also determines the power quality of the converter and cannot be reduced arbitrarily. Therefore, further study of PWM's impact is needed after the study of PWM's impact on switching losses.

3.3 PWM and Ripple

Power electronics converters are not ideal voltage sources or current sources, but generate discrete voltage pulses or current pulses. There is an obvious difference between output voltage/current and the reference. Therefore, output current/voltage ripple is generated as by-products. Taking voltage source converter as an example, Fig. 3.3 shows the phase-leg and the load. The terminal voltage of the phase-leg is pulse voltage generated by PWM, but load voltage is continuous. If the load is motor, the load voltage will be the back-EMF. If the load is grid, the load voltage will be grid voltage. Based on volt-seconds balancing principle, the average value of the pulse voltage will be the same as the load voltage in each switching cycle. However, during each switching cycle, pulse voltage V_s will have obvious difference from V_a. This difference will drop on the output inductor and generate current ripple. This ripple will impact the system vibration, noise and electrical performance, which is usually within limit.

The slope of current in the inductor (di/dt) is proportional to the voltage on the inductor. Since the difference between the pulse voltage and the load voltage is added on the inductor, each pulse sector will be with a linear current ripple, and the full ripple current in each switching cycle will be with zig-zag waveform. On understanding of this principle, current ripple can be effectively predicted. For simple single-phase inverter, the switching combination is simple and the current ripple can be predicted easily. But for three-phase and multiphase converter, current ripple will be impacted by complex switching actions and the logic is also complex. For

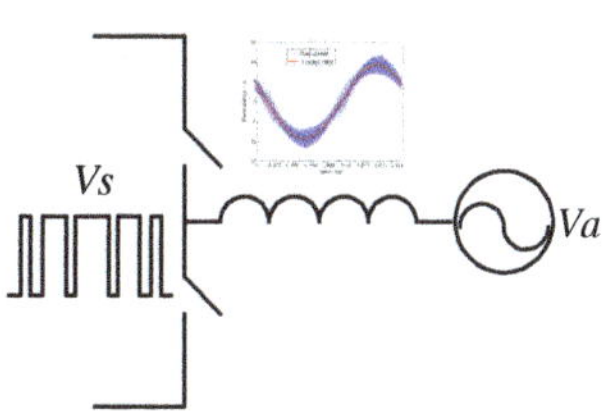

Fig. 3.3 Typical phase-leg and load of voltage source converter

different load connections, current ripple will be different too. Chapter 4 of this book will introduce the current ripple prediction method.

Voltage source converter generates pulse voltage to the load and current ripple. Correspondingly, current source converter will have different characteristics. Figure 3.4 shows a typical current source converter and its load. DC side inductor will keep the DC side with current source characteristics. The output terminals are generating pulse current and the three-phase output capacitors are with continuous (sinusoidal) voltage. Since the load is with three-phase continuous current, the error between output current pulses and sinusoidal load current will be charging the three-phase capacitors and generating voltage ripples based on sinusoidal voltage. Therefore, PWM will bring voltage ripple in the AC side for current source converter instead of current ripple.

PWM impacts not only AC side current/voltage ripple but also the DC side ripple. Figure 3.5 shows the relationship between AC side current and DC-link current in a voltage source converter. When the upper switch of one phase-leg is turned on, the corresponding phase current will flow into the positive DC bus. Then, the DC-link current will follow (3.3). S_a, S_b and S_c are the three-phase switching functions whose values are 0 (off) or 1 (on). The DC-link current is chopped current pulses generated by PWM. The DC-link current will directly influence the DC-link capacitor design or the battery design. Obviously, PWM is a key technology for DC-link design.

$$i_{dc} = S_a i_a + S_b i_b + S_c i_c \tag{3.3}$$

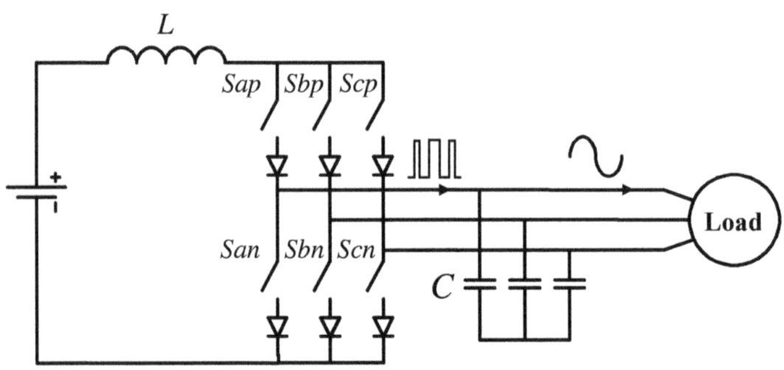

Fig. 3.4 Typical current source converter and load

Fig. 3.5 DC-link current analysis of voltage source converter

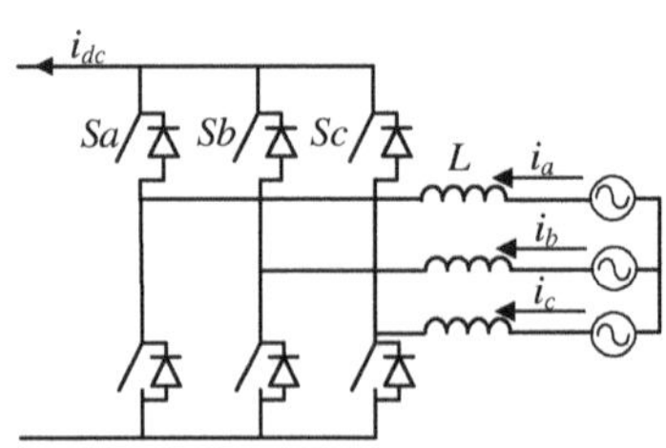

3.4 PWM and Electromagnetic Interference (EMI)

In mathematics, step function can be with infinite bandwidth in frequency domain theoretically. Because power electronics devices' switching function is close to step function for current or voltage, it brings wide bandwidth voltage or current response to the system. Especially, the high-frequency components in the response will conduct as electromagnetic interference (EMI); even higher frequency components will propagate as electromagnetic wave. That is the cause of EMI in the power electronics system. In conventional electrical or electronics circuit, EMI sources are usually the electronics components. Since the voltage and current variation in power electronics devices is much larger than microelectronics devices, EMI in power electronics circuit is worse and brings concern to electromagnetic compatibility (EMC) and the corresponding standard. Generally, if the conducted EMI can be mitigated, radiated EMI can be mitigated too. This book is mainly for conducted EMI study. With different EMI standards, conducted EMI's range will be different. Usually, the low range is not less than 10 kHz and the high range is not more than 30 MHz.

Taking voltage source converter as an example, Fig. 3.6 shows the EMI paths. Conducted EMI can be divided into differential-mode (DM) EMI and common-mode (CM) EMI. DM EMI will conduct between phases in the converter and will be in opposite directions in the DC bus. CM EMI will conduct together in the same direction in the AC bus to the load and through the parasitics capacitors to the ground. It will be in the same direction in the DC bus.

Based on the EMI analysis in Fig. 3.6, EMI equivalent circuits can be derived in Fig. 3.7. Figure 3.7a is the equivalent circuit for AC side EMI. The switching voltage of the output point of each phase-leg can be modeled as EMI source, which can generate EMI through three-phase inductors and load voltage sources to the AC neutral point O_2. The equivalent grounding impedance is between O_1 and O_2. With the noise source model (V_{aO1}, V_{bO1}, V_{cO1}) and the impedance in each phase, the EMI current in each phase can be derived, that is DM EMI. The sum of the three switching voltage sources and the CM impedance will determine the CM EMI current. Figure 3.7b is the EMI in the DC side. DC bus current is the sum of the product between three-phase currents and the switching functions (S_a, S_b, S_c).

Based on the EMI principle in the AC side and DC side, it can be found that the EMI noise source is directly determined by switching functions. Generally, EMI

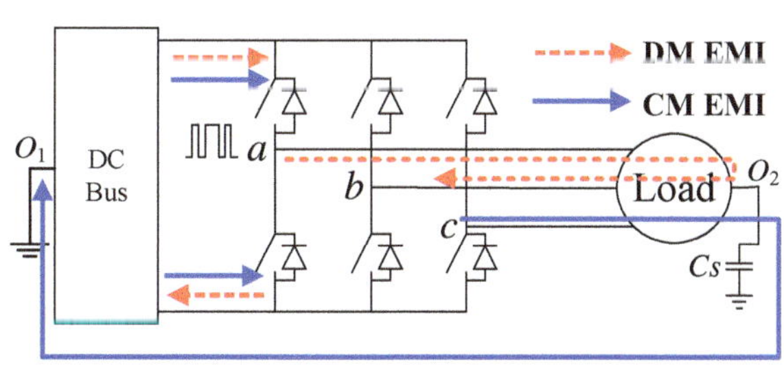

Fig. 3.6 EMI in voltage source converters

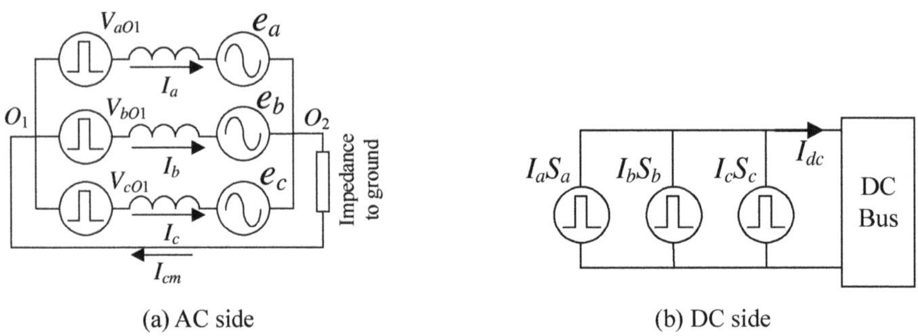

Fig. 3.7 EMI of voltage source converter in the AC and DC side

noise source can be divided into two major parts in frequency domain. One part is the current and voltage step response, which determines the high-frequency EMI. The time scale of this part is usually below μs. The other part is the combination of pulses, which is usually with time scale above μs. The first part is influenced by power electronics device characteristics and gate drivers, while the second part is influenced by PWM. For conventional PWM with constant switching frequency, EMI noise spectrum will concentrate near the harmonics of constant switching frequency and make the EMI below MHz the main switching harmonics [2].

Compared with DM EMI, CM EMI is more special. CM EMI is highly related to the nonideal characteristics of the converter system including parasitics, and the CM EMI source is determined by all the switching actions together. In Fig. 3.7a, the CM voltage is determined by the sum of three-phase switching voltages. It is highly related to the modulation schemes. In motor drive application, CM EMI not only contributes to the full system EMI, but also impacts the insulation and bearing in the motor and the reliability of the system. Mitigation of CM EMI is a very practical issue.

3.5 Overview of Spread Spectrum PWM

In Chap. 2, it has been mentioned that conventional PWM technologies have a few important freedoms which are not well utilized. The major reason is that the focusing point of the previous study on PWM is the fundamental frequency components and low-order harmonics, while the influence of switching frequency and higher frequency range is lacking. This chapter introduced the three major impacts of PWM: switching losses, voltage/current ripple and EMI. Aiming at the improvement for these performances, freedoms in PWM should be utilized. Before building up the prediction model for PWM in the next chapter, several spread spectrum PWM strategies are going to be introduced in this chapter, including random PWM

(RPWM) [3–6], programmed PWM [7–9] and periodic PWM [10–12]. It does not need prediction model, but can improve some performances as well.

Conventional PWM strategies have the following common characteristics: (1) fixed switching frequency; (2) PWM pulse is distributed symmetrically in each switching cycle; (3) the injected zero-sequence component is regular. This approach is simple for converter design, but brings the EMI near the harmonics of the constant switching frequency and makes the EMI peak to be high. EMI filter should be designed to mitigate the high EMI. One approach to improve is to spread the EMI noise in a wide range and the EMI peak can be reduced. In fact, assuring the volt-second balance in any switching cycle, the change of switching frequency, pulse position and zero-sequence component can achieve the "spread spectrum" effect on high-frequency harmonics and reduce the EMI peak. Focusing on reducing the peak value of high-frequency harmonic and EMI noise of the converter system by using PWM freedom, "spread spectrum PWM" has become a research hotspot in recent decades, which can be divided into RPWM, programmed PWM and periodic PWM. Figure 3.8 shows the comparison of EMI between conventional PWM and spread spectrum PWM. With conventional PWM, the EMI energy will be concentrated near the integer harmonics of the constant switching frequency and make the EMI peak to be high and easy to exceed the EMI standard. With spread spectrum PWM, energy will be distributed in a wider range in frequency domain. Although the main EMI energy is similar to conventional PWM, the EMI peak can be effectively reduced, which can make it easier to meet the EMI standard. Generally speaking, spread spectrum modulation mainly controls PWM pulse series, and little attention is paid to switching transient switching process. Therefore, in terms of EMI suppression effect, spread spectrum modulation is mainly aimed at the conduction EMI in the middle and low-frequency range, and has little influence on the high-frequency EMI. Compared with pulse position and zero-sequence component, PWM with variable switching frequency has better "spread spectrum" effect.

The main idea of RPWM is to randomize one PWM freedom or more, so as to achieve the effect of spread spectrum, and suppress EMI peak. Taking switching frequency freedom as an example, RPWM is realized through Fig. 3.9. In control software, a group of random numbers is set as rand[n]. An integer n is set up as

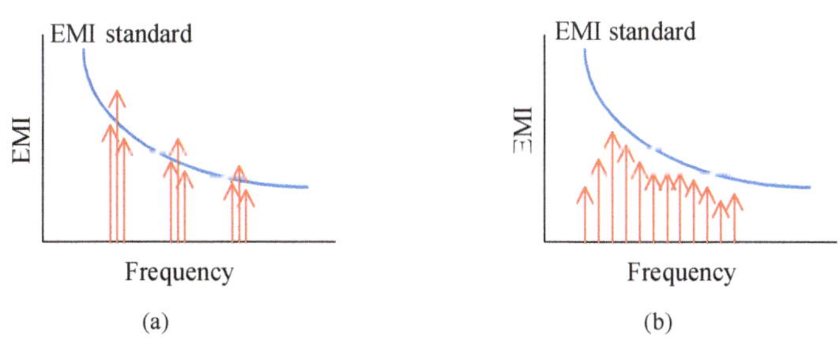

Fig. 3.8 EMI comparison between different PWM strategies. **a** conventional PWM, **b** spread spectrum PWM

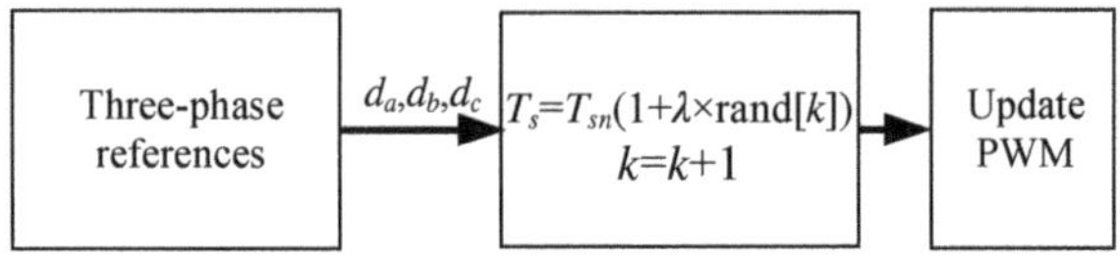

Fig. 3.9 RPWM generation principle

the quantity of random numbers to make sure randomization. Every number in rand[n] is a random number between -1 and 1. T_{sn} is a standard switching cycle. In order to control the switching frequency variation range, coefficient λ is designed. In every switching cycle, switching cycle T_s is updated and the carrier cycle is updated synchronously. Then, T_s is updated with the next random number in the next cycle. Therefore, the switching cycle will be randomly updated and the switching frequency is also varying in a controllable range. EMI will be distributed in a wider range than near the harmonics of constant switching frequency.

This part gives a real example to show the function of RPWM. Figure 3.10 shows the line voltage, DC-link voltage and phase current in a three-level rectifier with three PWM methods. Figure 3.10a, b shows 70 and 40 kHz constant switching frequency SVPWM, and Fig. 3.10c shows RPWM with switching frequency range between 55 and 70 kHz. It is clear that the waveforms in three cases are similar. The low-frequency spectra for the voltages and currents in three cases are also similar, indicating that the power quality has not been impacted obviously with RPWM. But the EMI in three cases shown in Fig. 3.11 has proved that with RPWM, the peak value is reduced from that of the cases with constant switching frequency PWM. EMI will be easier to meet the EMI standard with RPWM, proving the effect of RPWM for EMI mitigation.

The concept of programmed PWM is based on the decomposition of PWM signal, which depends on the formula of a given waveform and its statistical characteristics. Programmed PWM needs to establish the mathematical model of output voltage and carry out reasonable design for each switching position, as shown in Fig. 3.12. The goal of programmed PWM is to eliminate a set of harmonic components or to reduce the weighted square sum of a set of harmonic components. The technical difficulty of programmed PWM is to solve the nonlinear transcendental equation of its output voltage mathematical model. Generally, switching position can only be calculated offline, which is difficult to be solved online in real time. It is not easy

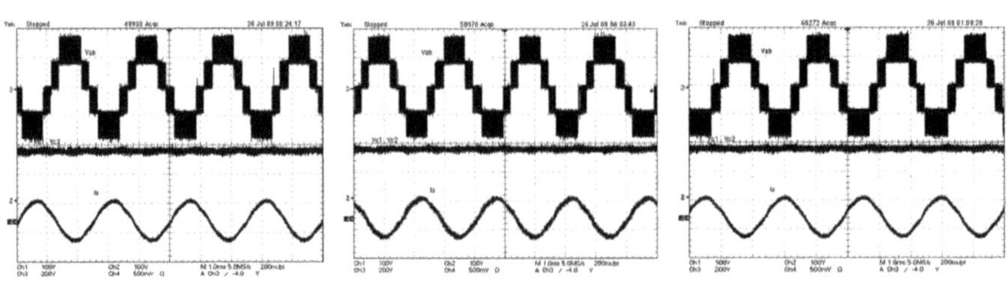

(a) 70 kHz SVPWM (b) 40 kHz SVPWM (c) 55~70 kHz RPWM

Fig. 3.10 Voltage and current waveforms for three different PWM methods [6]

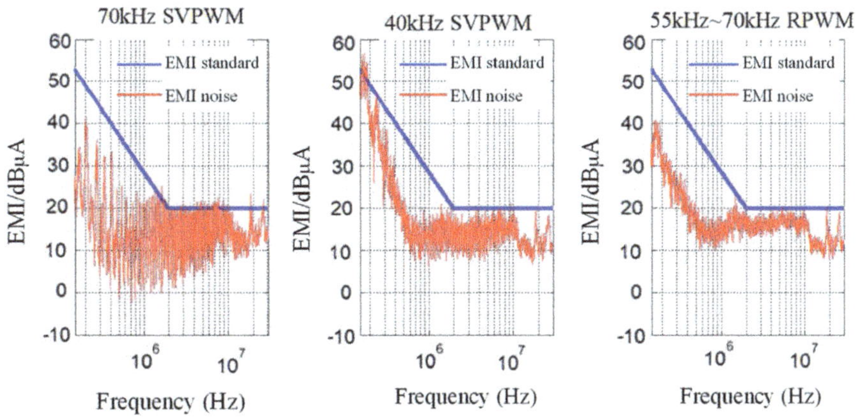

Fig. 3.11 Conducted EMI comparison between constant switching frequency PWM and RPWM [6]

Fig. 3.12 Output voltage waveform of programmed PWM

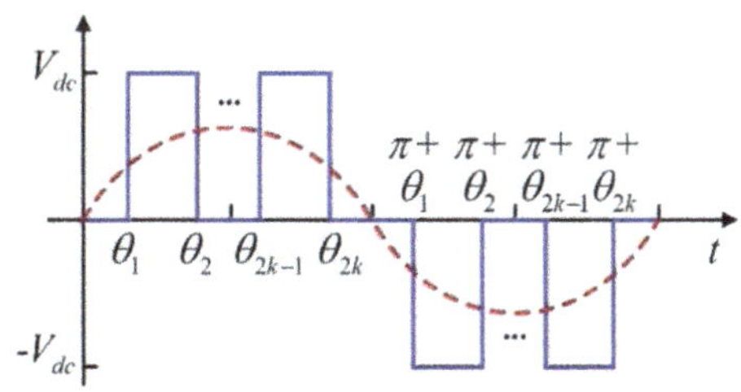

for programmed PWM to deal with a large number of harmonics, and the solved switching time cannot be accurately obtained.

Periodic PWM is a simple and controllable switching frequency modulation strategy, which mainly realizes the spread spectrum of high-frequency harmonic and EMI. Different from RPWM, the switching frequency function of periodic PWM is deterministic and changes periodically. Because the switching frequency function of periodic PWM is deterministic, its spread spectrum results can be repeatable. As shown in Fig. 3.13, switching frequency functions mainly include sinusoidal wave function, triangular wave function, sawtooth wave function and exponential wave function. Among the above switching frequency functions, the sinusoidal wave function has the worst spread spectrum effect. The exponential wave function and sawtooth wave function have the best spread spectrum effect, and the triangular wave function is relatively moderate. On the other hand, the exponential wave function is too complex to realize and analyze, while the sine wave function is much simpler to realize and analyze. Each switching frequency function has three control parameters: amplitude Δf, phase φ and frequency. After selecting the switching frequency function, the adjustment of the three control parameters will affect the final spread spectrum effect.

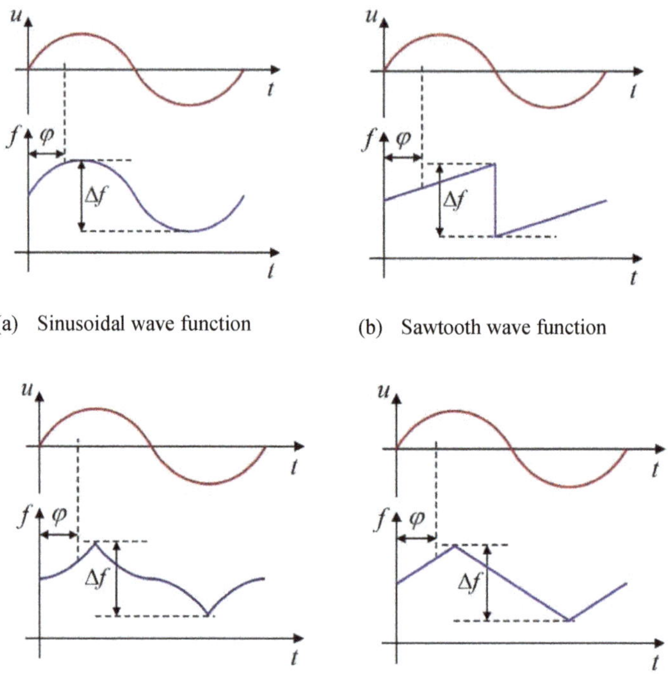

(a) Sinusoidal wave function (b) Sawtooth wave function

(c) Exponential wave function (d) Triangular wave function

Fig. 3.13 Periodic switching frequency function

Because of the effect of EMI mitigation of the above three kinds of spread spectrum modulation technologies, plenty of research work has been done in the past few decades. However, the above spread spectrum PWM has its own problems. By using PWM freedom, the spread spectrum of high-frequency harmonic and EMI noise is realized, but the time-domain characteristics (switching ripple) generated synchronously with PWM are ignored. Meanwhile, the improvement and controllability of switching losses are poor. That is because the above three kinds of spread spectrum PWM are not with prediction model. In the preface, "advanced PWM" is defined with prediction model as a major characteristic. Therefore, the above spread spectrum modulation technologies are just improved PWM. Model predictive variable switching frequency PWM will be introduced in the next few chapters.

3.6 Summary

This chapter has introduced three major impacts of PWM on the converter: switching losses, voltage/current ripple and EMI. Switching losses will accumulate together with PWM and influence the total losses and efficiency. The error between terminal

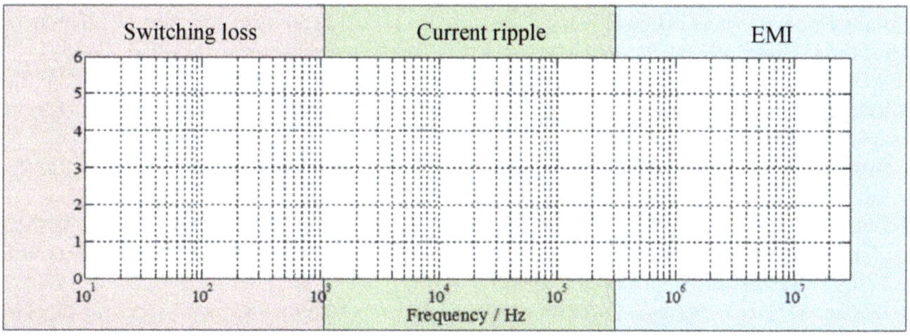

Fig. 3.14 Frequency range for switching loss, ripple and EMI

PWM voltage and continuous load voltage will generate current ripple and influence the input power ripple and torque ripple of the motor. PWM methods can also influence the switching EMI noise source and the EMI characteristics. These three major factors determine the converter's performance and are the major evaluation targets for PWM methods. This chapter has taken RPWM as an example to show a possible way to improve the PWM's performance. But RPWM is not with prediction model and relies on statistical results. The next few chapters will introduce the model prediction-based advanced PWM methods to further improve the converter performance[13].

Among the three major influence factors, the impacted frequency ranges are different, shown in Fig. 3.14. Switching loss is the effect of PWM accumulation and the time scale is long and the frequency is much lower than PWM. EMI is the high-frequency effect of PWM harmonics and the frequency is much higher than PWM. But voltage/current ripple is exactly synchronous with PWM in terms of frequency. Ripple can be the most proper target for prediction, which will be introduced in the next chapter.

References

1. Jiang D, Burgos R, Wang F, Boroyevich D (2012) Temperature-dependent characteristics of sic devices: performance evaluation and loss calculation. IEEE Trans Power Electron 27(2):1013–1024
2. Ma W, Zhang L, Meng J (2007) Electromagnetic compatibility of independent power system and power electronics equipment. Science Press (in Chinese)
3. Trzynadlowski AM, Kirlin RL, Legowski SF (1997) Space vector PWM technique with minimum switching losses and a variable pulse rate [for VSI]. IEEE Trans Ind Electron 44(2):173–181
4. Trzynadlowski AM, Wang Z, Nagashima JM, Stancu C, Zelechowski MH (2003) Comparative investigation of PWM techniques for a new drive for electric vehicles. IEEE Trans Ind Appl 39(5):1396–1403
5. Kirlin RL, Lascu C, Trzynadlowski AM (2011) Shaping the Noise spectrum in power electronic converters. IEEE Trans Ind Electron 58(7):2780–2788

6. Jiang D, Lai R, Wang F, Luo F, Wang S, Boroyevich D (2011) Study of conducted EMI reduction for three-phase active front-end rectifier. IEEE Trans Power Electron 26(12):3823–3831
7. Wang AC, Sanders SR (1993) Programmed pulsewidth modulated waveforms for electromagnetic interference mitigation in DC-DC converters. IEEE Trans Power Electron 8(4):596–605
8. Bowes SR, Clark PR (1995) Regular-sampled harmonic-elimination PWM control of inverter drives. IEEE Trans Power Electron 10(5):521–531
9. Dahidah MSA, Konstantinou G, Agelidis VG (2014) A review of multilevel selective harmonic elimination PWM: formulations, solving algorithms, implementation and applications. IEEE Trans Power Electron 30(8):4091–4106
10. De Caro D (2013) Optimal discontinuous frequency modulation for spread-spectrum clocking. IEEE Trans Electromagn Compat 55(5):891–900
11. Song M, Ahn S, Jung I et al (2012) Piecewise linear modulation technique for spread spectrum clock generation. IEEE Trans Very Large Scale Int (VLSI) Syst 21(7):1234–1245
12. Yeh ML, Liou WR, Hsieh HP et al (2009) An electromagnetic interference (EMI) reduced high-efficiency switching power amplifier. IEEE Trans Power Electron 25(3):710–718
13. Dong J, Wang F (2013) Variable switching frequency PWM for Three-Phase converters based on current ripple prediction. IEEE Trans Power Electron 28(11):4951–4961

Chapter 4
Current Ripple Prediction Model for Power Electronics Converter

In Chap. 3, the PWM effects on losses, current ripples and EMI of power electronics converter have been studied. In this chapter, current ripple is taken as the main research object for detailed analysis. First, the current ripple prediction model of single-phase inverter is introduced. Then, for the general three-phase voltage source converter, the current ripple prediction method based on the Thevenin equivalent circuit model is introduced. For multiphase converters, a more general current ripple prediction method is also introduced in this chapter. Considering the asymmetry of the application-like interior permanent magnet synchronous motor with varying phase-winding inductance, the current ripple prediction method using d–q transformation is also introduced. Based on AC side current ripple prediction, the real-time prediction model of DC-link current and the accurate calculation of RMS current stress on DC-link capacitor are introduced. Finally, the influence of nonideal characteristics on current ripple prediction and its countermeasures is analyzed. This chapter focuses on the voltage source inverters for the detailed analysis, and the related conclusions can also be applied to the rectifier topologies [1].

4.1 Current Ripple Prediction for Single-Phase Inverter

As described in Chap. 3, the power electronics converter realizes digital voltage (current) conversion by controlling the on/off state of switching devices. The terminal voltage (current) is a series of pulses with equal amplitude, while the load voltage (current) is continuous. In one switching cycle, the average value of pulse voltage (current) is balanced with the load voltage (current), and the instantaneous voltage (current) difference generates the corresponding current (voltage) ripple. Current (voltage) ripple is an important parameter to evaluate the performance of power electronics converter, and it is also the direct control target of PWM.

D. Jiang et al., *Advanced Pulse-Width-Modulation: With Freedom to Optimize Power Electronics Converters*, CPSS Power Electronics Series,
https://doi.org/10.1007/978-981-33-4385-6_4

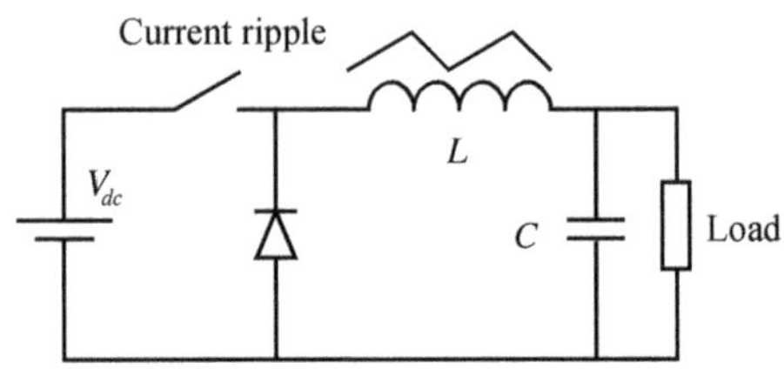

Fig. 4.1 Buck converter

From Fig. 4.1, with a simple buck converter, the inductor current is linearly varying with constant voltage. With PWM switching, the voltage on the inductor has two modes in each switching cycle. Then the current in the inductor behaves with triangular tooth waveform. The current slope in the tooth is determined by the terminal voltage of the converter, and can be predicted by the duty cycle. The duty ratio of switching devices in any switching period is constant in steady state, and the current ripple is consistent in each switching cycle.

However, for DC-AC and AC-DC converters, because of the varying AC side voltage and duty cycles during each line-cycle, the current ripple is not consistent in each switching cycle. Therefore, the current ripple prediction for AC-DC or DC-AC converters is more complicated. Figure 4.2 shows the single-phase DC-AC converter, which is widely used in household appliances with single-phase utility, such as in-home photovoltaic grid connection and plug-in hybrid electric vehicles. In the specific working condition, DC bus voltage is 400 V, reference sinusoidal voltage frequency is 50 Hz, carrier modulation ratio is 0.8 and PWM switching frequency is 10 kHz. The output current of single-phase inverter is shown in Fig. 4.3. The actual load current is superimposed by the fundamental current and current ripple. Figure 4.4 shows the components of current ripple in a fundamental period. It is not difficult to find that the distribution of output current ripple in single-phase inverter is extremely uneven, with the peak value of local current ripple reaching about 5 A, while the minimum current ripple is close to 0 A.

Figure 4.5 shows the schematic diagram of reference wave/carrier wave, output voltage/output average voltage and current ripple in two switching cycles. PWM

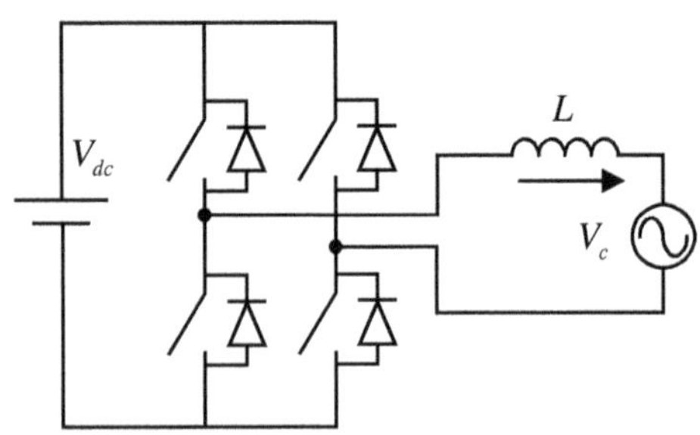

Fig. 4.2 Single-phase DC-AC converter

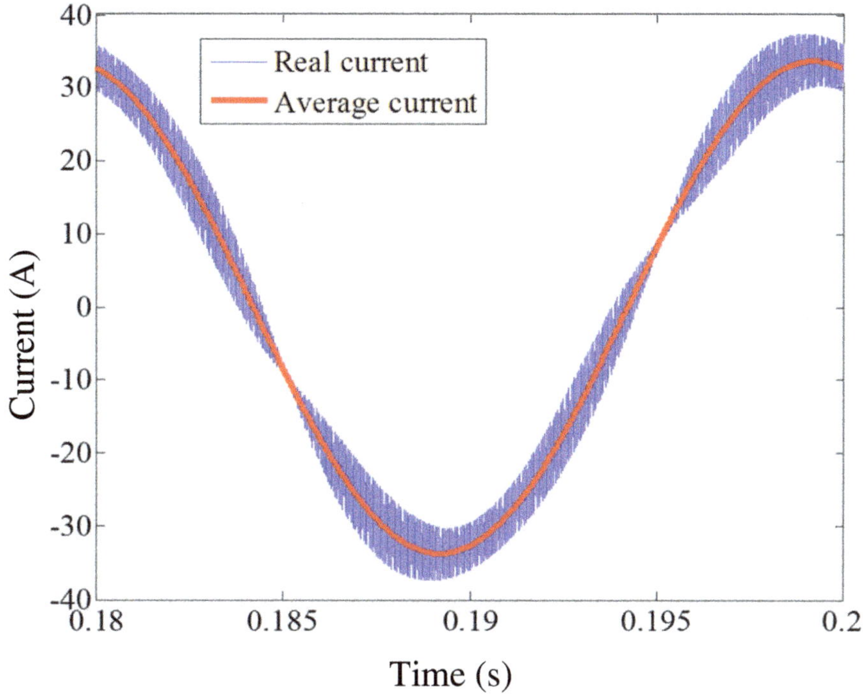

Fig. 4.3 Output current of single-phase DC-AC converter

Fig. 4.4 Current ripple of single-phase DC-AC converter

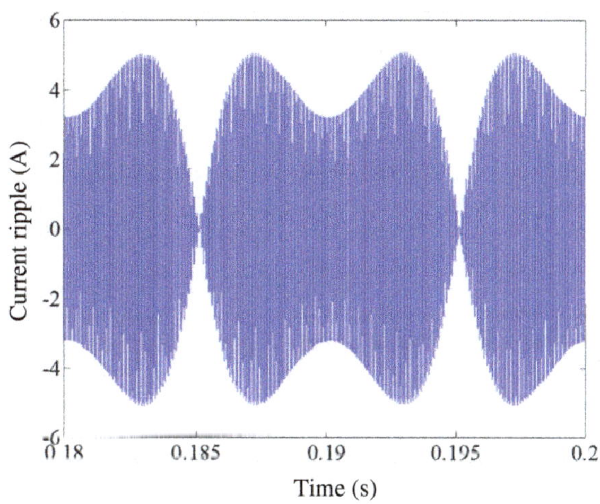

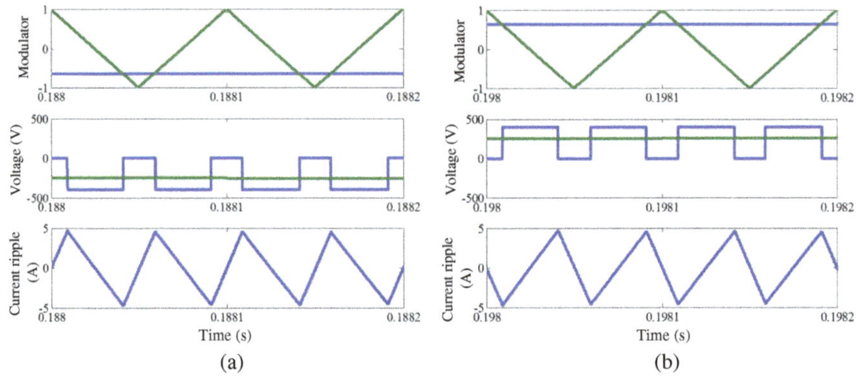

Fig. 4.5 Current ripple analysis of single-phase DC-AC converter: **a** negative half cycle, **b** positive half cycle

signal is generated by comparing reference wave with carrier wave. When the reference wave is located in the negative half cycle, as shown in Fig. 4.5a, the output voltage is switched between the negative DC-link voltage and the zero voltage. In the zero-voltage range, the difference between the output voltage and the average voltage is positive and the inductor current increases linearly. In the negative DC-link voltage range, the difference between the output voltage and the average voltage is negative, and the inductor current decreases linearly. When the reference wave is located in the positive half cycle, as shown in Fig. 4.5b, the output voltage is switched between the positive DC-link voltage and the zero voltage. In the positive DC-link voltage range, the difference between the output voltage and the average voltage is positive and the inductor current increases linearly. In the zero-voltage range, the difference between the output voltage and the average voltage is negative and the inductor current declines linearly. Therefore, there are two modes in each switching period and the current ripple behaves like triangle tooth curve.

In the PWM control process, the terminal voltage can be predicted by duty cycle, so the $\mathrm{d}i/\mathrm{d}t$ can also be predicted for on and off modes. With the modulation scheme, duty cycle for each switching period can be derived and utilized for $\mathrm{d}i/\mathrm{d}t$ and action time calculation. According to this principle, reference [2] has deduced the expressions of current ripple peak of single-phase inverter in positive and negative half cycle, as shown in (4.1) and (4.2), where V_{dc} is the DC bus voltage, L is the output inductance, T_s is the switching cycle, $d(t)$ is the real-time duty ratio and its variation range is -1 to 1.

$$\Delta I_{pk-pk} = \frac{[V_{dc} - d(t)V_{dc}]d(t)T_s}{L} = \frac{V_{dc}T_s}{L}[1 - d(t)]d(t) \quad (4.1)$$

$$\Delta I_{pk-pk} = \frac{[-V_{dc} - d(t)V_{dc}](-d(t))T_s}{L} = \frac{V_{dc}T_s}{L}[1 + d(t)]d(t) \quad (4.2)$$

Taking sinusoidal PWM (SPWM) as an example, its duty ratio $d(t) = m_a\cos(\omega t)$, where m_a is the modulation index. Combining (4.1) and (4.2), the unified expression of the peak value of current ripple in each switching cycle can be obtained as:

$$\Delta I_{pk-pk} = \frac{V_{dc}T_s}{L}[1 - m_a|\cos(\omega t)|]m_a \cos(\omega t) \quad (4.3)$$

From (4.3), it can be concluded that the current ripple distribution is correlated with the phase angle. When the phase angle is π/2, the peak value of current ripple reaches to the minimum value of 0. When the phase angle is arccos($0.5/m_a$), the peak value of the current ripple reaches the maximum value of $V_{dc}T_s/(4L)$. Figure 4.6 shows the comparison between the calculated peak value of current ripple and the simulation result. It can be seen that the predicted value of current ripple matches well with the envelope of simulated current ripple, which proves the correctness of the prediction method.

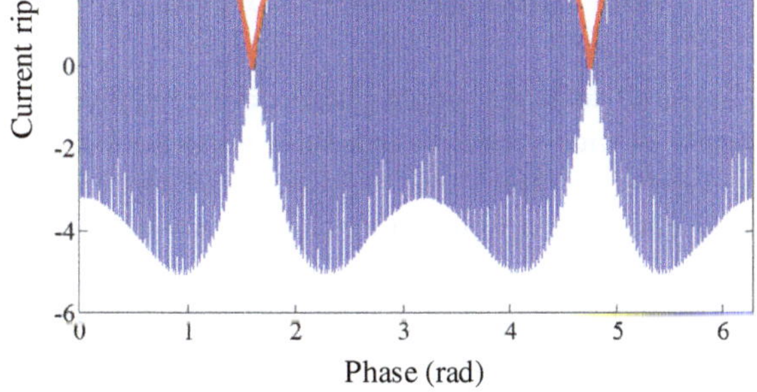

Fig. 4.6 Comparison between prediction and simulation current ripple

Furthermore, the RMS value of the current ripple in one switching cycle can be obtained, as shown in (4.4). Equation (4.5) is the calculation expression of the RMS value of current ripple in a fundamental period, which can be simplified to (4.6).

$$\Delta I_{\text{rms\_}Sub} = \frac{\Delta I_{\text{pk}-pk}}{2\sqrt{3}} = \frac{V_{dc}T_s}{2\sqrt{3}L}[1 - m_a|\cos(\omega t)|]m_a\cos(\omega t) \tag{4.4}$$

$$\begin{aligned}\Delta I_{\text{rms}} &= \sqrt{\frac{1}{\pi}\int_0^{\pi}\Delta I_{rms\_Sub}^2 d\omega t}\\ &= \sqrt{\frac{1}{\pi}\int_0^{\pi}(\frac{V_{dc}T_s}{2\sqrt{3}L})^2[1 - m_a|\cos(\omega t)|]^2 m_a^2\cos^2(\omega t)d\omega t}\end{aligned} \tag{4.5}$$

$$\Delta I_{\text{rms}} = \frac{1}{\sqrt{12\pi}}\frac{V_{dc}T_s}{L}m_a\sqrt{\frac{3\pi}{8}m_a^2 - \frac{8}{3}m_a + \frac{\pi}{2}} \tag{4.6}$$

According to (4.3) and (4.4), the peak value and RMS value of current ripple are inversely proportional to the switching frequency in each switching cycle. Meanwhile, the current ripple of single-phase inverter is unevenly distributed in the fundamental period. With the same requirement of current ripple peak or RMS value, the switching frequency can be reduced to lower the switching losses. Thanks to the switching frequency variation, the EMI noise will be much broader in the frequency domain, reducing the EMI peak value. But for three-phase and even multiphase voltage source inverter, the current ripple analysis is more complex, and very little real-time current ripple prediction method can be found [3–6]. In the next two parts, real-time current ripple prediction methods will be introduced.

4.2 Current Ripple Prediction for Three-Phase VSI: Thevenin Equivalent Circuit [7, 8]

Compared with the single-phase power supply in household, three-phase power supply is widely used in industrial applications, so three-phase power electronics converters are more popular in industrial applications. Figure 4.7 shows the equivalent circuit of a typical three-phase two-level VSI. The load is replaced by the

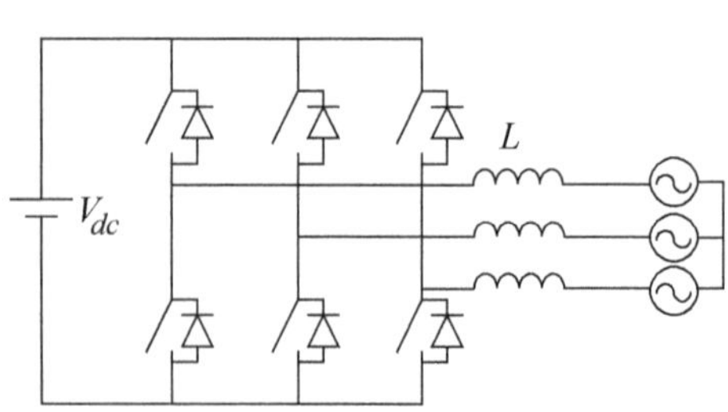

Fig. 4.7 Three-phase PWM converter and load

Y-connected three-phase AC sinusoidal voltage source, which can represent the back-EMF of AC motors or grid voltage in the grid-connected inverter. The difference between the terminal voltage of the inverter and the continuous load voltage is dropped on three-phase filter, generating the current ripple. For three-phase converters, d_a, d_b and d_c are the duty cycles of the three phases varying from 0 to 1. The "quasi-duty cycle" d_x' which varies from -1 to 1 can be defined as $d_x' = 2d_x - 1$ ($x = a, b, c$). The average output voltage is generated as shown in (4.7) and can represent the AC side fundamental voltage in steady state.s

$$\begin{cases} V_a = d_a' \dfrac{V_{dc}}{2} \\ V_b = d_b' \dfrac{V_{dc}}{2} \\ V_c = d_c' \dfrac{V_{dc}}{2} \end{cases} \tag{4.7}$$

The current ripple analysis of single-phase inverters in Sect. 4.1 can be similarly applied to the current ripple analysis of three-phase inverters. That is, under each voltage vector, the voltage dropped on the inductor is considered as fixed and the current ripple changes linearly. However, the switching state of three-phase inverter is more complicated than that of single-phase inverter. In the single-phase inverter, there are only two different switching states in each switching cycle, while the three-phase inverter has eight different switching states. The switching action of each bridge arm in the three-phase converter will affect the current ripple. For the three-phase converter, it is difficult to derive analytical expressions such as (4.4) or (4.6), but it does not mean that the current ripple of the three-phase converter cannot be predicted. There are eight different voltage vectors in a three-phase converter; each of them corresponds to one certain voltage combination and can change the pattern of current ripple. Figure 4.8 shows the equivalent circuit of the voltage vector 000, which means phase-a, phase-b and phase-c are all connected to negative DC bus. To analyze the current ripple in phase-a, the Thevenin equivalent circuit for phase-b and phase-c is derived. With Kirchhoff's voltage law, the voltage on the inductance of

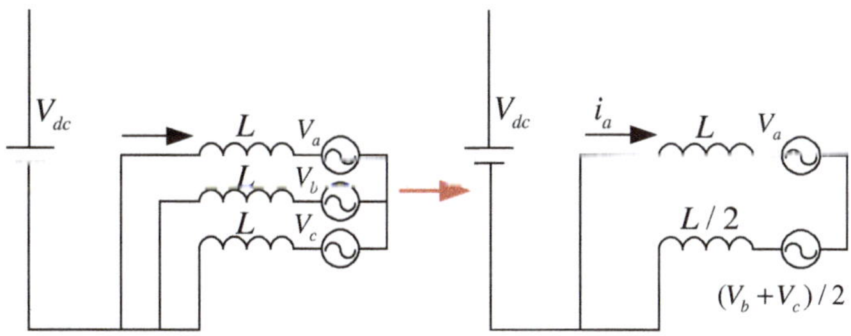

Fig. 4.8 Equivalent circuit and Thevenin equivalent circuit of voltage vector 000

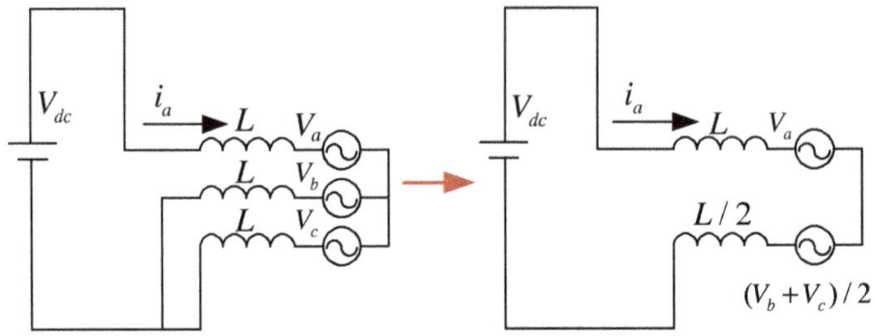

Fig. 4.9 Equivalent circuit and Thevenin equivalent circuit of voltage vector 100

phase-a can be simply obtained. Meanwhile, according to the principle of balancing the load voltage and the output average voltage of the converter, substituting (4.7), the slope of the current ripple on the phase inductor can be expressed as (4.8).

$$\frac{di_a}{dt} = \frac{V_{dc}}{3L}(\frac{d'_b + d'_c}{2} - d'_a) \tag{4.8}$$

Figure 4.8 shows a typical equivalent circuit diagram in the state of zero vector. If active vector is adopted, such as 100 vector, the corresponding equivalent circuit is shown in Fig. 4.9. Both phase-b and phase-c are connected to negative DC bus and phase-a is connected to positive DC bus. To analyze the current ripple in phase-a, the Thevenin equivalent circuit should contain not only phase-b and phase-c but also the DC-link, as shown in Fig. 4.9. The slope of current ripple can be derived in (4.9).

$$\frac{di_a}{dt} = \frac{2V_{dc}}{3L}\left(1 + \frac{d'_b + d'_c}{4} - \frac{d'_a}{2}\right) \tag{4.9}$$

Similarly, the Thevenin equivalent circuit can be applied to all voltage vectors of three-phase two-level converter, and eight groups of different equivalent circuits are obtained, as shown in Fig. 4.10.

Combining the Thevenin equivalent circuits of eight voltage vectors, the current ripple slope expression corresponding to each voltage vector can be achieved, as shown in Table 4.1.

With SVPWM, the switching state has seven zones in each switching cycle, as shown in Fig. 4.11, with a typical sequence. In each zone, a certain voltage combination is generated. The AC load voltage is treated as constant, so the voltage dropping on the output inductor will also be constant. The current ripple variation in one switching cycle will also have seven linear periods.

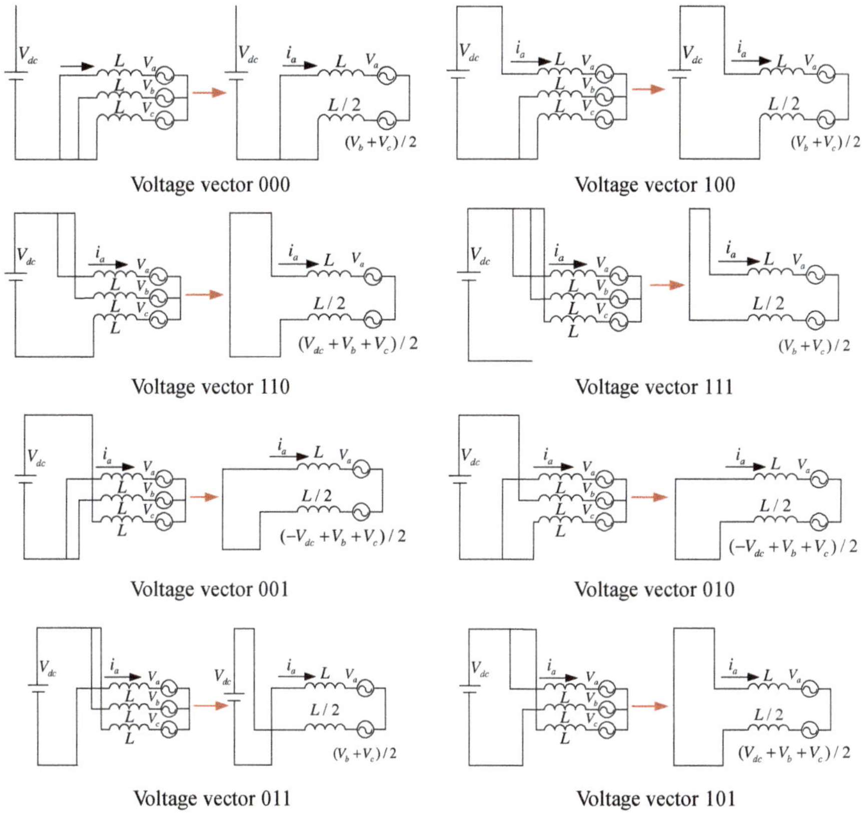

Fig. 4.10 Equivalent circuit and Thevenin equivalent circuit of all voltage vectors

Based on Fig. 4.11, the slope of current ripple in each zone can be predicted by combining the d*i*/d*t* corresponding to each voltage vector and the acting time, as shown in Fig. 4.12, taking the switching state sequence in Fig. 4.11 as an example. In the first zone, the voltage vector 000 is loaded on the inverter. The equivalent circuit is shown in Fig. 4.10a. With the expression in Table 4.1, the current ripple slope can be found as k_1. The corresponding vector of the second zone is 100. Similarly, the slope of phase-a current is k_2 in Table 4.1, and the current ripple inflection point values of the first and second zones are obtained from (4.10). According to the symmetry of SVPWM, the current ripple in the whole switching cycle can be finally obtained, as shown in Fig. 4.12. According to the current ripple in Fig. 4.12, the peak value of the current ripple in the switching cycle is max(|*x*|, |*y*|). The root mean square (RMS) value of current ripple can also be obtained from Fig. 4.12, as shown in Eq. (4.11).

Table 4.1 Current ripple slopes for eight voltage vectors

Voltage vector	Current ripple slope	Voltage vector	Current ripple slope
000	$\frac{di_a}{dt} = \frac{V_{dc}}{3L}(\frac{d'_b+d'_c}{2} - d'_a)$	001	$\frac{di_a}{dt} = \frac{V_{dc}}{3L}(\frac{d'_b+d'_c}{2} - 1 - d'_a)$
100	$\frac{di_a}{dt} = \frac{2V_{dc}}{3L}(1 + \frac{d'_b+d'_c}{4} - \frac{d'_a}{2})$	010	$\frac{di_a}{dt} = \frac{V_{dc}}{3L}(\frac{d'_b+d'_c}{2} - 1 - d'_a)$
110	$\frac{di_a}{dt} = \frac{2V_{dc}}{3L}(\frac{1}{2} + \frac{d'_b+d'_c}{4} - \frac{d'_a}{2})$	011	$\frac{di_a}{dt} = \frac{2V_{dc}}{3L}(\frac{d'_b+d'_c}{4} - 1 - \frac{d'_a}{2})$
111	$\frac{di_a}{dt} = \frac{V_{dc}}{3L}(\frac{d'_b+d'_c}{2} - d'_a)$	101	$\frac{di_a}{dt} = \frac{2V_{dc}}{3L}(\frac{1}{2} + \frac{d'_b+d'_c}{4} - \frac{d'_a}{2})$

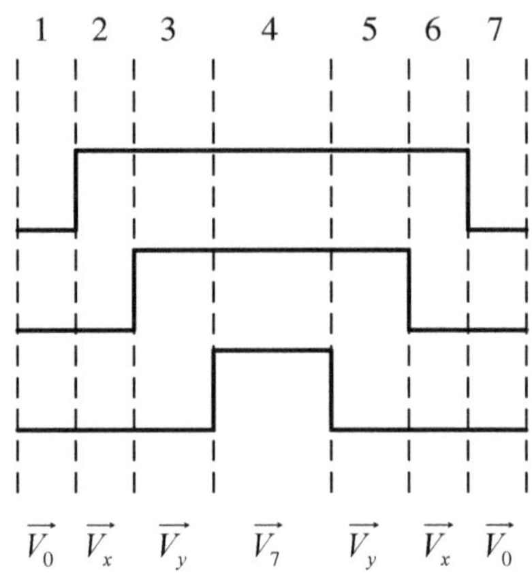

Fig. 4.11 Seven periods in one switching cycle with SVPWM

$$\begin{cases} x = k_1 \cdot \frac{t_0}{4} \\ y = k_1 \cdot \frac{t_0}{4} + k_2 \cdot \frac{t_1}{2} \end{cases} \tag{4.10}$$

$$\Delta I_{rms} = \sqrt{\frac{t_0}{T_s}\frac{x^2}{3} + \frac{t_1}{T_s}\frac{x^2 + xy + y^2}{3} + \frac{t_2}{T_s}\frac{x^2 - xy + y^2}{3}} \tag{4.11}$$

For DPWM, the current ripple will be different. In each switching cycle, there are five zones, and the slope of current ripple corresponding to each vector is also shown in Table 4.1. According to the two classical cases of DPWM_Min and DPWM_Max,

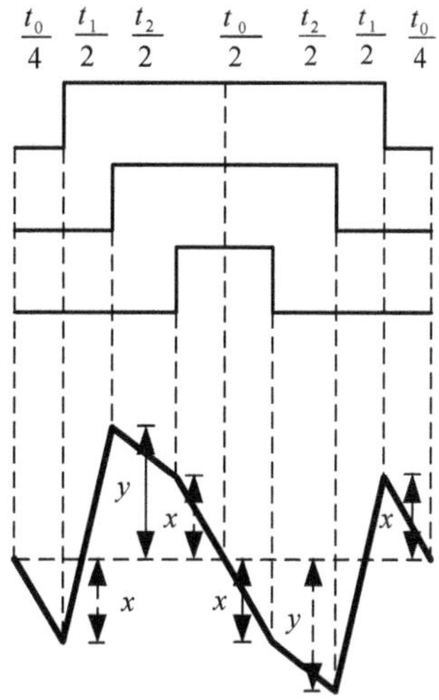

Fig. 4.12 Current ripple in one switching cycle with SVPWM

the current ripple in one switching cycle is shown in Fig. 4.13. Similar to the analysis of SVPWM, the current ripple of DPWM can also be obtained from (4.12). In each switching cycle, the peak value of current ripple corresponding to DPWM is max(|x|, |y|), and the RMS value is obtained by Eq. (4.13).

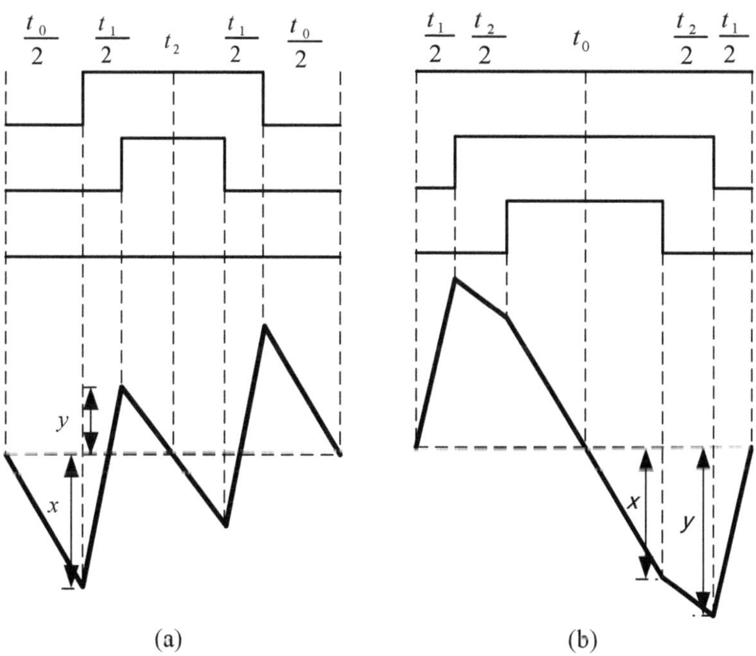

Fig. 4.13 Current ripple in one switching cycle with DPWM: **a** DPWM_Min, **b** DPWM_Max

$$\begin{cases} x = k_1 \cdot \frac{t_0}{2} \\ y = \begin{cases} k_1 \cdot \frac{t_0}{2} + k_2 \cdot \frac{t_1}{2} & \text{(DPWM\_Min)} \\ k_1 \cdot \frac{t_0}{2} + k_2 \cdot \frac{t_2}{2} & \text{(DPWM\_Max)} \end{cases} \end{cases} \tag{4.12}$$

$$\Delta I_{rms} = \begin{cases} \sqrt{\frac{t_0}{T_s}\frac{x^2}{3} + \frac{t_2}{T_s}\frac{y^2}{3} + \frac{t_1}{T_s}\frac{x^2 + xy + y^2}{3}} & \text{(DPWM\_Min)} \\ \sqrt{\frac{t_0}{T_s}\frac{x^2}{3} + \frac{t_1}{T_s}\frac{y^2}{3} + \frac{t_2}{T_s}\frac{x^2 + xy + y^2}{3}} & \text{(DPWM\_Max)} \end{cases} \tag{4.13}$$

In order to verify the current ripple prediction method described in this section, the MATLAB/SIMULINK simulation waveform and theoretical predicted result are compared and analyzed in Figs. 4.14, 4.15, 4.16, 4.17, 4.18 and 4.19 with SVPWM, having $V_{dc} = 400$ V, $L = 0.5$ mH, and $f_s = 20$ kHz. Phase-a current is shown in Fig. 4.14, and the fundamental wave value is obtained through FFT as shown in the thick line. The current ripple can be obtained by subtracting the fundamental wave value from the current simulation results, as shown in Fig. 4.15. It can be seen that the simulation results are highly consistent with the theoretical prediction results regardless of the overall current ripple (within a linear period) or the zoom-in detail comparison. If the peak value of current ripple predicted by the theory and the simulation results in each switching cycle are placed in the same figure, as shown in Fig. 4.16, the predicted peak value of current ripple and the envelope of simulation results match perfectly.

Figures 4.17, 4.18 and 4.19 show the comparison analysis using DPWM. Figure 4.17 shows the phase-a current waveform and its fundamental (average) value. After subtraction, the simulation result of current ripple is obtained, which is in good agreement with the prediction result, as shown in Fig. 4.18. Figure 4.19 shows

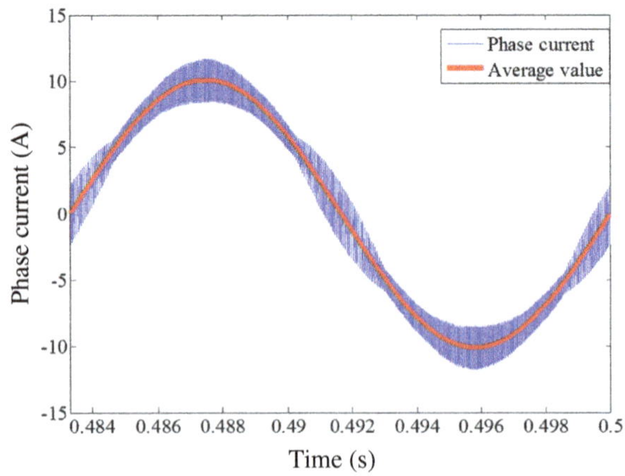

Fig. 4.14 Phase-a current with SVPWM

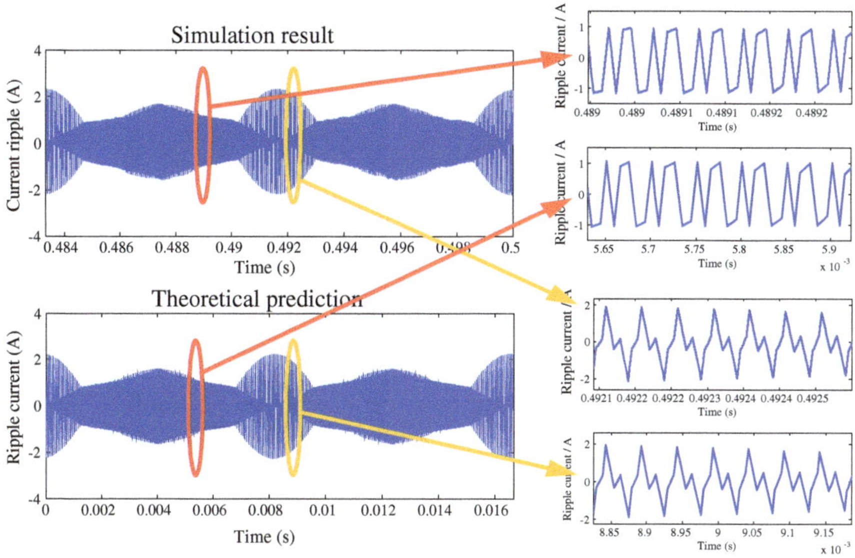

Fig. 4.15 Current ripple comparison with SVPWM

Fig. 4.16 Comparison between simulation result and predicted peak value with SVPWM

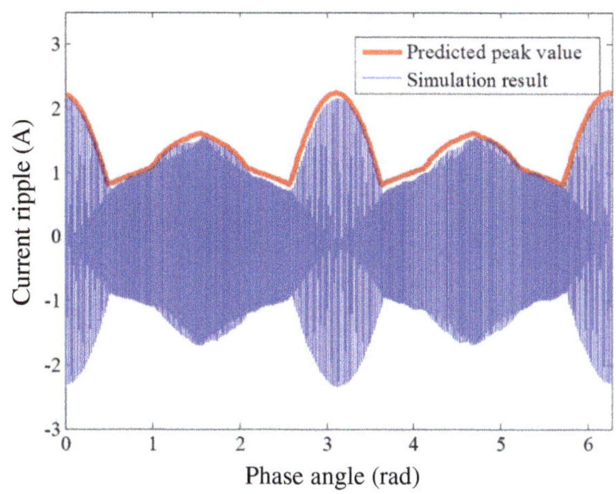

the comparison between the predicted current ripple peak value and the simulated current ripple. It can be seen that the predicted peak value matches well with the envelope of the simulated current ripple. By comparing the current ripple of DPWM and SVPWM, it can be seen that the peak value of the current ripple of DPWM is higher than that of SVPWM under the same condition, because of the decrease in the equivalent switching frequency.

Based on simulation, the experiment is further compared, having $V_{dc} = 240$ V, $L = 0.8$ mH, and $f_s = 6.25$ kHz. The phase current and its fundamental wave values

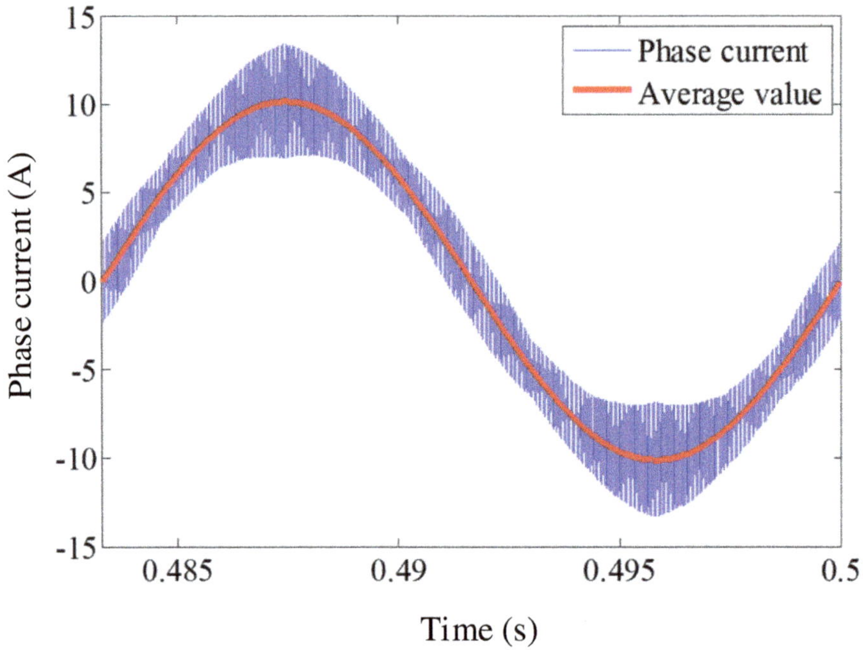

Fig. 4.17 Phase-a current with DPWM

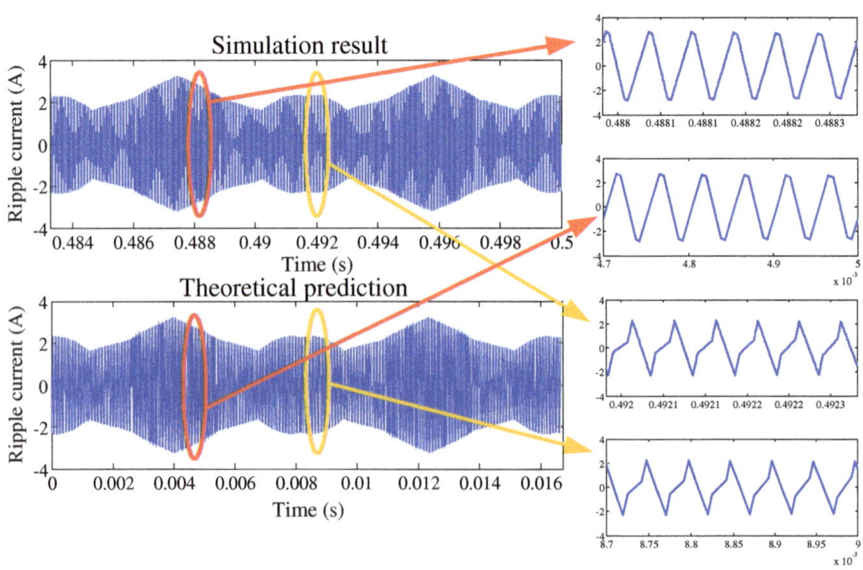

Fig. 4.18 Current ripple comparison with DPWM

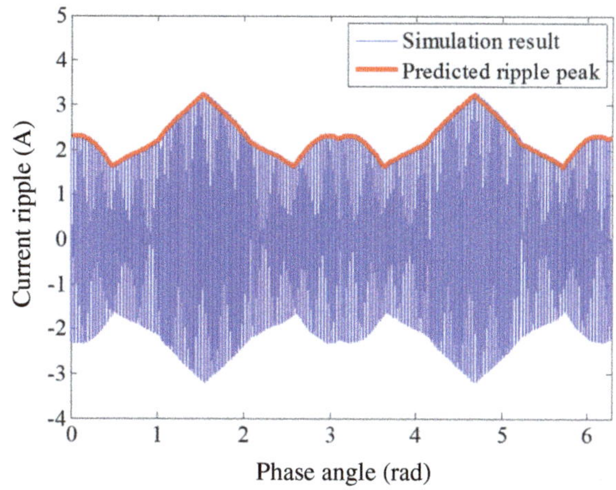

Fig. 4.19 Comparison between simulation result and predicted peak value with DPWM

using SVPWM and DPWM are shown in Figs. 4.20 and 4.21. The actual current is subtracted from the fundamental current, and the ripple current obtained is shown in Figs. 4.22 and 4.23. It can be seen that, no matter SVPWM or DPWM, the predicted current ripple peak value is close to the current ripple envelope of the experimental results. Compared with the simulation results, the prediction error of the experimental results is larger, because it is necessary to consider the nonideal conditions including the change of inductance parameters in the experiment.

In this section, the current ripple prediction method is derived using the Thevenin equivalent circuit of eight switch vectors of three-phase converter, and the accuracy and effectiveness of the prediction method are verified by simulation and experimental comparison. Current ripple prediction method can be effectively applied in the design of variable switching frequency PWM, which will be introduced in the

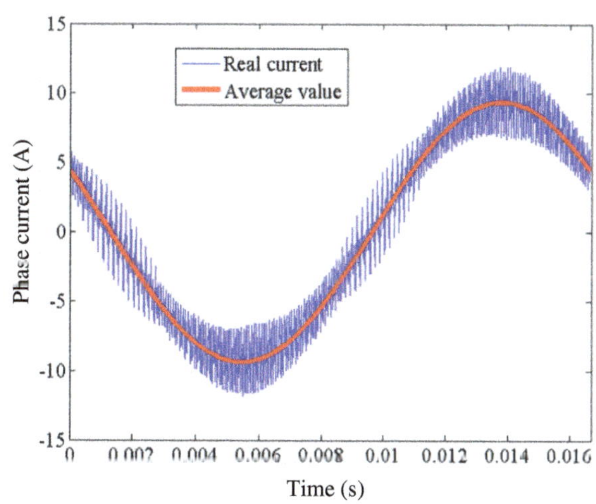

Fig. 4.20 Experimental phase-a current with SVPWM

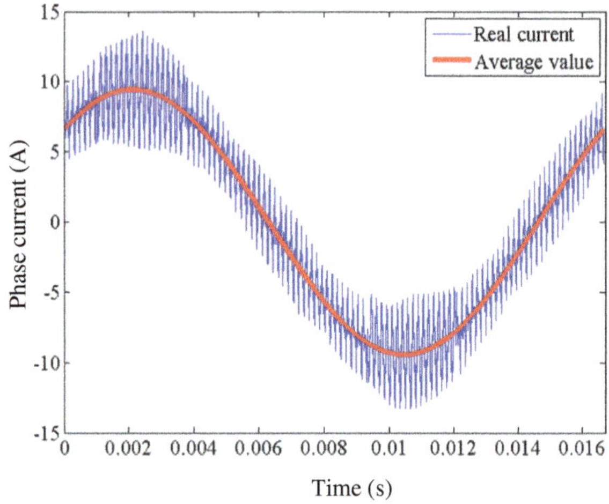

Fig. 4.21 Experimental phase-a current with DPWM

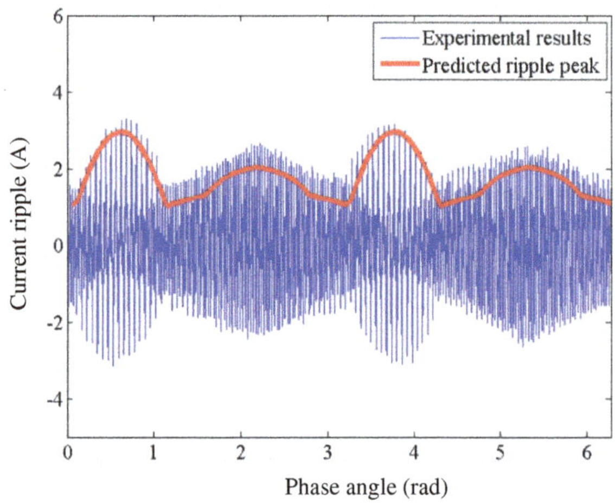

Fig. 4.22 Experimental current ripple comparison with SVPWM

next chapter. However, this approach has limitations, and a more general approach is described in Sect. 4.3.

4.3 Current Ripple Prediction for Multiphase VSI [9]

The current ripple prediction method based on the Thevenin equivalent circuit introduced in Sect. 4.2 is relatively suitable for three-phase converters, but has obvious defects for complex multiphase inverters. The total number of switching vectors for two-level n-phase voltage converters is 2^n, so for multiphase converters, the number of voltage vectors will increase sharply. For example, five-phase converters have 32

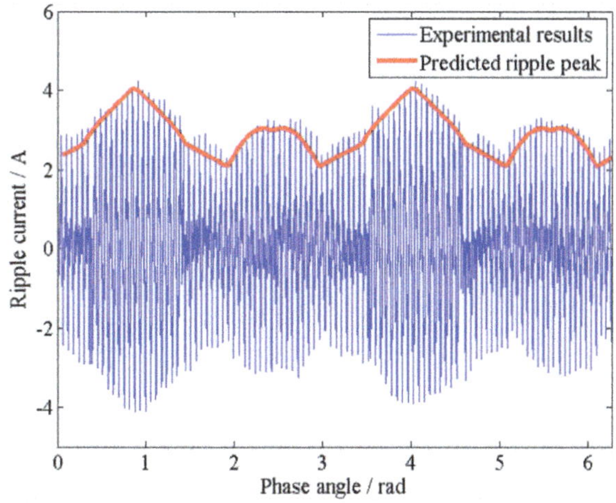

Fig. 4.23 Experimental current ripple comparison with DPWM

V vectors and six-phase converters have 64 V vectors. It will be very difficult to use the Thevenin equivalent circuit for each switch vector in Sect. 4.2 and get the analytical expression of the slope of current ripple, which will increase the computational burden of the controller hardware.

This section introduces a more general current ripple prediction method for multiphase converters. It is not necessary to calculate the Thevenin equivalent circuit for each voltage vector, but to solve the current ripple through the general equivalent circuit. This method will greatly simplify the prediction of current ripple in multiphase converters.

Figure 4.24 shows the equivalent circuit of typical n-phase converter. n terminals of n-phase converter $B_1, B_2, \ldots, B_n$ are connected to n-phase equivalent voltage source loads through n inductors $L_1, L_2, \ldots, L_n$. Similar to the three-phase converter shown in Fig. 4.7, n equivalent voltage sources V_{m1}–V_{mn} represent the average (fundamental) voltage of n-phase load. O_1 and O_2 represent the midpoints of the DC and AC sides.

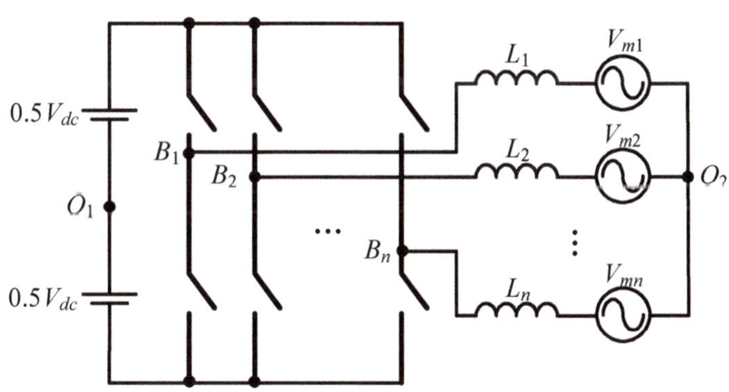

Fig. 4.24 The equivalent circuit of n-phase converter

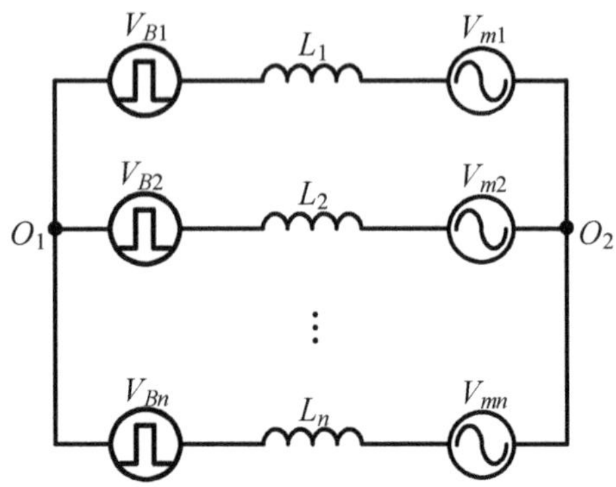

Fig. 4.25 The simplified equivalent circuit of n-phase converter

In Fig. 4.24, the terminal voltage of each phase is an impulse voltage sequence switched to the positive and negative DC-link voltage determined by PWM. Therefore, the equivalent circuit of the n-phase voltage source converter can be simplified into the form of Fig. 4.25, where V_{B1}–V_{Bn} represent the output pulse voltages.

In Fig. 4.25, taking any phase for analysis, such as phase-k ($1 \leq k \leq n$), the obtained circuit is shown in Fig. 4.26. Since the voltage at the midpoint O_2 on the AC side is not the same as that at the midpoint O_1 on the DC side, the voltage difference between O_1 and O_2 needs to be obtained to close the single-phase loop. According to symmetry, the common-mode voltage of n branches in Fig. 4.25 is equal to the voltage between O_2 and O_1, which includes the common-mode voltage V_{cm} of n impulse voltages V_{B1}–V_{Bn} and the common-mode voltage V_{l_cm} of n average voltage sources V_{m1}–V_{mn}. With V_{cm} and V_{l_cm}, the kth branch is closed. Since each voltage component in Fig. 4.26 can be obtained by the duty ratio of the n-phase converter, the voltage dropped on the inductor can also be obtained by the duty ratio. This general equivalent circuit is applicable to each phase of n-phase converter, without deriving Thevenin equivalent circuits for each switch vector.

In Fig. 4.26, the average voltage V_{mk} and V_{l_cm} in each switching cycle can be considered as constants, determined by duty cycle, while the switching voltage V_{Bk} and V_{cm} are step variations. In each switching cycle, V_{mk} is equal to duty cycle $d_k{'}$ multiplied by half of the DC-link voltage ($V_{mk} = d_k{'} V_{dc}/2$). The common-mode average voltage V_{l_cm} is a non-negligible variable, which is equal to the product of the common-mode duty ratio ($d_{cm}{'}$) and half of the DC-link voltage. For example, for SVPWM, $d_{cm}{'} = 1/2\ (d_{max}{'} + d_{min}{'})$. This is zero only for SPWM. Figure 4.27 shows the V_{Bk} and V_{cm} over a switching period. Figure 4.27a shows that V_{Bk} switches

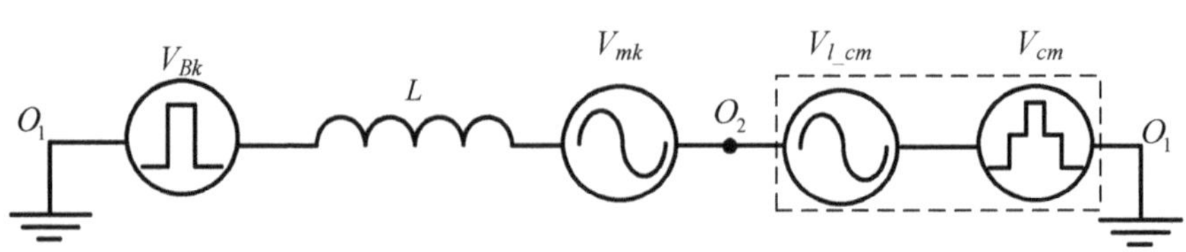

Fig. 4.26 Single-phase (kth) equivalent circuit′

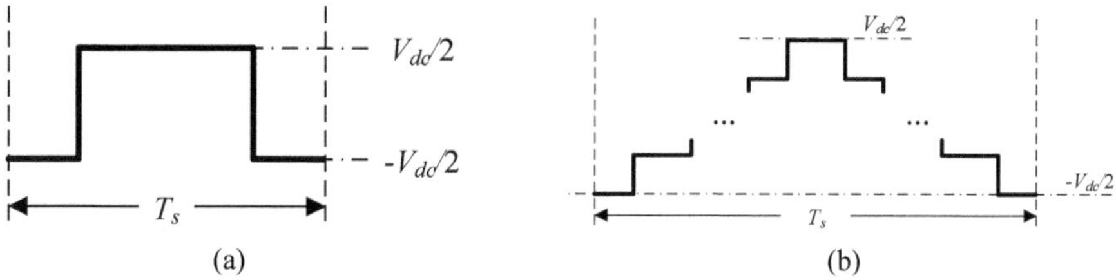

Fig. 4.27 Switching voltage in the *k*-phase equivalent circuit: **a** converter output voltage, **b** converter common-mode voltage

between $V_{dc}/2$ and $-V_{dc}/2$. Figure 4.27b shows that V_{cm} has $n + 1$ steps in each switching cycle, and the voltage difference of each step is V_{dc}/n.

Therefore, the voltage equation in Fig. 4.26 can be described as (4.14). In Eq. (4.14), $(d_k' + d_{cm}')V_{dc}/2$ is a constant, and V_{cm} is a voltage of $2n + 2$ zones as shown in Fig. 4.27b, which is symmetrical with respect to the midpoint. The common-mode voltage in *m*th zone ($0 \le m \le n$) is (4.15). V_{Bk} is a two-level voltage of three segments, as shown in Fig. 4.27a, which can be expressed in Eq. (4.16) in each switching cycle. For simplicity, define $d_k'' = d_k' + d_{cm}'$.

$$L\frac{di}{dt} = V_{Bk} - V_{mk} - V_{cm} - V_{l\_cm} = V_{Bk} - (d_k' + d_{cm}')\frac{V_{dc}}{2} - V_{cm} \tag{4.14}$$

$$V_{cm} = \left(-\frac{V_{dc}}{2} + m \times \frac{V_{dc}}{n}\right) \tag{4.15}$$

$$V_{Bk} = \begin{cases} \frac{V_{dc}}{2}; \left(\frac{1}{2} - \frac{d_k}{2}\right)T_s < t < \left(\frac{1}{2} + \frac{d_k}{2}\right)T_s \\ -\frac{V_{dc}}{2}; \; t \le \left(\frac{1}{2} - \frac{d_k}{2}\right)T_s \; or \; t \ge \left(\frac{1}{2} + \frac{d_k}{2}\right)T_s \end{cases} \tag{4.16}$$

In the *n*-phase converter, each switching period can be divided into $2n + 2$ zones, which includes two consecutive zones in the center, and distributes symmetrically. The current ripple is shown in Fig. 4.28. Each zone corresponds to a linear current ripple. The current ripple slope is determined by (4.14). The current ripple in zone *k* is symmetric with that in zone $(2n + 1 - k)$. Therefore, the current ripple analysis can be simplified to half of the switching cycle.

Without losing generality, it is assumed that the duty cycle order is $d_n \le d_{n-1} \le \cdots d_1$. The duty cycle of phase-*m* is d_m. Then, the action time of *m*th zone in the switching cycle is determined by Eq. (4.17). Finally, the action time of *n*th zone, $t_n = d_n T_s$, and the action time of 0th zone is $t_0 = (1 - d_1)T_s$. Due to impulse symmetry, t_m is divided by two for each period in Fig. 4.28.

$$t_m - d_m T_s - d_{m+1} T_s \tag{4.17}$$

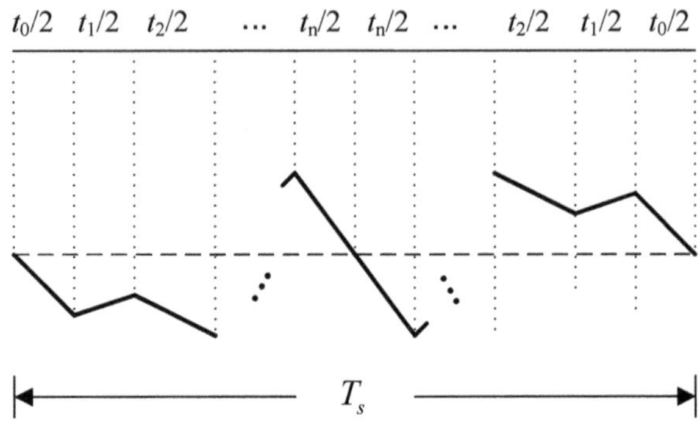

Fig. 4.28 Current ripple in one switching cycle for *n*-phase converter

According to the expressions of (4.15) and (4.16), the slope of phase-*k* current ripple in *m*th zone is determined by (4.18), and can be further simplified to (4.19).

$$L\frac{di}{dt}=\begin{cases}-\frac{V_{dc}}{2}-\left(-\frac{V_{dc}}{2}+m\times\frac{V_{dc}}{n}\right)-d_k''\frac{V_{dc}}{2}, & k>m\\ \frac{V_{dc}}{2}-\left(-\frac{V_{dc}}{2}+m\times\frac{V_{dc}}{n}\right)-d_k''\frac{V_{dc}}{2}, & k\le m\end{cases} \tag{4.18}$$

$$L\frac{di}{dt}=\begin{cases}-m\times\frac{V_{dc}}{n}-d_k''\frac{V_{dc}}{2}, & k>m\\ \left(1-\frac{m}{n}\right)V_{dc}-d_k''\frac{V_{dc}}{2}, & k\le m\end{cases} \tag{4.19}$$

Taking three-phase converter as an example, $n=3$, $d_3\le d_2\le d_1$. For phase-a, k is equal to 1.

(1) In the 0th zone of phase-a ($m=0$), $Ldi/dt=-d_1''\times V_{dc}/2$;
(2) In the first zone ($m=1$), $Ldi/dt=(1-1/n-d_1''/2)V_{dc}=(2/3-d_1''/2)V_{dc}$;
(3) In the second zone ($m=2$), $Ldi/dt=(1-2/n-d_1''/2)V_{dc}=V_{dc}\,(1/3-d_1''/2)$;
(4) In the third zone ($m=3$), $Ldi/dt=(1-3/n-d_1''/2)V_{dc}=-d_1''\times V_{dc}/2$.

The last four zones are symmetric with the first four zones. The expression of d*i*/d*t* is consistent with the results in Sect. 4.2.

This method can be easily extended to multiphase converters. Taking the five-phase ($n=5$) converter as an example, there are 12 zones in each switching cycle. The current ripple analysis of phase-2 ($k=2$) is as follows: in the first half of the switching cycle, there are six zones with $m=0$–5. When $m<2$, the slope of current ripple is in accordance with the first formula in (4.19). When $m\ge 2$, the slope of current ripple is in accordance with the second formula in (4.19). Considering the symmetry between the first half switching period and the second half switching period, the slope of current ripple in all 12 zones is shown in Table 4.2.

The current ripple of the five-phase converter can be predicted according to the slope of current ripple in 12 zones and the action time of each zone. Figure 4.29

Table 4.2 The slope of phase-2 current ripple for five-phase converter

	Zone 0	Zone 1	Zone 2	Zone 3
Time	½$(1 - d_1)T_s$	½$(d_1 - d_2)T_s$	½$(d_2 - d_3)T_s$	½$(d_3 - d_4)T_s$
di/dt	$-\frac{d_2''}{2}\frac{V_{dc}}{L}$	$(-\frac{1}{5} - \frac{d_2''}{2})\frac{V_{dc}}{L}$	$(\frac{3}{5} - \frac{d_2''}{2})\frac{V_{dc}}{L}$	$(\frac{2}{5} - \frac{d_2''}{2})\frac{V_{dc}}{L}$
	Zone 4	Zone 5	Zone 6	Zone 7
Time	½$(d_4 - d_5)T_s$	½$d_5\ T_s$	½$d_5\ T_s$	½$(d_4 - d_5)T_s$
di/dt	$(\frac{1}{5} - \frac{d_2''}{2})\frac{V_{dc}}{L}$	$-\frac{d_2''}{2}\frac{V_{dc}}{L}$	$-\frac{d_2''}{2}\frac{V_{dc}}{L}$	$(\frac{1}{5} - \frac{d_2''}{2})\frac{V_{dc}}{L}$
	Zone 8	Zone 9	Zone 10	Zone 11
Time	½$(d_3 - d_4)T_s$	½$(d_2 - d_3)T_s$	½$(d_1 - d_2)T_s$	½$(1 - d_1)T_s$
di/dt	$(\frac{2}{5} - \frac{d_2''}{2})\frac{V_{dc}}{L}$	$(\frac{3}{5} - \frac{d_2''}{2})\frac{V_{dc}}{L}$	$(-\frac{1}{5} - \frac{d_2''}{2})\frac{V_{dc}}{L}$	$-\frac{d_2''}{2}\frac{V_{dc}}{L}$

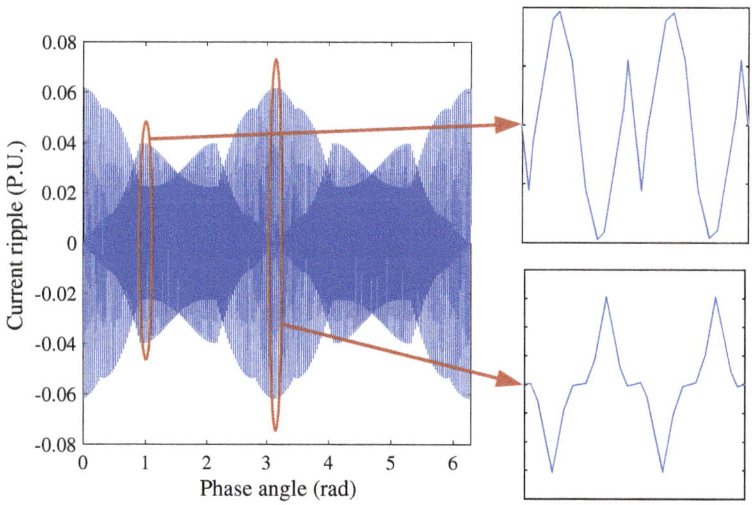

Fig. 4.29 Current ripple in one switching cycle for five-phase converter

shows the prediction results of the current ripple of a typical five-phase converter, including a full image of the whole fundamental wave period and zoom-in details. Similar to the current ripple of three-phase converter, the distribution of current ripple of five-phase converter is not uniform. With $t \cdot V_{dc}/L$ as the base value, the current ripple peak varies between 0.04 and 0.06 P.U.

A six-phase converter is taken as an example, that is, $n = 6$, and phase-3 (d_3'') is considered. In formula (4.19), $k = 3$. Each switching cycle has 14 zones. In the first half of the switching cycle, $m = 0$–6. When $m < 3$, the slope of current ripple follows the first formula (4.19). When $m \geq 3$, the slope of current ripple follows the second formula (4.19). The seven zones of the latter half of the switching cycle are symmetric with the first half of the switching cycle, so the expression of the current

Table 4.3 The slope of phase-3 current ripple for six-phase converter

	Zone 0	Zone 1	Zone 2	Zone 3	Zone 4
Time	½$(1-d_1)T_s$	½$(d_1-d_2)T_s$	½$(d_2-d_3)T_s$	½$(d_3-d_4)T_s$	½$(d_4-d_5)T_s$
di/dt	$-\frac{d_3''}{2}\frac{V_{dc}}{L}$	$(-\frac{1}{6}-\frac{d_3''}{2})\frac{V_{dc}}{L}$	$(-\frac{2}{6}-\frac{d_3''}{2})\frac{V_{dc}}{L}$	$(\frac{3}{6}-\frac{d_3''}{2})\frac{V_{dc}}{L}$	$(\frac{2}{6}-\frac{d_3''}{2})\frac{V_{dc}}{L}$
	Zone 5	Zone 6	Zone 7	Zone 8	Zone 9
Time	½$(d_5-d_6)T_s$	½$d_6\ T_s$	½$d_6\ T_s$	½$(d_5-d_6)T_s$	½$(d_4-d_5)T_s$
di/dt	$(\frac{1}{6}-\frac{d_3''}{2})\frac{V_{dc}}{L}$	$-\frac{d_3''}{2}\frac{V_{dc}}{L}$	$-\frac{d_3''}{2}\frac{V_{dc}}{L}$	$(\frac{1}{6}-\frac{d_3''}{2})\frac{V_{dc}}{L}$	$(\frac{2}{6}-\frac{d_3''}{2})\frac{V_{dc}}{L}$
	Zone 10	Zone 11	Zone 12	Zone 13	
Time	½$(d_3-d_4)T_s$	½$(d_2-d_3)T_s$	½$(d_1-d_2)T_s$	½$(1-d_1)T_s$	
di/dt	$(\frac{3}{6}-\frac{d_3''}{2})\frac{V_{dc}}{L}$	$(-\frac{2}{6}-\frac{d_3''}{2})\frac{V_{dc}}{L}$	$(-\frac{1}{6}-\frac{d_3''}{2})\frac{V_{dc}}{L}$	$-\frac{d_3''}{2}\frac{V_{dc}}{L}$	

ripple slope in the 14 zones of the whole switching cycle can be obtained, as shown in Table 4.3. Then, the prediction results of output current ripple of the six-phase converter are obtained, as shown in Fig. 4.30.

Similarly, current ripple prediction results can be compared with simulation results and experimental results, as shown in Figs. 4.31 and 4.32. In Fig. 4.31, for the five-phase and six-phase converters, the envelopes of the predicted peak value of current ripple and the simulated current ripple match well. In Fig. 4.32, for the five-phase and six-phase converters, the envelopes of the predicted peak value of current ripple and the experimental results are basically identical. The nonideal characteristics of the system, including the nonlinear inductance, should be considered in the experiment, so the relative error is larger than that of the simulation. The comparison results verify

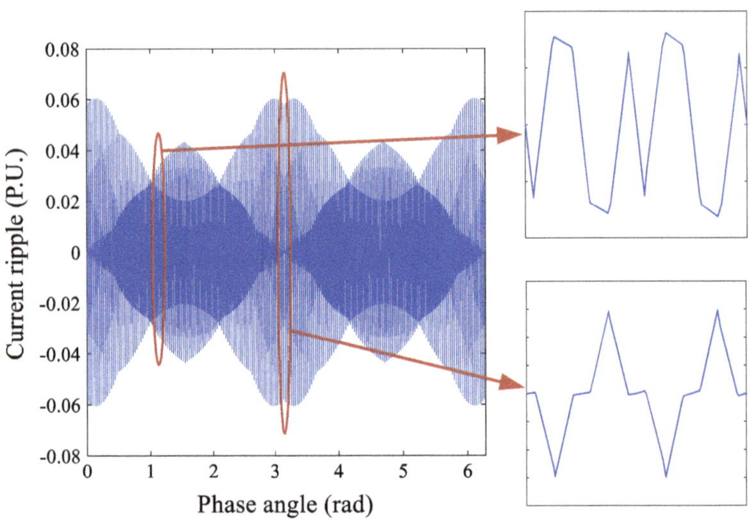

Fig. 4.30 Current ripple in one switching cycle for six-phase converter

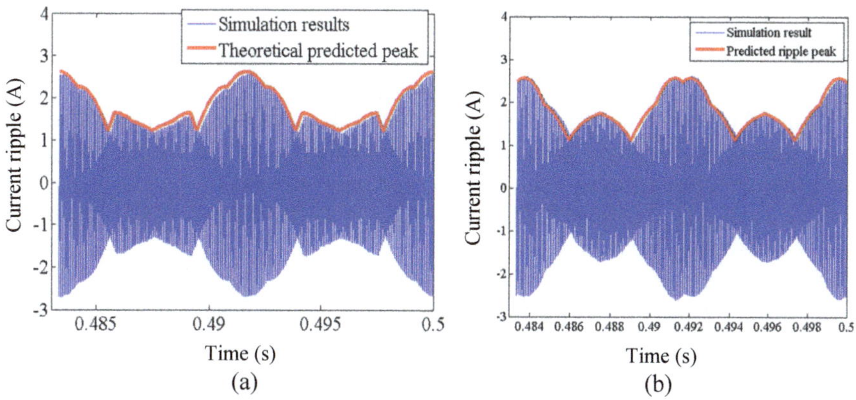

Fig. 4.31 Simulated current ripple comparison: **a** for five-phase converter, **b** for six-phase converter

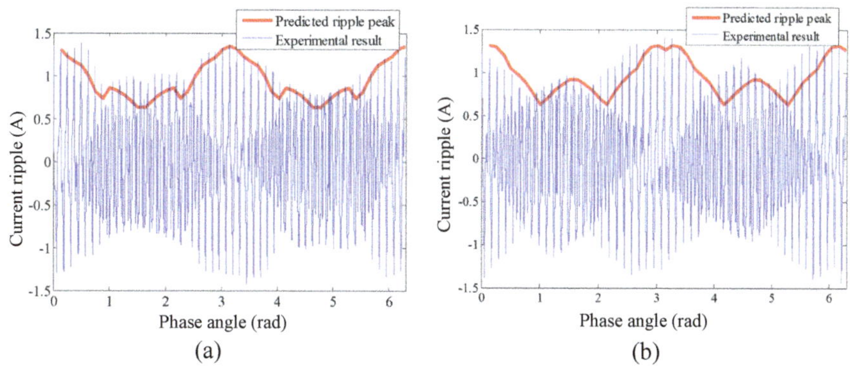

Fig. 4.32 Experimental current ripple comparison: **a** for five-phase converter, **b** for six-phase converter

the accuracy of current ripple prediction method for general multiphase converters. This current ripple prediction method is simpler than the method in Sect. 4.2, and is especially suitable for multiphase converters.

4.4 Current Ripple Prediction Using *d-q* Transformation [10]

The current ripple prediction methods introduced in Sects. 4.2 and 4.3 are based on symmetric and constant inductance-based load. In practical applications, the prediction method is applicable to grid-connected inverter/rectifier with three-phase independent symmetrical inductors and nonsalient pole motor drive with little mutual inductance. However, these methods are no longer applicable to three-phase asymmetric inductance-based load, such as salient pole motor drive, and so on. In the salient pole motor drive, due to the difference of *d-q* axis inductance, three-phase inductance actually changes with electrical angle, so the current ripple prediction method based on constant inductance is no longer applicable.

Taking the salient permanent magnet synchronous motor as an example, the flux equation in the three-phase *a-b-c* coordinate system is shown in (4.20). Self-inductance of each phase is with time-varying component:

$L_{aa} = L_{ls} + L_{0s} + L_{2s}\cos 2\theta$, $L_{bb} = L_{ls} + L_{0s} + L_{2s}\cos 2(\theta - 2\pi/3)$, $L_{cc} = L_{ls} + L_{0s} + L_{2s}\cos 2(\theta + 2\pi/3)$.

Mutual inductance is time-varying, such as

$L_{ab} = L_{ba} = L_{0s}/2 + L_{2s}\cos 2(\theta - \pi/3)$, $L_{ac} = L_{ca} = L_{0s}/2 + L_{2s}\cos 2(\theta + \pi/3)$, $L_{bc} = L_{cb} = -L_{0s}/2 + L_{2s}\cos 2(\theta + \pi)$. Among them, L_{ls}, L_{0s} and L_{2s} are, respectively, the leakage inductance, DC and AC components of excitation inductance of each phase.

$$\vec{\lambda}_{abc} = \begin{bmatrix} L_{aa}, L_{ab}, L_{ac} \\ L_{ba}, L_{bb}, L_{bc} \\ L_{ca}, L_{cb}, L_{cc} \end{bmatrix} \times \vec{i}_{abc} + \lambda_m \begin{bmatrix} \cos\theta \\ \cos(\theta - 2\pi/3) \\ \cos(\theta + 2\pi/3) \end{bmatrix} \tag{4.20}$$

It can be seen that it will be very complicated for salient pole motor to calculate the slope of current ripple in *a-b-c* coordinate system through the voltage equation in Sects. 4.2 and 4.3. It is necessary to consider not only the change of self-inductance with time but also the influence of mutual inductance. However, by converting the inductance into the *d-q* coordinate, the direct axis and quadrature axis inductance can be obtained as follows:

$$L_d = L_{ls} + L_{md} = L_{ls} + \frac{3}{2}(L_{0s} + L_{2s}) \tag{4.21}$$

$$L_q = L_{ls} + L_{mq} = L_{ls} + \frac{3}{2}(L_{0s} - L_{2s}) \tag{4.22}$$

In this way, the inductance becomes constant in *d-q* coordinates. It can be inspired that solving the current ripple of the salient pole motor in the *d-q* coordinate system can avoid the problem of time-varying inductance in the *a-b-c* coordinate system.

According to these ideas, the current ripple prediction method for asymmetric inductance load in the *d-q* coordinate system is obtained, which can be divided into the following two steps:

Step 1: Solve the current ripple Δi_d and Δi_q in *d-q* coordinate system.

Through Park transformation, the voltage is converted from *a-b-c* coordinate system to *d-q* coordinate system, as shown in (4.23).

$$\begin{bmatrix} v_d \\ v_q \\ v_0 \end{bmatrix} = T_{abc-dq0} \begin{bmatrix} v_{an} \\ v_{bn} \\ v_{cn} \end{bmatrix} \tag{4.23}$$

Similar to Sect. 4.2, voltage in *d-q* axis also contains the average component V_d, V_q and ripple component ΔV_d and ΔV_q which can be obtained by (4.23). For Eqs. (4.24) and (4.25) of permanent magnet synchronous motor in the *d-q* coordinate system, Eqs. (4.26) and (4.27) of average voltage and Eqs. (4.28) and (4.29) of ripple voltage component can also be obtained. In Eqs. (4.28) and (4.29), the resistance voltage drop and cross-coupling term are ignored, which can be simplified to (4.30) and (4.31). Therefore, according to the difference between the switching voltage and average voltage (ripple component Δv_d, Δv_q), current ripple Δi_d and Δi_q can be achieved.

$$v_d = R_s i_d + L_d \frac{di_d}{dt} - \omega L_q i_q \tag{4.24}$$

$$v_q = R_s i_q + L_q \frac{di_q}{dt} + \omega (L_d i_d + \lambda_m) \tag{4.25}$$

$$V_d = R_s I_d + L_d \frac{dI_d}{dt} - \omega L_q I_q \tag{4.26}$$

$$V_q = R_s I_q + L_q \frac{dI_q}{dt} + \omega (L_d I_d + \lambda_m) \tag{4.27}$$

$$\Delta v_d = R_s \Delta i_d + L_d \frac{d\Delta i_d}{dt} - \omega L_q \Delta i_q \tag{4.28}$$

$$\Delta v_q = R_s \Delta i_q + L_q \frac{d\Delta i_q}{dt} + \omega L_d \Delta i_d \tag{4.29}$$

$$\Delta v_d = L_d \frac{d\Delta i_d}{dt} \tag{4.30}$$

$$\Delta v_q = L_q \frac{d\Delta i_q}{dt} \tag{4.31}$$

Step 2: Current ripple is converted from *d-q* coordinate system to *a-b-c* coordinate system.

The three-phase current is converted from *d-q* coordinates to *a-b-c* coordinates as shown in Eq. (4.32). For the current ripple, Eq. (4.33) can be obtained. In the previous step, piece-wise linear prediction of current ripple has been achieved by using the model of fixed inductance under *d-q* coordinates, so current ripple segments in *a-b-c* coordinate system can also be obtained through (4.33). Like the current ripple in Sects. 4.2 and 4.3, the current ripple in asymmetric inductance is also divided into corresponding segments in accordance with PWM in each switching cycle. For example, for seven-zone SVPWM, the current ripple should be with seven segments in each switching cycle. The difference is that the d*i*/d*t* for each segment is not given by the fixed inductance, but by the variable inductance.

$$\begin{bmatrix} i_{as} \\ i_{bs} \\ i_{cs} \end{bmatrix} = T_{dq0-abc} \begin{bmatrix} i_d \\ i_q \\ i_0 \end{bmatrix} \tag{4.32}$$

$$\begin{bmatrix} \Delta i_{as} \\ \Delta i_{bs} \\ \Delta i_{cs} \end{bmatrix} = T_{dq0-abc} \begin{bmatrix} \Delta i_d \\ \Delta i_q \\ \Delta i_0 \end{bmatrix} \tag{4.33}$$

4.5 DC-Link Current Prediction [11]

The prediction method of current ripple on AC side will be helpful for real-time prediction of DC-link current and accurate calculation of RMS value. In the design of voltage source converter, the DC-link current is an important and complex parameter. The DC-link current is closely related to PWM strategy, which is essentially the high-frequency current obtained by PWM. The DC-link current will continuously charge and discharge the DC-link capacitor, generating corresponding charge and discharge losses on the DC-link capacitor. At the same time, the DC voltage ripple is generated by the charging and discharging process. If the battery is used on the DC-link, the DC-link current will continuously charge and discharge the battery, affecting the thermal effect and reliability of the battery.

DC-link current is closely related to PWM, which determines the switching action. The method of calculating DC-link current through PWM is shown in Fig. 4.33. DC-link current is obtained by AC side current and switching action. When the upper switch (S_a, S_b, S_c) of each phase is turned on, the current (i_a, i_b, i_c) of the corresponding phase will flow into the DC-link through the upper switch. Therefore, the DC-link current can be shown in Eq. (4.34), where S_a, S_b and S_c are 1 (on) or 0 (off). On this basis, the typical waveform of DC-link current in a switching cycle can be obtained, as shown in Fig. 4.34. When it is located in 111 or 000 vector, the total DC-link current is 0. When the voltage vector is 100, $S_a = 1$, $S_b = 0$, $S_c = 0$, the DC-link current is equal to i_a. When the voltage vector is 110, $S_a = 1$, $S_b = 1$, S_c

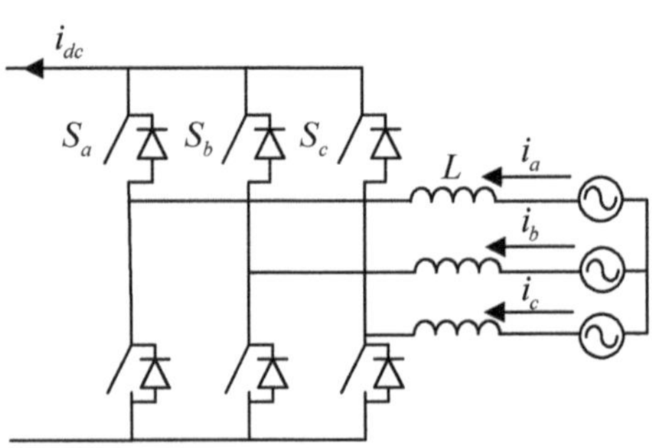

Fig. 4.33 AC side and DC-link current of three-phase voltage source inverter

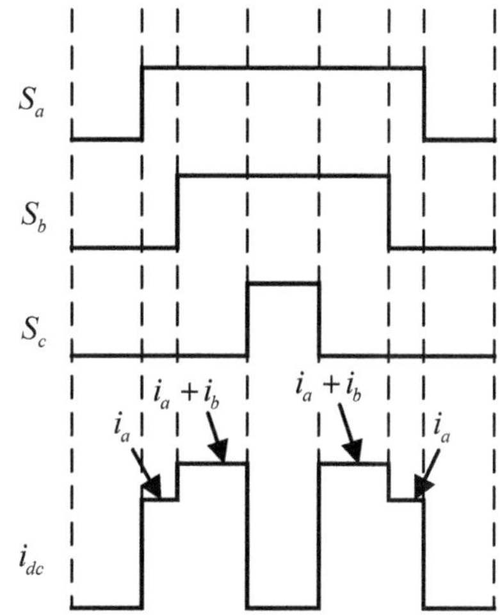

Fig. 4.34 DC-link current in one switching cycle

= 0, the DC-link current is equal to $i_a + i_b$ (or $-i_c$). When this kind of calculation is done for each switching period, the DC-link current can be obtained through the three-phase AC side current and the modulation strategy.

$$i_{dc} = S_a i_a + S_b i_b + S_c i_c \tag{4.34}$$

In most of DC-link current analysis, phase currents i_a, i_b and i_c are generally considered to be standard sinusoidal waves, which are treated as constants during each switching cycle. However, in the calculation of DC-link current in Fig. 4.34, if only the fundamental (average) component of the AC side current is considered and the current ripple is ignored, the calculated results of DC-link current will have obvious errors. In Sects. 4.1, 4.2, 4.3 and 4.4, it is known that, due to the effect of PWM, AC side current has switching current ripple synchronizing with PWM, and these current ripples will also be injected into the DC-link along with the switching action. Therefore, the AC side current ripple needs to be considered in the calculation of DC-link current. As shown in Fig. 4.35, the waveform of the DC-link current in a switching cycle with AC side current ripple considered is different from that shown in Fig. 4.34. In each zone, the AC side current is no longer regarded as a fixed value, but contains a value with obvious slope. Therefore, the DC-link current is no longer a chopped rectangle, but a trapezoid obtained by chopping.

Figure 4.36 shows the DC-link current calculation diagram considering the AC side current ripple. In the controller, the three-phase AC fundamental currents i_{a1}, i_{b1} and i_{c1} can be obtained through *d*-*q* transformation with the average *d*-*q* currents. Meanwhile, by using the current ripple prediction method in Sects. 4.2 or 4.3, three-phase ripple currents i_{ar}, i_{br} and i_{cr} can be obtained with duty cycles d_a, d_b and d_c in each switching period. By adding the fundamental current and current ripple, the three-phase full current can be achieved. Finally, the DC-link current can be calculated by (4.34).

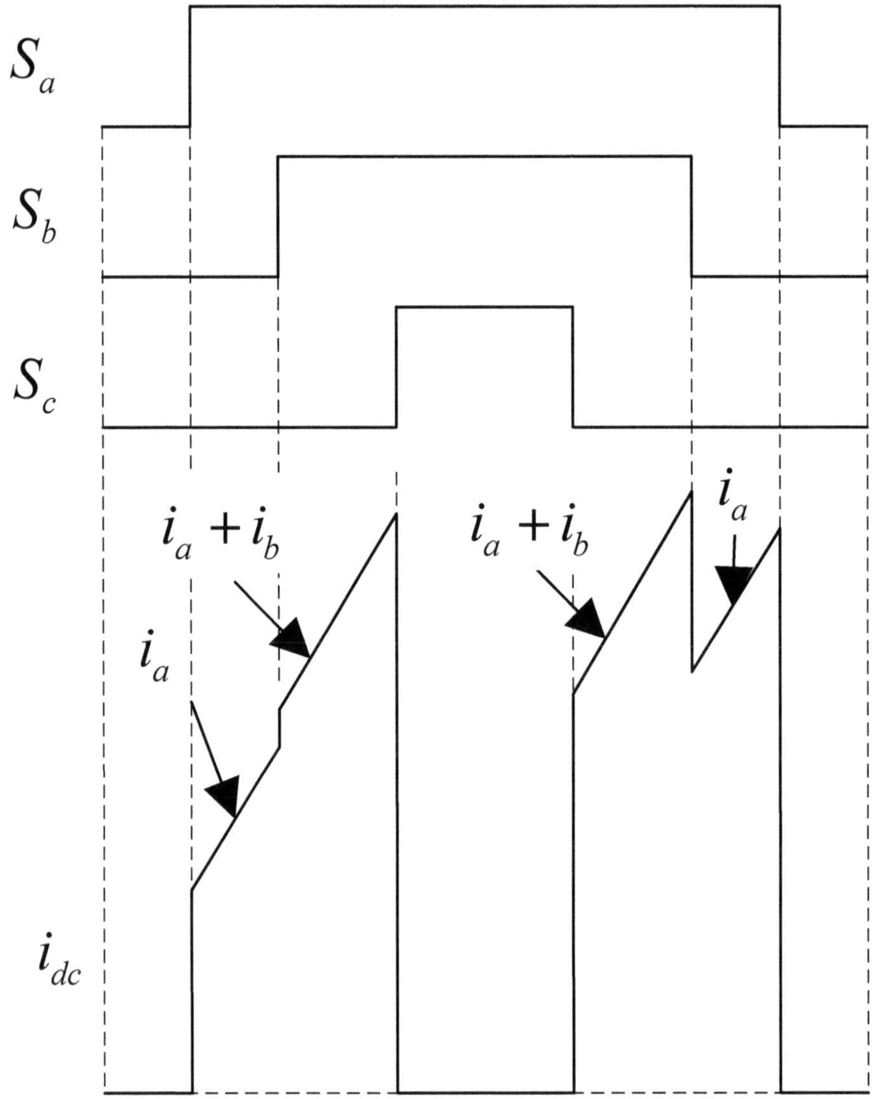

Fig. 4.35 DC-link current in one switching cycle considering AC side current ripple

The DC-link current comparison between simulation results and prediction results is shown in Figs. 4.37 and 4.38. Figure 4.37 shows the comparison results without considering the AC side current ripple. During the whole fundamental wave period, it can be seen that the envelope of the simulation results is about 20% higher than that of the predicted results and the zoom-in detail is shown in Fig. 4.37b. Because the AC side current ripple is not considered in the DC-link current calculation but only the fundamental value is adopted, which is obviously different from the actual

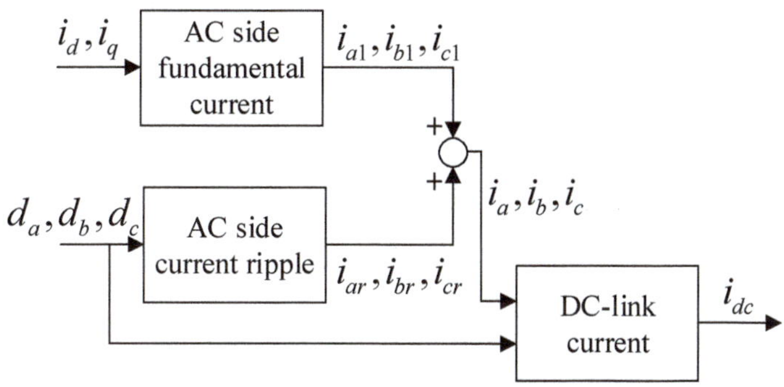

Fig. 4.36 DC-link current calculation considering AC side current ripple

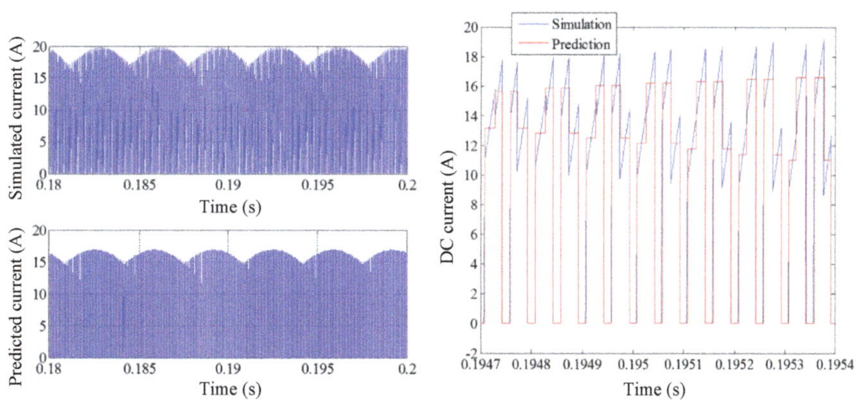

Fig. 4.37 DC-link current comparison without considering AC side current ripple: **a** fundamental period, **b** zoom-in detail

simulation result, the peak value of DC-link current is smaller than that of the simulation results. Figure 4.38 shows the comparison results considering AC side current ripple. Figure 4.38a shows the comparison results during the whole fundamental wave period. It can be seen that the simulated DC-link current is basically consistent with the predicted envelope. As can be seen from the zoom-in detail shown in Fig. 4.38b, the calculation result matches well with the simulation, proving the accuracy of the DC-link current calculation considering the AC side current ripple shown in Fig. 4.36. Figure 4.39 shows the spectrum comparison of DC-link current. Figure 4.39a shows that the spectrum of DC-link current prediction results without considering AC side current ripple is significantly different from simulation results, while Fig. 4.39b shows that the spectrum of DC-link current prediction results considering AC side current ripple is closer to simulation results. The prediction method

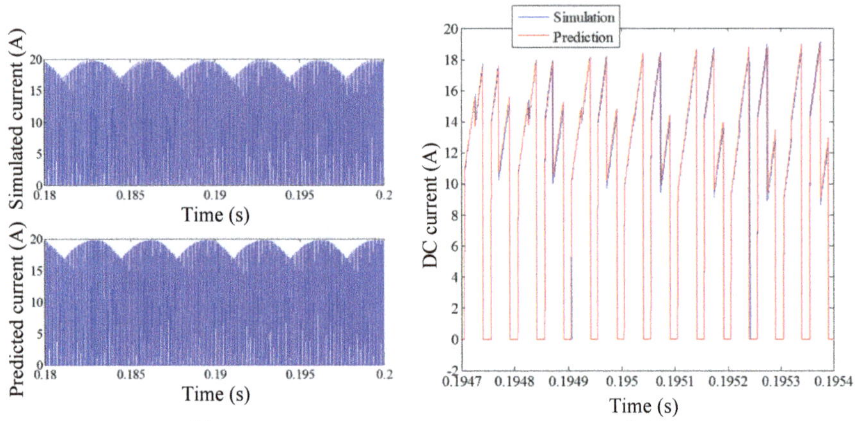

(a) (b)

Fig. 4.38 DC-link current comparison considering AC side current ripple: **a** fundamental period, **b** zoom-in detail

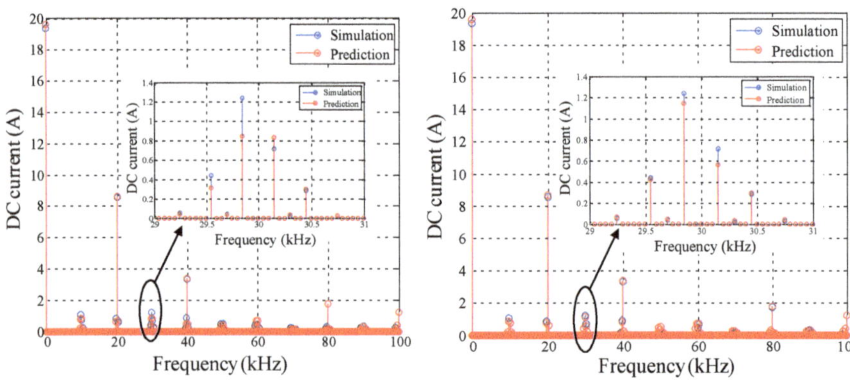

(a) (b)

Fig. 4.39 DC-link current spectrum comparison: **a** without considering AC side current ripple, **b** considering AC side current ripple

shown in Fig. 4.36 provides a more precise mathematical tool for calculating DC-link capacitance or DC battery losses in the frequency domain.

The experimental results are shown in Fig. 4.40, which are respectively the predicted DC-link current and the experimental DC-link current within a whole fundamental period. It can be seen that the amplitude and envelope of both the predicted and experimental results match well, which further proves the accuracy of the prediction.

In the design of inverter, the rigorous calculation of DC-link RMS current is helpful to the design and optimization of DC-link capacitor. Based on the DC-link current

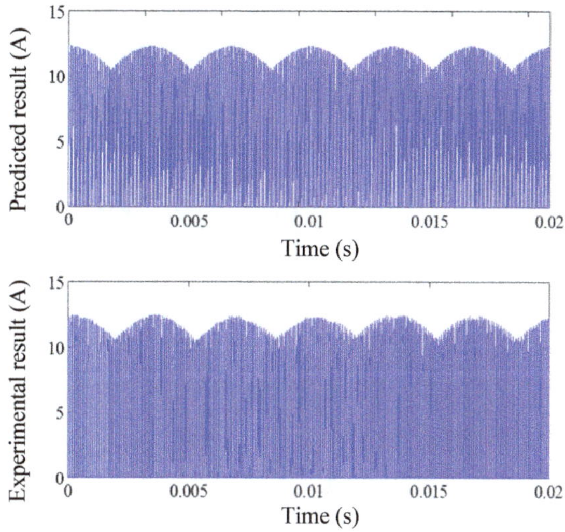

Fig. 4.40 DC-link current comparison between prediction and experiment

prediction method, the RMS value of DC-link current can be calculated accurately from the perspective of space vector. Because of the symmetry of the DC-link current, only one-sixth of the fundamental period is considered. Figure 4.41 shows the space vector diagram of two-level inverter and the reference vectors of voltage and current. The fundamental components of three-phase current on the AC side are shown in (4.35). SVPWM and DPWM use two adjacent effective voltage vectors to synthesize the target voltage vector. As can be seen from Fig. 4.41, the reference voltage is located in $\theta \in (\pi/3, 2\pi/3)$, and the voltage vector $\overrightarrow{V_0} - \overrightarrow{V_3} - \overrightarrow{V_2} - \overrightarrow{V_7} - \overrightarrow{V_2} - \overrightarrow{V_3} - \overrightarrow{V_0}$ will be adopted. The action time of each vector is shown in (4.36). Without losing

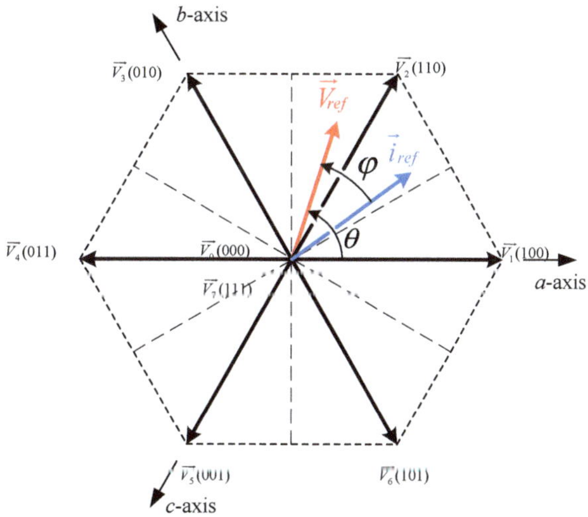

Fig. 4.41 Vectors of inverter output voltage and current

generality, the parameter k_z is defined to allocate zero vector time, for example: for DPWM_Max, $k_z = 1$; for DPWM_Min, $k_z = 0$. In fact, the AC side current ripple vector can be expressed as the time integral of the difference between the inverter output voltage vector and the target reference vector, and then divided by the output filter inductance L, as shown in (4.37).

$$\begin{cases} i_{a\_avg} = I_N \cos(\theta + \varphi) \\ i_{b\_avg} = I_N \cos(\theta - 2\pi/3 + \varphi) \\ i_{c\_avg} = I_N \cos(\theta + 2\pi/3 + \varphi) \end{cases} \tag{4.35}$$

$$\begin{cases} \lambda_{(010)} = \frac{\sqrt{3}}{2} m T_s \sin(\theta - \frac{\pi}{3}) \\ \lambda_{(110)} = \frac{\sqrt{3}}{2} m T_s \sin(\theta + \frac{\pi}{3}) \\ \lambda_{(111)} = k_z(T_s - \lambda_{(010)} - \lambda_{(110)}) \\ \lambda_{(000)} = (1 - k_z)(T_s - \lambda_{(010)} - \lambda_{(110)}) \end{cases} \tag{4.36}$$

$$\vec{i}_{ripple} = \frac{1}{L} \int (\vec{V}_x - \vec{V}_{ref}) dt \tag{4.37}$$

where I_N represents the peak value of AC side fundamental current, θ is the power factor angle, i_{a_avg}, i_{b_avg} and i_{c_avg} are the three-phase fundamental currents, m is modulation ratio ($m = 2V_m/V_{dc}$), V_m is the amplitude of reference voltage, $\lambda_{(010)}$, $\lambda_{(110)}$, $\lambda_{(111)}$ and $\lambda_{(000)}$ are the action time of voltage vectors $\vec{V}_3$, $\vec{V}_2$, $\vec{V}_7$, $\vec{V}_0$, $\vec{V}_x$ is the output voltage vector, $\vec{i}_{ripple}$ is the current ripple vector.

As mentioned above, the DC-link current is superimposed by the AC side fundamental current and the current ripple, as shown in Fig. 4.42. In one switching cycle, considering the symmetry of current ripple, the average value of DC-link current is only determined by the AC side fundamental current, which can be expressed as Eq. (4.38) and simplified to (4.39). As can be seen from Eq. (4.39), the average DC-link current is only determined by the peak value of the AC side fundamental current, the modulation index and the power factor, and is not affected by the AC side current ripple. It is assumed that the DC-link capacitor absorbs all the high-frequency components of the DC-link current. Similarly, the expression of the RMS current flowing through the DC-link capacitor during one switching cycle can be written as (4.40). In Eq. (4.40), the RMS current is composed of the AC side fundamental current and the current ripple. Among them, the RMS value of current ripple on AC side within one switching cycle has been obtained in (4.13), where the switching function should be considered in the calculation of current ripple. As shown in Fig. 4.42b, the switching function acts on the selection of the AC side current ripple. For example, when the inverter outputs the switching vector $\vec{V}_3$, the phase-b current ripple is superimposed on the DC-link current. The RMS value of current ripple can be expressed as (4.41), where y_{1_b}, y_{2_b}, y_{1_c} and y_{2_c} are the turning points of current ripple in phase-b and

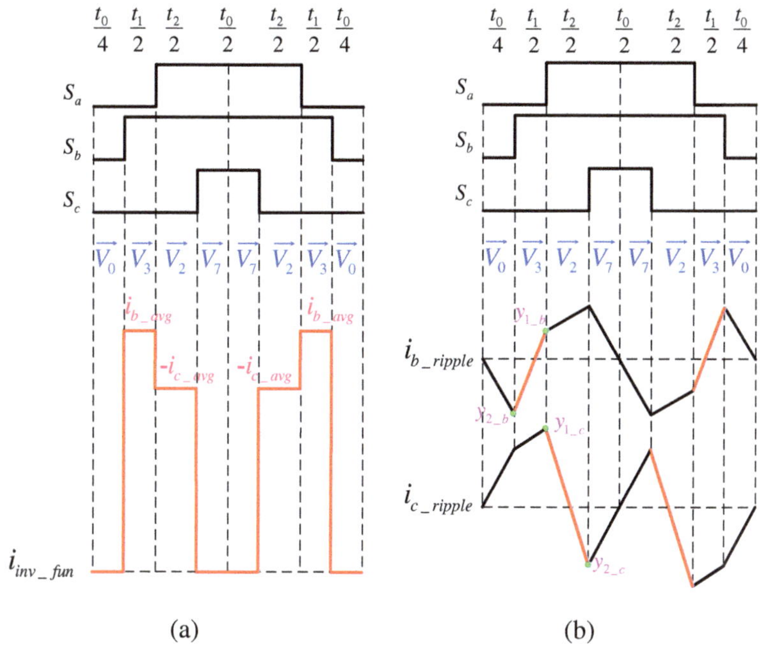

Fig. 4.42 DC-link current in one switching cycle: **a** fundamental current, **b** current ripple

phase-c, which can be obtained by the current ripple prediction method [12].

$$i_{avg} = \frac{\lambda_{(010)}}{T_s} i_{b\_avg} + \frac{\lambda_{(110)}}{T_s}(-i_{c\_avg}) \tag{4.38}$$

$$i_{avg} = \frac{3}{4} I_N m \cos \varphi \tag{4.39}$$

$$i_{cap\_rms} = \sqrt{[\frac{\lambda_{(010)}}{T_s} i^2_{b\_avg} + \frac{\lambda_{(110)}}{T_s} i^2_{c\_avg} - i^2_{avg}] + i^2_{ripple\_rms}} \tag{4.40}$$

$$i_{ripple\_rms} = \sqrt{\frac{\lambda_{(010)}}{T_s} \frac{y^2_{1\_b} + y^2_{2\_b} + y_{1\_b} y_{2\_b}}{3} + \frac{\lambda_{(110)}}{T_s} \frac{y^2_{1\_c} + y^2_{2\_c} + y_{1\_c} y_{2\_c}}{3}} \tag{4.41}$$

With $\theta \in (\pi/3, 2\pi/3)$, the RMS value of DC-link current flowing into the capacitor can be calculated by integrating θ into (4.40), as shown in (4.42). As can be seen from (4.43), the RMS value of current ripple is relatively complex, which is jointly determined by DC-link voltage, modulation index, switching period, filter inductance and zero vector distribution parameter, and is not affected by power factor.

$$I_{cap,rms} = \sqrt{I_N^2 m[\frac{\sqrt{3}}{4\pi} + (\frac{\sqrt{3}}{\pi} - \frac{9}{16}m)\cos^2\varphi] + I_{ripple,rms}^2} \tag{4.42}$$

$$\begin{aligned} I_{ripple,rms}^2 &= \frac{3}{\pi}\int_{\frac{\pi}{3}}^{\frac{2\pi}{3}} i_{ripple\_rms}^2 dt = f(V_{dc}, m, T_s, L, k_z) \\ &= \frac{3\sqrt{3}m^3 V_{dc}^2 T_s^2}{64\pi L^2}\left\{\begin{array}{l} \left[\frac{91}{80}k_z^2 - \frac{91}{80}k_z + \frac{179}{480}\right]m^2 \\ +\left[-(\frac{\sqrt{3}}{3}\pi + \frac{15}{16})k_z^2 + \right. \\ \left.(\frac{\sqrt{3}}{3}\pi + \frac{15}{16})k_z - \frac{3}{32} - \frac{7\sqrt{3}}{48}\pi\right]m \\ +(\frac{5}{3}k_z^2 - \frac{5}{3}k_z + \frac{29}{54}) \end{array}\right\} \end{aligned} \tag{4.43}$$

Keeping $I_{N,rms} = 36/\sqrt{2}$ A, $V_{dc} = 400$ V, $L = 0.5$ mH, for fair comparison, the switching frequency of SVPWM is 5 kHz and that of DPWM is 7.5 kHz. Figure 4.43 shows the 3D curve of the RMS value of DC-link capacitor current. When the power factor is higher, the RMS value of current is larger, which is usually the normal operation condition of AC motors. When designing DC-link capacitance, the maximum RMS value of the surface can provide guidance.

In order to verify the correctness of the calculation of the DC-link capacitor current, taking SVPWM as an example, the simulation is carried out in MATLAB/Simulink and compared with the theoretical calculation results. The simulation parameters are: $V_{dc} = 400$ V, $L = 0.2$ mH, $I_N = 20$ A, and the switching frequency varies from 5 to 40 kHz. As shown in Fig. 4.44, the calculated results of AC side current ripple are in good agreement with the simulation results. However,

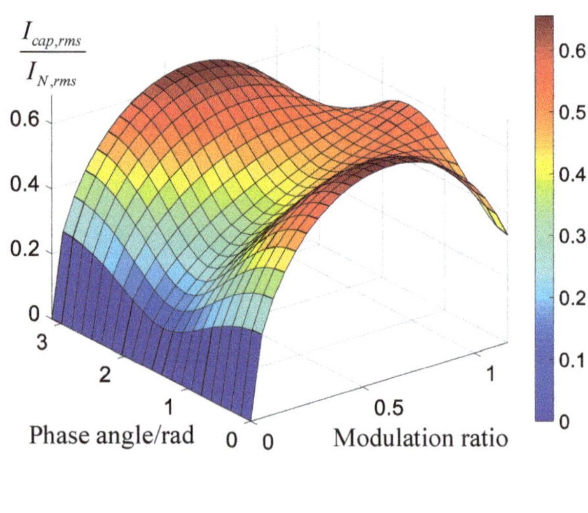

Fig. 4.43 DC-link capacitor RMS current

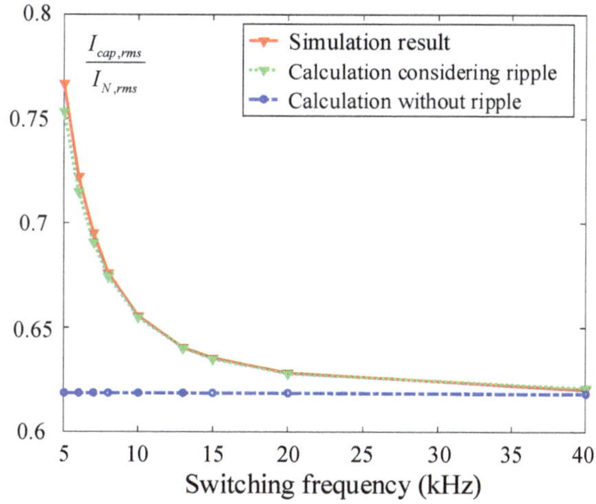

Fig. 4.44 Comparison of DC-link capacitor RMS current (SVPWM)

in the previous analysis and calculation, it is generally considered that the calculation of the RMS value of DC-link capacitor current has nothing to do with the switching frequency. As the proportion of AC side current ripple increases, the calculation result without considering ripple will cause a large error.

4.6 The Impact of Nonideal Characteristics on Current Ripple Prediction

In Sects. 4.1, 4.2, 4.3, 4.4 and 4.5, the current ripple prediction model is based on ideal circuit model without considering the practical issues in real power electronics hardware. In real applications, nonideal characteristics should be considered. In real hardware, circuit is more complex than the modeled circuit like Fig. 4.7, a typical issue is the existence of common-mode loop by parasitic parameters. Also, the components in the circuit are not ideal labels of "*R, L, C*", a typical issue is the variation of inductance by magnetic performance. In this section, the nonideal issues for current ripple prediction are discussed with two major examples.

4.6.1 Common-Mode Loop [13]

In a real three-phase symmetry inverter system, the conducted current components can be classified as common-mode and differential-mode components (CM and DM) according to their conduction paths. With the PWM operation, the CM current flows into the ground system through the parasitic capacitors and CM EMI emission is introduced. However, in the previous analysis on current ripple, the CM component is usually ignored in a system without neutral wire. In addition, the parasitic parameters in semiconductor devices and passive elements play an important role in the nonideal behavior of the current ripple prediction.

As discussed in Sect. 4.3, a general single-phase model has been proposed for current ripple prediction as shown in Fig. 4.26. In each switching cycle, seven zones are divided by six edges of three-phase PWM signals, as shown in Fig. 4.12. The action time of each sector can be achieved with three-phase duty cycles, and the slope of current ripple in each zone is determined by output inductor (L), inverter terminal voltage (V_{Bk}), output average voltage (V_{mk}), CM voltage of the n terminal voltage (V_{cm}) and the n output average voltage (V_{l_cm}). In the previous calculation of AC-side current ripple, V_{cm} is simply treated as the sum of three-phase terminal voltages of inverter, where V_{cm} is equal to $(V_{AN} + V_{BN} + V_{CN})/3$. It means that there is no CM circulation for CM component of current ripple expressed as (4.44). However, the CM component could not be ignored if a real CM circuit is introduced. The AC side equivalent circuit is shown in Fig. 4.45, and the three-phase terminal voltages are modeled as pulse voltage with reference to midpoint of DC-link (V_{AN}, V_{BN}, V_{CN}). Z_p represents the parasitic parameters contributing to the AC side CM circuit, such as phase-to-ground stray capacitance, series inductance and resistance.

$$i_{cm} = i_a + i_b + i_c \tag{4.44}$$

In fact, the CM current flows into the ground system through the parasitic capacitors and CM EMI emission is introduced, thus the impact of CM circuit on current ripple should be evaluated. If the CM circuit is taken into consideration, the CM voltage should be expressed as (4.45). Based on the AC side equivalent circuit, (4.46) gives its frequency-domain expression. It is clear that the real CM voltage is not ideally equal to $(V_{AN} + V_{BN} + V_{CN})/3$, unless the impedance of Z_p is high enough.

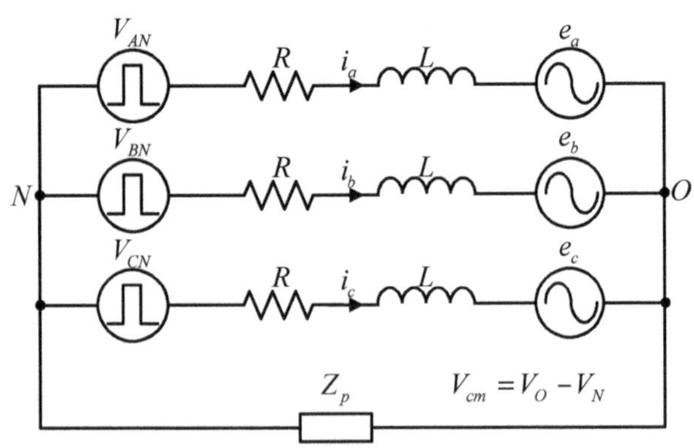

Fig. 4.45 Equivalent circuit of three-phase inverter considering CM circuit

$$V_{cm} = \frac{V_{AN} + V_{BN} + V_{CN}}{3} - \frac{R}{3} i_{cm} - \frac{L}{3} \frac{di_{cm}}{dt} \tag{4.45}$$

$$\frac{V_{cm}}{(V_{AN}(j\omega) + V_{BN}(j\omega) + V_{CN}(j\omega))/3} = \frac{Z_p(j\omega)}{R/3 + j\omega L/3 + Z_p(j\omega)} \tag{4.46}$$

Usually, the $Z_p(j\omega)$ is with high impedance in a normal inverter system, without impacting on the low-frequency harmonic current (in tens of Hz or hundreds of Hz). However, for a PWM system, the harmonic current distributes mainly to the integer multiple of switching frequency, where errors cannot be ignored if the resonant frequency in CM circuit is close to the frequency of the dominant harmonics of current ripple. Due to the parasitic parameters in the CM circuit, it is not possible to accurately calculate the CM voltage in time domain. Thus, with the conventional current ripple prediction model, this part of error caused by actual CM circuit is hard to amend.

Related experiments are carried out in an IGBT-based inverter with three-phase *L-R* load. With the neutral point of the load connected to the ground through a 1-m-long cable and series noninductive capacitor, the AC side CM circuit is investigated. When switching frequency is set as 20 kHz, DC voltage is 200 V and line inductance is 0.5 mH, the THD of output current increases along with the CM capacitance as shown in Fig. 4.46, because of the decreasing resonant frequency as shown in Fig. 4.47. Compared with the inverter system without connected CM capacitor, an

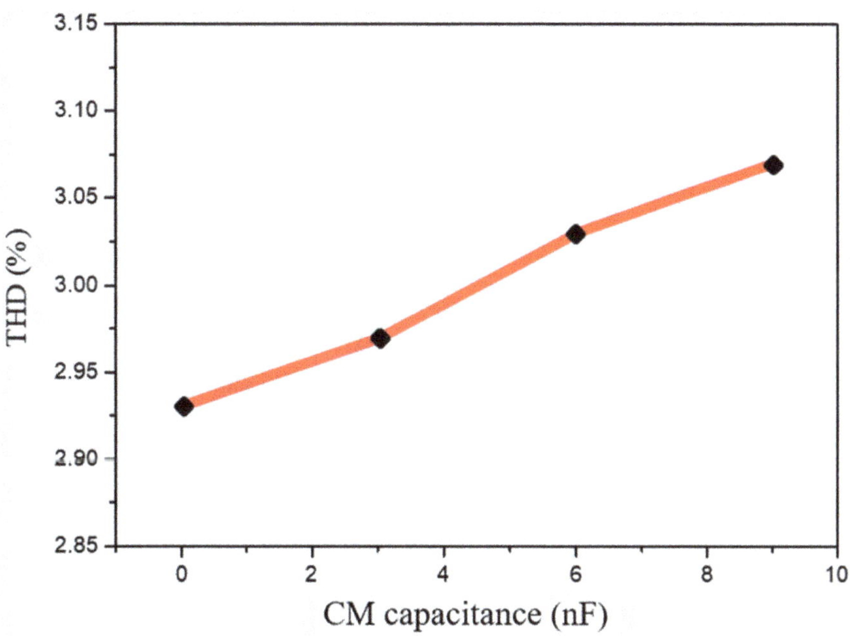

Fig. 4.46 THD of output current ($f_s = 20$ kHz)

Fig. 4.47 Impedance of CM circuit

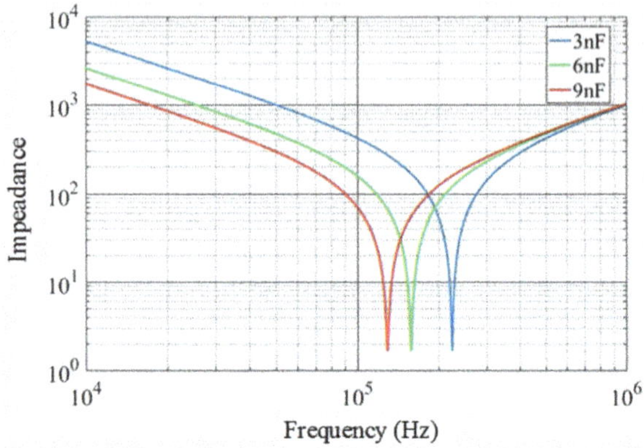

error parameter is defined as (4.47) for evaluating the impact of CM circuit. When the noninductive capacitance is set as 9 nF, it can be seen that the parasitic capacitor artificially added to the AC side CM circuit brings resonance and ringing in the current ripple in the time domain, as shown in Fig. 4.48. In addition, it can be found that the error caused by the CM circuit is increasing along with the switching frequency, as shown in Fig. 4.49. Thus, the resonance frequency introduced by AC side CM circuit should be much larger than the dominant harmonics frequency of current ripple, with the target for error reduction.

$$k = \frac{|\sum_{i=1}^{N} R_{rms\_a}(i) - \sum_{i=1}^{N} R_{rms\_a\_0nF}(i)|}{\sum_{i=1}^{N} R_{rms\_a\_0nF}(i)} \times 100\% \tag{4.47}$$

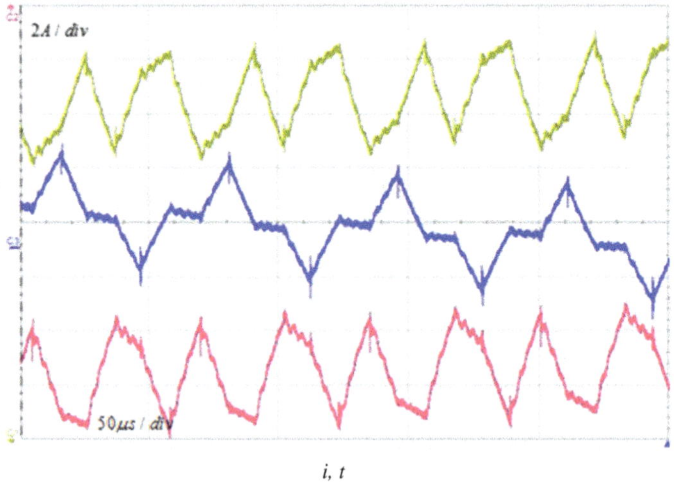

Fig. 4.48 Phase current with CM capacitor

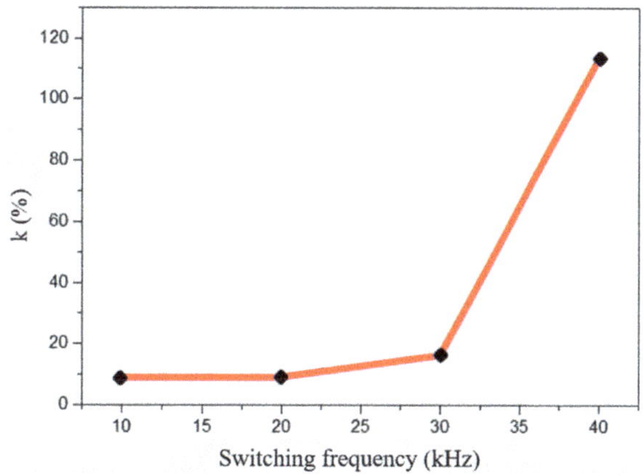

Fig. 4.49 Error of RMS value ($C_p = 9$ nF)

4.6.2 *Inductance Variation [14]*

On the other hand, passive filters are always needed to meet the standard in harmonic mitigation, where the inductance is not a constant value if the load current reaches the saturation limit determined by the core material. When the effective inductance of the filter inductor or windings varies in a wide range, both the output fundamental current and current ripple are influenced. Thus, inductance variation should be taken into consideration on current ripple prediction. The topology of a two-level three-phase VSI system is presented in Fig. 4.50, where three-phase inductors are inserted in the AC-side. It is noted that the inductance of line inductor varies over a wide range when AC current changes in a period.

Observed from Fig. 4.50, the inductance variation is taken into account. V_{KN} ($K =$ A, B, C) is the terminal voltage of three-phase, L_k ($k =$ a, b, c) is the line inductance and R_k is the series resistance, and V_{ON} is defined as the common-mode voltage. According to the KVL law, the voltage equations can be written for three-phase as:

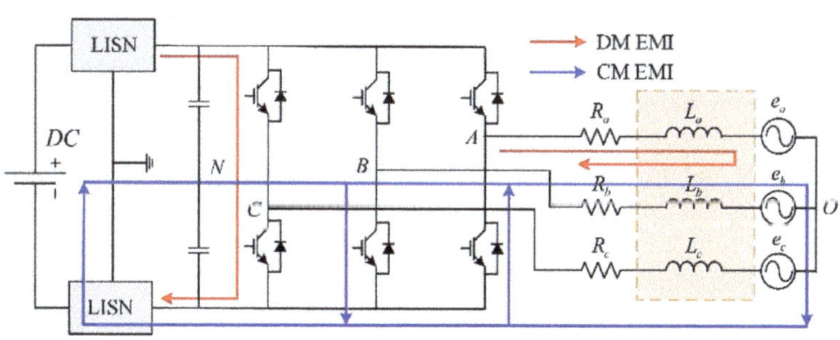

Fig. 4.50 Equivalent circuit for a two-level VSI with varying inductance

$$\begin{cases} V_{AN}(t) - V_{ON}(t) = Ri_a(t) + \dfrac{d\psi_a}{dt} + e_a(t) \\ V_{BN}(t) - V_{ON}(t) = Ri_b(t) + \dfrac{d\psi_b}{dt} + e_b(t) \\ V_{CN}(t) - V_{ON}(t) = Ri_c(t) + \dfrac{d\psi_c}{dt} + e_c(t) \end{cases} \tag{4.48}$$

where ψ_k stands for the excitation flux linkage in the line inductor, and the second term on the right side accounts for the inductance voltage drop of AC current, which can be expressed as:

$$\begin{cases} \dfrac{d\psi_a}{dt} = L_a\dfrac{di_a}{dt} + i_a\dfrac{dL_a}{di_a}\dfrac{di_a}{dt} = (L_a + i_a\dfrac{dL_a}{di_a})\dfrac{di_a}{dt} \\ \dfrac{d\psi_b}{dt} = L_b\dfrac{di_b}{dt} + i_b\dfrac{dL_b}{di_b}\dfrac{di_b}{dt} = (L_b + i_b\dfrac{dL_b}{di_b})\dfrac{di_b}{dt} \\ \dfrac{d\psi_c}{dt} = L_c\dfrac{di_c}{dt} + i_c\dfrac{dL_c}{di_c}\dfrac{di_c}{dt} = (L_c + i_c\dfrac{dL_c}{di_c})\dfrac{di_c}{dt} \end{cases} \tag{4.49}$$

Generally, the inductance L_k is considered to be a constant value, which means that variation of the inductance with the bias current can be ignored, and $i_k dL_k/di_k$ is always kept to be zero. In practice, the inductance L_k will decrease when AC current reaches or goes beyond the saturation limit of the line inductor. Having this fact in mind, an artificial inductance L_k^* can be defined as:

$$L_k^* = L_k + i_k\frac{dL_k}{di_k} \tag{4.50}$$

Equation (4.51) describes how the inductance variation is introduced under AC current flow. It can be observed that:

(1) L_k^* is the artificial inductance for phase-k and it consists of two terms. The first term L_k is the effective inductance, which is a function of permeability of the magnetic core related to the bias current through the inductor; the second term $i_k \mathrm{d}L_k/\mathrm{d}i_k$ stands for the product of the bias current and the derivative of inductance to bias current.
(2) Once the effective inductance is not a constant value, the variation of effective inductance and the bias current, which are involved in the artificial inductance, will then influence the current ripple analysis.
(3) L_k^* is still a function of bias current through the inductor. Both the two terms change along with bias current and it is worth noting that L_k^* is just determined by the bias current if the saturation characteristic of the inductor has been designed.

Substituting (4.49) and (4.50) into (4.48), the following equations are then obtained:

$$\begin{cases} L_b^* L_c^* = [Ri_a(t) + L_a^* \frac{di_a}{dt} + e_a(t)] L_b^* L_c^* \\ [V_{BN}(t) - V_{ON}(t)] L_a^* L_c^* = [Ri_b(t) + L_b^* \frac{di_b}{dt} + e_b(t)] L_a^* L_c^* \\ [V_{CN}(t) - V_{ON}(t)] L_a^* L_b^* = [Ri_c(t) + L_c^* \frac{di_c}{dt} + e_c(t)] L_a^* L_b^* \end{cases} \quad (4.51)$$

Ignoring the resistance voltage drop, the instantaneous CM voltage can be deduced as:

$$V_{ON}(t) = \frac{L_b^* L_c^* V_{AN}(t) + L_a^* L_c^* V_{BN}(t) + L_a^* L_b^* V_{CN}(t)}{L_b^* L_c^* + L_a^* L_c^* + L_a^* L_b^*} - \frac{L_b^* L_c^* e_a(t) + L_a^* L_c^* e_b(t) + L_a^* L_b^* e_c(t)}{L_b^* L_c^* + L_a^* L_c^* + L_a^* L_b^*} \quad (4.52)$$

Considering the inductance saturation characteristics, the values of L_a^*, L_b^* and L_c^* are different because three-phase AC current differs, and the practical CM voltage is a function of three-phase artificial inductance, terminal voltage and load voltage. Specifically, it may be observed from (4.52) that the CM voltage can be simplified as $[V_{AN}(t) + V_{BN}(t) + V_{CN}(t)]/3$ if the artificial inductance is regarded as an unsaturated value. Therefore, the normal CM voltage definition of a three-phase VSI is just a special case, in which artificial inductance is treated to be a constant value.

Due to the symmetry of three-phase VSI, phase-a is studied for current ripple analysis, taking the inductance variation into account. Without losing generality, averaging (4.48) for phase-a over a switching period T_s leads to:

$$\overline{V_{AN}}(T_s) - \overline{V_{ON}}(T_s) = R\overline{i_a}(T_s) + L_a^* \frac{\Delta i_a}{T_s} + \overline{e_a}(T_s) \quad (4.53)$$

Here, $\overline{V_{AN}}(T_s)$, $\overline{V_{ON}}(T_s)$, $\overline{i_a}(T_s)$ and $\overline{e_a}(T_s)$ are the average value of $V_{AN}(t)$, $V_{ON}(t)$, $i_a(t)$ and $e_a(t)$ over a switching period, respectively. Δi_a represents the current change of phase-a over a switching period. Substituting (4.49) and (4.50) into (4.48), and subtracting (4.53) we obtain:

$$\widetilde{v}_a(t) = R[i_a(t) - \overline{i_a}(T_s)] + L_a^*[\frac{di_a}{dt} - \frac{\Delta i_a}{T_s}] + [e_a(t) - \overline{e_a}(T_s)] \quad (4.54)$$

where $\widetilde{v}_a(t)$ is defined as the difference between instantaneous $v_a(t)$ and average voltage $\overline{v}_a(T_s)$:

$$\widetilde{v}_a(t) = v_a(t) - \overline{v}_a(T_s) = [V_{AN}(t) - V_{ON}(t)] - [\overline{V_{AN}}(T_s) - \overline{V_{ON}}(T_s)] \quad (4.55)$$

In Eq. (4.54), the first term on the right side accounts for the resistance voltage drop of current ripple and the third term is the deviation between the actual load voltage and its average voltage over one switching period; both of them can be negligible [12]. Thus, it can be simplified as:

$$\widetilde{v}_a(t) \cong L_a^*\left[\frac{di_a}{dt} - \frac{\Delta i_a}{T_s}\right] \tag{4.56}$$

By integrating (4.56) in subperiod with $t \in [0, T_s]$, the instantaneous current ripple can be defined as:

$$i_{a\_ripple}(t) = i_a(t) - \frac{t\Delta i_a}{T_s} \cong \frac{1}{L_a^*}\int_0^t \widetilde{v}_a(t)dt \tag{4.57}$$

where $i_a(t)$ is the full current of phase-a and $t\Delta i_a/T_s$ stands for the fundamental current changing with time in one switching period. Thus, the slope of phase-*k* current ripple is:

$$\frac{di_{k\_ripple}}{dt} = \frac{1}{L_k^*}[V_{KN}(t) - V_{ON}(t) - \overline{V_{KN}}(T_s) + \overline{V_{ON}}(T_s)] \tag{4.58}$$

Here, L_k^* and $V_{ON}(t)$ are the artificial inductance and CM voltage, respectively. $V_{KN}(t)$ is the terminal voltage of phase-*k*, switched between the positive and negative DC-link voltage. $\overline{V_{KN}}(T_s)$ is the average value of terminal voltage of phase-*k* ($k =$ a, b, c), which is a constant value over one switching period by regular sampling. $\overline{V_{ON}}(T_s)$ is the average value of instantaneous CM voltage over one switching period.

Without losing generality, Fig. 4.51 shows the switching pulse pattern of three-phase VSI in SVPWM technique. There are seven zones corresponding to seven

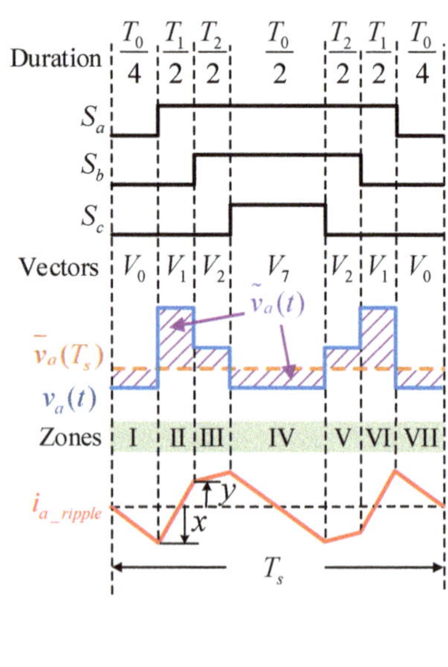

Fig. 4.51 Current ripple of phase-a in one switching cycle with SVPWM

voltage vectors in one switching period. Within each zone, terminal voltage $V_{KN}(t)$ and instantaneous CM voltage $V_{ON}(t)$ can be obtained and treated as constant. In addition, over one switching period, the average value of three-phase terminal voltage and instantaneous CM voltage can be calculated as:

$$\begin{cases} \overline{V_{AN}}(T_s) = \dfrac{V_{dc}(T_1 + T_2)}{2T_s} \\ \overline{V_{BN}}(T_s) = \dfrac{V_{dc}(T_2 - T_1)}{2T_s} \\ \overline{V_{CN}}(T_s) = \dfrac{-V_{dc}(T_1 + T_2)}{2T_s} \end{cases} \tag{4.59}$$

$$\begin{aligned} \overline{V_{ON}}(T_s) = & \frac{L_b^*L_c^*V_{AN}(T_1) + L_a^*L_c^*V_{BN}(T_1) + L_a^*L_b^*V_{CN}(T_1)}{L_b^*L_c^* + L_a^*L_c^* + L_a^*L_b^*}\frac{T_1}{T_s} \\ & + \frac{L_b^*L_c^*V_{AN}(T_2) + L_a^*L_c^*V_{BN}(T_2) + L_a^*L_b^*V_{CN}(T_2)}{L_b^*L_c^* + L_a^*L_c^* + L_a^*L_b^*}\frac{T_2}{T_s} \\ & - \frac{L_b^*L_c^*\overline{e_a}(T_s) + L_a^*L_c^*\overline{e_b}(T_s) + L_a^*L_b^*\overline{e_c}(T_s)}{L_b^*L_c^* + L_a^*L_c^* + L_a^*L_b^*} \end{aligned} \tag{4.60}$$

where $V_{KN}(T_1)$ ($K = A, B, C$) is the terminal voltage of phase-k in vector V_1, and $V_{KN}(T_2)$ is the terminal voltage of phase-k in vector V_2. $\overline{e_k}(T_s)$ represents the average value of three-phase load voltage over one switching period and the deviation of $\overline{e_k}(T_s)$ from $e_k(t)$ can be negligible. Thus, the last term in (4.52) and (4.60) can be considered equal, and can cancel each other in (4.58) for current ripple calculation. It means that no additional load voltage sensors are necessary for current ripple calculation.

The validity of current ripple analysis has been verified through dynamic simulations of a three-phase VSI feeding into *L-C-R* load in MATLAB/Simulink, having $V_{dc} = 200$ V, $m = 0.7$, $R = 5\ \Omega$, $C = 35\ \mu$F. A model of nonlinear inductor has been used here, which is determined by the current vector and magnetic flux vector, using smooth interpolation method.

Figure 4.52 shows the measured inductance variation with bias current, where the effective inductance of line inductor is about 720 μH under low injected current and drops to 500 μH around the current peak (14 A). As it may be observed from Fig. 4.53 that under the sinusoidal current flow, high bias current will cause the inductance to decrease and a significant difference ($i_a dL_a/di_a$) between the effective inductance (L_a) and the artificial inductance (L_a^*) can be observed. It can be seen that the maximum inductance error caused by $i_a dL_a/di_a$ is more than 27%. As can be seen from Fig. 4.54, the practical CM voltage is not equal to $[V_{AN}(t) + V_{BN}(t) + V_{CN}(t)]/3$ in some zones because of the inductance variation, but matches well with the predicted result proposed in this section.

Figure 4.55 shows the comparison of current ripple between simulation and prediction for phase-a in different cases. From the zoom-in details of current ripple

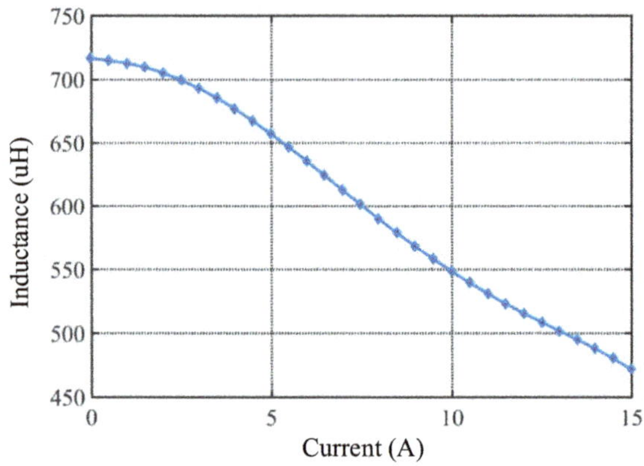

Fig. 4.52 Effective inductance variation with bias current

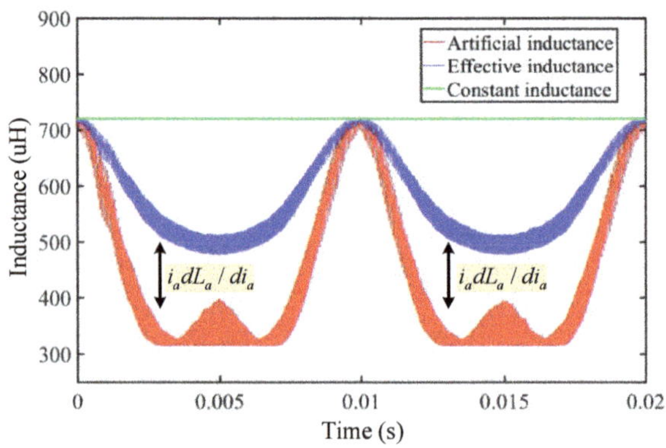

Fig. 4.53 The inductance variation under sinusoidal current flow

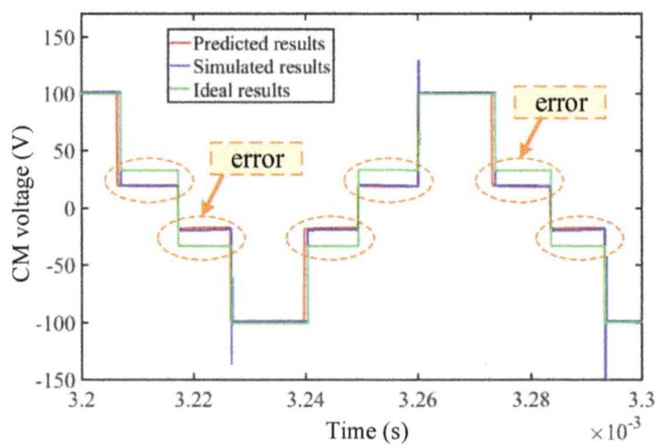

Fig. 4.54 CM voltage comparison

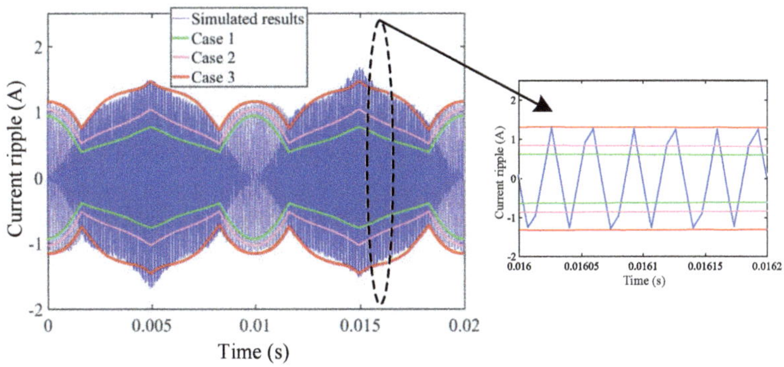

Fig. 4.55 Simulation results: current ripple comparison of phase-a

in several switching cycles for normal SVPWM, it is clear that the current ripple is reset in each switching cycle. Case 1 stands for the predicted peak-to-peak value of current ripple, with a constant inductance (720 μH); in case 2, the varying effective inductance (L_a) is considered to calculate the peak-to-peak value of current ripple, but ignoring the influence of $i_a \mathrm{d}L_a/\mathrm{d}i_a$; Case 3 represents the predicted results considering the inductance variation of L_a^* proposed here. Compared with the simulated current ripple, significant errors can be observed in both case 1 and case 2. Though the effective inductance variation is taken into account in case 2, the prediction error is still significant. Using the artificial inductance L_k^* and the discussed prediction method, the predicted peak-to-peak value of current ripple matches well with the simulated current ripple, which demonstrates the previous current ripple analysis considering the varying inductance.

4.7 Summary

In this chapter, the influence of PWM strategy on current ripple and its prediction model have been introduced. First, the current ripple prediction method of the simplest single-phase converter is introduced: the slope of current ripple is calculated by the difference between the terminal voltage of the inverter and the average load voltage, and the current ripple is predicted by considering the action time together. The analysis method is also the basis of current ripple prediction of multiphase converters. Compared with single-phase converter, three-phase converters are obviously more complex. In Sect. 4.2, the current ripple prediction method of three-phase converters based on the Thevenin equivalent circuit is introduced, and the validity is verified by simulation and experiment comparison. When the phase number of the converter increases further, it becomes very complicated to calculate the current ripple for each voltage vector. Therefore, in Sect. 4.3, a current ripple prediction method based on the general single-phase model is introduced. Compared with the method in Sect. 4.2, the method in Sect. 4.3 does not need to find the equivalent circuit for each voltage vector, but adopts the general equivalent circuit, which is simpler and more suitable for multiphase converters. To solve the problem of asymmetric inductance in salient pole motor, Sect. 4.4 introduces a current ripple prediction method based on *d-q* coordinate system. Section 4.5 applies current ripple prediction effectively to the prediction of DC-link current, and deduces and calculates the RMS value of DC-link capacitor current. The current ripple prediction should also consider nonideal effects in the system. Section 4.6 mainly analyzes the influence of common-mode circuit and nonlinear inductance of inverter system on current ripple prediction. This chapter introduces the theory, simulation and experiment of current ripple prediction, which lays a foundation for the further design of advanced PWM to control the freedom of switching frequency to optimize converter performance.

References

1. Holmes DG, Lipo TA (2003) Pulse width modulation for power converters-principle and practice. IEEE Press
2. Mao X, Ayyanar R, Krishnamurthy HK (2009) Optimal variable switching frequency scheme for reducing switching loss in single-phase inverters based on time-domain ripple analysis. IEEE Trans Power Electron 24(4):991–1001
3. Zhao D, Hari V, Narayanan G, Ayyanar R (2010) Space-vector-based hybrid pulsewidth modulation techniques for reduced harmonic distortion and switching loss. IEEE Trans Power Electron 20, 25(3):760–774
4. Dujic D, Jones M, Levi E, Prieto J, Barrero F (2011) Switching ripple characteristics of space vector PWM schemes for five-phase two-level voltage source inverters—part 1: flux harmonic distortion factors. IEEE Trans Ind Electron 58(7):2789–2798
5. Jones M, Dujic D, Levi E, Prieto J, Barrero F (2011) Switching ripple characteristics of space vector PWM schemes for five-phase two-level voltage source inverters—part 2: current ripple. IEEE Trans Ind Electron 58(7):2799–2808
6. Holtz J, Beyer B (1994) Optimal pulse width modulation for AC servos and low-cost industrial drives. IEEE Trans Ind Appl 30(4):1039–1047
7. Jiang D, Wang F (2013) Variable switching frequency PWM for three-phase converters based on current ripple prediction. IEEE Trans Power Electron 28(11):4951–4961
8. Jiang D, Wang F (2014) Current ripple prediction for three-phase PWM converters. IEEE Trans Ind Appl 50(1):531–538
9. Jiang D, Wang F (2014) A general current ripple prediction method for the multiphase voltage source converter. IEEE Trans Power Electron 29(6):2643–2648
10. Yang F, Taylor AR, Bai H, Cheng B, Khan AA (2015) Using d–q transformation to vary the switching frequency for interior permanent magnet synchronous motor drive systems. IEEE Trans Transp Electrific 1(3):277–286
11. Li Q, Jiang D (2018) DC-link current analysis of three-phase two-level VSI considering AC current ripple. IET Power Electrons 11(1):202–211
12. Grandi G, Loncarski J, Dordevic O (2015) Analysis and comparison of peak-to-peak current ripple in two-level and multilevel PWM inverters. IEEE Trans Ind Electron 62(5):2721–2730
13. Li Q, Jiang D, Chen J, Zhang Y (2019) Impact of common-mode circuit on variable switching frequency PWM strategy in voltage source inverters. In: Proceedings of IEEE applied power electronics conference and exposition, pp 932–937
14. Li Q, Jiang D, Zhang Y (2019) Analysis and calculation of current ripple considering inductance saturation and its application to variable switching frequency PWM. IEEE Trans Power Electron 34(12):12262–12273

Chapter 5
Model Predictive VSFPWM

In Chap. 4 the real-time prediction method of current ripple is introduced . Based on the real-time prediction model, the traditional open-loop PWM method can be optimized. It is called model predictive PWM in this book. Among them, switching frequency is a typical controllable freedom in PWM of power electronics converter. This chapter mainly introduces the variable switching frequency PWM (VSFPWM) based on current ripple prediction model: VSFPWM based on current ripple peak value and VSFPWM based on current ripple RMS value. In addition, VSFPWM based on other optimization objectives will also be briefly introduced, including torque ripple of motor and DC-link voltage ripple in the PWM rectifier. The quantified impact of switching frequency's distribution on the EMI has been studied and unified distributed PWM (UDPWM) and normal distributed PWM (NDPWM) have been proposed which can further optimize EMI. Then, two fundamental questions for VSFPWM have been answered: the variation of switching frequency's impact on low-frequency harmonics and feedback control. Finally, in addition to the freedom of switching frequency, pulse position is also an important freedom in PWM.

5.1 Model Predictive PWM

The traditional PWM strategy is an open-loop driving process: the reference voltage/current signal is converted into duty ratio, and the final output is in the form of pulse, without using the correlation prediction model to optimize the system performance. However, the PWM process is directly related to the system current ripple, voltage ripple and other physical variables, so it is feasible to establish the real-time prediction model of related physical variables. Different from the traditional PWM strategy, the model predictive PWM (MPP), based on the real-time prediction model of related physical variable, makes full use of the freedom of PWM and can effectively improve the switching losses and conducted EMI of the system under the

D. Jiang et al., *Advanced Pulse-Width-Modulation: With Freedom to Optimize Power Electronics Converters*, CPSS Power Electronics Series,
https://doi.org/10.1007/978-981-33-4385-6_5

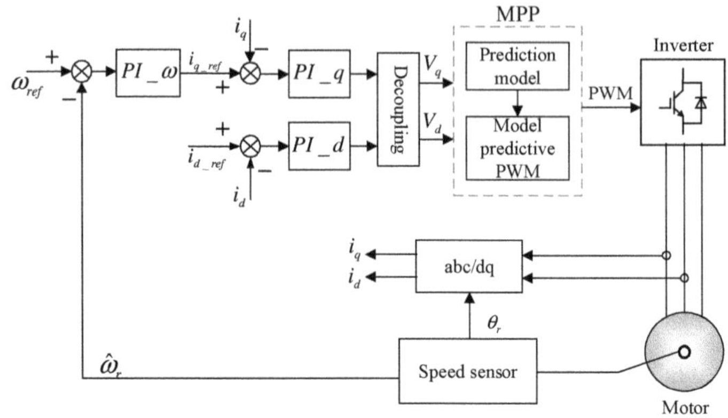

Fig. 5.1 AC motor drive control diagram with model predictive PWM

same conditions as the relevant standards of the system. Figure 5.1 shows a typical application model of PWM converter schematic diagram, with AC motor drive as an example. The control structure of the converter is exactly the same as that of the ordinary control system. The main difference is that the duty ratio is no longer directly converted into the gate drive signal in the PWM module, but is optimized according to the prediction model.

Model predictive PWM has two important positive characteristics, one is to improve the performance of the inverter by using the freedom of PWM, and the other is the freedom control must be based on the prediction model. Although the random PWM introduced in Chap. 3 uses the freedom of switching frequency to improve conducted EMI, its control method is random and not realized according to the prediction model, so it does not belong to the scope of model predictive PWM. At the same time, random PWM does not use prediction model to control the switching frequency, but relies on statistical results, and cannot effectively control the current ripple and switching loss of the inverter, which has application limitations.

The model predictive control (MPC) has been very popular in recent years. Figure 5.2 shows a model predictive control block diagram taking motor drive as an example. Different from ordinary vector control, MPC does not control the motor torque through the current loop. Instead, it directly takes the current reference as input and selects the appropriate output voltage vector through the prediction model, to realize the optimal control of the motor. It is also suitable for other applications, including grid-connected inverters. The prediction model of the control object is directly used for the synthesis of output vectors, and the optimization objective can also select different parameters according to the application [1–4].

Although the prediction model is applied in both MPP and MPC, they are essentially different. First, without inner loop (current loop), MPC is implemented by using the method of direct vector synthesis to achieve the current (torque) control. However, the MPP is based on the traditional control method, including the outer loop and inner loop (current loop). The reference voltage in MPP is generated from

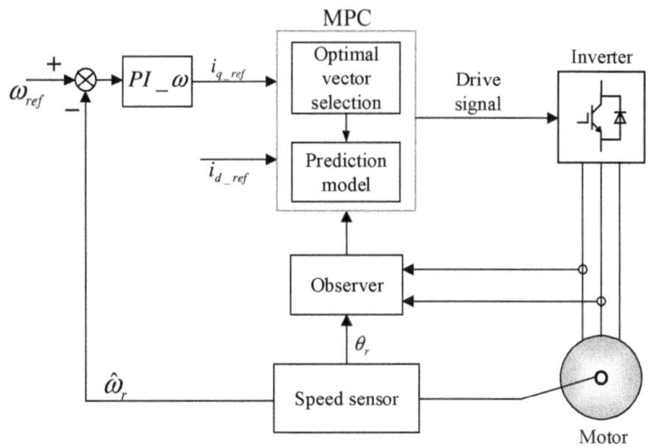

Fig. 5.2 AC motor drive control diagram with model predictive control

Table 5.1 Comparison between MPC and MPP

	MPC	MPP
Pulse generation	Directly generated	Reference voltage combination
Control target	Can directly control torque and current	Torque and current transfer control through voltage
Control approach	Based on nonlinear control architecture	Based on conventional vector control architecture, only change PWM generation

the current loop, and achieve the optimization through PWM. The object of MPC is to control torque and current directly, while the MPP controls torque and current indirectly by controlling output voltage. MPC can be considered as a nonlinear control architecture, which changes the traditional control structure. However, MPP is based on the traditional control method, which only changes the way of PWM generation. The correlation comparison is shown in Table 5.1.

The above analysis is used to compare and explain the difference between MPP and MPC. The core content of this book is to improve PWM through prediction model, without too much description of the control method. Based on the prediction model introduced in the previous chapter, this chapter optimizes PWM method by using switching frequency and other freedoms.

5.2 The Architecture of Variable Switching Frequency PWM

The most typical model predictive PWM in this chapter is variable switching frequency PWM (VSFPWM). The variable switching frequency PWM defined in this chapter is compared with the traditional constant switching frequency PWM

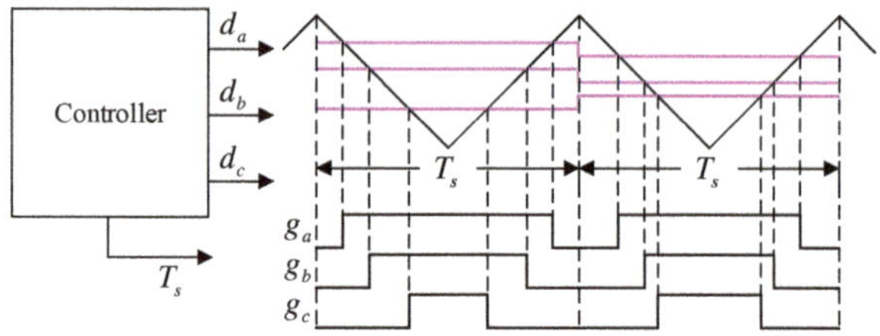

Fig. 5.3 The diagram of variable switching frequency PWM

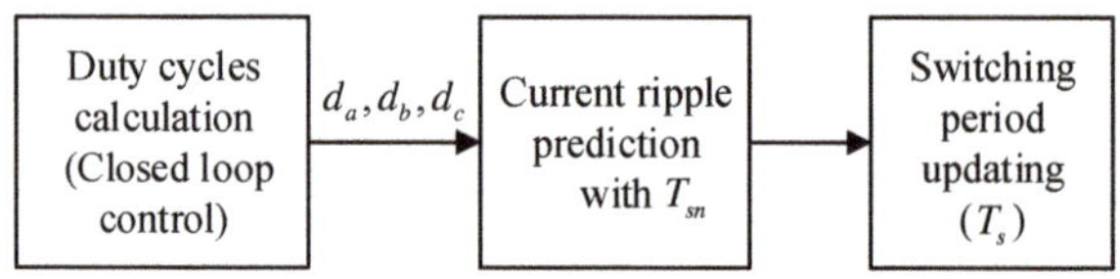

Fig. 5.4 Switching period updating process

(CSFPWM) [5–7]. VSFPWM can control the switching frequency in each switching cycle. Similar to random PWM introduced in Sect. 3.5, the core idea of variable switching frequency PWM is to update the switching frequency cycle by cycle in real time. Figure 5.3 shows the schematic diagram of variable switching frequency PWM. The CSFPWM updates the duty ratios (d_a, d_b, d_c) once in each switching cycle and compares them with the carrier with fixed frequency to finally send out the pulse. VSFPWM not only updates the duty ratios (d_a, d_b, d_c) once in each switching cycle but also updates the switching cycle T_s. That is, the frequency of triangular carrier is also updated in real time [8–13].

Different from the random PWM introduced in Sect. 3.5, the switching frequency variation of VSFPWM introduced in this chapter does not depend on random number, but is based on the current ripple prediction method introduced in Chap. 4. Switching frequency updating process is shown in Fig. 5.4. In each interruption cycle, duty cycles are calculated from the controller and delivered into the current ripple prediction module. The current ripple is predicted based on the fixed switching period T_{sn}, and the switching period T_s is updated in real time according to the requirement of current ripple.

Based on Figs. 5.3 and 5.4, the block diagram of VSFPWM control is obtained, as shown in Fig. 5.5. In the controller, duty cycles (d_a, d_b, d_c) are calculated in every cycle, and current ripple prediction module reads duty cycles after receiving a sampling signal. Current ripple can be predicted in each switching cycle. Then, based on the current ripple peak value requirement, switching period T_s is updated. A sampling signal will be generated when a complete carrier waveform has been sent to the comparator, and the VSFPWM control will enter the next update.

Assuming the predicted peak value of current ripple based on T_{sN} is $I_{ripple\_predicted}$, as the peak value of current ripple is proportional to the switching period, I_{ripple_update} can be obtained by using new switching period T_s, as shown in (5.1). If the new current ripple (I_{ripple_update}) is required to be equal to the current ripple limit ($I_{ripple\_require}$), then the switching period T_s should meet (5.2). Therefore, the real-time control of current ripple can be realized by controlling PWM switching frequency.

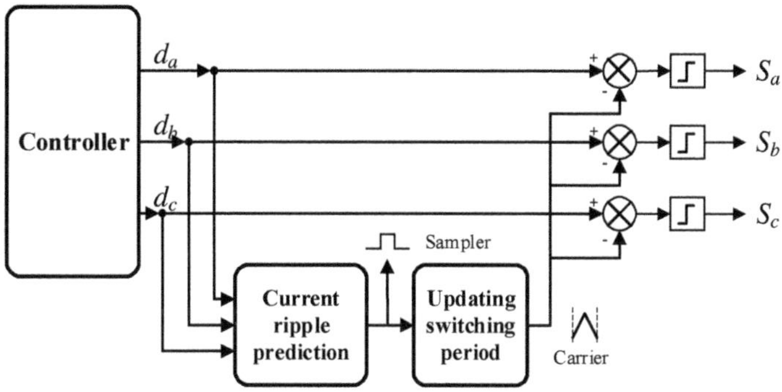

Fig. 5.5 VSFPWM control diagram

$$I_{ripple\_update} = I_{ripple\_predicted} \times \frac{T_s}{T_{sN}} \tag{5.1}$$

$$T_s = T_{sN} \times \frac{I_{ripple\_require}}{I_{ripple\_predicted}} \tag{5.2}$$

For example, it is assumed that the nominal switching frequency is 20 kHz ($T_{sN} =$ 50 μs) and the required current ripple ($I_{ripple\_require}$) is 4 A. In one certain switching cycle, the maximum current ripple with T_{sN} is predicted to be 2 A. Therefore, certain switching cycle could be updated to be $T_s = 100$ μs and the maximum current ripple is controlled to be as the required value of 4 A.

For current ripple, there are mainly two requirements: peak value and RMS value. Sections 5.3 and 5.4 will design corresponding VSFPWM strategies for these two requirements, respectively, and carry out simulation and experimental verification. In addition, there are other requirements related to current ripple in the application of power electronics converter, and the corresponding VSFPWM can also be designed, which is introduced in Sect. 5.5.

5.3 VSFPWM for Current Ripple Peak Value Control (VSFPWM1)

In many applications, the current ripple peak value of power electronics converter is required. The larger the current ripple peak value is, there will be more serious current distortions. In Chap. 4, the current ripple prediction model has told us that the current ripple caused by the switching action of the power electronics converter is directly related to the duty cycle at the switching time. In AC-DC or DC-AC converters, the duty cycle varies with time and is determined by the modulation function, which means that the peak value of current ripple varies in different switching cycles.

In this section, a unified simulation model is adopted. The circuit is shown in Fig. 5.6, with 400 V DC-link voltage, and the three-phase inverter is connected to L-type filter, feeding three-phase load connected in "Y" type. The system adopts SVPWM method with $m = 0.78$, and the peak value of current ripple is required below 1.4 A.

In order to meet the requirements of current ripple peak value, the inverter in Fig. 5.4 needs to adopt a constant switching frequency of 33.7 kHz, and the current ripple is shown in Fig. 5.7. As you can see, the current ripple is controlled within the ± 1.4 A range. But the current ripple distribution is not uniform with time. The peak value of current ripple has obvious change within one fundamental period. In fact, there are several switching cycles where the peak value of current ripple is equal to the maximum value of current ripple required by the standard, and the

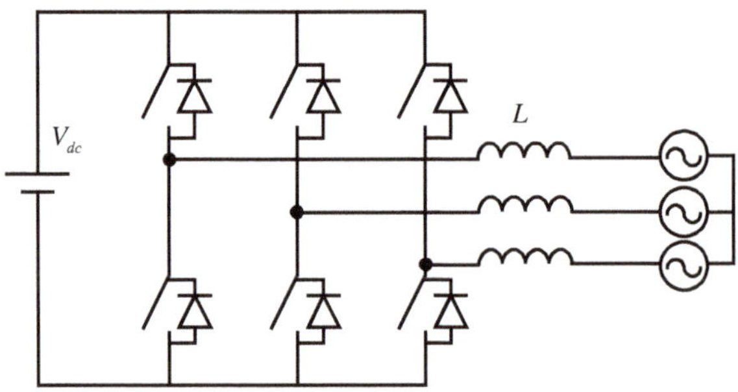

Fig. 5.6 Three-phase inverter with AC load

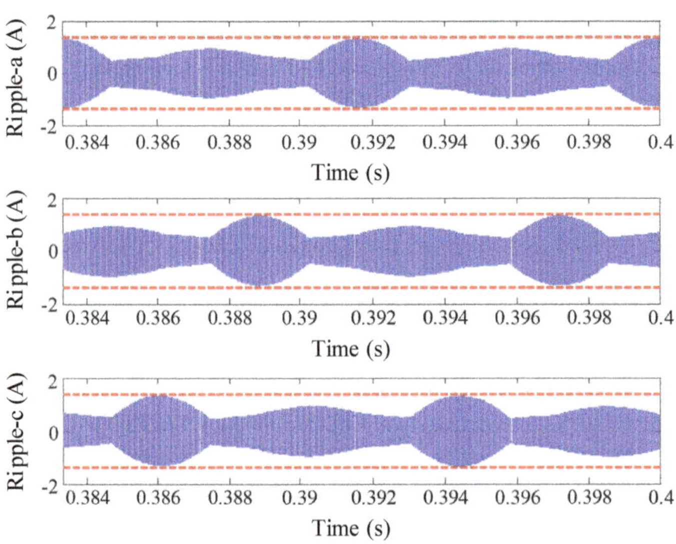

Fig. 5.7 Three-phase current ripple with constant switching frequency (33.7 kHz)

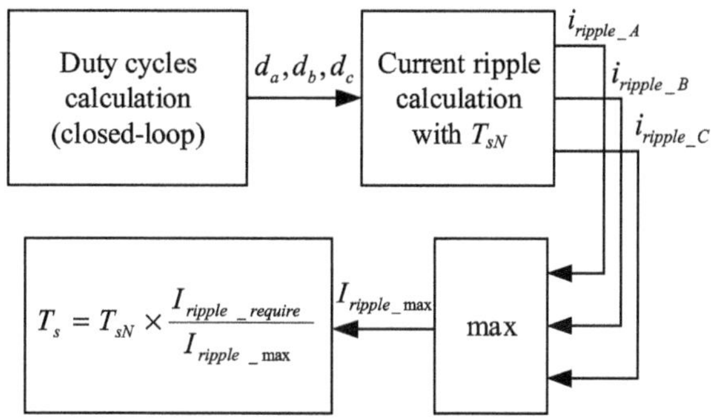

Fig. 5.8 VSFPWM control diagram based on current ripple peak value

current ripple in most switching cycles is less than the standard, even less than half. However, because SVPWM with constant switching frequency is adopted, the switching frequency is always designed according to the maximum current ripple value, resulting in excessive switching loss. According to the prediction model of current ripple, the current ripple can be controlled more effectively by changing the switching frequency in real time, and the switching loss can be further reduced.

Figure 5.8 shows the diagram of VSFPWM based on current ripple peak value. In each interruption period, the controller calculates the output three-phase duty cycles d_a, d_b and d_c, which are delivered to the current ripple prediction module. The prediction module of current ripple is based on the analysis in Chap. 4. The three-phase current ripple in the switching period is obtained through the prediction module, and its maximum value is I_{ripple_max}. This maximum value of current ripple is delivered to the switching cycle update module, which updates the switching cycle according to the rule of (5.2). In this way, the maximum value of the three-phase current ripple in this switching cycle is exactly equal to the standard requirement of the current ripple, which ensures that the current ripple in each switching cycle is not greater than the standard requirement, and the maximum value of the current ripple in just one phase is equal to the standard requirement.

The switching frequency of VSFPWM based on current ripple peak value under the same requirement is shown in Fig. 5.9. It can be seen that the switching frequency varies from 33.7 to 23 kHz within a fundamental period for six periodic changes. Compared with constant switching frequency of 33.7 kHz, the equivalent switching frequency of VSFPWM is decreased by 15%. In this case, the current ripple is shown in Fig. 5.10. The current ripple is still controlled within the requirement by the dotted line. Compared with Fig. 5.7, the current ripple of VSFPWM touches the dotted line for more times. That is, the peak value of three-phase current ripple in each switching cycle is exactly equal to the upper limit of current ripple requirement. VSFPWM makes better use of the requirement of current ripple peak value and realizes the decrease of average switching frequency.

The equivalent switching frequency decrease in VSFPWM can directly realize the switching loss reduction. In Chap. 3, the total switching loss E_{sw} over a fundamental period is shown in Eq. (5.3), which is the sum of the switching losses of three phases

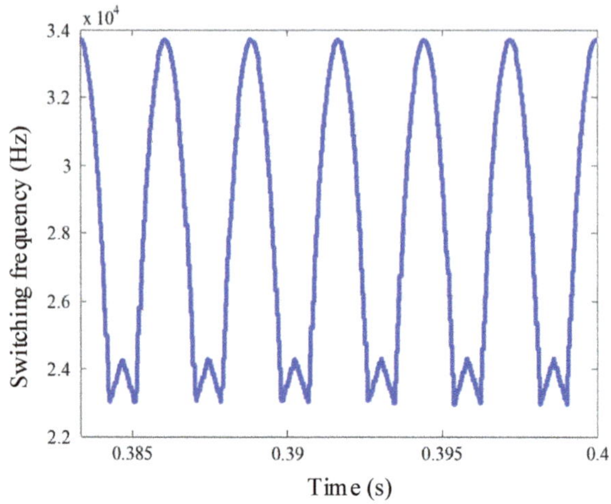

Fig. 5.9 Switching frequency of VSFPWM based on current ripple peak value

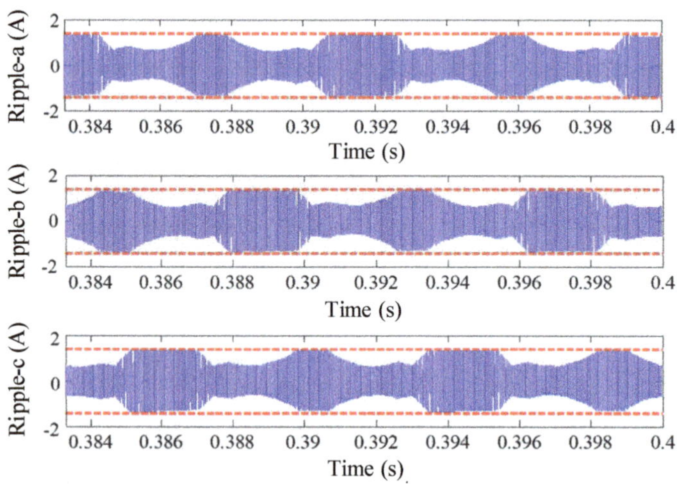

Fig. 5.10 Three-phase current ripple of VSFPWM based on current ripple peak value

in each switching period. In Eq. (5.3), unit switching loss e_{unit} is a fixed value, the current is basically treated as unchanged and the effect of reducing the equivalent switching frequency is the decrease of integral number N, which makes the total switching losses decrease synchronously. The switching losses of VSFPWM and CSFPWM are compared in (5.4).

$$E_{sw} = \sum_{x=a,b,c} \sum_{n=1}^{N} E_{sw,x}(n) = \sum_{x=a,b,c} \sum_{n=1}^{N} |i_x(n)| e_{unit} \tag{5.3}$$

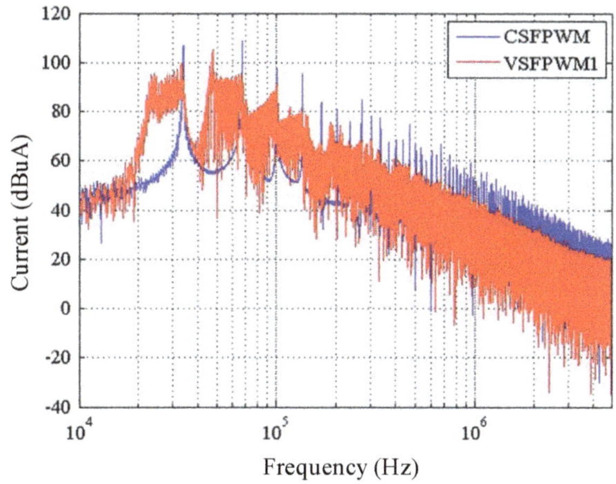

Fig. 5.11 EMI comparison between CSFPWM and VSFPWM

$$\frac{E_{sw\_VSFPWM}}{E_{sw\_CSFPWM}} = \frac{\sum_{n=1}^{N1} |i(n)|}{\sum_{n=1}^{N2} |i(n)|} \tag{5.4}$$

Another advantage of VSFPWM is the suppression of EMI. Due to the variation of switching frequency, EMI is no longer concentrated on the integer multiple of constant switching frequency, so that EMI peak is suppressed. The effect is similar to random PWM. EMI simulation comparison is shown in Fig. 5.11. It can be seen that the EMI of CSFPWM presents periodic peak in frequency domain according to integer multiple of switching frequency. The amplitude of EMI oscillation of VSFPWM is much smaller. Although the average energy of the two is similar, the peak EMI of VSFPWM is more than 10 dB smaller than that of CSFPWM. Therefore, VSFPWM based on current ripple prediction can reduce switching losses and suppress EMI.

For VSFPWM, it is necessary to pay attention to the influence of switching frequency variation on low-order current harmonics. Figure 5.12 shows low-order harmonics comparison between CSFPWM and VSFPWM. When the fundamental current is 10 A, the low-order harmonics difference between the two can be negligible. That is to say, VSFPWM does not worsen the low-order harmonics of the inverter. This issue will be further discussed in Sect. 5.7.

Based on the simulation analysis, further experimental research is carried out. In the experiment, DC-link voltage is set to 150 V and output inductance is 800 μH. The maximum value of current ripple is defined as 2.5 A. In order to meet the current ripple requirement of 2.5 A, for SVPWM with constant switching frequency, the switching frequency must reach 8.66 kHz. When PWM with variable switching frequency is adopted, the switching frequency is shown in Fig. 5.13, which varies from 5.75 to 8.66 kHz. Compared with CSFPWM, the equivalent switching frequency

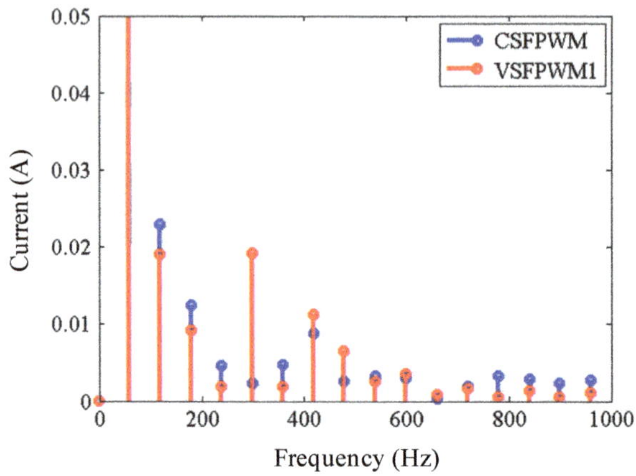

Fig. 5.12 Low-order current harmonic comparison between CSFPWM and VSFPWM

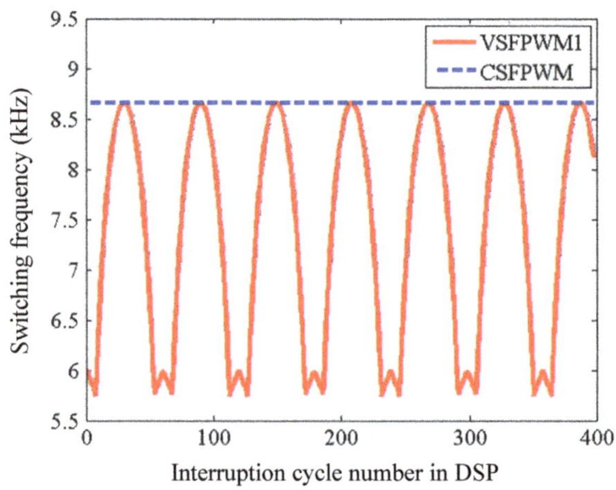

Fig. 5.13 Switching frequency comparison between CSFPWM and VSFPWM (experimental results)

of VSFPWM decreases by 14%. The output current ripple is shown in Fig. 5.14. Both CSFPWM and VSFPWM can keep the current ripple within the peak range of 2.5 A. Compared with the simulation results, the experimental results are more affected by the nonideal characteristics of circuit parameters. The frequency-domain analysis is shown in Fig. 5.15. Similar to the simulation results, there is no obvious deterioration on the low-order harmonics for VSFPWM, but the peak value of EMI is effectively suppressed.

It is worth pointing out that there is only one controllable variable of switching frequency PWM, so there can only be one control target. Therefore, the VSFPWM based on the current ripple peak value takes the maximum value of three-phase current ripple as the control target to make it equal to the requirement. If the current ripple peak value is controlled only for one phase, the current ripple of the other two phases will be distorted. Figure 5.16 shows the VSFPWM that controls the peak value of phase-a current ripple. It can be seen that phase current keeps a constant

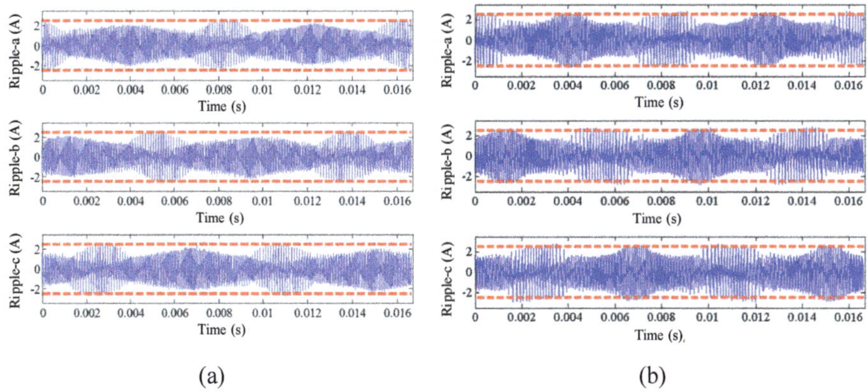

Fig. 5.14 Three-phase current ripple: **a** CSFPWM (8.66 kHz), **b** VSFPWM (5.75–8.66 kHz)

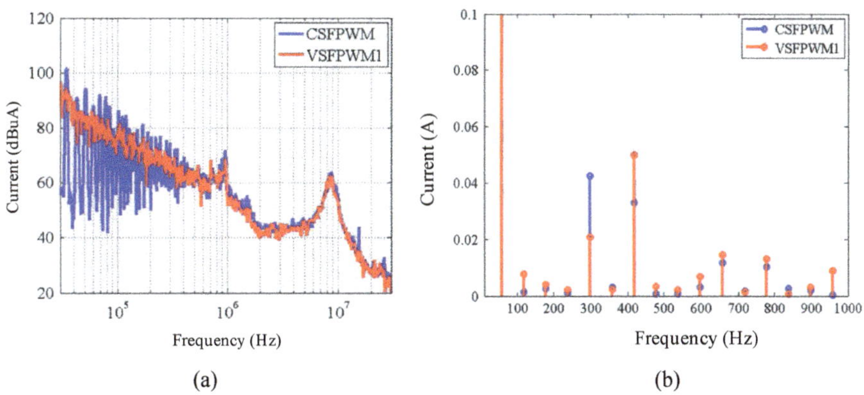

Fig. 5.15 Frequency-domain comparison: **a** EMI, **b** low-order harmonics

Fig. 5.16 Three-phase current with VSFPWM only controlling phase-a current ripple

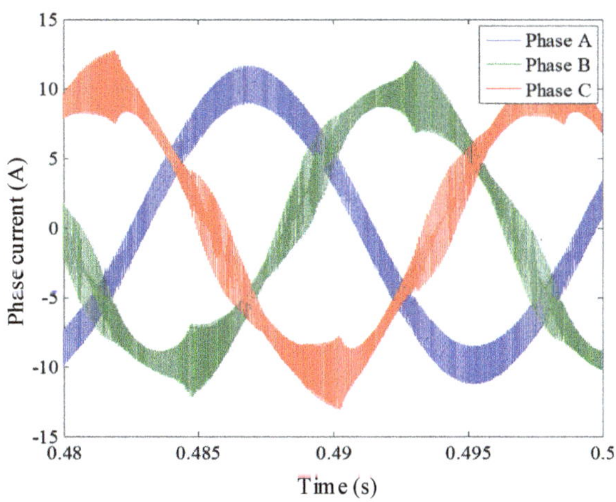

ripple amplitude, but the current ripple of phase-b and phase-c is worse. Therefore, this method is not recommended. If there is more than one control target, the required control variables will be increased accordingly, which will be described in the next chapter.

5.4 VSFPWM for Current Ripple RMS Value Control (VSFPWM2)

VSFPWM based on current ripple peak value makes good use of the uneven distribution of current ripple, and can reduce switching frequency and EMI at the same time. In many applications, the requirement of current ripple is not the peak value, but the RMS value. That is, considering the ohm loss of the current ripple to the system, the RMS value of current ripple is defined. Different from the VSFPWM based on the current ripple peak value, when the current ripple requirement is based on the RMS value of long time scale, although the VSFPWM can achieve the switching frequency variation, the average switching frequency reduction is very limited. That is, the switching losses reduction for VSFPWM based on current ripple RMS value will not be obvious. However, it is still possible to design VSFPWM to ensure the RMS value requirement of current ripple and to improve the EMI. Compared with the random PWM, such VSFPWM has better control on current ripple.

Figure 5.17 shows the diagram of VSFPWM based on current ripple RMS value. Similar to Fig. 5.8, the three-phase duty cycles d_a, d_b and d_c are first obtained through closed-loop control, and then delivered to the current ripple prediction module. Different from Fig. 5.8, the peak value of current ripple in this switching period is no longer predicted, but is based on the RMS value of current ripple I_{ra}, I_{rb} and I_{rc}. Then, the RMS value of the three-phase current ripple is calculated again to obtain the average RMS value, which is sent to the switching cycle update module. The switching period T_s is updated according to the requirement of current ripple RMS value so that the average RMS value of three-phase current ripple is equal to the required RMS value of current ripple $I_{rms\_require}$.

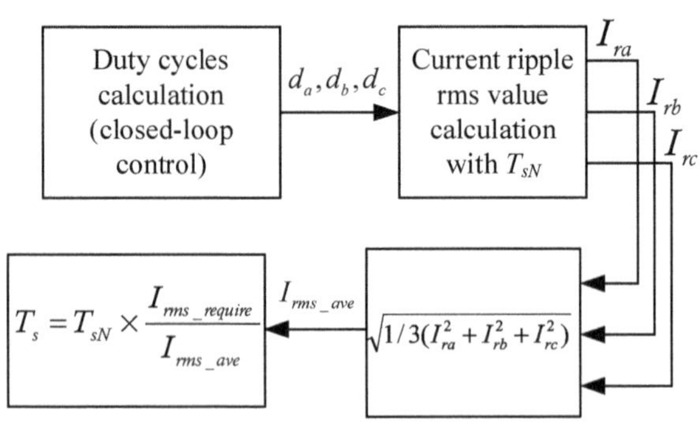

Fig. 5.17 The diagram of VSFPWM based on current ripple RMS value

In accordance with the simulation conditions in Sect. 5.2, the RMS value requirement of current ripple is set as 1 A. The switching frequency of CSFPWM should be set to 16.1 kHz, and the switching frequency of VSFPWM based on current ripple RMS value is shown in Fig. 5.18, which varies from 13.2 to 18.3 kHz. It can be seen that, in terms of average switching frequency, VSFPWM based on current ripple RMS value has no obvious advantage over CSFPWM, and the improvement of switching losses is very limited. However, the requirement is met as the RMS value of three-phase current ripple is well controlled as 1 A. In addition, the advantage of VSFPWM in EMI suppression can still be achieved, as shown in Fig. 5.19. EMI noise peak reaches an attenuation of more than 10 dB.

Both VSFPWM1 and VSFPWM2 can effectively reduce the oscillation of EMI in frequency domain, thus reduce the peak value of EMI. Under the requirements of

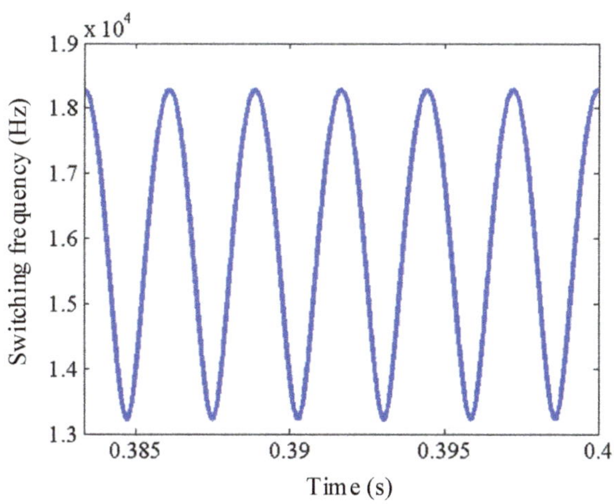

Fig. 5.18 The switching frequency of VSFPWM based on current ripple RMS value

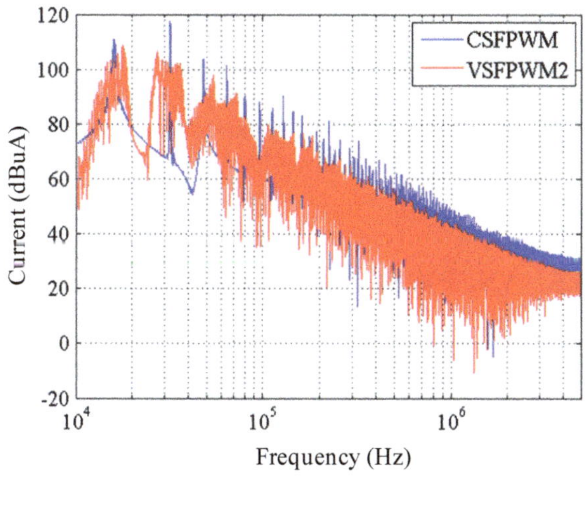

Fig. 5.19 EMI comparison between CSFPWM and VSFPWM2

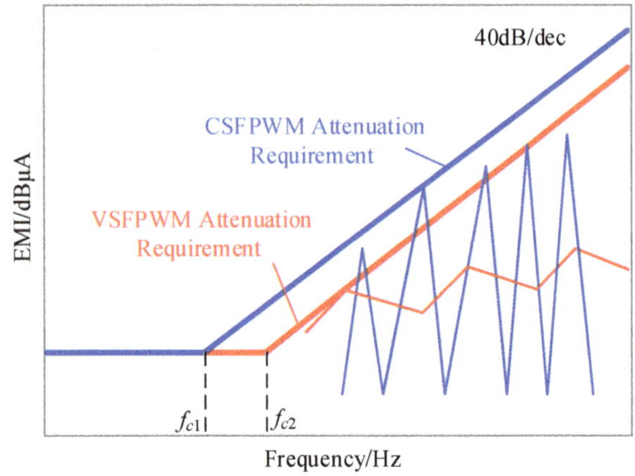

Fig. 5.20 The effect of CSFPWM and VSFPWM on EMI filter

EMI standard, it is equivalent to reducing the demand of EMI attenuation. Figure 5.20 shows the effect of VSFPWM and CSPFWM on EMI filter. An *L-C* filter is taken as an example. Ideally, the filter can be considered to have an attenuation of 40 dB/dec above the corner frequency f_c. According to EMI attenuation demand, corresponding corner frequency f_c can be designed. Compared with CSFPWM, the average attenuation demand has no obvious change, but the maximum attenuation demand has been suppressed. Therefore, the corner frequency f_{c2} of *L-C* filter designed by VSFPWM can be greater than f_{c1} of *L-C* filter designed by CSFPWM. Because the other parameters are the same in two cases, the larger corner frequency can get the smaller filter volume and weight, thus improving the power density of the system.

The VSFPWMs introduced in Sects. 5.3 and 5.4 are optimized PWM methods for AC side output current because AC side current ripple is directly determined by the modulation mode. In many applications, the optimization target is not AC current ripple, but is closely related to modulation method. In the design of VSFPWM, the only available freedom is the switching frequency, so there is only one optimization objective. For example, the maximum value of three-phase current ripple in Sect. 5.2 or the total RMS value of three-phase current ripple in Sect. 5.3. Therefore, the VSFPWM for other optimization objectives can only have one optimization target. In next section, VSFPWM is extended to torque ripple control of AC motors and voltage ripple control of two-level rectifier.

5.5 VSFPWM Based on Other Optimization Objectives

5.5.1 VSFPWM for Torque Ripple Control

In modern AC drive system, torque ripple is an important parameter for the system performance. Torque ripple will influence the vibration and acoustic noise of the motor system. In many applications, this is associated with the quality of the system. There are many reasons for motor torque ripple. The major cause for torque ripple usually comes from two parts: motor structure and PWM strategy of the inverter. The torque ripple caused by motor structure is usually with fundamental frequency and its harmonics. The solution is usually dealing with motor structure inherently. However, for the torque ripple caused by PWM of the motor drive, it is usually associated with switching frequency and its harmonics. The latter is more related to the acoustic noise. The solution is usually with the modification of PWM method.

PWM torque ripple is directly determined by PWM current ripple. Because the back-EMF of the motor is continuous and sinusoidal voltage, but the terminal voltage of the inverter phase-leg is pulse voltage switched between the positive and negative DC bus, the difference between the two voltages is dropped on the winding inductor and it generates the PWM current ripple. Since the electromagnetic torque is generated by the winding current, PWM current ripple will generate PWM torque ripple for the motor. Due to the nonuniform distribution of current ripple, the PWM torque ripple of the motor will also appear as nonuniform distribution. Similar to current ripple, if CSFPWM strategy is adopted, switching frequency needs to be designed for the worst point of torque ripple, resulting in unnecessary increase of switching losses. At the same time, the torque ripple is concentrated around the integer multiple of the switching frequency, which will make the peak value of vibration noise larger and cause audible noise.

Similar to the previous idea of designing VSFPWM for the current ripple control, VSFPWM can also achieve uniform control of PWM torque ripple. The key is to get the relationship between PWM torque ripple and PWM method. Figure 5.21 shows a simple PWM torque ripple algorithm. That is, through coordinate transformation, the three-phase current ripple prediction results are converted to the *d-q* coordinate system, and the current ripples i_{ripple_d} and i_{ripple_q} in the *d-q* coordinate system are obtained. Through the electromagnetic torque calculation formula in *d-q* coordinate system, PWM torque ripple $\Delta T_{predict}$ is achieved.

Based on the torque ripple prediction method shown in Fig. 5.21, the control block diagram of VSFPWM based on torque ripple control can be obtained, as shown in Fig. 5.22. With the real-time prediction of PWM torque ripple in each switching cycle, the maximum torque ripple with constant switching period T_{sN} is calculated. Then the updated switching cycle T_s is calculated where the PWM torque ripple is proportional to switching period. The updated switching period will be employed to

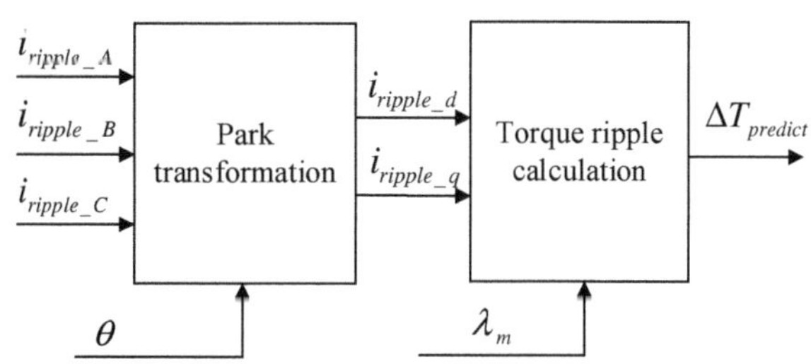

Fig. 5.21 PWM torque ripple prediction method

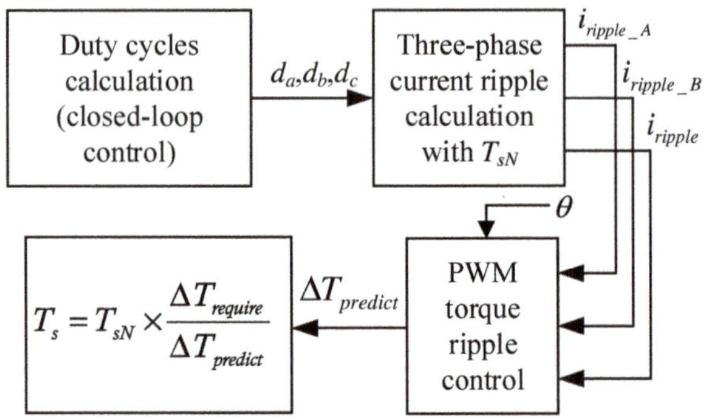

Fig. 5.22 VSFPWM diagram for torque ripple control

make the maximum PWM torque ripple equal to the requirement in each switching cycle. In this way, VSFPWM based on PWM torque pulsation control is realized.

In the simulation example of this section, the CSFPWM achieves the peak torque ripple requirement at 2.5 kHz switching frequency. With VSFPWM, switching frequency varies from 1.8 to 2.5 kHz as shown in Fig. 5.23, effectively reducing switching losses. The torque ripple comparison is shown in Fig. 5.24. The torque ripple peaks of VSPFWM and CSFPWM are basically consistent, except that VSFPWM achieves uniform torque distribution, while the torque ripple amplitude of CSFPWM varies.

Figure 5.25 shows the torque ripple spectrum comparison. Using CSFPWM, the torque ripple spectrum concentrates on integral multiple of switching frequency, and there is an obvious peak near 5 kHz. However, VSFPWM spreads the torque ripple spectrum in a wider range, and the peak value decreases greatly. Because 5 kHz is the frequency within the range of human hearing, CSFPWM brings obvious auditory noise, while VSFPWM can suppress the noise to some extent.

A permanent-magnet synchronous motor has been selected as load to verify the analytical method for PWM torque ripple. In this experiment, traditional vector control method is used to control the permanent magnet synchronous machine

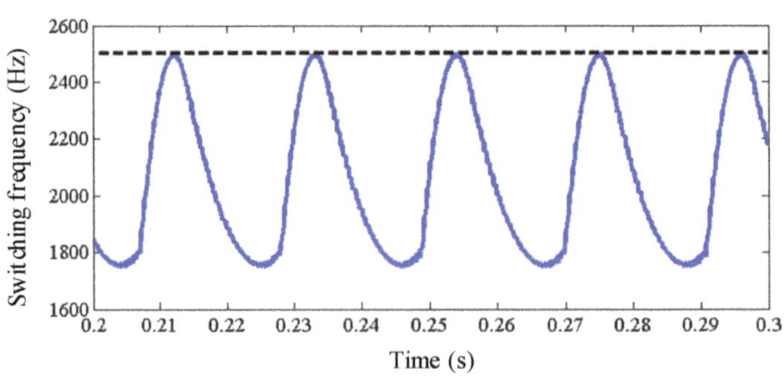

Fig. 5.23 Switching frequency comparison between CSFPWM and VSFPWM for torque ripple control

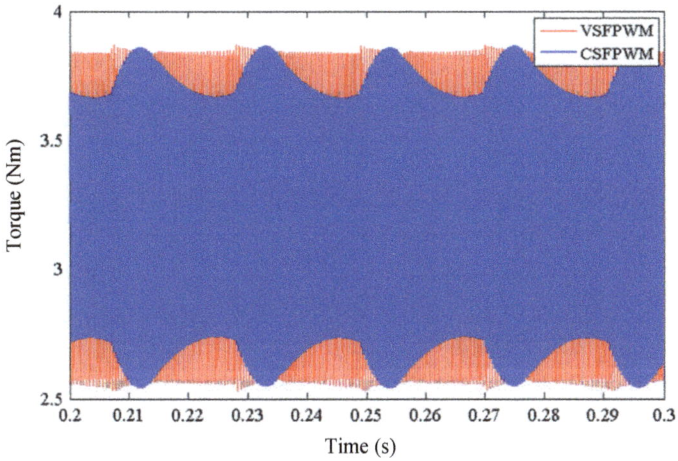

Fig. 5.24 PWM torque ripple comparison

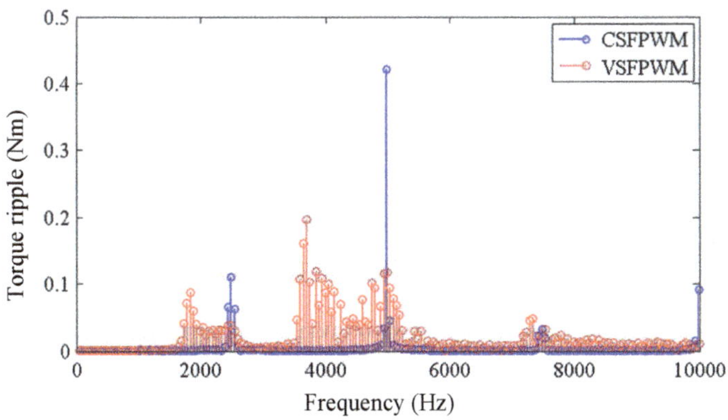

Fig. 5.25 Torque ripple spectrum comparison

(PMSM). The machine is initially driven by an inverter with 15.2 kHz constant switching frequency, and then the program is turned to VSFPWM algorithm.

With VSFPWM, switching period T_s is updated continuously and meets the criterion that PWM torque ripple peak value is controlled according to the limitation ($\Delta T_{require}$) in each switching cycle. PWM torque ripples are compared in Fig. 5.26, and it can be observed that VSFPWM and CSFPWM have the same PWM torque ripple peak value, but torque ripple distribution for VSFPWM is more homogeneous. FFT analysis of the PWM torque ripple is acquired in Fig. 5.27. The VSFPWM control has a wider range of PWM torque ripple harmonics, reducing the ripple peak value in CSFPWM significantly. From Fig. 5.28, the switching frequency for VSFPWM varies from 12.6 to 15.2 kHz resulting in smaller average switching frequency and less switching losses directly. The obvious reduction in EMI noise

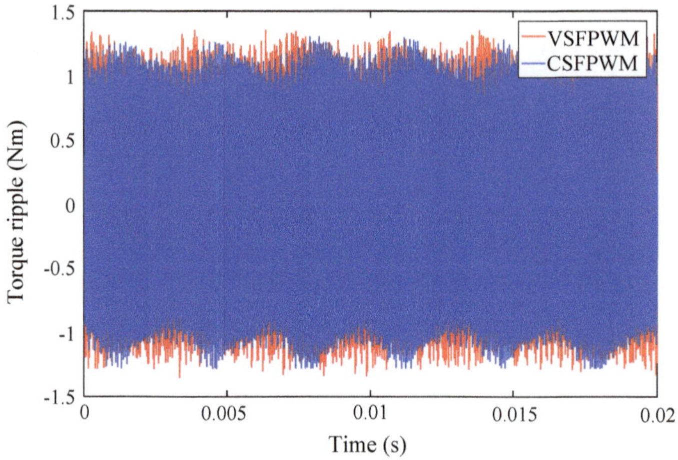

Fig. 5.26 Torque ripple comparison

Fig. 5.27 Torque ripple spectrum comparison

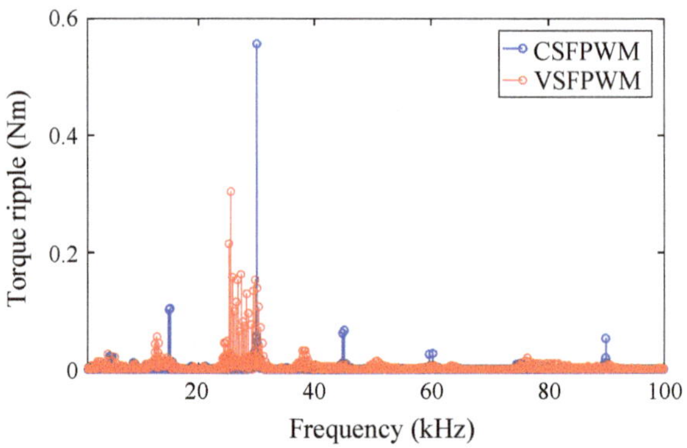

Fig. 5.28 Switching frequency comparison

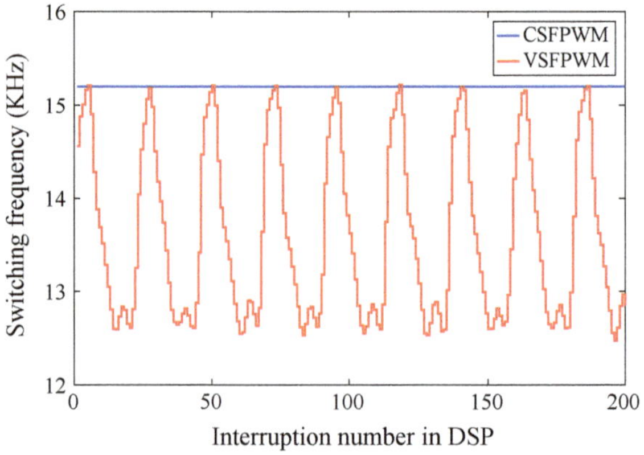

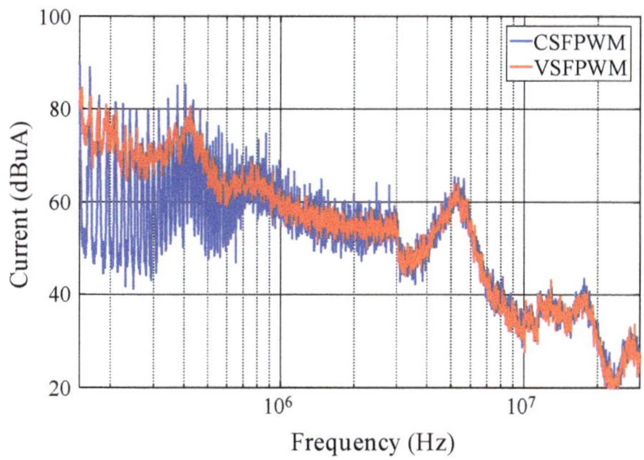

Fig. 5.29 EMI comparison

peak value can be observed in Fig. 5.29, which illustrates VSFPWM has a better EMI performance with respect to CSFPWM.

5.5.2 VSFPWM for Voltage Ripple Control

VSFPWM has been proposed for three-phase PWM converters with the target of AC side current ripple initially, and extended work has been carried out for PWM torque ripple control. Voltage ripple in the DC-link capacitor is a key factor for the design of DC-link capacitor in voltage source rectifier. Many researches have been done for optimization of DC-link capacitor. In the view of PWM, the variation of DC-link voltage ripple gives an opportunity to make variation of switching frequency which can control the DC-link voltage ripple distribution. Then, with the same DC-link voltage ripple requirement, the average switching frequency can be reduced and the switching losses can be reduced. Also, the variation of switching frequency can reduce the EMI noise of the PWM rectifier.

However, the physical principles of DC-link voltage ripple and AC side current ripple are different. Without understanding the DC-link voltage ripple model which is associated with PWM, the DC-link voltage ripple based VSFPWM is not possible. The active-front-end (AFE) rectifier is considered here as shown in Fig. 5.30, which consists of a three-phase two-level converter, three inductors inserted in the AC side and a capacitor connected to the DC-link. The LISN is used here to prevent external conductive noise on the power system, and the EMI path mode is also depicted in Fig. 5.30. The voltage-oriented current control method is adopted based on the revolving *d-q* reference frame. For the output current control, the terminal voltage of the converter is adjusted according to the Eq. (5.4). Figure 5.31 gives the basic control structure of a two-level rectifier, including the voltage and current double-loop control, as well as PWM module. The grid voltage vector is oriented to *d* axis and the *q*-axis with phase locking loop (PLL) tracking controller. In *d-q* axis, coupling

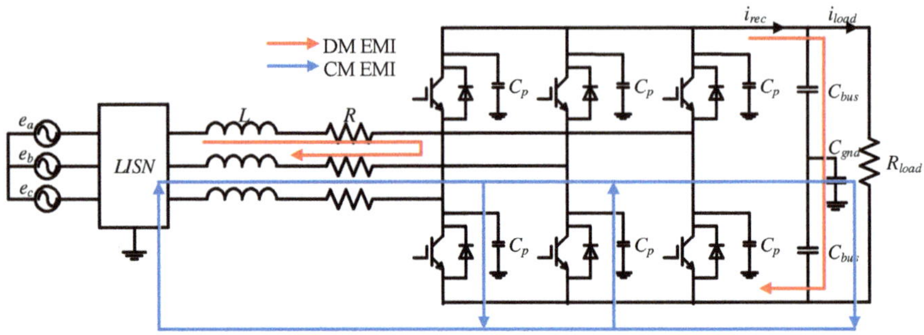

Fig. 5.30 Equivalent circuit for AFE rectifier as well as EMI mode path

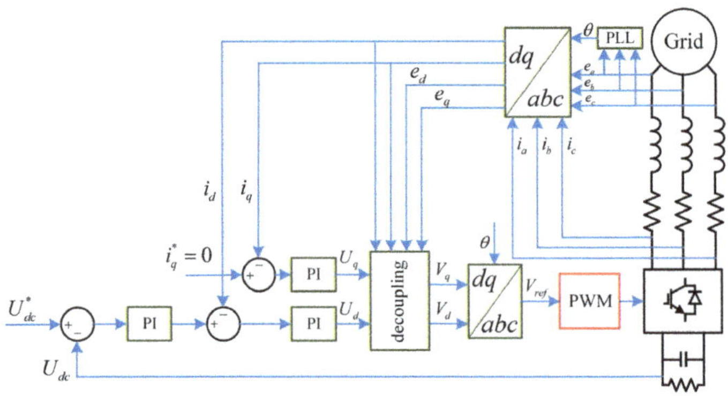

Fig. 5.31 Control structure of AFE rectifier

terms are added to the u_d and u_q, respectively, in the internal current loop, using two conventional proportional integral (PI) controllers to regulate the output current. For DC-link voltage control, a voltage PI controller is used here, and delivers the *d-q* axis current reference. The unity power factor is analyzed in this paper, thus $i_q{}^*$ is set to 0.

$$\begin{cases} V_d = -L\frac{di_d}{dt} - Ri_d + \omega L i_q + e_d \\ V_q = -L\frac{di_q}{dt} - Ri_q - \omega L i_d + e_q \end{cases} \tag{5.4}$$

With typical SVPWM, Fig. 5.32a shows the DC-link current in one switching cycle, a function of switching state and AC side current determined by the AC side voltage/current vectors, as discussed in Sect. 4.5. By computing the integral of the current difference through the capacitor in each sector, the DC-link voltage ripple is obtained Eq. (5.5). The DC-link current is a ladder-like distribution if AC current ripple is ignored, and voltage ripple peak is simply proportional to the switching period (T_s), as shown in Fig. 5.32b.

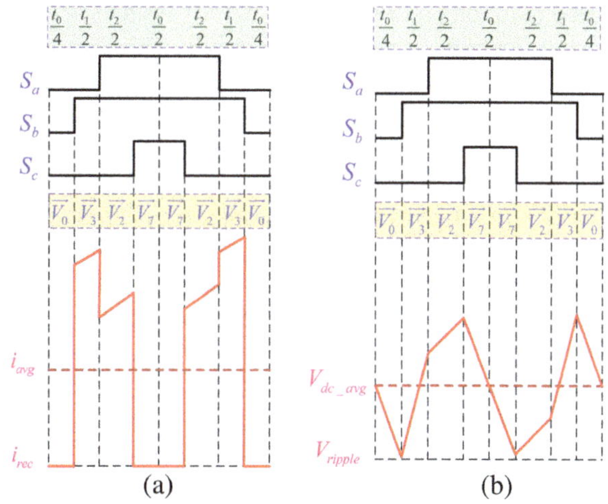

Fig. 5.32 DC-link current/voltage ripple in one switching frequency: a DC-link current, b DC-link voltage ripple

$$V_{ripple} = \frac{\int (i_{rec} - i_{load})dt}{C_{bus}} \tag{5.5}$$

where i_{rec} is the DC-link current and i_{load} is the load current, C_{bus} is the DC-link capacitance and V_{ripple} is the DC-link voltage ripple.

In order to verify the prediction method, an AFE rectifier has been simulated by MATLAB/Simulink, having $V_{dc} = 200$ V, $f_s = 10$ kHz and $C_{bus} = 17.5$ μF. With typical SVPWM, Fig. 5.33 shows the DC-link voltage ripple comparison between the predicted and the simulated results. In a fundamental period, the DC-link voltage ripple is rising and falling with the peak value (± 5.5 V), matching well with the predicted envelope, which illustrates the correctness of prediction method. It can be observed that the voltage ripple reaches the peak value only in several time points. The requirement of DC-link voltage ripple for an AFE rectifier is usually with the peak

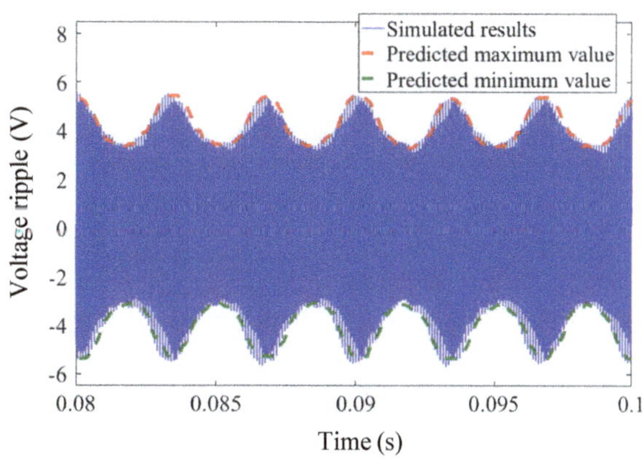

Fig. 5.33 Comparison between the prediction and the simulation results of the DC-link voltage ripple in one fundamental period

value requirement. So, with (± 5.5 V) voltage ripple requirement, the performance is over-qualified in most of the periods, which means the switching frequency could be reduced in those periods, thus the space limited by the peak value is not fully utilized with CSFPWM. The periodic variation of DC-link voltage ripple can be well predicted by the proposed method in real time, which provides a tool for VSFPWM for DC-link voltage ripple control.

Figure 5.34 gives the voltage ripple prediction process in detail. The duty cycles in three phases and the *d-q* axis current in the controller are sent to the prediction block to estimate the DC-link voltage ripple V_{ripple} with the method in Chap. 4 with nominal switching cycle T_{SN}. With the predicted maximum voltage ripple (V_{ripple_max}) of one switching cycle, the updating switching period T_S can be calculated and acted by means of controlling the slope of updating triangular carrier. In that way, the peak value of actual DC-link voltage ripple in this switching cycle will be exactly equal to the requirement. By continuously doing that, the DC-link voltage ripple will be controlled equally with the required value and variable T_S.

VSFPWM and CSFPWM have been respectively carried out with the target of DC-link voltage ripple in MATLAB/Simulink. The requirement of voltage ripple is sure to be ± 5.5 V. Figure 5.35 shows the DC-link voltage ripple comparison between VSFPWM and CSFPWM. Both of them can be controlled within the identical limitation (± 5.5 V). Compared with the voltage ripple of CSFPWM, voltage ripple of VSFPWM can reach the requirement almost in every switching cycle. It means that the switching frequency is decreased in the area, where the voltage ripple peak for CSFPWM is lower than the requirement. Figure 5.36 shows the comparison of switching frequency. Compared with the classical CSFPWM whose switching frequency is always a constant (10 kHz), switching frequency of VSFPWM varies between 6 and 10 kHz resulting in 20% average switching frequency reduction.

The performance of VSFPWM strategy is carried out in an AFE rectifier platform and the physical map of each part of the rectifier system is presented in Fig. 5.37. In this experiment, the control program is computed based on the DSP of TMS320F28335. When the DC-link voltage has achieved the voltage reference

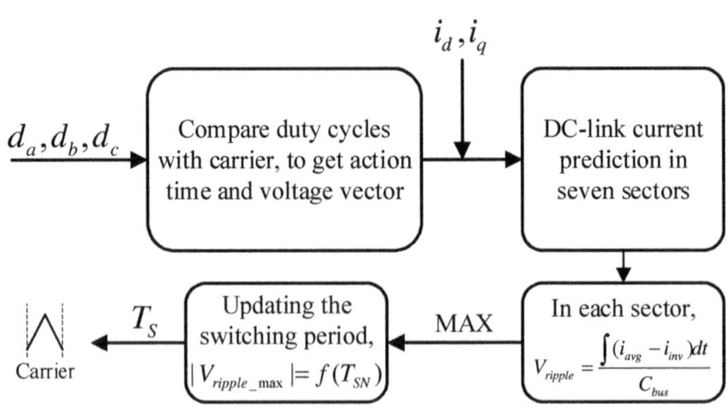

Fig. 5.34 The diagram of VSFPWM for voltage ripple control

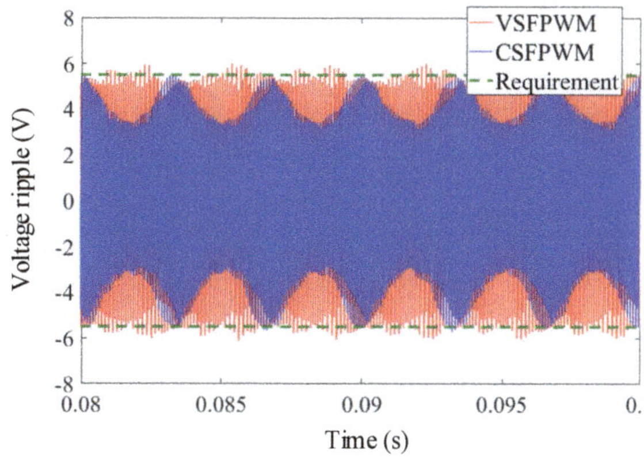

Fig. 5.35 DC-link voltage ripple comparison

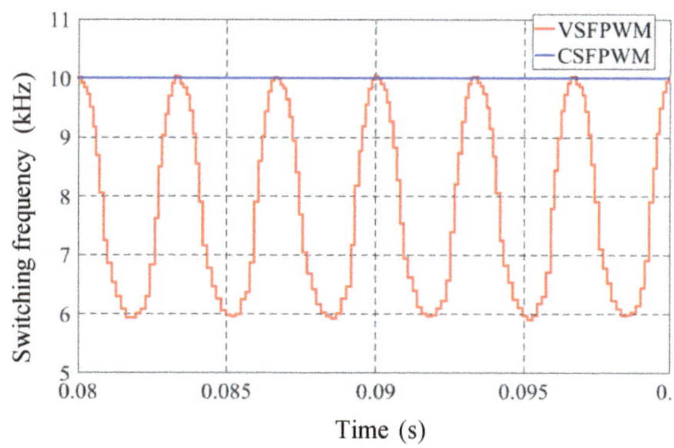

Fig. 5.36 Switching frequency comparison

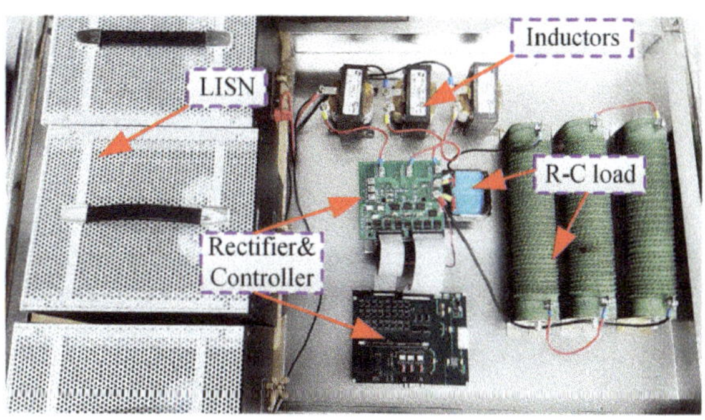

Fig. 5.37 Picture of experimental setup

with CSFPWM, the control program is turned into VSFPWM strategy. During the implementation process of CSFPWM, a switching frequency of 10 kHz is used for traditional SVPWM, and the requirement of DC-link voltage ripple can be obtained (± 5.5 V) as a control target for the later VSFPWM.

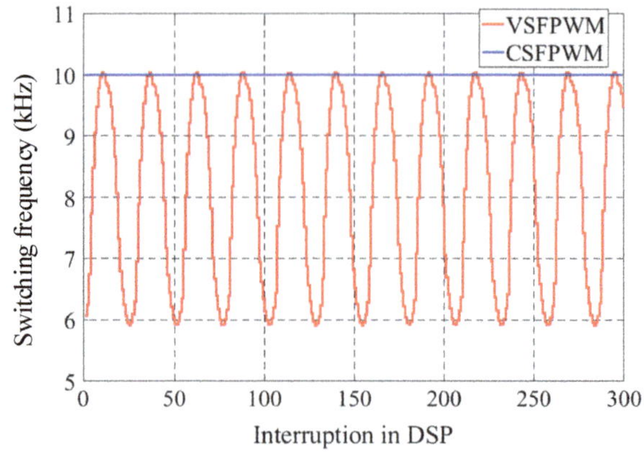

Fig. 5.38 Experimental result: switching frequency comparison

In this case, the switching frequency is stored in the DSP interruption, as compared in Fig. 5.38. The switching frequency of VSFPWM varies from 6 to 10 kHz, always below the selected constant switching frequency. 20% reduction in the average switching frequency can be achieved with VSFPWM, so the switching losses are directly reduced. The comparison of DC-link voltage ripple has been depicted in Fig. 5.39 as well as enlarged detail view. With respect to CSFPWM, the ripple for VSFPWM is still controlled within the limitation, but nearly covering all areas between the dash lines. Since variable switching frequencies are obtained in VSFPWM continuously, its current harmonics will not concentrate on the multiple switching frequency shown in CSFPWM. From Fig. 5.40, the measured EMI can be suppressed with VSFPWM, and a reduction of 8 dB can be seen in the 150 kHz to 500 kHz range.

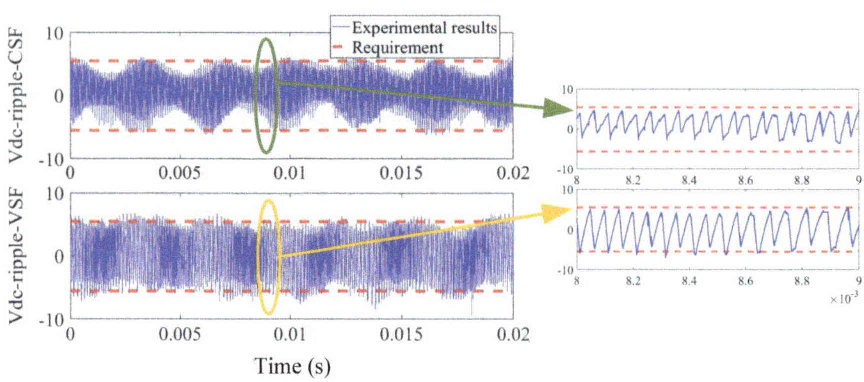

Fig. 5.39 Experimental result: DC-link voltage ripple comparison

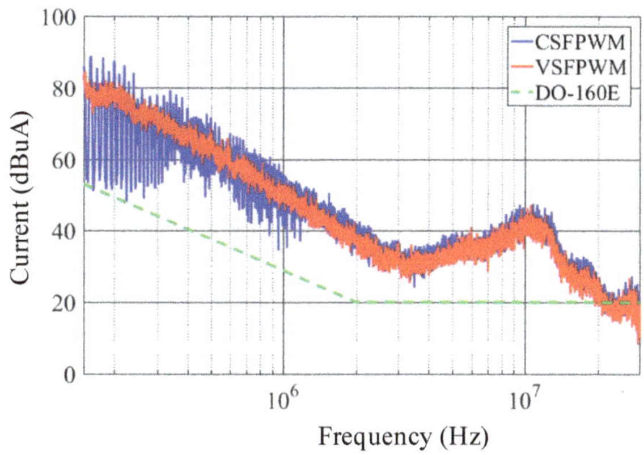

Fig. 5.40 Experimental result: EMI comparison

5.6 Switching Frequency Distribution's Further Optimization [23, 24]

In the previous sections, VSFPWM has been developed based on prediction model. With periodic variation of switching frequency, EMI will be spread to a wider range in frequency domain and reduced. But what kind of variation can obtain the best EMI reduction effect? This section will answer this question [19, 20].

The harmonic energy of traditional CSFPWM mainly distributes near the carrier harmonics, which eventually manifests as the obvious EMI spikes in spectrum because of the fixed switching frequency. Therefore, the freedom of switching frequency can be used to improve the distribution of the spectrum and reduce the conducted EMI spikes for CSFPWM. As discussed in Chap. 3, random PWM is able to reduce EMI by randomly varying the switching frequency. Nevertheless, the unpredictability of current ripple is difficult for the design of power electronics converters. In addition, the real random signals for RPWM are difficult to generate in practical applications.

The previous sections have introduced the VSFPWM that changes switching frequency according to current ripple prediction model. Although VSFPWM can reduce conducted EMI, many EMI spikes still exist in spectrum, which makes the conducted EMI reduction for VSFPWM not as good as expected. Hence, this section will provide comprehensive analysis of the above phenomenon, and then the PWM strategies based on the uniform or normal distribution of switching frequency are proposed to eliminate the spikes in spectrum, further reducing the conducted EMI.

For VSFPWM, the variation of switching frequency depends on the current ripple, whose characteristic is associated with modulation index. Consequently, the waveforms of switching frequency vary considerably with the change of modulation index. As shown in Fig. 5.41a and b, the variation ranges of switching frequency are different with each other: the lower limit of switching frequency can reach 21.5 kHz in Fig. 5.41b whereas only 33 kHz in Fig. 5.41a. The shapes of the switching frequency waveforms are also different in addition to the variation range. Therefore,

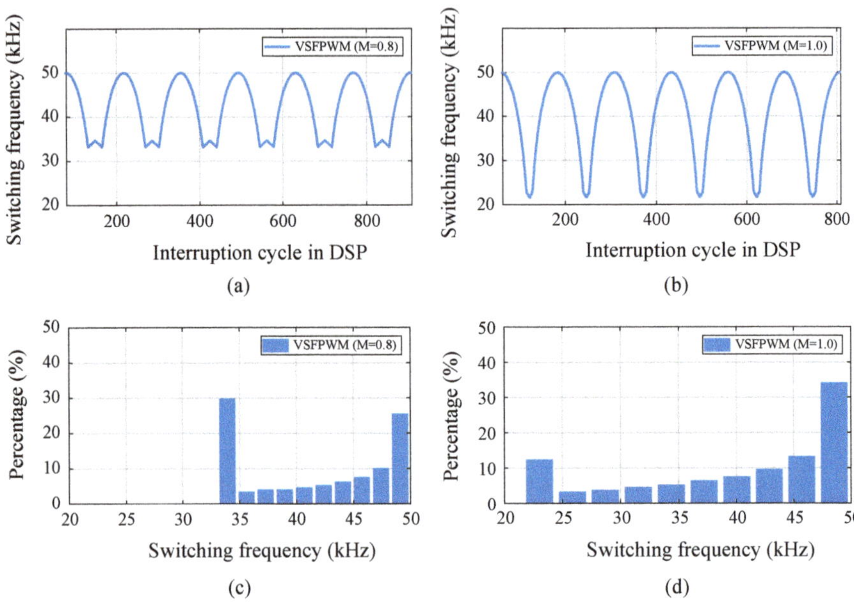

Fig. 5.41 **a** Switching frequency waveform for modulation index 0.8, **b** switching frequency waveform for modulation index 1.0, **c** switching frequency distribution for modulation index 0.8, **d** switching frequency distribution for modulation index 1.0

Fig. 5.41c and d further shows the statistical distribution of the switching frequency for Fig. 5.41a and b, respectively. The common feature of the two cases is the nonuniform distribution of switching frequency. Switching frequency is mainly distributed around 50 kHz where the switching frequency nearly accounts for 30% so that the EMI spikes still exist around the carrier harmonics of 50 kHz.

Figure 5.42 shows the conducted EMI comparison between CSFPWM and VSFPWM for the cases in Fig. 5.41. VSFPWM can suppress conducted EMI to a certain level, but there are still many EMI spikes in the spectrum. As analyzed above, obvious EMI spikes marked by arrows in Fig. 5.42 exist at 150 and 200 kHz, which are the third and the fourth carrier harmonics of 50 kHz, respectively. For the case of modulation index 0.8, the proportion of switching frequency 34 kHz is 30%, so the EMI spikes appear at 238 and 272 kHz marked by yellow arrows in Fig. 5.42a, which are the seventh and eighth carrier harmonics of 34 kHz, respectively. For the case of modulation index 1.0, the proportion of 22.5 kHz in Fig. 5.41d is only 10%, so there is no distinct spike of its carrier harmonics in Fig. 5.42b. Similarly, since the proportions of other switching frequencies are not large, thus there is no corresponding EMI spike in the spectrum.

RPWM has a good effect in reducing conducted EMI without these kinds of regional EMI spikes, and the random number is generally based on uniform distribution or normal distribution. Taking the RPWM with uniform distribution for example, Fig. 5.43 is the switching frequency of RPWM based on a uniformly distributed

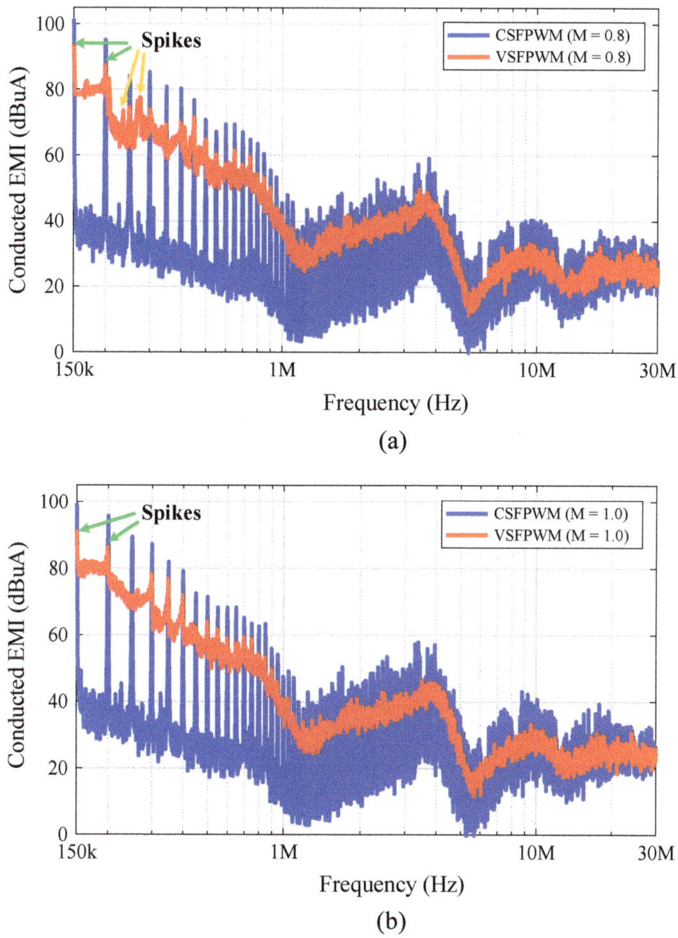

Fig. 5.42 Conducted EMI comparison between CSFPWM and VSFPWM with different modulation indexes: **a** 0.8, **b** 1.0

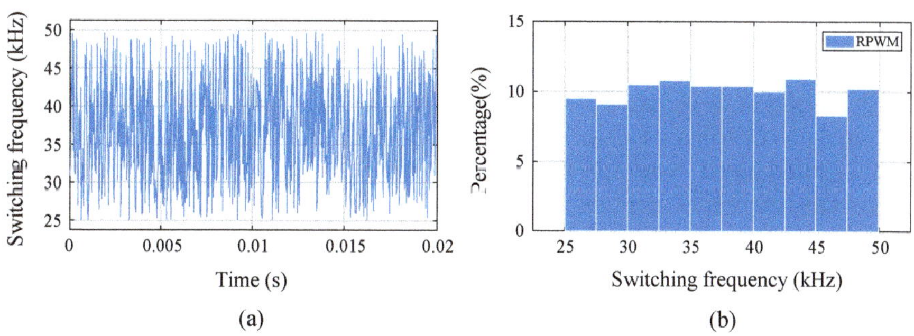

Fig. 5.43 Switching frequency for RPWM: **a** waveform in a line cycle, **b** distribution density

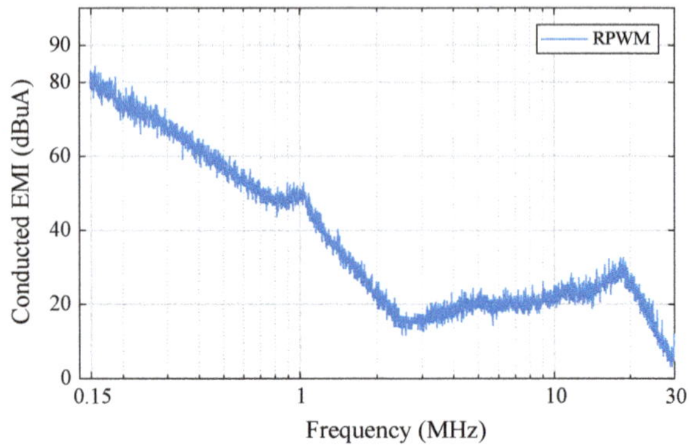

Fig. 5.44 Conducted EMI for RPWM with uniform distribution of switching frequency

pseudo-random number. The switching frequency is random in the time domain, and the statistical distribution approximately follows a uniform distribution. Figure 5.44 shows the conducted EMI of RPWM, and there are almost no obvious EMI spikes in the spectrum.

In summary, the EMI spikes in spectrum are caused by the excessively concentrated distribution of switching frequency. That is to say, the statistical distribution of switching frequency plays an important role in reducing conducted EMI. If the distribution of switching frequency is uniform distribution or normal distribution like RPWM, the regional EMI spikes can be eliminated to achieve more conducted EMI reduction in the spectrum.

This part will introduce novel PWM strategies to eliminate the EMI spikes in spectrum based on the uniform or normal distribution of switching frequency, and thus the proposed strategy is named as uniform distribution pulse width modulation (UDPWM) or normal distribution pulse width modulation (NDPWM). The following is arranged from three aspects. First, the differential equations of the switching frequency that make the switching frequency satisfy any distribution characteristics are derived. Second, the limited conditions of the switching frequency are proposed to optimize the performance. Finally, the switching frequency function waveforms of the two distributions of uniform distribution and normal distribution are discussed.

A Relationship between switching frequency and distribution

In order to obtain the equation of switching frequency, the analysis between switching frequency and distribution is performed as follows. It is assumed that θ is the electrical angle of three-phase system, $f(\theta)$ is the switching frequency in respect to angle θ, $f'(\theta)$ is the differential function, and D[$f(\theta)$] is the distribution density of switching frequency.

(1) The relationship between switching frequency $f(\theta)$ and distribution density $D[f(\theta)]$

Figure 5.45 shows a waveform of switching frequency with linear change. Taking points P and Q for example, since switching frequency f_P is lower than switching

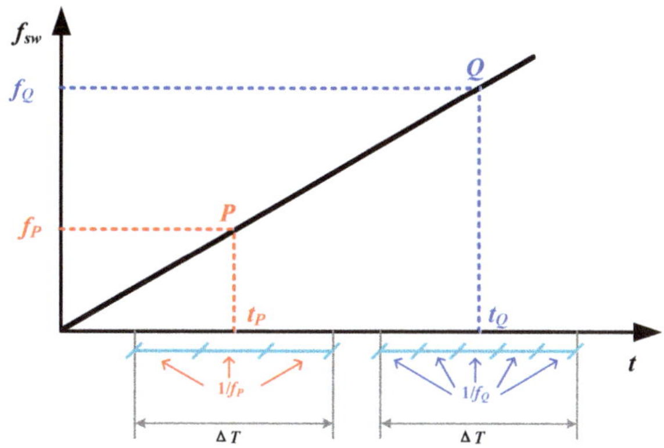

Fig. 5.45 Analysis for the distribution of switching frequency when the switching frequency changes linearly

frequency f_Q, the carrier period $1/f_P$ is longer than the carrier period $1/f_Q$. Assuming that the switching frequency is constant during a certain period of time ΔT, then the number of sampling points at switching frequency f_P is less than that of f_Q. That is to say, when the switching frequency is linearly varying, the higher the switching frequency, the more are the sampling points, that is, higher will be the distribution density of that switching frequency, as expressed in Eq. (5.6).

$$D[f(\theta)] \propto f(\theta) \tag{5.6}$$

(2) The relationship between the differential function $f'(\theta)$ and distribution density $D[f(\theta)]$

$|f'(\theta)|$ indicates the speed at which the switching frequency changes. Figure 5.46 presents the situation of different rates of changes at the same switching frequency. Assuming that the sampling time is t_P, then the next sampling time is $t_P + 1/f_P$. If the switching frequency changes slowly, the switching frequency at the next sampling instant is f_{Q1}. That switching frequency belongs to the frequency interval

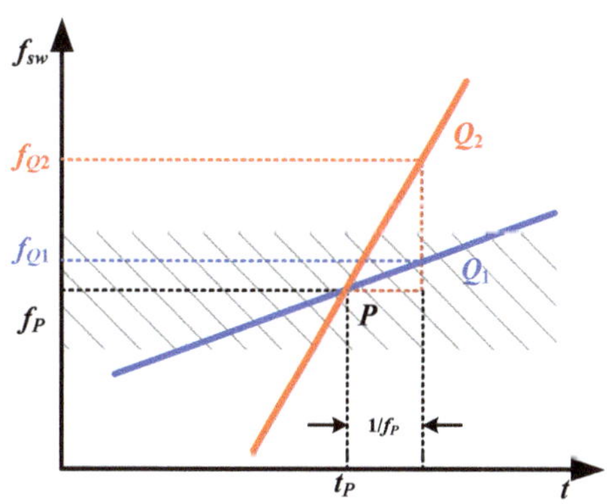

Fig. 5.46 Analysis for the distribution of switching frequency with different slopes

centered on switching frequency f_P, as the shadow part in Fig. 5.46, so that the number of sampling points for that frequency interval increases. Inversely, if the switching frequency changes fast, the switching frequency at the next sampling instant is f_{Q2}. That switching frequency does not belong to the frequency interval centered on switching frequency f_P so that the number of sampling points for that frequency interval decreases. That is to say, for a certain switching frequency, the faster the switching frequency changes, the less are sampling points, that is, less is the distribution of that switching frequency, as expressed in Eq. (5.7).

$$D[f(\theta)] \propto \frac{1}{|f'(\theta)|} \tag{5.7}$$

In summary, to make the distribution characteristic of the function of the switching frequency coincide with the probability density function $D[f(\theta)]$, switching frequency $f(\theta)$ obeys:

$$k \cdot \frac{f(\theta)}{|f'(\theta)|} = D[f(\theta)] \tag{5.8}$$

where k is the scale factor.

B Limitations for the variable switching frequency

Equation (5.8) is a first-order differential equation that requires two boundary conditions to determine the integral constant and the coefficient k. The boundary conditions are determined based on the waveform characteristics of the switching frequency function such as periodicity, symmetry and range of variation, and those characteristics of the switching frequency can be used to optimize other performances of the converters.

(1) Monotonicity of switching frequency

The differential equation of switching frequency reflects the microscopic change trend of the switching frequency, and the monotonicity of switching frequency wave reflects the macroscopic change trend of the switching frequency. The monotonicity is also the degree of freedom that can be used to optimize the performance of the converter. Therefore, it is used to reduce the switching losses. In general, the larger the current flows through the switching devices, greater will be the loss caused by a single switching action. When the switching frequency and the magnitude of the current are inversely varied, the switching loss can be effectively reduced.

For a single-phase converter, the relationship between the monotonicity of the switching frequency and the sinusoidal current waveform is shown in Fig. 5.47a. To apply the relationship to a three-phase system, it is necessary to comprehensively consider the magnitude of the three-phase current. Then, the variation of the maximum value of the three-phase current is considered to determine the change in the switching frequency. Figure 5.47b shows the three-phase currents and their

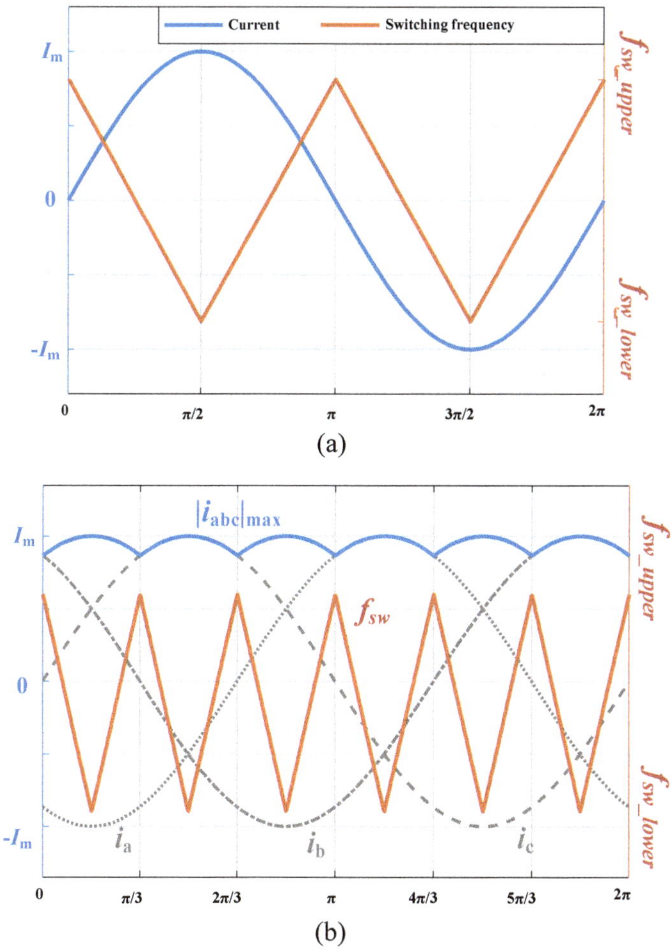

Fig. 5.47 Relationship between current and switching frequency: **a** single phase, **b** three phases (Take triangle wave for example)

maximum absolute value, that is, a six-pulse wave in a line cycle. Therefore, the switching frequency in triangular wave has the same periodicity and symmetry but opposite monotonicity as the current in blue line. In other words, the bigger the current, the lower is the switching frequency. The switching loss has a positive correlation with current, and the adverse variation between current and switching frequency can decrease the switching loss.

(2) Variable range of switching frequency

Now, the basic characteristics for the waveform of switching frequency have been determined as shown in Fig. 5.47. The waveform function needs to be solved between 0 and π/6 radians, and then the switching frequency waveform in the line cycle can be obtained according to the periodicity and symmetry. Therefore, the upper limit and lower limit of the variable switching frequency constitute the boundary condition in Eq. (5.8). The analytic formula of the arbitrarily distributed switching frequency can be solved by combining (5.8) and (5.9).

$$\begin{cases} f(0)=f_{sw\_upper} \\ f\left(\frac{\pi}{6}\right)=f_{sw\_lower} \end{cases} \tag{5.9}$$

The switching frequency range in practical applications should be flexibly selected according to the specific situation. The maximum switching frequency is limited by the switching speed, heat dissipation and efficiency of the power devices. The upper limit of the switching frequency of all spread spectrum modulation methods in this section is selected as the switching frequency of CSFPWM. The lowest switching frequency affects the design of the power filter and controller, and generally cannot be chosen to be too low. The lower limit of the switching frequency of all spread spectrum modulation methods in this section is chosen to be half of the upper limit to achieve better spread spectrum effect. For example, when the switching frequency is varied between 25 and 50 kHz, the harmonic energy is just scattered to each frequency without causing some frequency bands to have no spread spectrum effect as shown in Fig. 5.48a. Conversely, if the lower limit of the switching frequency is higher than 25 kHz, then there will be no harmonic energy being allocated to some frequency band as shown in Fig. 5.48b.

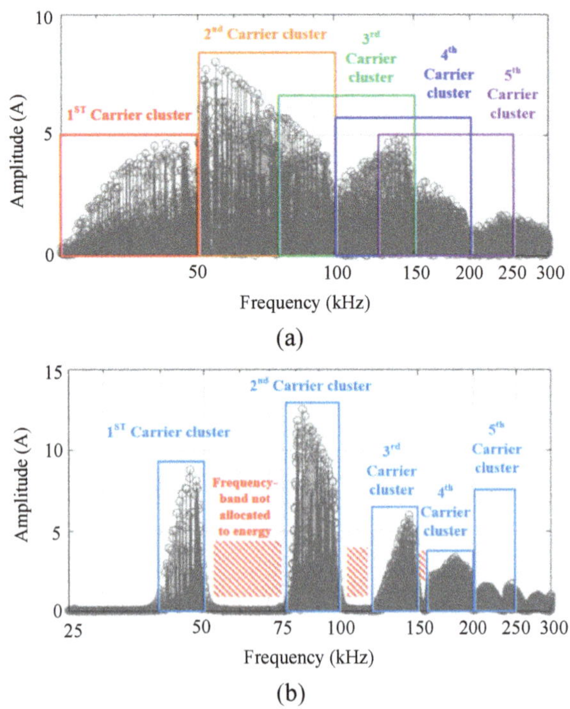

Fig. 5.48 The effect of spread spectrum contrast: **a** 25–50 kHz, **b** 40–50 kHz

C Implementation of UDPWM and NDPWM

On bringing the probable density of the desired distribution into (5.8), we can get the analytical expression of the switching frequency. The differential equations of the two distributions are (5.10), and the remaining work is purely mathematical derivation. The analytic solution of uniform distribution is (5.11). The function of the normal distribution is too complicated to obtain the analytical solution directly, so the numerical method can be used to solve the approximate solution.

$$k \cdot \frac{f(\theta)}{|f'(\theta)|} = \frac{1}{f_{sw\_upper} - f_{sw\_lower}} \tag{5.10a}$$

$$k \cdot \frac{f(\theta)}{|f'(\theta)|} = \frac{1}{\sqrt{2\pi\sigma^2}} \exp\left[-\frac{(f(\theta)-\mu)^2}{2\sigma^2}\right] \tag{5.10b}$$

$$f(\theta) = \exp\left[\frac{6}{\pi}\ln\left(\frac{f_{sw\_lower}}{f_{sw\_upper}}\right)\cdot\theta + \ln\left(f_{sw\_upper}\right)\right] \tag{5.11}$$

Figure 5.49 plots the time-domain waveforms of the switching frequency of the spread spectrum modulation based on the distribution characteristics, which changes

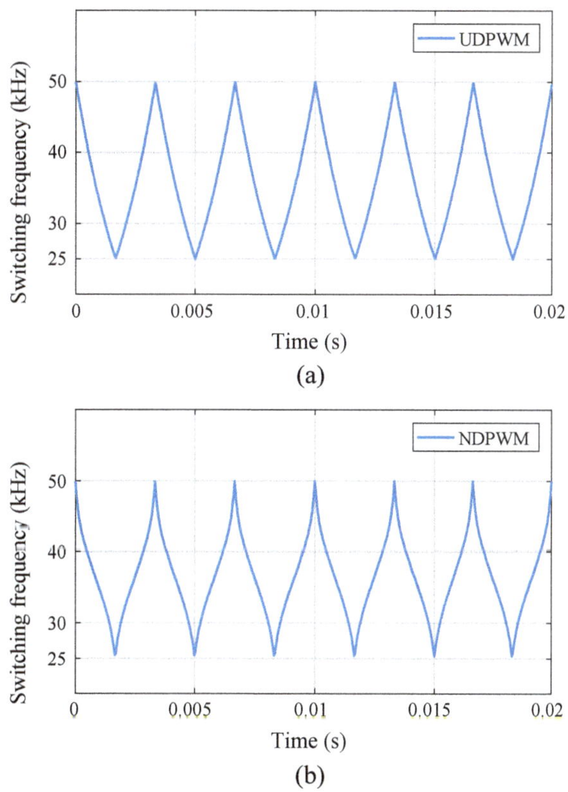

Fig. 5.49 Switching frequency waves for distribution-based modulation: **a** uniform, **b** normal

the rate of change at different switching frequencies to obtain the expected distribution law. In order to verify the correctness of the analytical expressions of the solved switching frequency, Fig. 5.50 further gives the statistical characteristics of the switching frequency obtained from the experiments. At the same time, the red lines of Fig. 5.50 indicate the expected distribution characteristics, and it is clear that the statistical characteristics of the individual figures agree well with that of anticipation. That is, the obtained switching frequency expression can comply with a specific distribution.

Figure 5.51 shows the block diagram of the implementation for UDPWM or NDPWM, which consists of two parts: the left part calculates the relevant physical quantities for the next carrier cycle (the $(k + 1)$th cycle), and the right part generates the PWM signals for the current carrier cycle (the kth cycle). For the left part, the control algorithm calculates the three-phase voltage references and the electrical angle θ of the three-phase system. Then, the angle θ will be transformed into the range between 0 and $\pi/6$ radians according to the periodicity and symmetry of

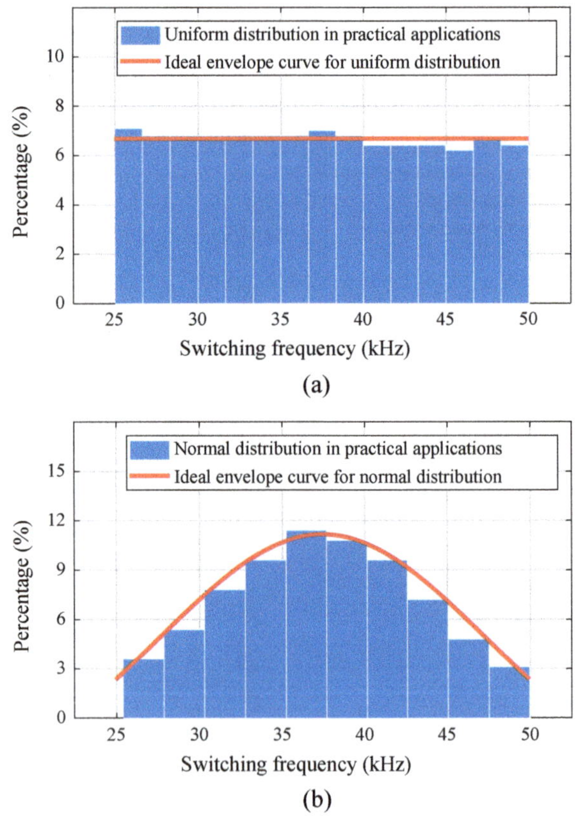

Fig. 5.50 Distribution characteristics for distribution-based modulation: **a** uniform, **b** normal

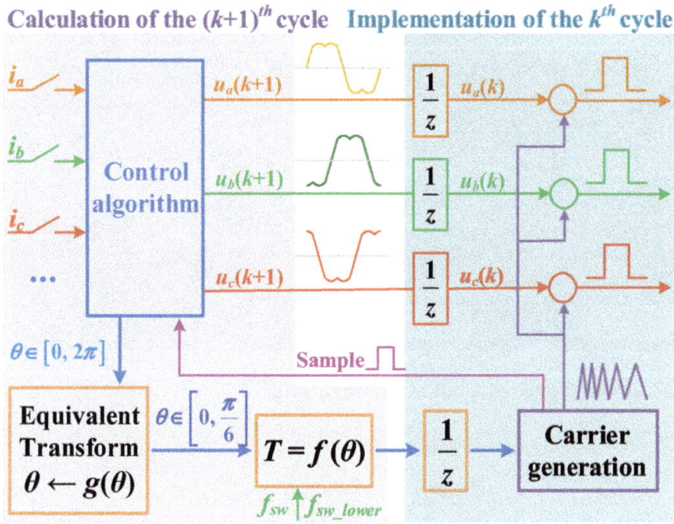

Fig. 5.51 Block diagram of proposed UDPWM or NDPWM strategy

switching frequency waveform. Finally, the transformed angle θ together with predefined parameters f_{sw} and f_{sw_lower} are substituted into waveform function to calculate the carrier period. The three-phase references and carrier period acquired from the above-mentioned calculation process are used for the $(k + 1)$th carrier cycle. For the right part, the three-phase references are compared with the carrier to generate the PWM signals of the kth carrier period. A trigger signal is generated when the carrier generation module completes the count of the kth switching cycle. On the one hand, the trigger signal informs the controller to perform the calculation for the $(k + 2)$th switching cycle; on the other hand, the trigger signal informs PWM generation part to load the calculated three-phase references and carrier period of the $(k + 1)$th carrier cycle to generate PWM signals.

In order to verify the validity of the proposed UDPWM methods, experiments are carried out based on the two-level converter prototype shown in Fig. 5.52. The switching devices of converter are SiC JEFTs (IJW120R100T1) provided by Infineon, which are suitable for high switching frequency. The circuit parameters and operating conditions are listed in Table 5.2.

Figure 5.53 shows the emission level of conducted EMI for the three PWM methods with different modulation indexes. The envelope curve of conducted EMI for UDPWM is the lowest, and there is no obvious EMI spike in spectrum for UDPWM. The worst points such as 150 and 200 kHz that are important for EMI filter design achieve great attenuation approximate to 20 dB compared with traditional CSFPWM. Although the VSFPWM can reduce conducted EMI, the EMI spike at 150 kHz is still dominating in spectrum. For example, the conducted EMI at 150 kHz only decreases 7.1 dB compared with CSFPWM and remains large spike in Fig. 5.53a. However, the UDPWM further decreases the conducted EMI at 150 kHz about 14.8 dB compared

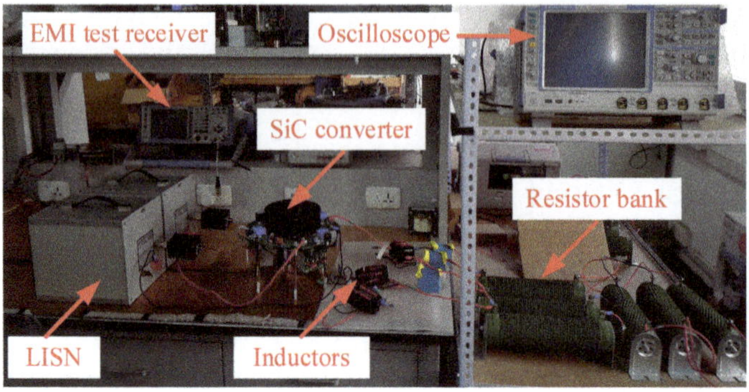

Fig. 5.52 SiC converter prototype and experimental setup

Table 5.2 The experimental parameters

Symbol	Parameters	Value
V_{dc}	DC voltage	200 V
f_{sw} (p.u.)	Constant switching frequency	50 kHz
f_{sw_lower}	Lower limit of switching frequency	25 kHz
f_0	Fundamental frequency	50 Hz
L	Filtering inductance	0.5 mH

with VSFPWM and eliminates the EMI spike in Fig. 5.53a. In both Fig. 5.53b and c, UDPWM also effectively decreases conducted EMI about 20 dB compared with CSFPWM.

Moreover, Fig. 5.54 shows the three-dimensional maps of conducted EMI for the four PWM methods with the variation of modulation index. Since the PWM methods mainly determine the spectrum in low-frequency band, Fig. 5.54 only shows the conducted EMI between 150 kHz and 1 MHz for clear observation of spectral characteristics. Compared with the spectra of CSFPWM and VSFPWM in Fig. 5.54a and b, the EMI spectrum of UDPWM in Fig. 5.54c is almost a plane rather than undulating surfaces with steep peaks and valleys, which demonstrates that UDPWM is more effective to disperse the concentrated EMI in spectrum. Figure 5.54d plots the EMI spectrum for NDPWM, which also eliminates EMI spikes like NDPWM. Further, the color bars on the right side of Fig. can also prove that UDPWM and NDPWM have the ability to reduce conducted EMI by 20 dB over the full range of modulation index. Therefore, Figs. 5.53 and 5.54 validate the predominance of the proposed UDPWM and NDPWM methods in reducing conducted EMI.5.54

The ability to reduce switching loss is another advantage of UDPWM. Since the lower limit of VSFPWM is not fixed, and the switching loss of UDPWM and VSFPWM cannot be fairly compared, Fig. 5.55 plots the efficiency of the inverter in different load power from 400 W to 2 kW with the same lower limit. Compared with CSFPWM and VSFPWM, UDPWM can make efficiency of inverter increase by 0.30 and 0.12%, respectively.

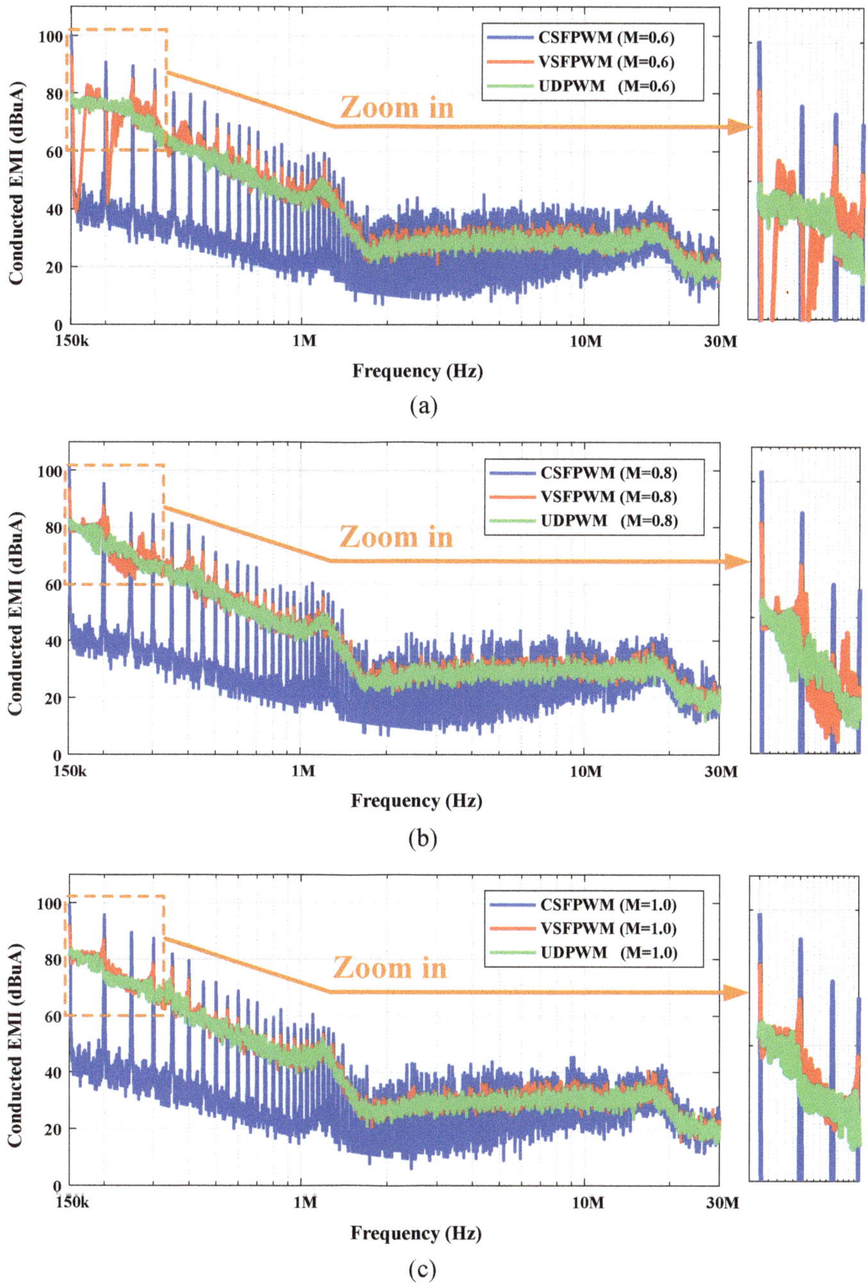

Fig. 5.53 Emission level of conducted EMI at different modulation indexes: **a** 0.6, **b** 0.8, **c** 1.0

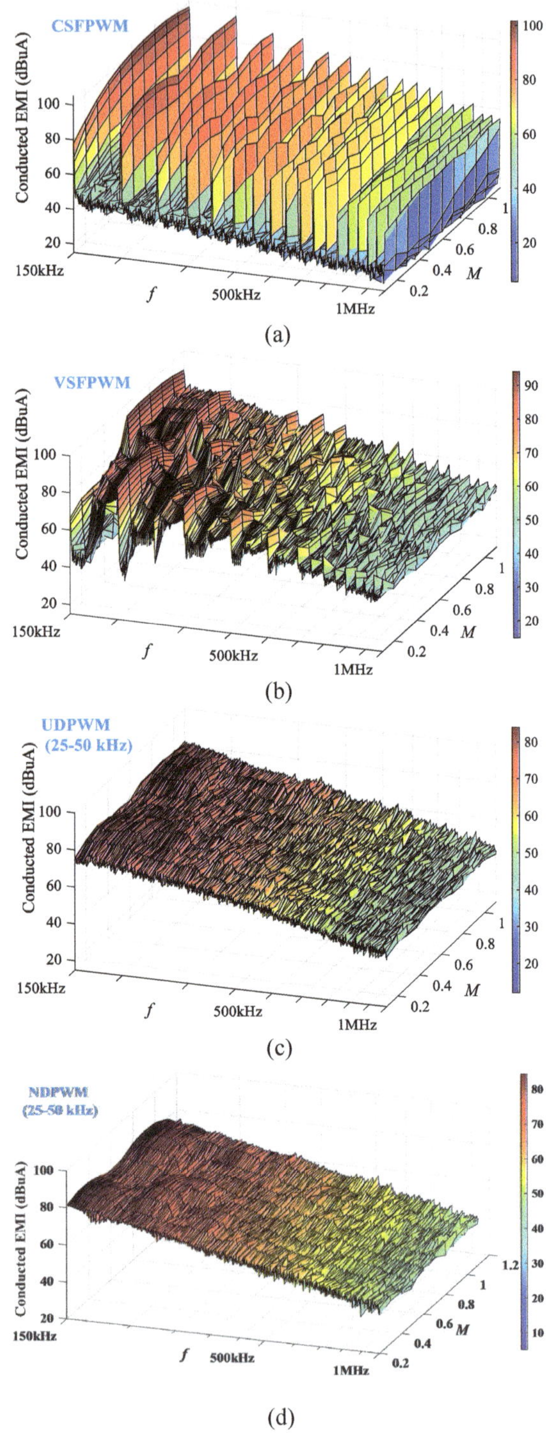

Fig. 5.54 Conducted EMI spectrum with the variation of modulation index: **a** CSFPWM, **b** VSFPWM, **c** UDPWM, **d** NDPWM

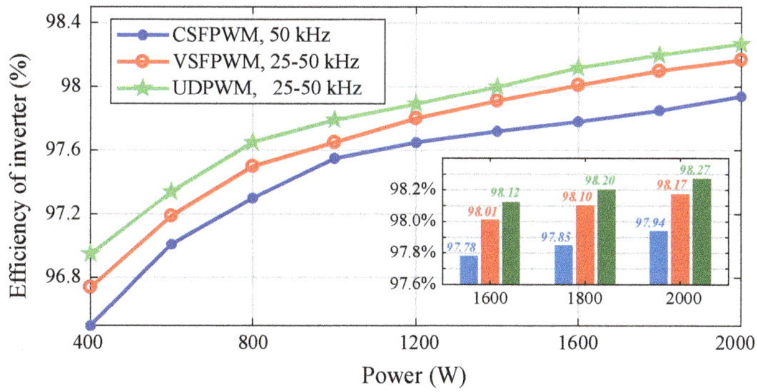

Fig. 5.55 Efficiency of inverter for CSFPWM, VSFPWM and UDPWM in different power (experimental results)

This section comprehensively investigates the phenomenon that the VSFPWM scheme cannot effectively reduce the conducted EMI, and the reason is owing to the excessively concentrated distribution of switching frequency. In order to eliminate the EMI spikes in spectrum and further reduce the conducted EMI, novel PWM strategies called UDPWM and NDPWM have been proposed for converters. The design principles are based on the uniform distribution of switching frequency and less switching loss. The analytical formula of switching frequency in time domain is deduced in detail. Experimental results indicate that UDPWM can reduce conducted EMI by 20 dB compared with CSFPWM, and all obvious spikes in spectrum are eliminated compared with VSFPWM. This modulation strategy is very suitable for the high switching frequency converters with wide band-gap devices to reduce conducted EMI. The switching loss of UDPWM can also be reduced to improve the efficiency of the inverter. However, the price of the reduction of conducted EMI and switching loss is the bigger THD. Also, compared with model predictive VSFPWM, novel switching frequency-distributed NDPWM and UDPWM cannot realize real-time control of ripple in the converter.

In Sects. 5.1–5.6, different VSFPWM methods have been discussed with a common law: the switching frequency variation with periodic law. The introduction of this periodic variation of switching frequency has two major concerns:

(1) Will it bring low-frequency harmonics to the system?
(2) Will it interfere with feedback control?

These two questions will be answered in next two sections.

5.7 The Spectrum Calculation and Analysis of VSFPWM [21]

Compared with the traditional CSFPWM, the VSFPWM can be used to effectively reduce the interference peaks at switching frequency and its harmonics. Actually, the switching function of the proposed VSFPWM and its further optimized method is periodic. In previous sections, the spectrum characteristics of CSFPWM and VSFPWM are mainly compared via FFT conversion; thus an analytical approach at the theoretical level is still lacking. In this section, spectrum-quantified expressions for both CSFPWM and VSFPWM will be derived through the method of double Fourier series which has been briefly introduced in Chap. 2.

According to the principle of the double Fourier series, the output phase voltage U_o(t) for SPWM is a function of the following two periodic time variables:

$$\begin{cases} x(t) = \omega_c t + \theta_c \\ y(t) = \omega_o t + \theta_o \end{cases} \tag{5.12}$$

where ω_c is the carrier angular frequency, θ_c is the initial phase angle of the carrier, ω_o is the fundamental angular frequency, and θ_0 is the initial phase angle of the fundamental wave. Furthermore, U_o(t) can be decomposed into:

$$\begin{aligned} U_o(x, y) = & \frac{A_{00}}{2} + \sum_{n=1}^{\infty} \{A_{0n} \cos ny + B_{0n} \sin ny\} \\ & + \sum_{m=1}^{\infty} \{A_{m0} \cos mx + B_{m0} \sin mx\} \\ & + \sum_{m=1}^{\infty} \sum_{\substack{n=-\infty \\ (n \neq 0)}}^{\infty} \{A_{mn} \cos(mx + ny) + B_{mn} \sin(mx + ny)\} \end{aligned} \tag{5.13}$$

In the above equation, the first term is DC bias, the second term is baseband harmonic component (the fundamental component when $n = 1$), the third term is carrier harmonic component, and the fourth term is side-band harmonic component, where the coefficient is obtained by:

$$C_{mn} = A_{mn} + jB_{mn} = \frac{1}{2\pi^2} \int_{-\pi}^{\pi} \int_{-\pi}^{\pi} f(x, y) e^{j(mx+ny)} dxdy \tag{5.14}$$

Figure 5.56 shows the natural sampling PWM with triangular carriers. In every carrier, the switching time can be expressed as:

When $f(x,y)$ switches from 0 to $2V_{dc}$:

$$x = 2\pi p - \frac{\pi}{2}(1 + M \cos y)\ p = 0, 1, 2, \ldots, \infty \tag{5.15}$$

When $f(x,y)$ switches from $2V_{dc}$ to 0:

$$x = 2\pi p + \frac{\pi}{2}(1 + M \cos y)\ p = 0, 1, 2, \ldots, \infty \tag{5.16}$$

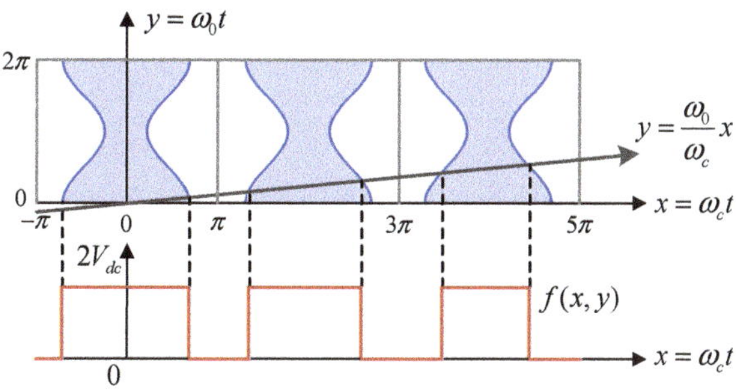

Fig. 5.56 Natural sampling PWM with triangular carriers

After determining the upper limit and lower limit of the integral, it can be calculated as:

$$
\begin{aligned}
U_o(t) = & V_{dc} + V_{dc} M \cos(\omega_o t + \theta_o) \\
& + \frac{4V_{dc}}{\pi} \sum_{m=1}^{\infty} \frac{1}{m} J_0(m\frac{\pi}{2}M) \sin[m\frac{\pi}{2}] \cos[m(\omega_c t + \theta_c)] \\
& + \frac{4V_{dc}}{\pi} \sum_{m=1}^{\infty} \sum_{\substack{n=-\infty \\ (n \neq 0)}}^{\infty} \frac{1}{m} J_n(m\frac{\pi}{2}M) \sin[(m+n)\frac{\pi}{2}] \cos[m(\omega_c t + \theta_c) + n(\omega_o t + \theta_o)]
\end{aligned}
\tag{5.17}
$$

With the DC offset removed, the harmonic components of the output phase voltage U_o(t) are expressed. In the triangular carrier natural sampling SPWM, there are no baseband harmonic components except the fundamental component. The odd-order side-band harmonic components around the odd-order carrier frequency and the even-order carrier harmonic components are completely eliminated by the $\sin[(m + n)/2]$ term in the above formula.

If the regular sampling PWM with triangular carriers is used as shown in Fig. 5.57, the switching time can be expressed as:

When $f(x,y)$ switches from 0 to $2V_{dc}$:

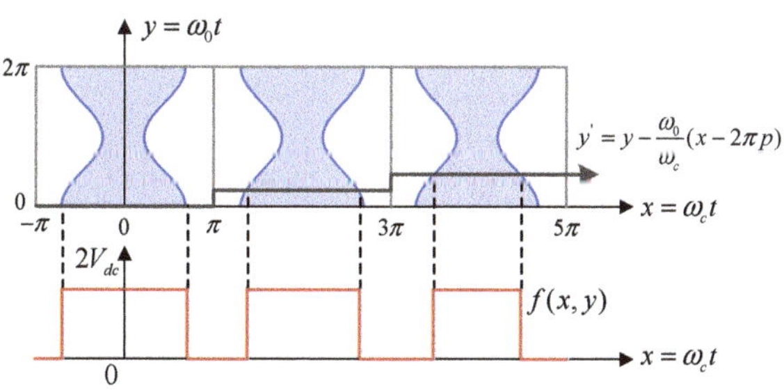

Fig. 5.57 Regular sampling PWM with triangular carriers

$$x = 2\pi p - \frac{\pi}{2}(1 + M\cos y')\ p = 0, 1, 2, \ldots, \infty \tag{5.18}$$

When $f(x,y)$ switches from $2V_{dc}$ to 0:

$$x = 2\pi p + \frac{\pi}{2}(1 + M\cos y')\ p = 0, 1, 2, \ldots, \infty \tag{5.19}$$

Here,

$$y' = y - \frac{\omega_o}{\omega_c}(x - 2\pi p) \tag{5.20}$$

After determining the upper limit and lower limit of the integral, it can be calculated as:

$$\begin{aligned} U_o(t) = V_{dc} &+ \frac{4V_{dc}}{\pi}\sum_{n=1}^{\infty}\frac{1}{(n\frac{\omega_o}{\omega_c})}J_n(n\frac{\omega_o}{\omega_c}\frac{\pi}{2}M)\sin[n(1+\frac{\omega_o}{\omega_c})\frac{\pi}{2}]\cos n(\omega_o t + \theta_o) \\ &+ \frac{4V_{dc}}{\pi}\sum_{m=1}^{\infty}\frac{1}{m}J_0(m\frac{\pi}{2}M)\sin m\frac{\pi}{2}\cos m(\omega_c t + \theta_c) \\ &+ \frac{4V_{dc}}{\pi}\sum_{m=1}^{\infty}\sum_{\substack{n=-\infty \\ (n\neq 0)}}^{\infty}\frac{1}{(m + n\frac{\omega_o}{\omega_c})}\left\{\begin{array}{l} J_n[(m + n\frac{\omega_o}{\omega_c})\frac{\pi}{2}M]\sin[(m + n\frac{\omega_o}{\omega_c} + n)\frac{\pi}{2}] \\ \cdot\cos[m(\omega_c t + \theta_c) + n(\omega_o t + \theta_o)] \end{array}\right\} \end{aligned} \tag{5.21}$$

With the DC offset removed, the harmonic components of the output phase voltage $U_o(t)$ are also expressed. In the regular sampling SPWM of triangular carriers, there are other baseband harmonic components except the fundamental component. The odd-order side-band harmonic components around the odd-order carrier frequency and the odd-order side-band harmonic components around the odd-order carrier frequency cannot be eliminated completely.

By comparing the Eqs. (5.17) and (5.21), the difference between the natural sampling and the regular sampling is the impact on the term $n\omega_o/\omega_c$ on C_{mn}. Figure 5.58 shows the switching harmonics comparison between the natural sampling and the regular sampling, with $\omega_o/\omega_c = 0.005$. It can be found that the impact on the term $n\omega_o/\omega_c$ on C_{mn} can be negligible when ω_o/ω_c is high. In digital controller, the regular sampling is often used.

In order to simplify the analysis, the initial phase angle of carrier and fundamental wave is set to zero, and the switching frequency function under VSFPWM control can be expressed as:

$$f_c(t) = f_{c0} + \Delta f_c v_m(t) \tag{5.22}$$

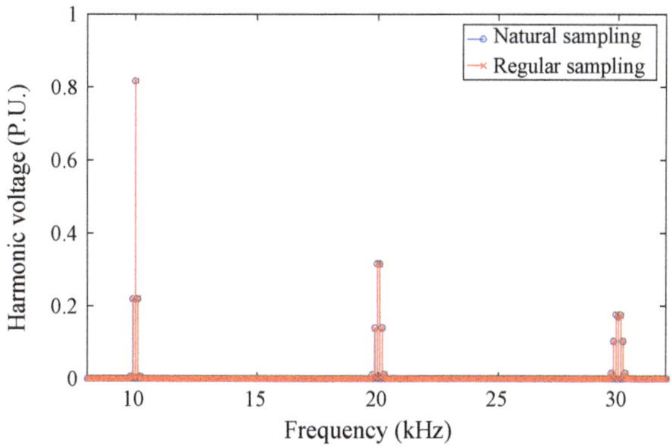

Fig. 5.58 Switching harmonics comparison between the natural sampling and the regular sampling

Here, f_{c0} is the center frequency, Δf_c is the switching frequency range, and the $v_m(t)$ is periodic signal of switch frequency function. Moreover, $v_m(t)$ can be performed by Fourier transform:

$$v_m(t) = \sum_{k=1}^{\infty} B_k \sin(2\pi k f_m t + \theta_k) \tag{5.23}$$

where f_m is the changing frequency of the switching function, and B_k and θ_k are the amplitude and phase of the kth harmonic.

Considering the high carrier ratio, the impact of the term $n\omega_o/\omega_c$ on C_{mn} can be negligible and calculation analysis can be simplified. While calculating C_{mn} under VSFPWM control, $\omega_c = 2\pi f_{c0}$. By substituting (5.22) into (5.21), the output phase voltage can be obtained as:

$$\begin{aligned} U_o(t) &= real\left\{\sum_{m=0}^{\infty}\sum_{n=-\infty}^{\infty} A_{mn} e^{j2\pi[mf_{c0}t + m\Delta f_c \int_0^t v_m(\tau)d\tau + nf_0 t]}\right\} \\ &= real\left\{\sum_{m=0}^{\infty}\sum_{n=-\infty}^{\infty} A_{mn} e^{j[2\pi m f_{c0}t + \varphi_{mk} - \sum_{k=1}^{\infty} \frac{m\Delta f_c B_k}{kf_m}\cos(2\pi k f_m t + \theta_k) + 2\pi n f_0 t]}\right\} \\ &= real\left\{\sum_{m=0}^{\infty}\sum_{n=-\infty}^{\infty} A_{mn} e^{j\varphi_{mk}} e^{j2\pi(mf_{c0}t + nf_0 t)} \cdot \prod_{k=1}^{\infty} e^{-j[\frac{m\Delta f_c B_k}{kf_m}\cos(2\pi k f_m t + \theta_k)]}\right\} \end{aligned} \tag{5.24}$$

Here,

$$A_{mn} = \frac{4V_{dc}}{(m + n\frac{\omega_o}{\omega_c})\pi} J_n\left[\left(m + n\frac{\omega_o}{\omega_c}\right)\frac{\pi}{2}M\right]\sin\left[\left(m + n\frac{\omega_o}{\omega_c} + n\right)\frac{\pi}{2}\right] \tag{5.25}$$

$$\varphi_{mk} = \sum_{k=1}^{\infty} \frac{m\Delta f_c B_k \cos\theta_k}{kf_m} \tag{5.26}$$

Using the Jacobi-Anger expansion, it can be simplified as:

$$U_o(t) = real\left\{\sum_{m=0}^{\infty}\sum_{n=-\infty}^{\infty} A_{mn}e^{j\varphi_{mk}}e^{j2\pi(mf_{c0}t+nf_0t)}\cdot\prod_{k=1}^{\infty}\sum_{r=-\infty}^{\infty} j^{-r}J_r\left(\frac{m\Delta f_c B_k}{kf_m}\right)e^{jr(2\pi kf_m t+\theta_k)}\right\} \tag{5.27}$$

Let $l = kr$,

$$U_o(t) = real\left\{\sum_{m=0}^{\infty}\sum_{n=-\infty}^{\infty} A_{mn}e^{j\varphi_{mk}}e^{j2\pi(mf_{c0}t+nf_0t)}\cdot\sum_{l=-\infty}^{\infty} A_l e^{j(2\pi lf_m t+\varphi_{kr})}\right\} \tag{5.28}$$

Here, A_l represents the final amplitude after merging every possible combinations of exp($j2\pi\, lf_m t$). Specially, a simple sinusoidal switching function is considered and it can be directly written as:

$$A_l = J_r\left(\frac{m\Delta f_c B_k}{kf_m}\right) \tag{5.29}$$

To facilitate observation and analysis, the above formula is further simplified as:

$$U_o(t) = real\left\{\sum_{l=-\infty}^{\infty}\sum_{m=0}^{\infty}\sum_{n=-\infty}^{\infty} A_{mn}A_l e^{j2\pi(mf_{c0}+nf_0+lf_m)t}e^{j(\varphi_{mk}+\varphi_{kr})}\right\} \tag{5.30}$$

In the above formula, $A_{mn}A_l$ is the harmonic amplitude at $(mf_{c0} + nf_0 + lf_m)$ frequency, and $A_{mn}A_l < A_{mn}$. Therefore, under VSFPWM control, harmonic distribution will be more extensive and harmonic amplitude will be effectively suppressed. If $m = 0$, $n > 0$, nf_0 frequency harmonic phase is $2\pi nf_0$, which is the same as CSFPWM. That is to say, if the effect of VSFWM on A_{mn} is ignored, VSFPWM has no effect on the baseband harmonic components of the output voltage.

To verify the correctness of spectrum quantified expressions for VSFPWM, the simulation results have been compared. The switching frequency harmonics of VSFPWM is shown in, Fig. 5.59 and the first 10 harmonics are considered for analytic calculation here. Figures 5.60 and 5.61, respectively, show the baseband harmonics and the carrier harmonics comparison between the analytic calculation and the simulation results. It can be found that both the analytic calculation and the simulation results match well, which illustrates the correctness of spectrum-quantified expressions for VSFPWM. In addition, VSFPWM nearly has no effect on the baseband harmonic components of the output voltage; for carrier harmonic components, VSFPWM can spread the harmonic distribution and effectively suppress the harmonic amplitude.

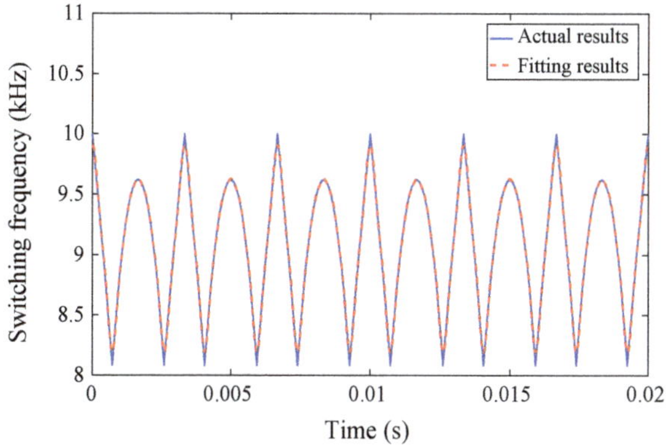

Fig. 5.59 Switching frequency of VSFPWM

Fig. 5.60 Baseband harmonic comparison

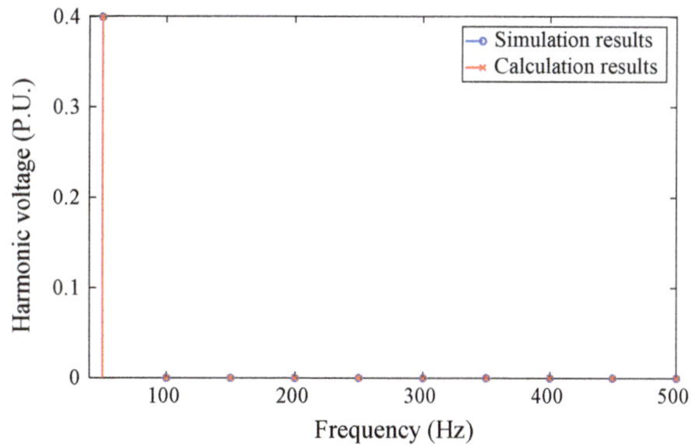

Fig. 5.61 Carrier harmonic comparison

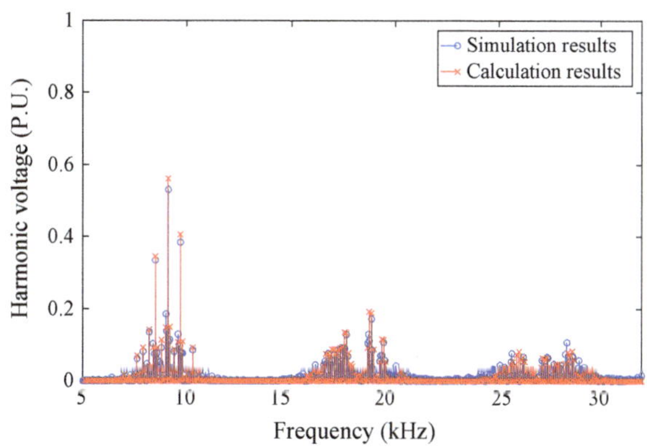

Therefore, the conclusion in this section is that the periodic variation of switching frequency will not impact on the low-frequency harmonics of the output current.

5.8 The Impact of VSFPWM on *d-q* Current Control [22]

From the point of view of modulation, the VSFPWM nearly has no effect on the base-band harmonic components. However, it may cause some deterioration in digital *d-q* current control of voltage source converters. In the case of proportional integral (PI) current regulator implemented with VSFPWM, the varying time-delay is accompanied by the frame rotation, which causes a varying phase-lag of output voltages, degrading the system performance. In this section, an analytical method based on complex coefficient transfer functions is developed to provide an accurate explanation of the relationship between the VSFPWM and the corresponding impacts. A compensation approach is then proposed to eliminate the varying phase-lag of output voltages when the PI current regulator with VSFPWM is employed.

The system of the three-phase grid-connected converter with *L*-type grid filter is considered here. Figure 5.62 shows the diagram of the basic PI current control with VSFPWM. The DC-link voltage is kept constant and its control is not discussed. The three-phase grid voltage is considered as ideal voltage source. In Fig. 5.62, $\vec{i}_{abc}$ and $\vec{V}_{gabc}$ are three-phase current and grid voltage defined as complex space vectors. ω_e denotes the electrical angular frequency. Using complex vectors, it can simplify the 2×2 multiple-input multiple-output model of a three-phase system into a single-input single-output model, with $\vec{V}_{dq} = V_d + jV_q$, $\vec{i}_{dq} = i_d + ji_q$ and $\vec{V}_{gdq} = V_{gd} + jV_{gq}$. Here, j is the complex sign. V_d and V_q are the converter output voltages. i_d and i_q are the *d-q* grid currents. V_{gd} and V_{gq} are the *d-q* grid voltages. In this paper, the grid voltage vector is oriented on q axis, thus $V_{gd} = 0$.

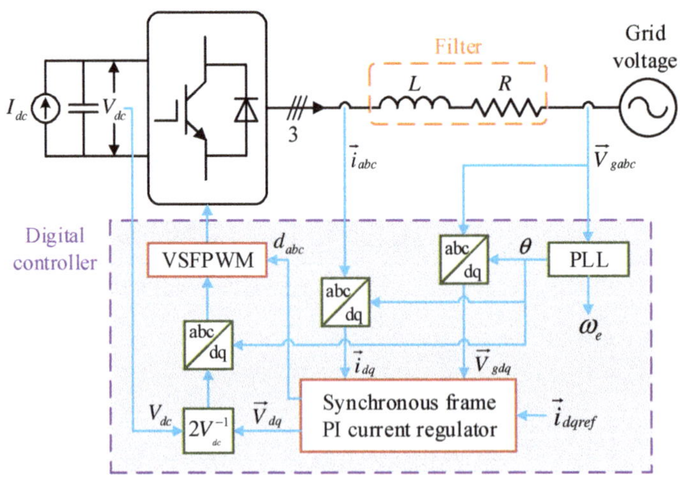

Fig. 5.62 System diagram of *d-q* current control with VSFPWM for grid-connected VSCs

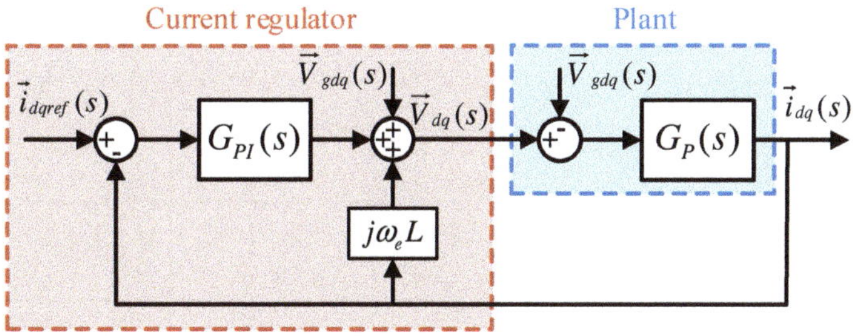

Fig. 5.63 Complex vector block diagram of synchronous frame PI current regulator with ICSF

To model the system, the dynamic of grid current in synchronous frame can be expressed with complex vectors as:

$$L\frac{d\vec{i}_{dq}}{dt} = \vec{V}_{dq} - R\vec{i}_{dq} - j\omega_e L\vec{i}_{dq} - \vec{V}_{gdq} \tag{5.31}$$

where L and R are the inductance and equivalent resistance of the grid filter. The relationship between the converter output voltages and the grid currents can be derived from (5.31) and expressed as a complex coefficient transfer function:

$$G_P(s) = \frac{\vec{i}_{dq}(s)}{\vec{V}_{dq}(s) - \vec{V}_{gdq}(s)} = \frac{1}{sL + R + j\omega_e L} \tag{5.32}$$

The term $j\omega_e L$ indicates that there are synchronous frequency axes crossing coupling in the plant, which means that the converter output voltages V_d and V_q not only control the corresponding axis current, but also affect the other axis current. If this term is decoupled from (5.32), the plant will become an RL load on synchronous frequency, enabling simple, fast and accurate current regulation. Usually, the inductor current state feedback (ICSF) is mostly applied to decouple the cross-coupling caused by the term $j\omega_e L$ in (5.32). Figure 5.63 shows the complex vector block diagram of synchronous frame PI current regulator with ICSF, and the measured grid voltages are used as feedforward signals.

The current references $\vec{i}_{dqref}(s)$ are set by the out-loop controller and the PI controller is expressed as:

$$G_{PI}(s) = K_p + \frac{K_i}{s} \tag{5.33}$$

Usually, the parameters of PI controller are tuned according to:

$$\begin{cases} K_p = \omega_c L \\ K_i = \omega_c R \end{cases} \tag{5.34}$$

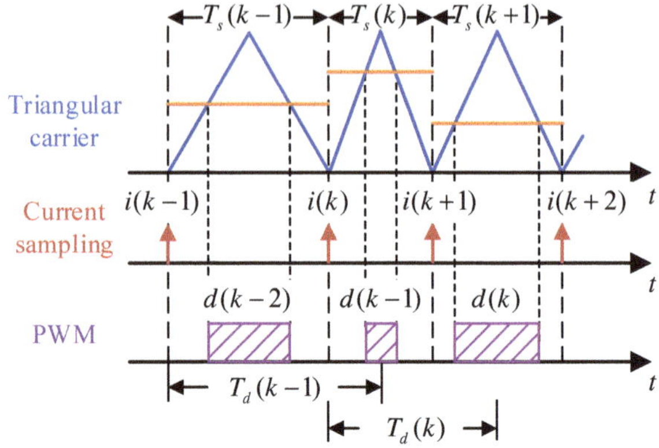

Fig. 5.64 Time sequence of current sampling, computation and VSFPWM output

where ω_c is the expected crossover frequency of the current loop gain. As can be noticed in Fig. 5.63, the converter is modeled as a unity gain without phase-lag. However, the time-delay always exists in the digital control system, which affects the converter modeling and finally degrades the performance of the *d*-*q* current control.

It is widely known that the calculation program in digital signal processor (DSP) should be completed in one sampling period and the sampling point should be set in the midpoint of the zero vector to avoid reading the current ripple component. Thus, the time-delay can be constantly viewed as one-and-a-half switching period for the symmetrical PWM with constant switching frequency, which updates the output voltage once per cycle. It is also intuitively known that the time-delay of VSFPWM is varying and highly related to the switching period updating process.

Figure 5.64 shows the typical time sequence of current sampling, computation and VSFPWM output, where $T_s(k)$ is the *k*th switching period, and $T_d(k)$ represents the *k*th time-delay after *k*th current sampling $i(k)$, including the *k*th switching period delay during the execution of control algorithm and the time-delay of PWM output in next switching period $T_s(k+1)$. For the constant switching frequency PWM (CSFPWM), the transfer function of time-delay can be easily expressed as $1/(1 + 1.5T_s s)$. Since the switching period is changing cycle by cycle in VSFPWM, the time-delay is no longer a constant. Theoretically, the original current loop cannot be treated as a linear system if the time-delay becomes a variable, which will make the analysis more complex.

The VSFPWM is usually with a high ratio of the switching frequency to the output frequency. It means that the variation of switching frequency hardly affects the system stability of a three-phase-connected VSI system with *L*-type filter. In addition, the effect of switching frequency variation on magnitude-frequency of time-delay process is very limited, and it can be considered as a small disturbance (D_{Tsdq}) to the corresponding current loop. D_{Tsdq} is introduced by the AC signals, and it can be ignored in the steady-state where the DC component is dominant. The determination of this small disturbance is out of the scope of a linear system, which is just a simplification for next analysis and is not here. The time-delay process

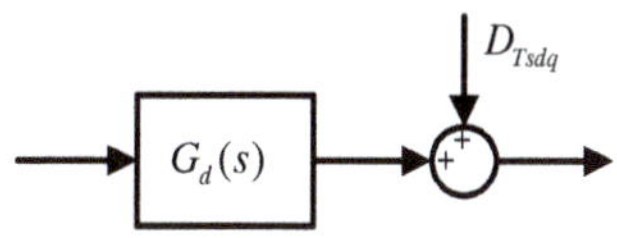

Fig. 5.65 Simplified presentation of time-delay in VSFPWM

of VSFPWM can be depicted as Fig. 5.65, where G_d(s) = 1/(1 + $T_{dn}s$) and T_{dn} represents the average time-delay in VSFPWM.

On the other hand, in the practical case of time sequence in Fig. 5.64, the synchronous reference frame rotates and the output voltage vector moves, as shown in Fig. 5.66. In Fig. 5.66, the rotation of synchronous reference frame is represented by $d^e(k\text{-}1)$, $d^e(k)$ and $d^e(k + 1)$, corresponding to each current sampling points. $\vec{V}_{dq}(k - 1)$ is the current regulator output based on the synchronous reference frame $d^e(k\text{-}1)$. Due to the time-delay in the digital control system, the voltage vector $\vec{V}_{dq}(k-1)$ is activated through PWM when the synchronous reference frame rotates between $d^e(k)$ and $d^e(k + 1)$. Thus, there is a phase-lag for voltage vector output caused by the synchronous reference frame rotation during the time-delay. With the assumption that the synchronous frequency ω_e is constant during the time-delay, the practical voltage vector output $\vec{V}_{dq}^{real}$ can be derived as:

$$\vec{V}_{dq}^{real} = e^{-j\omega_e T_d}\vec{V}_{dq} \approx (1 - j\omega_e T_d)\vec{V}_{dq} \tag{5.35}$$

The transfer function for phase-lag on voltage vector output is achieved by taking the Laplace transform of (5.35), to give:

$$\vec{V}_{dq}^{real}(s) \approx \vec{V}_{dq}(s) - L[j\omega_e T_d\vec{V}_{dq}] = \vec{V}_{dq}(s) - D_{rdq}(s) \tag{5.36}$$

where D_{rdq}(s) represents the Laplace transform of the second term $L[j\omega_e T_d\vec{V}_{dq}]$. Based on the above analysis, the complex vector block diagram considering the delay-time in VSFPWM can be illustrated in Fig. 5.67.

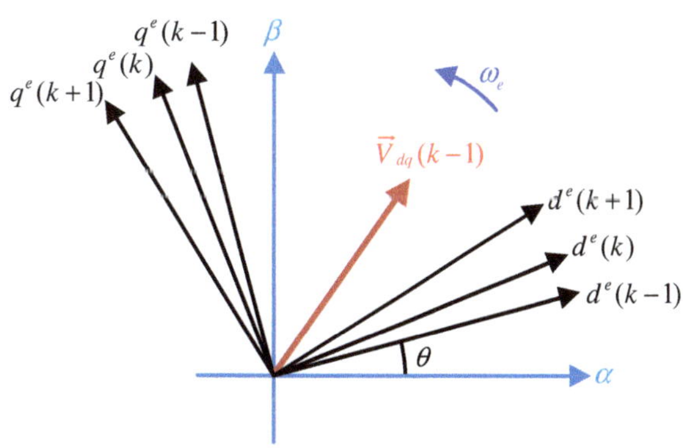

Fig. 5.66 Synchronous frame rotation corresponding to the time sequence

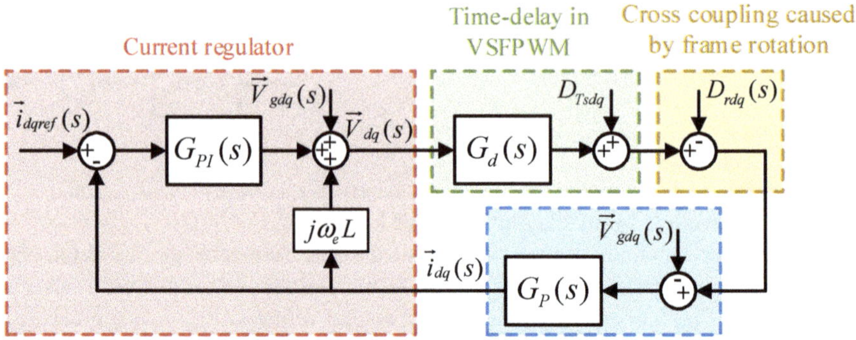

Fig. 5.67 Complex vector block diagram considering the time-delay in VSFPWM

From Fig. 5.67, it can be seen that the original decoupling performance of ICSF will be degraded due to the existing time-delay. It means that the original axes cross-coupling cannot be fully eliminated, but only be attenuated for the digital control system with VSFPWM. Focusing on Eq. (5.36), it can be seen that a new cross-coupling term D_{rdq}(s) appears, which is caused by the phase-lag in output voltage. Clearly, this cross-coupling term deteriorates the dynamic response of the current regulator. When the current regulator executes in the steady-state, D_{rdq}(s) is nearly equal to $j\omega_e \vec{V}_{dq} T_d(s)$. It means that the cross-coupling term D_{rdq}(s) is determined by the function of switching frequency in PWM, under steady-state. If the normal CSFPWM is used, T_d is constant ($1.5T_s$) and D_{rdq} is nearly DC signal. However, apart from the DC component, D_{rdq} is with rich AC component, introduced by the varying time-delay T_d(s) when VSFPWM is employed. Usually, the switching frequency variation of VSFPWM repeats six times in a fundamental period. If 50 Hz grid frequency is adopted in a three-phase grid-connected VSC, T_d(s) is with rich harmonic concentrated on 300 Hz. Consequently, with VSFPWM, low-amplitude oscillation of six times grid frequency caused by the new cross-coupling term may be excited in the *d*-*q* current control, even though the current regulator reaches the steady-state.

It is instructive to develop the closed-loop tracking error sensitivity caused by the new cross-coupling term D_{rdq}(s), as given by:

$$E_D(s) = \frac{0 - \vec{i}_{dq}(s)}{D_{rdq}(s)} = \frac{G'_P(s)}{1 + G_{PI}(s)G_d(s)G'_P(s)} \tag{5.37}$$

Equation (5.37) identifies the plant sensitivity to the disturbance of D_{rdq}(s). In Eq. (5.37), $G_P{}'$(s) is considered as $1/(R + sL)$, with the assumption that the axes cross-coupling caused by the plant can be attenuated to very small by ICSF. To minimize the steady-state errors caused by D_{rdq}(s), E_D(s) needs to approach zero in a wide frequency range.

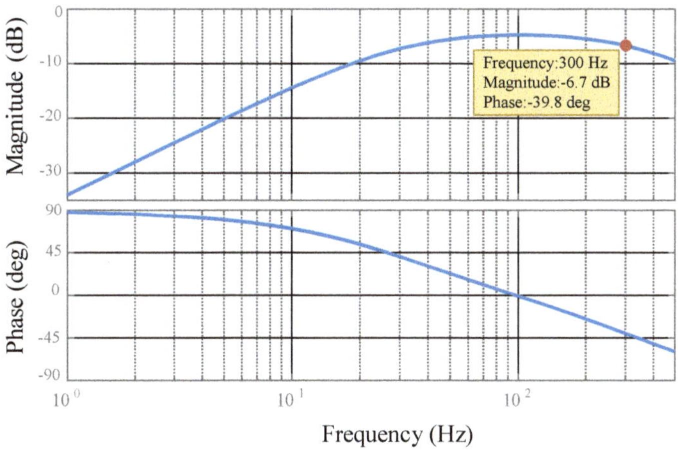

Fig. 5.68 Bode diagram of E_D(s)

Accordingly, it is convenient to depict the bode diagram of E_D(s) as shown in Fig. 5.68, with $\omega_c = 500\pi$ rad/s, $L = 1$ mH, $R = 0.2\ \Omega$ and $T_N = 8$ kHz. As discussed before, the disturbance of CSFPWM mainly locates at 0 Hz under steady-state, which can be fully eliminated because the gain of E_D(s) in $f = 0$ Hz is zero theoretically. It means that the impact of CSFPWM on *d*-*q* current control can be ignored when the system operates in steady-state. It is noted here that the gain in six times grid frequency ($f = 300$ Hz) has been marked in Fig. 5.68. Clearly, the gain of -6.7 dB is insufficient for eliminating the special harmonic caused by VSFPWM. To conclude, due to the varying time-delay in VSFPWM, fluctuations and harmonics concentrated on six times grid frequency are introduced by the cross-coupling term D_{rdq}(s), deteriorating the *d*-*q* current control.

To deal with the new cross-coupling issue caused by the synchronous frame rotation during varying time-delay in VSFPWM, a phase compensation term $e^{j\omega_e T_d}$ can be employed to compensate this varying phase-lag as expressed in (5.38). For the VSFPWM, T_d is not a constant, but it depends on the switching frequency in real time. In other words, the phase compensation term $e^{j\omega_e T_d}$ needs to be calculated and updated in every switching cycle if VSFPWM is used. Figure 5.69 shows the proposed compensation method for PI current regulator implemented with VSFPWM in k^{th} switching cycle. When PWM interruption is triggered, three-phase currents are sampled by the current sensors and then *d*-*q* currents are achieved by Park transformation. With the input of *d*-*q* currents $i_{d,q}(k)$, the control algorithm is processed, and then output voltage vector in next switching cycle $\vec{V}_{dq}(k+1)$ is obtained. On the other hand, VSFPWM algorithm is processed with the three-phase duty-cycles in the current switching cycle $d_{a,b,c}(k)$, and then next switching period is determined $T_N(k + 1)$. According to the time sequence in Fig. 5.66, the time-delay from the sampling point in current switching cycle to the voltage output point in next switching cycle is obtained with $T_d = T_N(k) + 0.5T_N(k + 1)$. Finally, the compensation term $e^{j\omega_e T_d}$

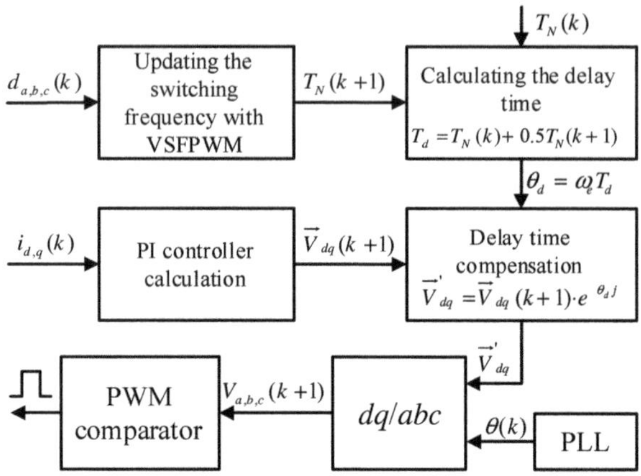

Fig. 5.69 Proposed compensation method for the varying phase-lag in VSFPWM

is added to the output voltage vector $\overrightarrow{V}_{dq}(k+1)$, which is realized by adding an advancing phase θ_d during the inverse Park transformation. In the next switching cycle, PWM signals are actuated by comparing the compensated reference voltages $V_{a,b,c}(k+1)$ with the triangular carrier, and the previous compensation method executes again.

$$\overrightarrow{V}_{dq}^{real} = e^{j\omega_e T_d} \cdot e^{-j\omega_e T_d} \overrightarrow{V}_{dq} = \overrightarrow{V}_{dq} \quad (5.38)$$

Experimental tests are implemented on a low-power laboratory prototype with the same topology shown in Fig. 5.62. The grid is emulated with a three-phase programmable AC voltage source. The VSC is connected to the grid through three-phase inductors and the DC-link voltage is kept constant by a DC voltage source. During the implementation process of CSFPWM, with a constant switching frequency of 10 kHz for the traditional SVPWM, the maximum peak-to-peak value of current ripple is 2.6 A, which is set as a control requirement for latter VSFPWM.

5.8.1 Steady-State Performance

First, the comparison of steady-state performance has been carried out, with the current reference $i^*_{dref} = 0$ A and $i^*_{qref} = 8$ A. It should be noted that the practical *d*-*q* current cannot be regulated as good as the simulations because of nonideal effects in real system, such as deadtime and sampling errors. In this case, the switching frequency of VSFPWM varies between 6.5 and 10 kHz to make the three-phase current ripple meet the peak-to-peak value requirement, resulting in a reduction of 19% in average switching frequency compared with the CSFPWM. From Fig. 5.70a and b, it can be observed that six times grid frequency (300 Hz) oscillations in i_d and i_q become larger if the VSFPWM is implemented without compensation. In Fig. 5.70b,

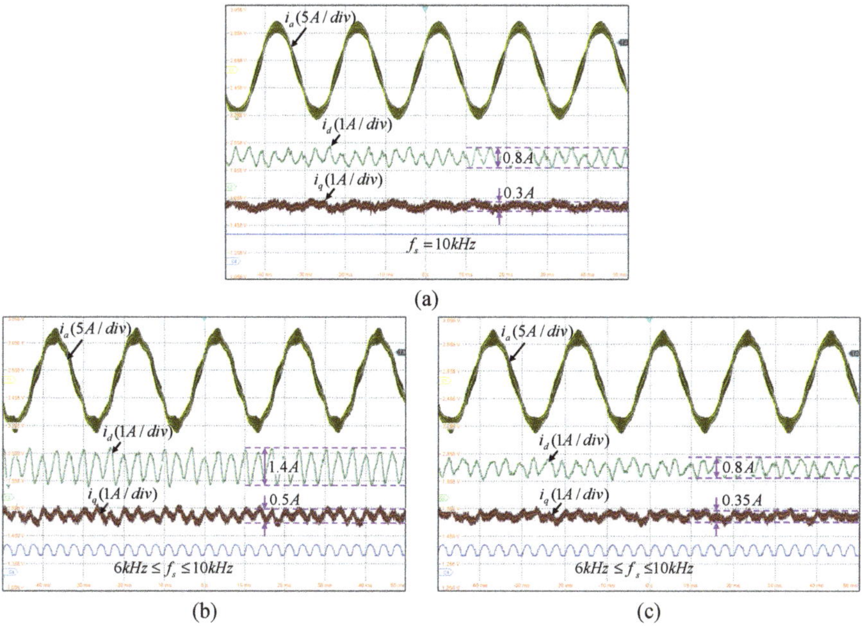

Fig. 5.70 Steady-state performance: **a** CSFPWM, **b** VSFPWM without compensation, **c** VSFPWM with compensation

the fluctuation ranges of i_d and i_q are, respectively, 1.4 and 0.5 A, increasing by 75 and 67% in *d-q* current oscillations compared with the CSFPWM. Also, a visible low-frequency distortion is introduced into the output current during VSFPWM implementation. After implementing the proposed compensation method, oscillations in *d-q* current for VSFPWM are significantly reduced to the same level of CSFPWM because the switching-period-related disturbance caused by the cross-coupling term $D_{rdq}(s)$ is removed, as shown in Fig. 5.70c. Moreover, the FFT analysis in *d-q* current for different methods is acquired in Fig. 5.71. It can be observed that the VSFPWM

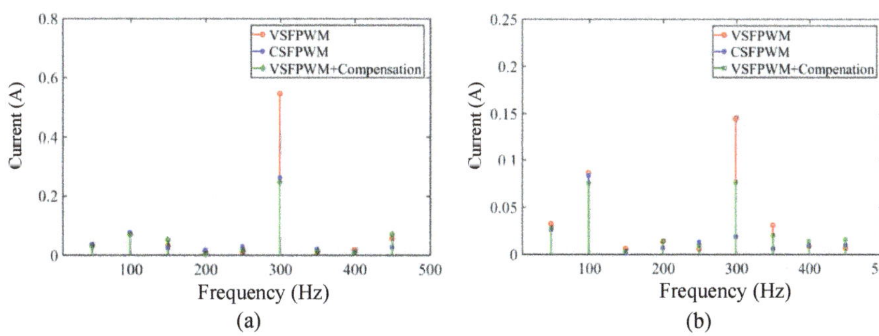

Fig. 5.71 *d-q* current comparison in frequency domain: **a** *d*-axis current, **b** *q*-axis current

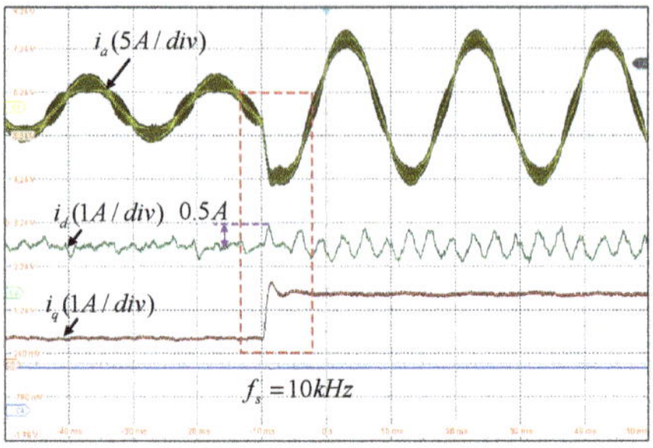

(a)

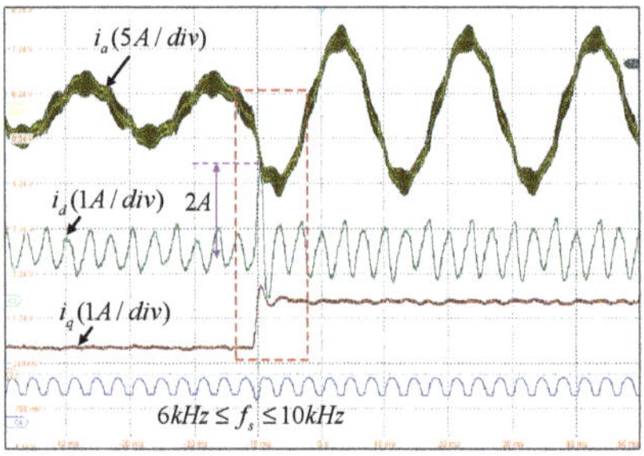

(b)

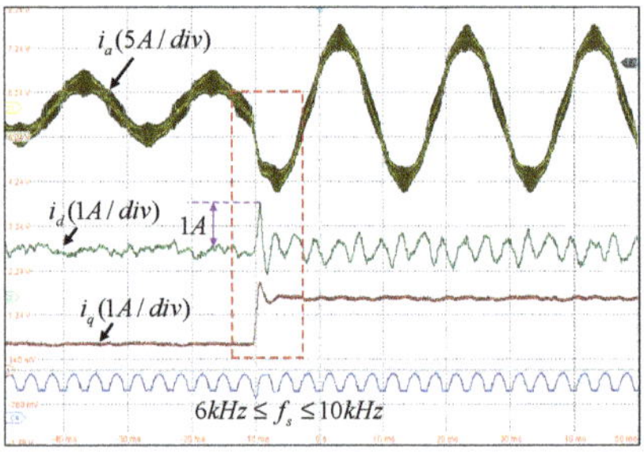

(c)

Fig. 5.72 Dynamic performance: **a** CSFPWM, **b** VSFPWM without compensation, **c** VSFPWM with compensation

mainly deteriorates the harmonic magnitude of six times grid frequency (300 Hz), which can be significantly improved by the proposed compensation method.

5.8.2 Dynamic Performance

Further tests are conducted on the transient responses of the grid-connected converter when the *q*-axis current reference is step changing from 3 to 8 A and *d*-axis current remains 0 A. Due to $V_q > V_d$, the *d*-axis current is more severely cross-coupled with the *q*-axis. In Fig. 5.72b, as i_q tracks the reference quickly, the i_d is severely disturbed and significantly deviates from its reference. The peak deviation of i_d in VSFPWM is about 2 A, increasing by 400% compared with the CSFPWM shown in Fig. 5.72a. As shown in Fig. 5.72c, it can be found that the better transient performances of *d*-*q*-axes currents can be achieved for VSFPWM when the proposed compensation method is adopted. After the compensation, the *d*-*q*-axes currents change rapidly to follow the reference signal with smaller overshoot, with the transient process finished within 4 ms.

In Sects. 5.7 and 5.8, two fundamental questions for VSFPWM have been answered:

(1) With double Fourier analysis for VSFPWM, the mathematical proof of reduction of switching harmonics by VSFPWM can be provided. With switching frequency obviously higher than the fundamental frequency, it is validated that the low-frequency harmonic is hardly impacted by VSFPWM.
(2) With feedback control in multiphase converter, VSFPWM will bring a variable delay in the control loop, which is with AC component (six times of fundamental frequency). Feedback control will react to this AC component. But with real-time compensation, it can be mitigated.

5.9 Pulse Position Control: Phase-Shift PWM [14]

Various variable switching frequency PWM methods introduced in Sects. 5.2–5.8 are actually the application of one freedom in PWM: switching frequency, to achieve the performance improvement of the converter system. As mentioned in the previous chapter introduction, besides switching frequency, pulse position is also an important freedom. In the model prediction PWM, how to effectively use the pulse position to improve the performance of the converter is also an important part. Similar to the VSFPWM, the utilization of pulse distribution also includes two main parts: one is how to realize the control of pulse position, and the other is how to predict the influence of the change of pulse position on the system performance (especially the current ripple). This section describes the work of these two parts.

First, how to realize arbitrary distribution of pulse position. As described in Chap. 2, in the widely used PWM controller, there are mainly two kinds of pulse distribution: symmetric PWM and side-aligned PWM. The former is obtained by symmetrical triangular carrier comparison and the latter by saw-tooth carrier comparison. In fact, both kinds of PWM are just special cases of pulse arbitrarily distributed PWM. Figure 5.73 shows examples of several pulse distributions. The first (a) shows the central symmetric PWM, which is realized by symmetrical triangular wave comparison. (b) and (c) show side-aligned PWM, which is realized by comparing with saw-tooth waves. In fact, the pulse can be distributed arbitrarily in this switching period without changing the duty ratio. In other words, it can be aligned on both sides to any position in the central symmetry without affecting the output average value, as shown in (d). More extremely, the pulse shift may exceed this period and re-enter from the other side. That is, the high level of the pulse is no longer a continuous part but is separated on both sides, as shown in (e).

As can be seen in Fig. 5.73, the arbitrary distribution of pulse is a freedom with a wide control range, and it is difficult to realize with the traditional carrier. Figure 5.74 shows the carrier comparison implementation of three-phase symmetric PWM (regular sampling). The three-phase adopts the same symmetric carrier, and updates the three-phase modulation signals (d_a, d_b and d_c) once in each switching cycle so that the three-phase pulses are symmetric. Taking phase-a as an example, in order to realize any distribution of the pulse in the switching cycle, carriers of phase-b and phase-c must be adjusted to realize the adjustable pulse position.

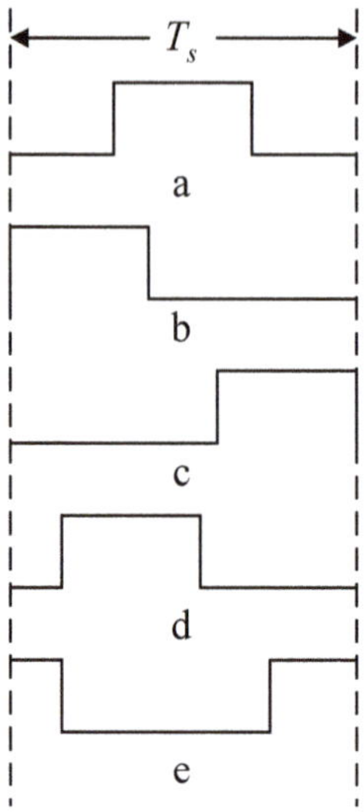

Fig. 5.73 Several examples of pulse distribution

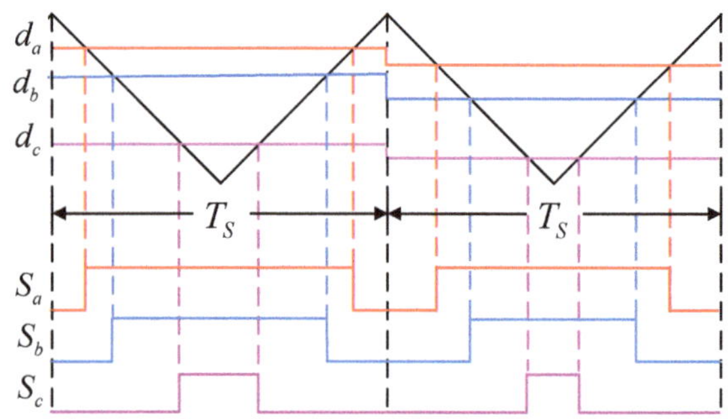

Fig. 5.74 Symmetric PWM generation

A typical adjustment method is to use asymmetric carriers as shown in Fig. 5.75. With this method, the carriers corresponding to phase-b and phase-c are no longer symmetric, but any triangular wave with a definite period. Then the corresponding pulses (S_b, S_c) are sent out by comparing with the corresponding modulation signals (d_b, d_c). At this point, compared with phase-a, phase-b and phase-c achieve pulse displacement while maintaining constant duty ratio. By controlling the vertex positions of phase-b and phase-c carriers, the displacement can also be controlled so that the pulse position can be controlled.

However, the approach shown in Fig. 5.75 has some limitations. Since the vertex displacement of the triangle carrier can only occur during the whole period, the pulse position can only be adjusted from left-most alignment to right-most alignment. The high level of the pulse is always continuous, which makes it impossible to realize the situation in Fig. 5.73e. In addition, the implementation of asymmetric triangular carrier in DSP is also complex. Therefore, this approach is not recommended.

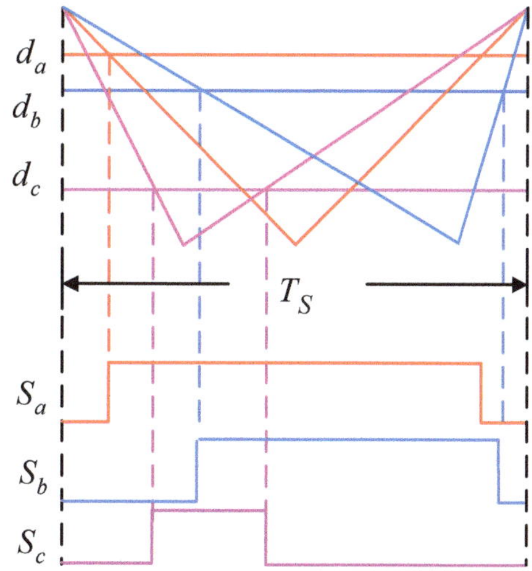

Fig. 5.75 Asymmetric carrier for pulse shift PWM

Another pulse position control method effectively improves these shortcomings, as shown in Fig. 5.76. This method is still implemented by symmetric carrier. The difference is that three phases no longer use the same carrier. The carriers of phase-b and phase-c can shift f_{s_b} and f_{s_c} arbitrarily ($0 < f_{s\_b}, f_{s\_c} < T_s$), and the phase-shift distance is exactly the displacement of the pulses S_b and S_c in reference to S_a. Since the displacement contains all the possibilities of 0–T_s, the pulse also contains all the possibilities of displacement. This also includes the "overflow" situation in Fig. 5.73e, which implements high-level segments on both sides of the switching cycle. Moreover, symmetric carrier is easy to be implemented in DSP counter, so this method is more recommended.

Another important task is to study the prediction model of arbitrary pulse distribution. Similar to the prediction model of VSFPWM, current ripple is synchronized

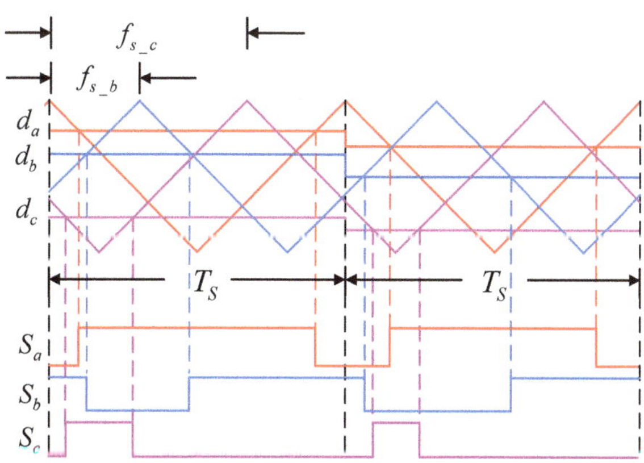

Fig. 5.76 Three-phase PWM with arbitrary pulse distribution by carrier phase-shift

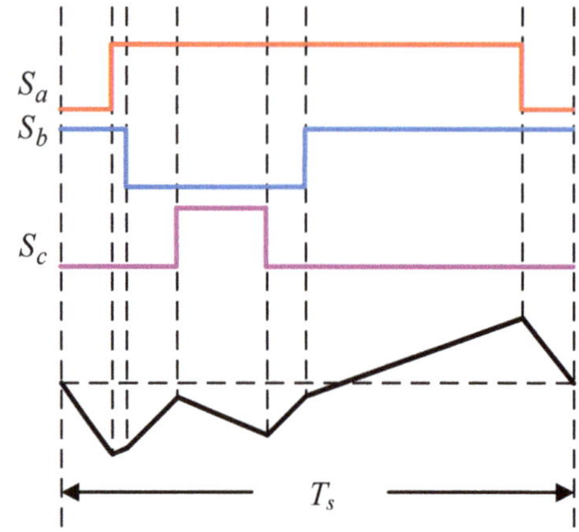

Fig. 5.77 Current ripple with phase-shift PWM

with PWM and easy to predict. However, compared with VSFPWM, it is more difficult to predict the current ripple using PWM of arbitrary pulse distribution. For phase-shift PWM, not only action time is arbitrarily controlled but also the original vector is changed and no longer determined by ordinary SVPWM. Taking the current ripple under arbitrary pulse distribution shown in Fig. 5.77 as an example, there are still seven zones corresponding to seven voltage vectors and seven segments of linear current ripple in one switching cycle. However, due to the phase-shift of phase-b and phase-c in reference to phase-a, not only the action time of each linear segment can be arbitrarily changed but also the sequence of voltage vector action has changed. Therefore, it is necessary to improve the current ripple prediction method.

Based on the current ripple diagram of arbitrary pulse distribution shown in Fig. 5.77, the current ripple prediction process shown in Fig. 5.78 can cope with

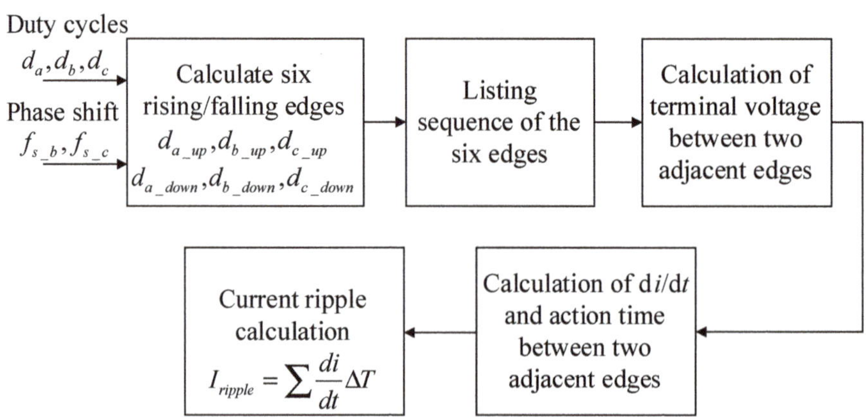

Fig. 5.78 Current ripple prediction process with phase-shift PWM

the PWM strategy of arbitrary pulse distribution. In each switching cycle, input the duty cycles d_a, d_b and d_c, and input the phase-shifting f_{s_b} and f_{s_c} of phase-b and phase-c relative to phase-a. Then, the rising edges (d_{a_up}, d_{b_up}, d_{c_up}) and falling edges (d_{a_down}, d_{b_down}, d_{c_down}) of the output pulse of three-phase can be obtained. The methods for calculating the six edges are shown in (5.39)–(5.44) (P.U. value), where if the calculated edge value is greater than 1, it means that it overflows after phase-shift and enters the period from the other side, and is minus one. The six edges obtained by such calculation are listed in the order from smallest to largest, as shown in t_1 –t_6 in Fig. 5.77. In this way, the action time between two adjacent edges can be calculated, and the voltage vector between two adjacent edges can be determined, so as to obtain the voltage dropped on the inductor. So, we can figure out the action time and d*i*/d*t* for each zone. By integrating d*i*/d*t* with time, the current ripple can be predicted in each switching cycle.

$$d_{a\_up} = (1 - d_a)/2 \tag{5.39}$$

$$d_{b\_up} = (1 - d_b)/2 + f_{s\_b} \tag{5.40}$$

$$d_{c\_up} = (1 - d_c)/2 + f_{s\_c} \tag{5.41}$$

$$d_{a\_down} = (1 + d_a)/2 \tag{5.42}$$

$$d_{b\_down} = (1 + d_b)/2 + f_{s\_b} \tag{5.43}$$

$$d_{c\_down} = (1 + d_c)/2 + f_{s\_c} \tag{5.44}$$

Taking the third zone in Fig. 5.77 as an example, the values of the six edges are first calculated. It is between two edges of t_2 and t_3, and the corresponding voltage vector of this segment is 100. Thus, the expression of d*i*/d*t* can be obtained through the equivalent circuit method in the previous prediction model.

To verify with a simple example, it is assumed that phase-b shifts by half of the switching cycle, namely, $f_{s\_b} = T_s/2$. The maximum and minimum current ripple values of each switching cycle can be obtained through the above prediction method. Meanwhile, three-phase current is obtained through simulation. Both current ripple and the prediction result are shown in Fig. 5.79a. The predicted maximum and minimum current ripple values coincide well with the envelope of the simulation result, which proves the validity of the prediction method. Experimental results (Fig. 5.79b) also illustrate that.

Although the phase-shift will deteriorate with fixed phase-shift, there is definitely an opportunity for improvement of the current ripple if the optimal phase-shift is predicted according to the current ripple in each switching cycle. That is, the method of maintaining adaptive phase-shift. However, compared with the linear relation that

Fig. 5.79 **a** Comparison of prediction and simulation ripple current with $f_{s\_b} = T_s/2$. **b** Experimental results with $f_{s\_b} = T_s/2$

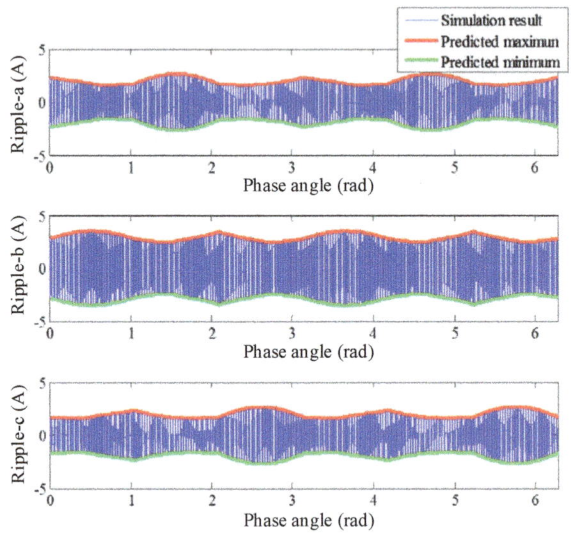

(a)

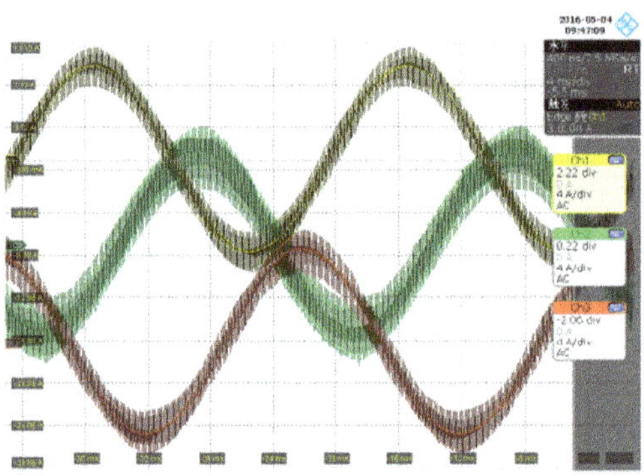

(b)

the current ripple is proportional to the switching period, the relationship between phase-shift and current ripple is highly nonlinear, and it is difficult to achieve real-time optimization. The work is still being explored.

Variable-pulse-position PWM can also be realized by carrier phase-shift. That is, change the pulse position in each switching period. However, as mentioned above, due to the complex relationship between current ripple and phase-shift, real-time control is difficult. A simple way is to shift phase randomly. Such PWM can improve EMI under the condition that the switching frequency remains unchanged. It is equivalent to random-pulse-position PWM [14, 15]. Figure 5.80 compares the current experimental waveforms of ordinary SVPWM and random phase-shift PWM. The deterioration of current ripple is predicted as before. The comparison of EMI is shown

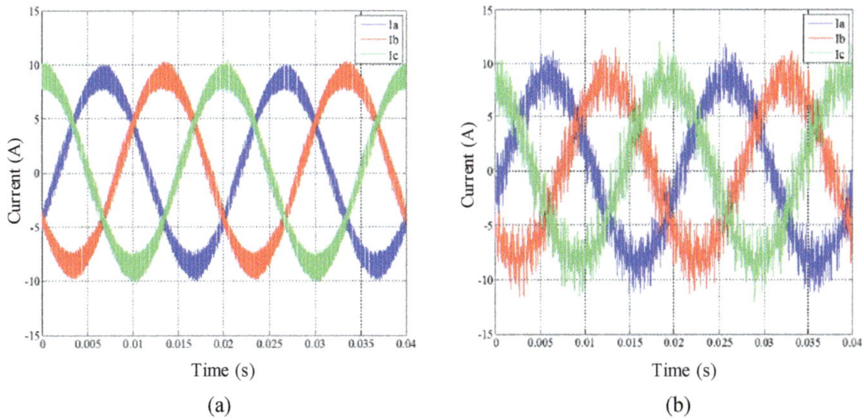

Fig. 5.80 Experimental result: **a** SVPWM, **b** random-pulse-position PWM

in Fig. 5.81, and the random phase-shift PWM slightly improves EMI compared with ordinary SVPWM. Since the switching frequency remains fixed, EMI improvement is limited.

Although the EMI improvement is limited, phase-shift PWM will have special advantages over common-mode voltage and common-mode EMI, which will be introduced in Chap. 7.

Fig. 5.81 EMI comparison

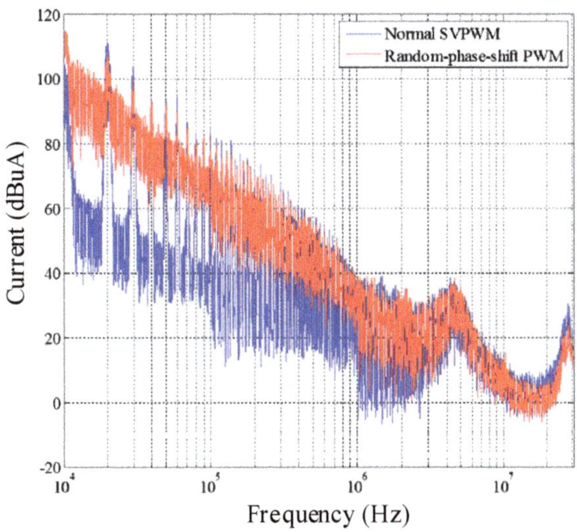

5.10 Summary

Based on current ripple prediction in Chap. 4, this chapter first introduces the concept and structure of model prediction PWM. This chapter mainly introduces the most typical model predictive PWM: variable switching frequency PWM. Different from the random PWM introduced in Chap. 3, the variable switching frequency PWM based on the prediction model changes the switching frequency in real time according to the needs of the optimization target, which is a predictive control method. The most direct applications are two kinds of VSFPWM methods with the control target of AC side current ripple. VSFPWM based on current ripple peak value can control the maximum value of three-phase current ripple in each switching cycle consistent with the ripple requirement, thereby reducing the average switching frequency, reducing the switching losses and inhibiting the peak value of EMI. VSFPWM based on current ripple RMS value has no obvious effect on the reduction of switching losses, but it can still suppress the EMI under the condition of guaranteeing the RMS value of current ripple. In addition, according to other control objectives, the corresponding VSFPWM strategy can also be designed according to the corresponding prediction model. VSFPWM is extended to torque ripple control of motor drives and voltage ripple control of two-level rectifier. It is worth pointing out that due to the VSFPWM only uses one control variable: switching frequency, so there can only be one control target. More control targets need more control variables. Deeper analysis for the switching frequency variation has been further introduced in this chapter. To better deal with the EMI problem, NDPWM and UDPWM have been proposed, which can further eliminate the EMI spikes in VSFPWM. With switching frequency variation, fundamental questions for harmonics and control performance need to be studied. Spectrum calculation and analysis of VSFPWM are presented here, and a conclusion can be given that VSFPWM nearly has few effects on the baseband harmonic components of the output voltage and can effectively suppress the carrier harmonics amplitude. Moreover, the impact of VSFPWM on *d*-*q* current control has been analyzed and a compensation method is also proposed. This chapter also introduces the improved PWM method using the freedom of pulse position, including carrier phase-shift method and current ripple prediction method. However, it is worth mentioning that the phase-shift PWM has a negative effect on current ripple, and the complex relationship between current ripple and phase-shift makes its application not as simple and direct as the application of switching frequency.

References

1. Rodriguez J, Cortes P (2012) Predictive control of power converters and electrical drives. Wiley-IEEE Press
2. Geyer T, Papafotiou G, Morari M (2009) Model predictive direct torque control—part i: concept, algorithm, and analysis. IEEE Trans Ind Electron 56(6):1894–1905

3. Geyer T, Oikonomou N, Papafotiou G, Kieferndorf FD (2012) Model predictive pulse pattern control. IEEE Trans Ind Appl 48(2):663–676
4. Zhang Y, Yang H (2015) Generalized two-vector-based model-predictive torque control of induction motor drives. IEEE Trans Power Electron 30(7):3818–3829
5. Holmes G, Lipo TA (2003) Pulse width modulation for power converters-principle and practice. IEEE Press
6. Blasko V (1997) Analysis of a hybrid PWM based on modified space-vector and triangle-comparison methods. IEEE Trans Ind Appl 33(3):756–764
7. Zhou K, Wang D (2002) Relationship between space-vector modulation and three-phase carrier-based PWM: a comprehensive analysis [three-phase inverters]. IEEE Trans Ind Electron 49(1):186–196
8. Mao X, Ayyanar R, Krishnamurthy HK (2009) Optimal variable switching frequency scheme for reducing switching loss in single-phase inverters based on time-domain ripple analysis. IEEE Trans Power Electron 24(4):991–1001
9. Jiang D, Wang F (2013) Variable switching frequency PWM for three-phase converters based on current ripple prediction. IEEE Trans Power Electron 28(11):4951–4961
10. Jiang D, Wang F (2014) Current ripple prediction for three-phase PWM converters. IEEE Trans Ind Appl 50(1):531–538
11. Jiang D, Wang F (2014) A general current ripple prediction method for the multiphase voltage source converter. IEEE Trans Power Electron 29(6):2643–2648
12. Hari VSSPK, Narayanan G (2016) Space-vector-based hybrid PWM technique to reduce peak-to-peak torque ripple in induction motor drives. IEEE Trans Ind Appl 52(2):1489–1499
13. Grandi G, Loncarski J, Dordevic O (2014) Analytical evaluation of output current ripple amplitude in three-phase three-level inverters. IET Power Electron 7(9):2258–2268
14. Jiang D, Shen Z, Qu R (2016) Phase-shift PWM for three-phase voltage source converters. In: Proceedings of european conference on power electronics and applications (EPE'16 ECCE Europe), pp 1–8
15. Bech MM, Blaabjerg F, Pedersen JK (2000) Random modulation techniques with fixed switching frequency for three-phase power converters. IEEE Trans Power Electron 15(4):753–761
16. Kirlin RL, Kwok S, Legowski S, Trzynadlowski AM (1994) Power spectra of a PWM inverter with randomized pulse position. IEEE Trans Power Electron 9(5):463–472
17. Li Q, Jiang D (2018) Variable switching frequency PWM strategy of two-level rectifier for DC-link voltage ripple control. IEEE Trans Power Electron 33(8):7193–7202
18. Jiang D, Li Q, Han X, Qu R (2016) Variable switching frequency PWM for torque ripple control of AC motors. In: Proceedings of the international conference on electrical machines and systems (ICEMS), pp 1–5
19. Huang J, Xiong R (2014) Study on modulating carrier frequency twice in SPWM single-phase inverter. IEEE Trans Power Electron 29(7):3384–3392
20. Holmes DG, Lipo TA, McGrath BP, Kong WY (2009) Optimized design of stationary frame three phase AC current regulators. IEEE Trans Power Electron 24(11):2417–2426
21. Li Q, Chen J, Jiang D (2020) Periodic variation in the effect of switching frequency on the harmonics of power electronic converters. Chinese Journal of Electrical Engineering 6(3):35–45
22. Li Q, Jiang D, Zhang Y, Liu Z (2020) The impact of VSFPWM on current control and a compensation method. IEEE Transactions on Power Electronics 36(3):3563–3572
23. Chen J, Jiang D, Sun W, Shen Z, Zhang Y (2020) A family of spread-spectrum modulation schemes based on distribution characteristics to reduce conducted EMI for power electronics converters. IEEE Transactions on Industry Applications 56(5):5142–5157
24. Chen J, Jiang D, Shen Z, Sun W, Fang Z (2019) Uniform distribution pulsewidth modulation strategy for three-phase converters to reduce conducted EMI and switching loss. IEEE Transactions on Industrial Electronics 67(8):6215–6226

Chapter 6
Advanced PWM Strategies for Complicated Topologies

In previous chapters, model predictive pulse width modulation (PWM) strategy is only applied to general two-level voltage source converters (VSCs). This chapter will discuss the complicated power electronics converters accompanied by corresponding modulation strategies. The advanced PWM strategies for three complicated topologies, including paralleled inverters, multilevel converters and current source converters (CSCs), are analyzed and developed to improve their performance one by one in the following parts.

6.1 Introduction to Complicated Topologies

In the previous chapters, the object of study is general three-phase two-level VSC, where there are only two switching states for each bridge leg. Therefore, the relationship between switching function and output voltage is readily accessible, and the relationship between switching function and output current can be deduced. According to these mathematical relationships, it is easy to develop modified PWM strategies to control the output voltage or current ripple of power converters, eventually improving the performance of the whole converters.

Although three-phase two-level converters are widely used to residential, commercial and industrial applications for simple structure, there are many limitations for this kind of converter yet. The freedom degree of output voltage is limited due to the binary switching states. At the same time, each power semiconductor must sustain the full DC side voltage and corresponding large current, and it is difficult to be applied to high-voltage high-power application for high stress. The high dv/dt of output voltage that jumps between positive bus and negative bus also results in huge voltage stress for load inductors and big current ripple at load side. For the applications of advanced PWM strategies, since the degree of freedom of two-level converters is limited, there is few chances to develop improved PWM methods for two-level topologies.

D. Jiang et al., *Advanced Pulse-Width-Modulation: With Freedom to Optimize Power Electronics Converters*, CPSS Power Electronics Series,
https://doi.org/10.1007/978-981-33-4385-6_6

In recent decades, the topologies of AC-DC or DC-AC power electronics converters have been developed rapidly based on the traditional two-level VSCs so that numerous new topologies with high-power, high-quality performance and more degrees of freedom have appeared, accompanied by kinds of modulation strategies. This chapter selects several typical complicated topologies to introduce, and the related modified PWM strategies will be developed in the following sections.

There are two main modified topologies in respect to regular three-phase two-level converters: one is paralleled structure as shown in Fig. 6.1a, and the other is series structure as shown in Fig. 6.1b. The paralleled structure can keep output voltage level unchanged, but increase power and control degrees of freedom through multiple converters in parallel, and distribute the load current through the addition in the number of bridge arms, which is suitable for large current applications. Similarly, the series structure can keep the number of bridge arms unchanged, increase power and control degrees of freedom through the addition in output voltage levels, and distribute the bus voltage through multiple power devices in series, which is suitable for high-voltage applications.

The basic form of paralleled structure is power electronics devices in parallel, which can increase the current capacity for the system without changing the topology structure of converter. It is necessary to pay more attention to keeping the synchronization of the turn-on and turn-off transient for the switches in parallel, and the current distributed in different devices should be balanced. However, there is no difference between the regular two-level converter and the converter with paralleled devices in terms of the control of converters. The increase of power devices fails to add the control degrees of freedom, and the output power quality has no space to improve, either. Therefore, it is worth studying the converters in parallel.

There are three solutions to make inverters in parallel as shown in Fig. 6.2. The first is the direct paralleled connection of loads as shown in Fig. 6.2a, which is realized through multiphase loads. Paralleled inverters are connected to each phase of loads to

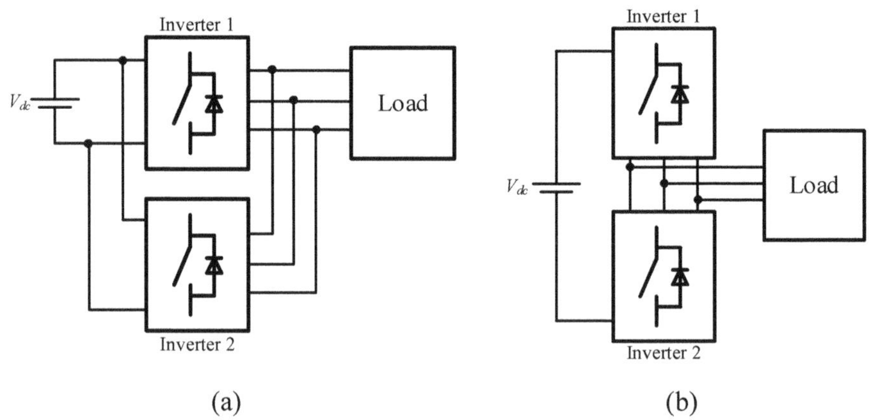

Fig. 6.1 The two main improved topologies: **a** paralleled structure, **b** series structure

(b)

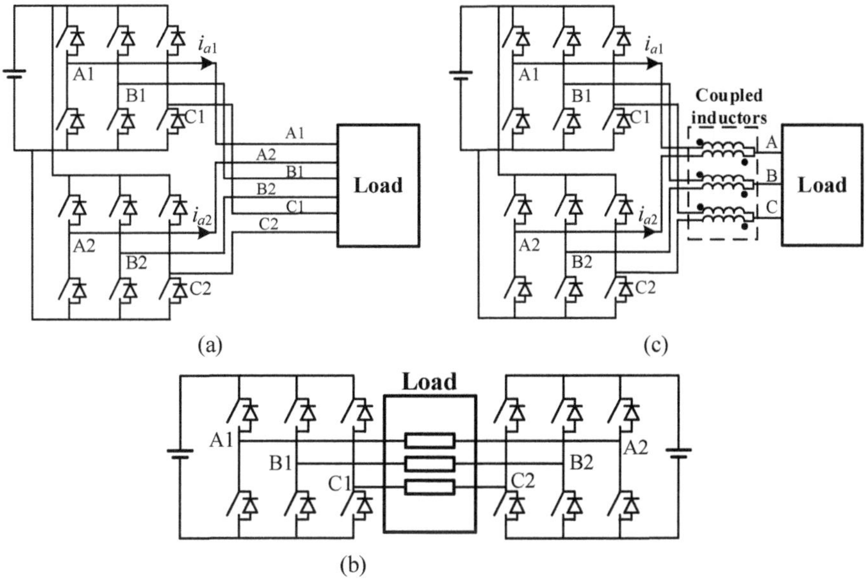

Fig. 6.2 Three typical structures for paralleled inverters: **a** multiphase load, **b** open-winding load, **c** inverters in parallel through coupled inductors for three-phase load

create the effect of load in parallel. The second is so-called open-winding as shown in Fig. 6.2b, where the loads in parallel are similar to the first one. However, the difference is that the structures still supply three-phase, and each winding is open to be supplied by a full H-bridge converter [1, 2]. The third utilizes coupled inductors to make inverters in parallel as shown in Fig. 6.2c, where the output poles of inverter 1# are connected with inverter 2# via three coupled inductors and the common terminals supply the three-phase loads [3, 4].

Among the three paralleled structures, there is a common issue that the traditional three-phase load such as motors must be modified to realize parallel connection for the first two topologies. This brings many challenges to the applications of regular three-phase load. However, the third paralleled structure is directly suitable for regular three-phase load, so this chapter mainly introduces the paralleled inverters whose structure is the same as the third solution as shown in Fig. 6.2c.

Analogous to paralleled structure, the simplest series structure is making power devices in series, which can increase the voltage level of the whole system without changing the topology of inverter. This method should pay attention to the synchronization of all devices in series during switching actions and the uniform blocking voltage distributed to each device. Nevertheless, there is no distinction between the topologies with devices in series and general two-level inverters in terms of control strategy. The addition of devices does not change the control freedom degree of system, and thus there is no space for improving the quality of output waveforms.

The study of the converters in series that are known as multilevel power electronics converters is more valuable.

In the past few decades, multilevel converters have derived many topologies and they have been successfully applied to high-voltage and high-power drive systems. Essentially, all topologies of multilevel converters achieve the multilevel of output pole voltage to improve the output power quality by means of the combination of switching states for devices. Based on the general voltage-source-based power electronics switching unit (i.e., single-bride module), the structures of multilevel converters develop a series of derivatives. Among them, the neutral point clamped (NPC) converters, flying capacitor (FC) converters and cascade of H-bridges are the three widely popular topologies for industrial applications.

Figure 6.3 takes three-level converters for example, which shows a typical NPC three-level inverter topology. For this structure, the positive bus of DC side connects with negative counterpart through two capacitors C_1 and C_2, so the neutral point (i.e., O) potential is equal to the half of DC voltage with reference to negative DC bus when inverters are operating on steady-state. Consequently, there are three kinds of voltage levels in DC side, that is, $+$, $-$ and 0. Each bridge leg consists of four bidirectional switches and two clamped diodes, and each bidirectional switch is composed of a controllable switching tube and an anti-paralleled diode. Taking phase-a for example, the output pole is connected to positive DC bus when S_{a1} and S_{a2} are on on-state; the output pole is connected to neutral point when S_{a2} and S_{a3} are on on-state, and the on/off-state of the clamped diodes is determined by the direction of current; the output pole is connected to negative DC bus when S_{a3} and S_{a4} are on on-state. Then the output pole voltage can switch among three kinds of voltage levels via PWM methods, thus controlling the output current.

Figure 6.3 shows the general NPC three-level inverter topology, and the converter topology to realize more voltage levels can be derived based on this structure. In addition, the voltage-clamped diode can be replaced by a controllable switching tube and an anti-paralleled diode to add the control freedom degree, which can effectively

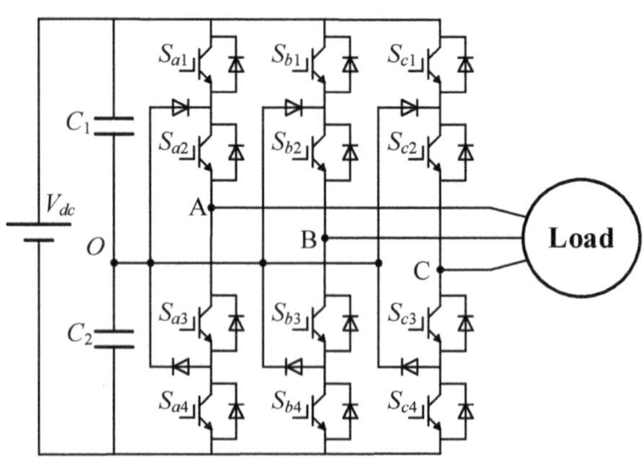

Fig. 6.3 The NPC three-level inverter

balance the switching loss between different devices. This topology is called as active neutral point clamped (ANPC) multilevel converters.

Another derivative of NPC multilevel converter is the T-type three-level converter, as shown in Fig. 6.4 [5], which is based on the regular two-level structure. However, the output pole is directly connected to neutral point of the DC bus via a bidirectional switch. Then, each output pole is connected to positive DC bus, neutral point and negative DC bus through the switches, resulting in three kinds of output voltage level through the combination of switches. Compared with the diode-clamped three-level converter, the concept of T-type three-level converter is more straightforward.

When the active switches of two-level bridge leg for the T-type are replaced by diodes, the bidirectional energy flow capability of the converter degrades into a unidirectional flow. However, that topology can still operate as a three-level rectifier, controlling current waveforms and power factor, which is the famous Vienna-type rectifier [6], as shown in Fig. 6.5. Both T-type converters and Vienna-type rectifiers can be classified as clamped multilevel converters.

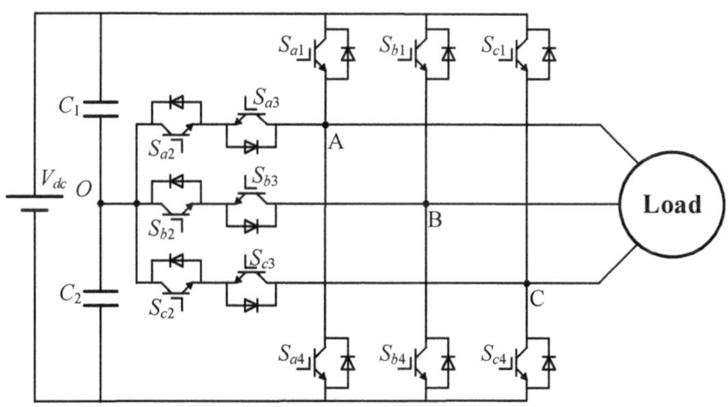

Fig. 6.4 The T-type three-level inverter

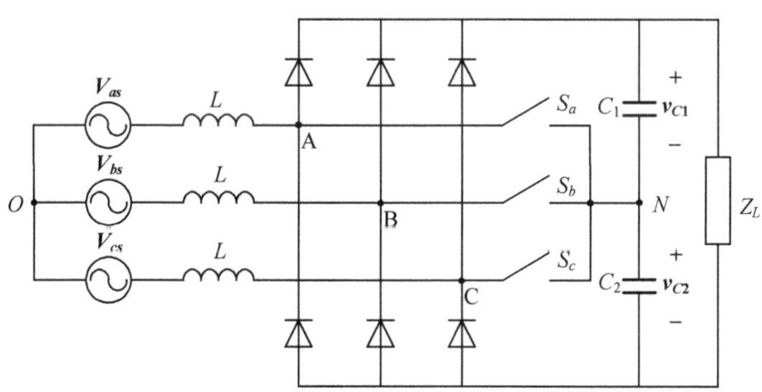

Fig. 6.5 Vienna-type three-level rectifier structure

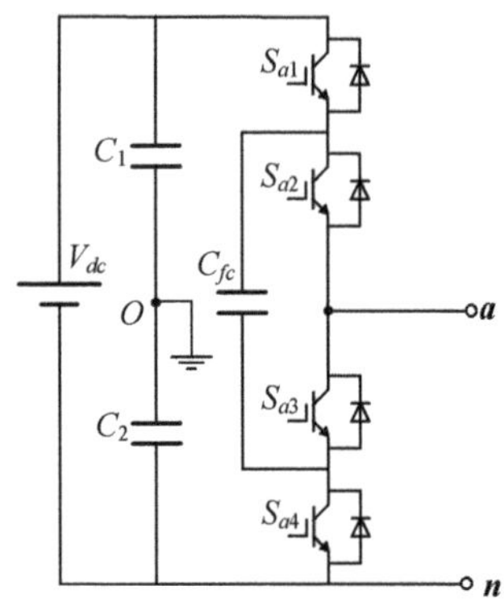

Fig. 6.6 A single-bridge-arm structure for flying capacitor three-level converter

Corresponding to the NPC multilevel converter, the capacitor-clamped multilevel power electronics converter is another kind of multilevel converter with the flying capacitor (FC) multilevel converter as representative. Figure 6.6 shows the flying capacitor structure for three-level converters. A separate capacitor is used to fly between S_{a2} and S_{a3} instead of a diode or a switch clamped to the neutral-point potential. The voltage of each capacitor is equal, so the output voltage can be switched between +, − and zero by switch combination. Since the number of clamped capacitors increases, the output voltage level can also be increased simultaneously.

Another popular multilevel topology is the H-bridge cascade structure, as shown in Fig. 6.7. The output voltage of each H-bridge has three modes: positive, negative and zero (freewheeling). Thus, the output voltage can be obtained by adding the pole voltage of each H-bridge to achieve multilevel output. This structure can be modularized and widely used in high-voltage and large-power variable-speed drive applications. Its disadvantage is that the DC side of each H-bridge converter requires

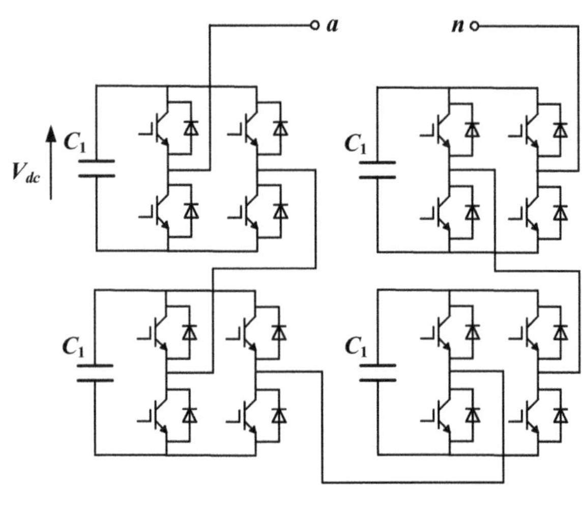

Fig. 6.7 H-bridge cascaded multilevel structure

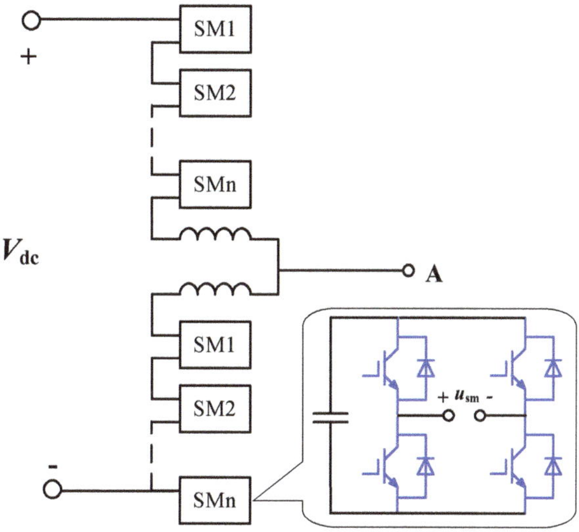

Fig. 6.8 Modular multilevel converter (MMC) structure

independent power supply control, so the requirements on the rectification side are complex, and the economical cost and bulk weight are also concerned. In recent years, a new generation of modular multilevel converter (MMC) [7] has been developed based on the H-bridge cascaded multilevel converter, as shown in Fig. 6.8. The topology does not require the independent power supply for each DC bus, but the voltage of submodule needs to be balanced. At present, this structure has begun to be applied in the fields of high-voltage direct current (HVDC) transmission and high-power drives.

The multilevel structure has more transformations than paralleled structure. This chapter is mainly aimed at introducing the modulation strategy of the most widely used three-level power electronics converter.

For both paralleled converters and multilevel converters, it is necessary to adopt a certain PWM method to generate output voltage effectively. Similar to the traditional three-phase two-level converter, the PWM strategies can also be divided into space vector PWM (SVPWM) and carrier-based PWM (CBPWM).

The SVPWM of a paralleled converter is similar to that of a traditional two-level converter. Each paralleled converter has eight switching vectors, and the six nonzero vectors form a regular hexagon, as shown in Fig. 6.9. Since each inverter can independently synthesize reference vectors V_1* and V_2*, the synthesis between V_1* and V_2* is the voltage vector output to the load. It is also worth noting that if the combined output voltages of the two converters are inconsistent, a voltage difference is generated between the two converters to generate a circulating current. The circulating current problem is a very noteworthy problem in paralleled converters, which will be discussed in detail in the subsequent chapters.

For the paralleled converter configuration in Fig. 6.2c, the contribution of the two converters to the load is consistent, so the space vectors of the two converters are identical. For the other two forms of parallel connection of the converter—multiphase

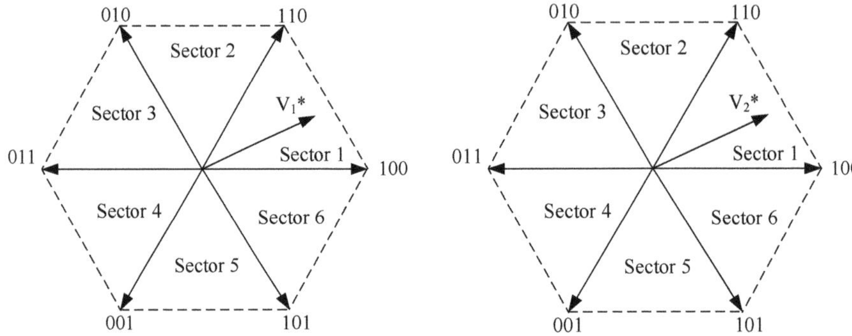

Fig. 6.9 Space vector synthesis distributed in two inverters

converter (Fig. 6.2a) and open-winding double-ended supply converter (Fig. 6.2b)—the converter's contribution to the load is no longer consistent, so the SVPWM is different from that of traditional three-phase converters. Taking a multiphase inverter as an example, since the spatial position difference between the phases is no longer 120°, the corresponding space vector will be different. Moreover, due to the addition of the degree of freedom, the selection of the synthesis reference voltage of the multiphase converter is also increasing. Figure 6.10 shows the space vector of a five-phase converter [8], which is divided into: fundamental plane shown in Fig. 6.10a and the third-harmonic plane shown in Fig. 6.10b. The same switching vector has different effects in the fundamental plane and the third-harmonic plane. For the synthesis of vector, the fundamental plane and the third-harmonic plane need to be solved simultaneously to obtain a unique solution of the duty cycle. The PWM

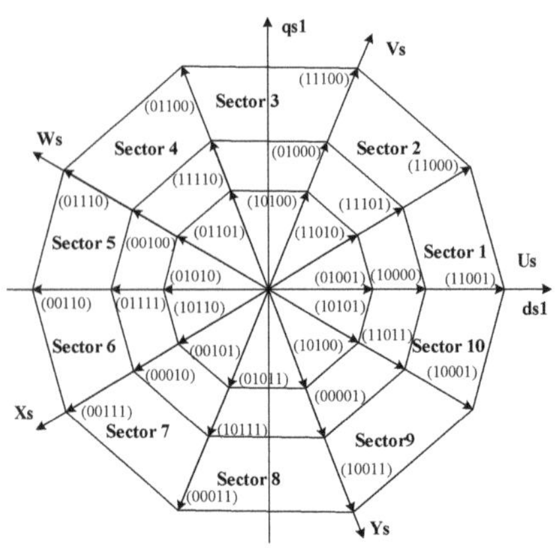

Fig. 6.10 Space vector of five-phase voltage converter: **a** fundamental plane, **b** third-harmonic plane [8]

strategies for multiphase converters and open-winding converters are not described in detail in this chapter.

The SVPWM of the paralleled converter also has a corresponding CBPWM strategy. The relationship between the SVPWM and the CBPWM is consistent with counterpart introduced in the previous chapter. However, there are more degrees of freedom available for paralleled converters, which will be covered in Sect. 6.2.

The number of space vectors of the multilevel converter increases exponentially compared with that of the traditional two-level converter. Taking a three-level converter as an example, the number of output voltage level for each phase increases from 2 to 3, so the three-phase three-level converter has a total of $3^3 = 27$ switching vectors. According to the output voltage of each bridge leg being positive (p), zero (o) and negative (n), the 27 switch vectors are placed on the three-phase space plane, as shown in Fig. 6.11. These vectors can be divided into four types: large vector, medium vector, small vector and zero vector according to the equivalent voltage amplitude. Some of the vectors are coincident in the space vector: for example, poo and onn are the same in terms of output, and the output effects of the three zero vectors ppp, ooo and nnn are also the same.

Different from the space vector of the three-phase two-level converter that divides the plane into six sectors, the space vectors of the three-level converter divide the plane into 24 sectors. The basic principle of vector synthesis is to select the three adjacent vectors of the sector in which the reference voltage vector is located to

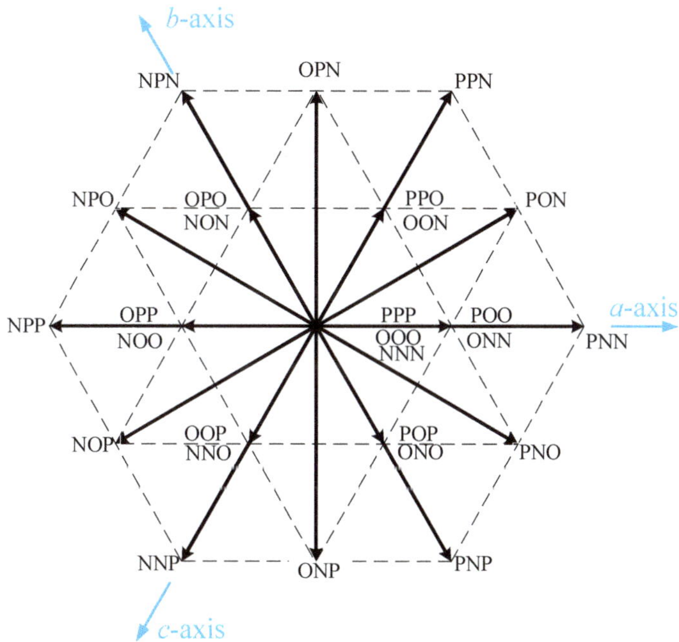

Fig. 6.11 Space vector of three-phase three-level converter

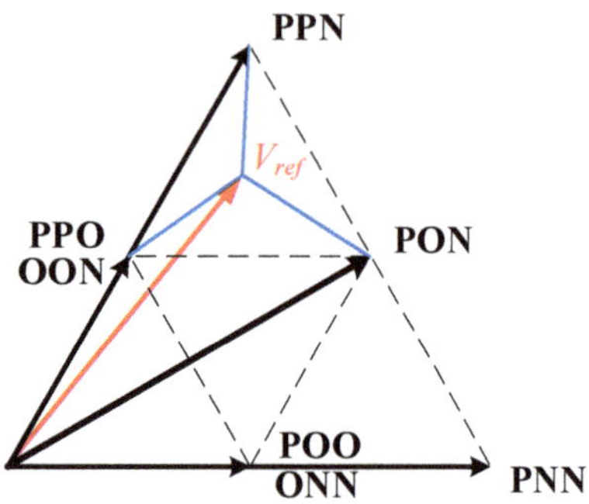

Fig. 6.12 Three-level space vector synthesis

achieve synthesis. Figure 6.12 shows a typical example: the reference vector V_{ref} is located in a small sector and is synthesized by three adjacent vectors ppn, pon and ppo/oon according to volt-seconds balance principle. Consequently, the difference between the output voltage and the reference voltage is the smallest, so the output current ripple is also optimal.

Similar to the two-level converter, the SVPWM of the three-level converter also has a corresponding CBPWM method. Figure 6.13 shows the carrier comparison implementation of the SVPWM for three-level converter. The three-phase modulation functions v_a, v_b and v_c are compared with the (upper and lower) two sets of triangular carriers. The upper carrier is between 0 and 1, and the lower carrier is between -1 and 0. When the modulation function is in the positive half cycle, the corresponding output phase voltage is determined to be a positive voltage or a zero voltage compared with the upper carrier; when the modulation function is in the negative half cycle, the corresponding desired output voltage is determined to be zero or a negative voltage compared with the lower carrier. Such an output phase voltage exhibits a three-level characteristic, and the output line voltage has a five-level structure. The same method in Chap. 2 can be used to prove the inherent consistency of CBPWM and SVPWM. The modulation function can also implement different PWM methods by adding various common-mode components into three-phase sinusoidal modulation waves.

The above presentation is mainly focused on the modified topology of general voltage-source converters by introducing paralleled and series structures. In the topology family of power electronics converters, there is also a current source converter corresponding to the voltage source converter. Figure 3.4 in Chap. 3 has briefly introduced the topology of the current source converter, that is, replacing the capacitor with a DC energy storage inductor so that the DC side forms a current source characteristic, and each bridge-arm is composed of a current unidirectional switch, which makes the output a rectangular pulse width current. This topology is also used as an example of a complicated topology in Sect. 6.4 of this chapter to introduce its PWM strategy.

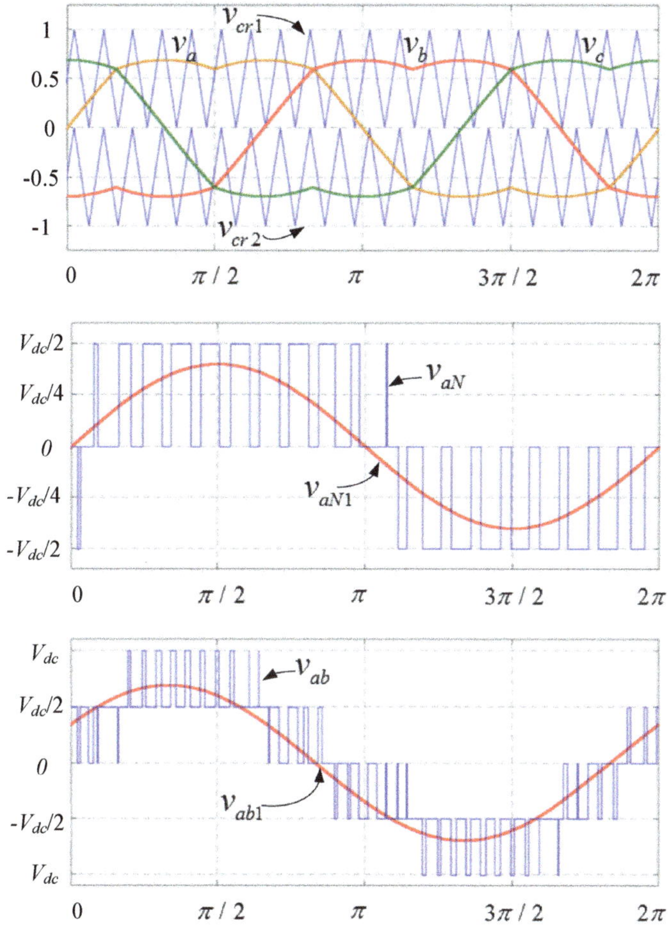

Fig. 6.13 Carrier comparison for three-level SVPWM

6.2 PWM Strategies for Paralleled Inverters [19, 28]

As described in Sect. 6.1, the modulation of paralleled converters can be implemented by synthesizing reference vectors separately or by means of CBPWM. The carrier phase-shift PWM described in this section is actually a way to improve the performance of PWM by using carrier comparison. The next chapter introduces the zero common-mode PWM of the paralleled inverter, which uses the idea of space vector synthesis. Since common-mode suppression is a separate chapter, the related content is not covered in this section.

The carrier phase-shift PWM described in this section is based on the paralleled structure shown in Fig. 6.2c. Then, this section will also show how to apply carrier phase-shift PWM to multisegment motors, which achieve paralleled effects through the motor structure. Finally, the circulating current issue is discussed in the last subsection.

6.2.1 Carrier Phase-Shift PWM for Paralleled Inverters

Figure 6.14 shows the most basic structure of carrier phase-shift [9]. For two paralleled power electronics converters, the corresponding carriers have the spacing of 2π in one switching cycle, and the carrier phase-shift diagram is shown in Fig. 6.14a. When the carriers of the two converters shift to angle κ, the corresponding switching time will also have the corresponding delay. There will be a difference between the output PWM voltages of the two converters. This difference produces two kinds of effects on the converters: after the coupled inductor, the common terminal output voltage will produce a cancellation effect based on the original PWM voltage, and the harmonics will be changed, as the output current i_{A} shown in Fig. 6.14b. The voltage difference between the two converters also creates a circulating current through the coupled inductor, flowing between the two converters, that is, the difference between i_{A1} and i_{A2} in Fig. 6.14b.

The analysis method described in Chap. 2 can perform a double Fourier transformation on the output voltage of the carrier phase-shift PWM. Assuming that the fundamental frequency is ω_0 and the carrier frequency is ω_{c}, the output voltage V_{A1N} in Fig. 6.14b is expressed as (6.1), which represents the nth side-band harmonic in the group of harmonics that are located around the mth carrier harmonic. C_{mn} is the harmonic amplitude. θ_c and θ_0 refer to the phase of carrier and fundamental component, respectively. When the carrier of the second converter shifts phase angle κ, the expression of its output voltage V_{A2N} can be expressed by (6.2).

$$V_{A1N(m,n)}(t) = C_{mn}\cos[(m\omega_c + n\omega_0)t + m\theta_c + n\theta_0] \tag{6.1}$$

$$V_{A2N(m,n)}(t) = C_{mn}\cos[(m\omega_c + n\omega_0)t + m(\theta_c + \kappa) + n\theta_0] \tag{6.2}$$

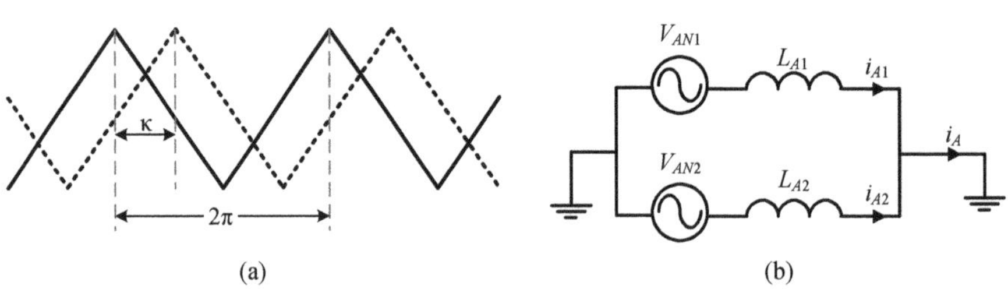

Fig. 6.14 Basic principle of carrier phase-shift of paralleled converter: **a** schematic diagram of carrier phase-shift, **b** simplified equivalent circuit

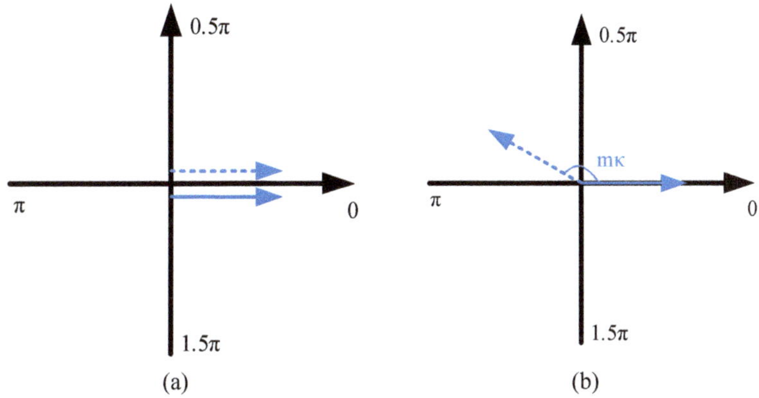

Fig. 6.15 Vector synthesis of the *m*th carrier harmonic of the output current for the paralleled converter: **a** without carrier phase-shift, **b** with carrier phase-shift angle κ [9]

Figure 6.15 shows the principle of the synthesis of the output pole voltage V_{AN} at the *m*th switching frequency. Figure 6.15a shows the case without carrier phase-shift: the corresponding vectors of the two voltage components are with the same phase, and the synthesized vector is equal to the algebraic superposition. Therefore, the corresponding harmonics of the output is the largest. Figure 6.15b shows the synthesis principle with phase-shift angle κ, where the vector of the harmonic component of the two voltage sources has a phase difference of $m\kappa$. The process is vector synthesis, and the output harmonics can be reduced compared with the former case.

Therefore, the principle of Fig. 6.15 can be used to design a suitable phase-shift angle κ to eliminate the corresponding harmonic component. The simplest example is to set $\kappa = \pi$, so that the first carrier harmonic is vector synthesis of two vectors with identical length and 180° phase difference, whose synthesis result is theoretically equal to zero. In addition, other odd-order carrier harmonics can also be eliminated. However, the two vectors will coincide at the even-order carrier harmonics, which is similar to the case without carrier phase-shift. Consequently, the output current of the paralleled converter will eliminate the odd-order carrier harmonics, greatly improving the quality of the output current. Similarly, the phase-shift angle κ can be set to $\pi/2$ to suppress the even-order carrier harmonics, and then the harmonic vector in Fig. 6.15b achieves cancellation when *m* is an even number. Therefore, different phase-shift angle κ can be set to meet the different requirements of harmonic elimination. The proportion of odd-order carrier harmonics and even-order carrier harmonics in the current harmonics varies with the modulation index. Figure 6.16 shows the THD attenuation rate with modulation index variation for different phase-shift angles. In the case where the modulation index is high, the phase-shift angle of $\kappa = \pi$ is used, and the harmonic achieves great improvement; with the phase-shift angle of $\kappa = \pi/2$, the improvement of harmonic is less. Therefore, there is an optimal phase-shift angle κ at different modulation indexes to optimize the output harmonics [9].

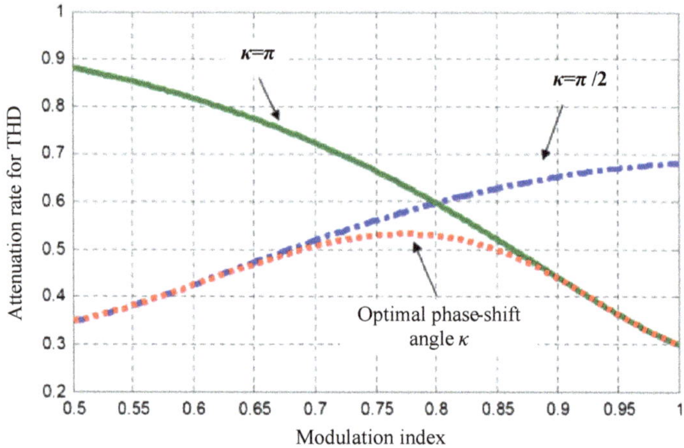

Fig. 6.16 THD attenuation rate with modulation index variation for different phase-shift angles κ [9]

In addition, due to the different characteristics of the load, the effect of carrier phase-shift strategy will be different. Especially for the circuit where there is a resonance point in the conducted loop, it is necessary to eliminate the harmonics near the resonance frequency. Therefore, the phase-shift angle will also be designed according to these requirements [10].

The experimental results can clearly reflect the harmonics elimination effect of carrier phase-shift strategy. Figure 6.7 shows the waveform of the AC side output current of the paralleled converters. Figure 6.17a shows the result without carrier phase-shift strategy, where there are abundant harmonics in the three-phase current, as shown in Fig. 6.18a. After adopting the carrier phase-shift strategy with $\kappa = \pi$, the ripple in the current shown in Fig. 6.17b is greatly reduced. The THD drops

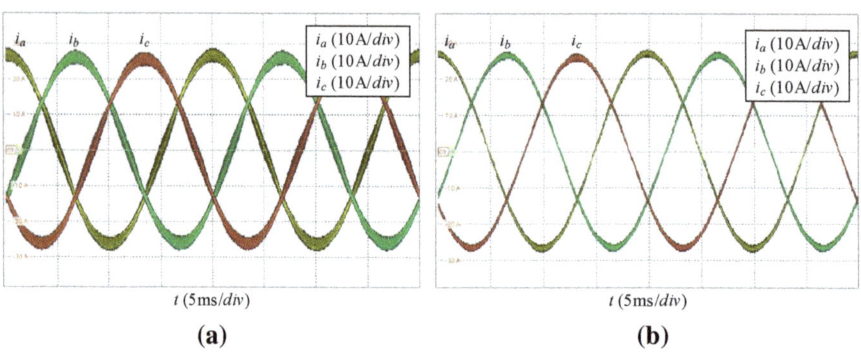

Fig. 6.17 Experimental results of paralleled converter output current: **a** without carrier phase-shift, **b** with carrier phase-shift angle κ

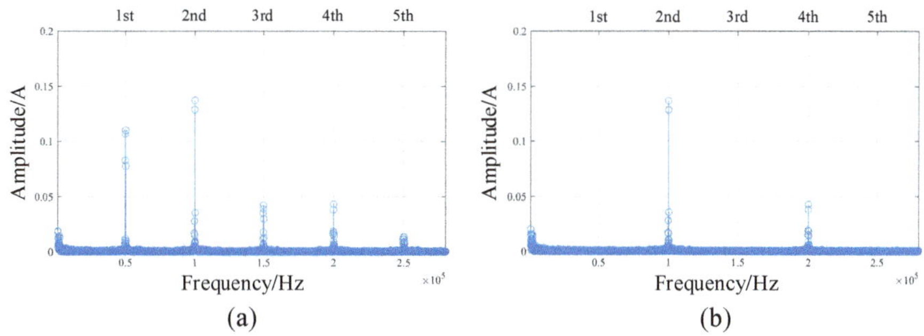

Fig. 6.18 Spectrum analysis for the output current of paralleled converter: **a** without carrier phase-shift, **b** with carrier phase-shift angle κ

from 2.28% to 1.47% but the fundamental current remains unchanged. As shown in Fig. 6.18b, the main odd-order carrier harmonics are suppressed.

It can be seen from the above analysis that the carrier phase-shift method can effectively suppress the specific carrier harmonics and electromagnetic interference, which is beneficial to the design of the system filter. However, to suppress the circulating current generated by the voltage difference between the two inverters, it is also necessary to increase the value of inductors, that is, the coupled inductors marked in Fig. 6.2c. From a mathematical point of view, the voltage difference obtained by subtracting (6.1) from (6.2) requires impedance suppression, otherwise the circulating current will be infinite. Fortunately, since the reference voltages of the two inverters are identical and only the phase of the carrier is transformed with the carrier phase-shift PWM scheme, so the difference between the voltages of the two inverters will be balanced in each switching cycle. In other words, the difference in voltage only presents at the frequency that is the integer times of switching frequency. Or, the difference between the instantaneous output voltages of the two inverters only produces the harmonics that are equal to or higher than the order of the switching frequency. Therefore, the coupled inductor only needs to suppress the circulating current with carrier harmonics, and the size and weight can be designed to be small. Figure 6.19 shows a typical structure of a coupled inductor, which consists of two U-shaped cores and is realized by reverse-phase winding. In this case, the flux generated by the circulating current between the two inverters will flow in the main loop of the core as shown in Fig. 6.19, and the flux independently generated by the respective currents will conduct through the leakage flux of Fig. 6.19. Therefore, the main inductance of the coupled inductor acts as a circulating current suppression inductor, and the leakage inductance has a smaller effect on the load. The methods of paralleled inverters for common-mode noise suppression and the corresponding coupled inductors are also described in Chap. 7.

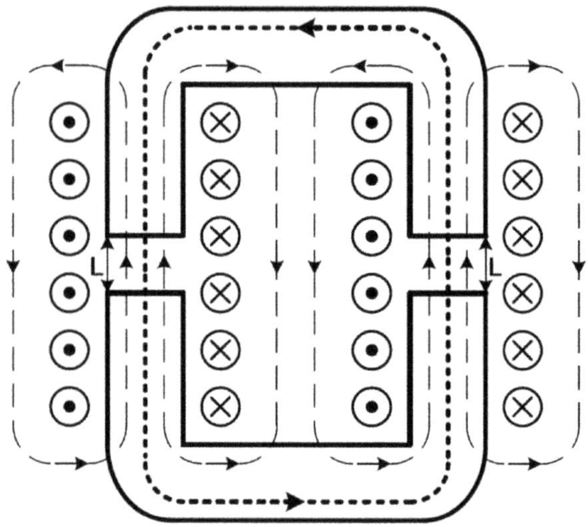

Fig. 6.19 The typical structure of coupled inductor [9]

6.2.2 *Carrier Phase-Shift PWM for Multisegment Motor*

All the above descriptions are the paralleled inverters supplied typical three-phase load with carrier phase-shift PWM scheme, which achieves the improvement of output harmonics and EMI by carrier phase-shift PWM to eliminate some harmonic components. This idea can also be applied to unconventional loads, that is, the load itself is in paralleled connection to improve the harmonic characteristics without the coupled inductors. Figure 6.20 shows an unconventional load: a multisegment three-phase motor. There are N three-phase modules in the motor, which are driven by N sets of three-phase windings through N drive circuits. This is actually equivalent to make the N inverters in parallel by the motor itself, and finally generate the accumulation of power and torque.

If the carrier phase-shift modulation scheme is applied to the multisegment motor with N inverters, the cancellation effect of the fluctuation of torque at carrier harmonics can be achieved. In fact, the main harmonic component of the electromagnetic torque of the AC motor is the second-order harmonic of the switching frequency, so the main cancellation target eliminates the torque ripple that is the second-order carrier harmonic. Figure 6.21 shows the carrier phase-shift method of the N drive inverters. Starting from the first carrier, there is a time difference of 1/2 N switching period between two adjacent carriers, that is, a phase difference of π/N, and then N sets of carrier are obtained. As a result, the current harmonics generated by the corresponding inverters of each two adjacent carriers will differ by 1/2 N switching period in the time domain. Then the second-order carrier harmonic component also has a phase difference of $2\pi/N$. Since each subtorque is directly generated by the current of corresponding inverter, the torque generated by the two adjacent modules also has a phase difference of $2\pi/N$ at the second-order carrier harmonic.

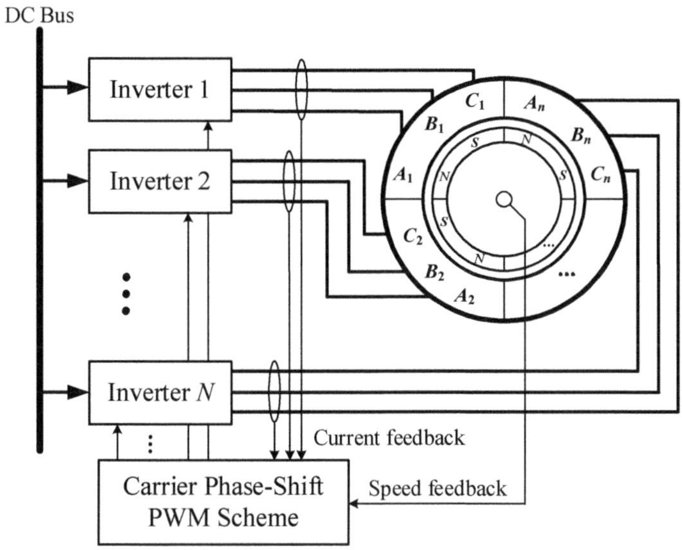

Fig. 6.20 The multisegment paralleled drive motor structure diagram

Fig. 6.21 The principle of carrier phase-shift scheme for multisegment motor

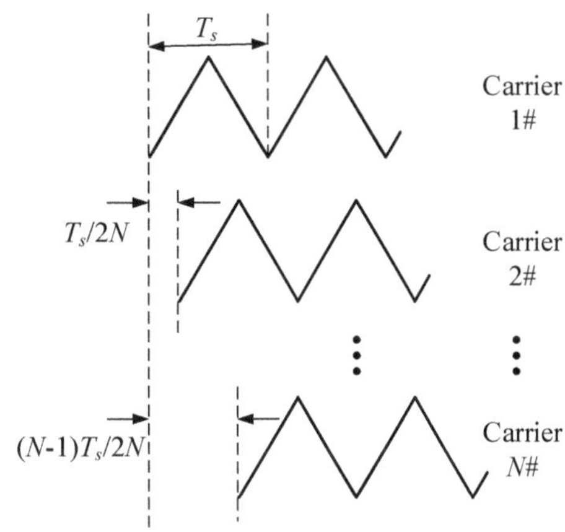

Figure 6.22 shows the principle of torque synthesis at second-order switching frequency component. The torque component of each module is with the same phase under the condition without carrier phase shift, so the synthetic torque is equal to the result of the algebraic addition, which is approximately N times of the torque of each module, as shown in Fig. 6.2a. When the carrier phase-shift scheme shown in Fig. 6.21 is used, the torque components generated by the adjacent modules at the second-order switching frequency is $2\pi/N$, and the combined torque will be theoretically equal to zero, as shown in Fig. 6.22b. Therefore, the torque ripple of motor will be significantly improved under the carrier phase-shift PWM.

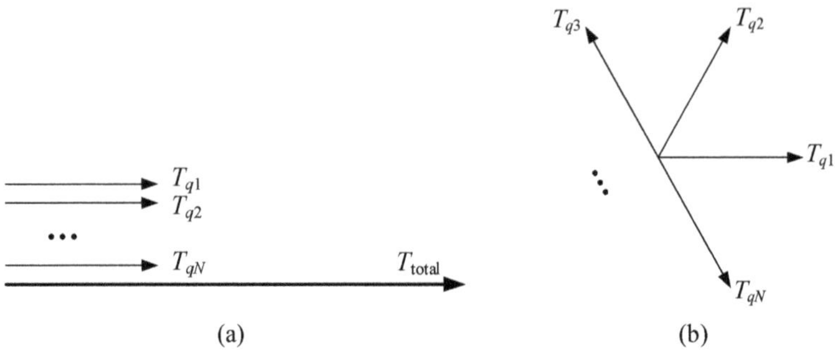

Fig. 6.22 Principle of torque ripple synthesis: **a** without carrier phase-shift, **b** with carrier phase-shift

6.2.3 An Example: Torque Ripple Reduction for Two-Segment PMSM

In Sect. 6.2.2, carrier phase-shift PWM, or interleaved PWM, has been applied in paralleled inverters with coupled inductors to three-phase load. A further idea is to parallel the inverters directly in the load motor and save the coupled inductors. This section presents a two-segment motor system powered by two inverters with carrier phase-shift PWM to fulfill the purpose of low-torque ripple and low vibration noise. The stator windings of this machine are divided into two segments, with each segment fed by one set of VSI, as illustrated in Fig. 6.23. Clearly, each inverter is controlled independently to provide three-phase current for the two segments, respectively. The total power and torque are shared by the two segments of the stator windings, too.

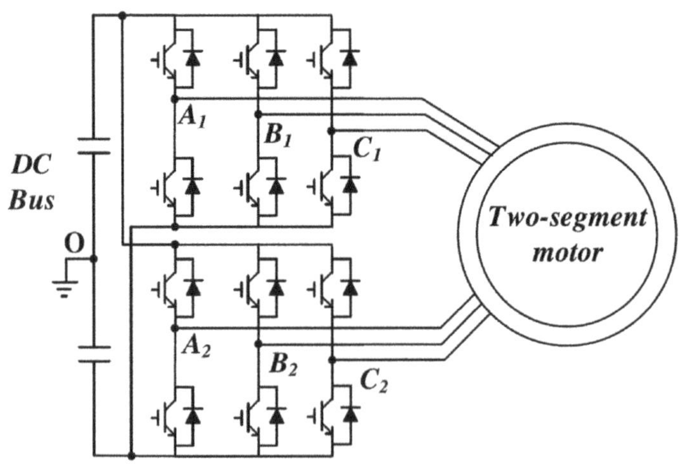

Fig. 6.23 Schematic of two-segment PMSM drive

Like the conventional paralleled drive system, the output phase voltage of two-segment motor drive system is expressed as (6.1) and (6.2) according to the double-integral Fourier analysis. Correspondingly, the analytical formula for three-phase output current can be derived as (6.3), (6.4), and (6.5), where C'_{mn} is the current harmonic amplitude, φ is the power factor angle. Since the phase voltage contains components of carrier-wave frequency harmonics, and, in turn, these frequency harmonics will result in three-phase current harmonics. In fact, these $(m\omega_c + n\omega_0)$ order current harmonics contribute greatly to high-frequency vibration noise. Through current harmonic component analysis in PMSM drive with SVPWM modulation strategy, it can be concluded that the main phase current harmonics concentrate on the first and the second switching frequency and their side-bands.

$$i_{A1(m,n)}(t) = C'_{mn}\cos[(m\omega_c + n\omega_0)t + m\theta_c + n\theta_0 - \varphi] \tag{6.3}$$

$$i_{B1(m,n)}(t) = C'_{mn}\cos\left[(m\omega_c + n\omega_0)t + m\theta_c + n\left(\theta_0 - 2\pi/3\right) - \varphi\right] \tag{6.4}$$

$$i_{C1(m,n)}(t) = C'_{mn}\cos\left[(m\omega_c + n\omega_0)t + m\theta_c + n\left(\theta_0 + 2\pi/3\right) - \varphi\right] \tag{6.5}$$

With the linear model of PMSM, the high-frequency torque ripple expression can be determined. First, the three-phase current harmonics can be turned into rotor reference frame by Park transformation (6.6), where θ_e is the rotor position. Then, the torque ripple in the surface-mounted PMSM is deducted as (6.7), where ψ_f is the flux linkage produced by permanent magnets.

$$\begin{pmatrix} i_{d(m,n)}(t) \\ i_{q(m,n)}(t) \end{pmatrix} = \frac{2}{3}\begin{pmatrix} \cos(\theta_e) & \cos\left(\theta_e - \frac{2\pi}{3}\right) & \cos\left(\theta_e + \frac{2\pi}{3}\right) \\ -\sin(\theta_e) & -\sin\left(\theta_e - \frac{2\pi}{3}\right) & -\sin\left(\theta_e + \frac{2\pi}{3}\right) \end{pmatrix} \cdot \begin{pmatrix} i_{A(m,n)}(t) \\ i_{B(m,n)}(t) \\ i_{C(m,n)}(t) \end{pmatrix} \tag{6.6}$$

$$T_e = \frac{3}{2}p\psi_f i_q = \frac{3}{2}p\psi_f \cdot C'_{mn}\sin[(m\omega_c + n\omega_0)t + m\theta_c + n\theta_0 - \phi \pm \theta_e] \tag{6.7}$$

The amplitude of switching frequency current ripple is generally dominated by both modulation index and torque angle, which are further determined by motor speed and load condition. The curved surfaces of the q-axis harmonic current versus motor speed and load, calculated from the double-integral Fourier analysis, are illustrated in Fig. 6.24, along with the torque ripple illustrated in Fig. 6.25. It can be seen that both the main component of q-axis current and the torque ripple harmonics in rotor frame concentrate around the second-order switching frequency harmonics and its side-bands. Therefore, the main target of torque ripple reduction is second-order switching frequency harmonics.

For the two-segment motor mentioned before, when control instructions for the two inverters are the same, the terminal phase current for the two groups of windings will also be the same, including the fundamental and switching harmonics. In

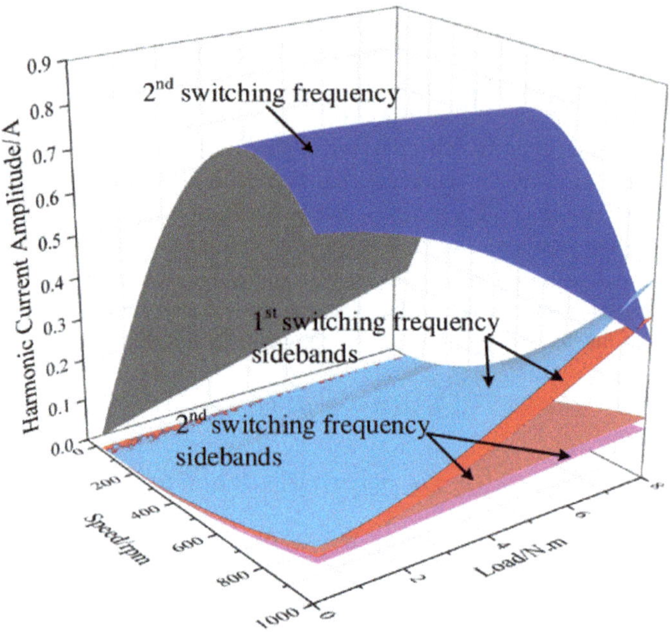

Fig. 6.24 Harmonic current calculation result in the full operation range

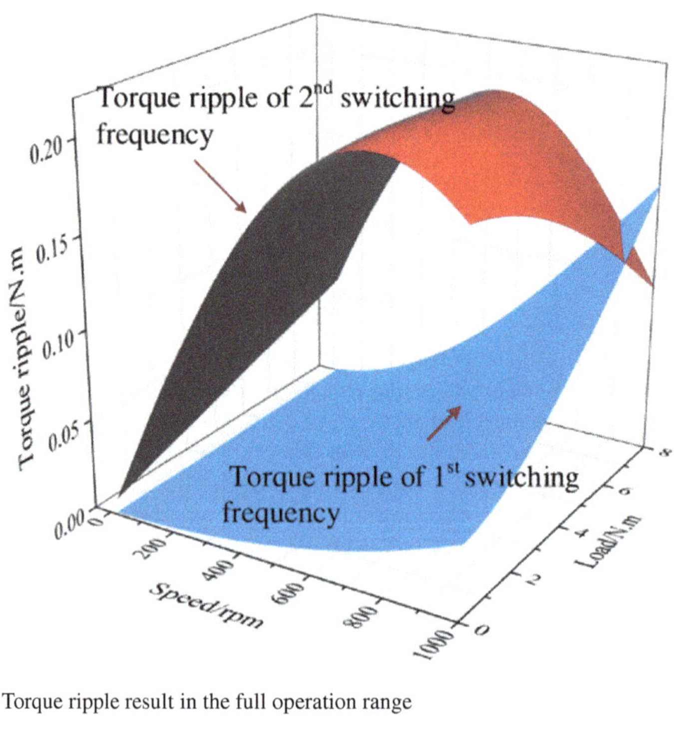

Fig. 6.25 Torque ripple result in the full operation range

Fig. 6.26 Carrier-wave phase-shift

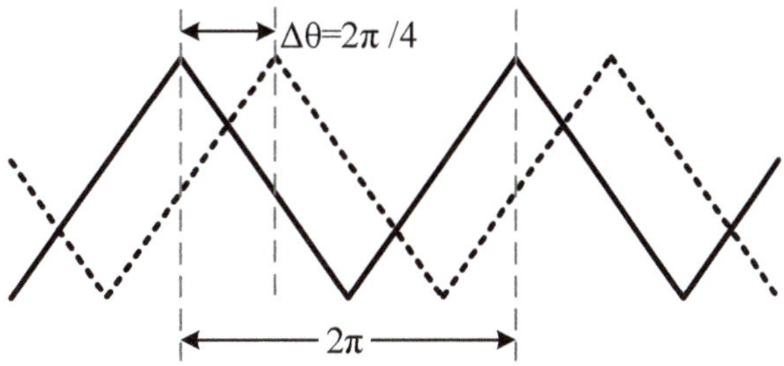

that case, the torque harmonics generated in two segments will be the same and the total torque ripple is the sum of these two parts which are added as a scalar function, similar to conventional PMSM. However, inspired by interleaving method in paralleled inverters, if the carrier phase of one of the two carriers is shifted by an angle $\Delta\theta = 2\pi/4$, as shown in Fig. 6.26, the corresponding phase output voltage and current harmonic of this inverter will have a phase-shift $m\cdot\Delta\theta$ in the initial phase term compared with the nonshift one. The expressions of phase voltage, current and torque are rewritten from (6.8) to (6.11). As a result, the amplitudes of second-order switching frequency torque harmonics are generated by two segments, respectively, with the value remaining the same but phase being shifted with π exactly, which is shown in Fig. 6.27. By vector summation (6.11), the second-order switching frequency harmonic and its side-bands in the total torque ripple will be eliminated completely and the other order harmonics will also attenuate simultaneously, thus greatly smoothing the output torque. Similarly, the radial magnetic force will also have this kind of cancellation of harmonics and the vibration together with the reduction of torque harmonics will be improved.

$$v'_{A(m,n)}(t) = C_{mn}\cos[(m\omega_c + n\omega_0)t + m(\theta_c + \Delta\theta) + n\theta_0] \tag{6.8}$$

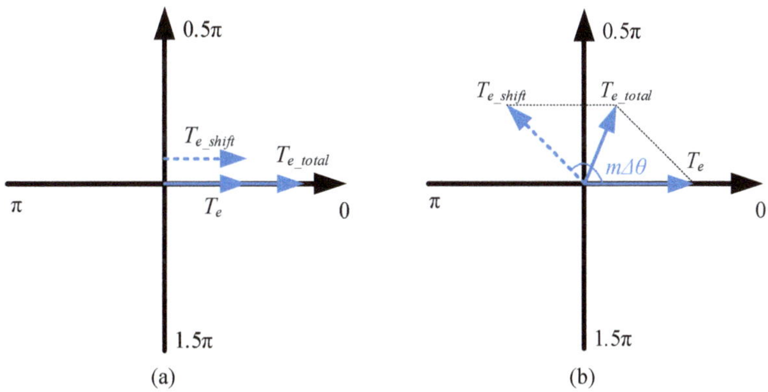

Fig. 6.27 Total torque harmonics (second order switching frequency): **a** without carrier phase shift, **b** with carrier phase-shift

$$i_{A(m,n)}(t) = C'_{mn}\cos[(m\omega_c + n\omega_0)t + m(\theta_c + \Delta\theta) + n\theta_0 - \varphi] \tag{6.9}$$

$$T_{e\_shift} = \frac{3}{2}p\psi_f \cdot C'_{mn}\sin[(m\omega_c + n\omega_0)t + m(\theta_c + \Delta\theta) + n\theta_0 - \varphi \pm \theta_e] \tag{6.10}$$

$$T_{e\_\text{total}} = T_{e\_\text{shift}} + T_e \tag{6.11}$$

It should be noted that for the major second-order harmonic torque ripple cancellation, carrier-wave phase has been shifted with $0.25T_s$, as shown in Fig. 6.28. The switching signal can be obtained through the comparison between the duty cycle and the carrier. Then these phase-shifted switching signals are given to two inverters to drive the motor.

Figure 6.29 shows the phase current between two units, including the full-scale and enlarged details in a few switching cycles. In Fig. 6.29a, when carrier phase has not been shifted, current ripples in two units are nearly the same. But in Fig. 6.29b, when carrier phase has been shifted, current ripples in two units are close to each other, validating the assumption that the two units share the same current and average torque. With the enlarged details, it can be seen that current harmonics phases have been shifted by 1/4 of the switching cycle. Because the major component of the current harmonics is the second-order harmonic of switching frequency, the current ripples in two units are close to opposite values.

Contrast of electromagnetic torque ripple before and after carrier phase-shift is illustrated in Fig. 6.30. It is clear that the average torques before and after carrier

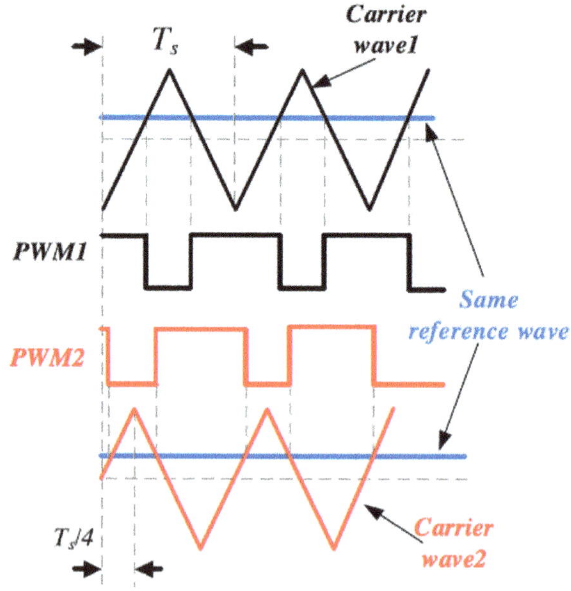

Fig. 6.28 Carrier-wave phase-shift with $T_s/4$

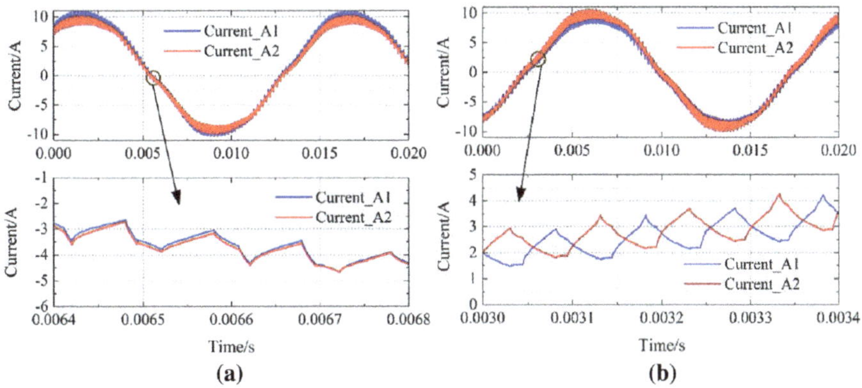

Fig. 6.29 Contrast of current ripple between two units: **a** without carrier phase-shift, **b** with carrier phase-shift

Fig. 6.30 Comparison of electromagnetic torque in time domain

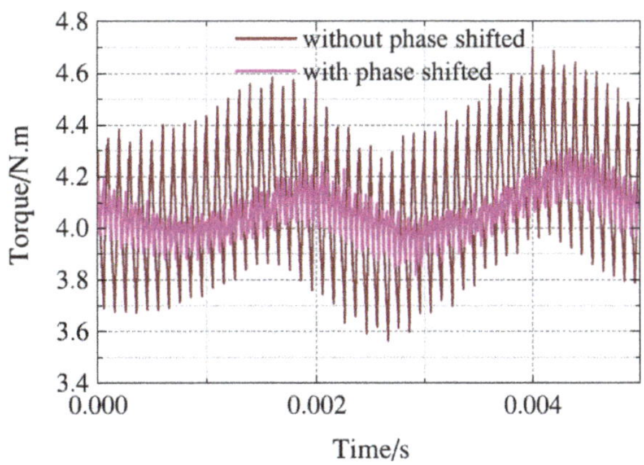

phase-shift are nearly the same but the high-frequency torque ripple after carrier phase-shift is significantly reduced by more than 20% of the case before carrier phase-shift. To further prove the effectiveness of the carrier phase-shift method, EMI test of the DC-bus current has been done with Rohde & Schwarz EMI analyzer. The test result is illustrated in Fig. 6.31, which shows the noise of switching has been greatly attenuated after carrier phase-shift, which is similar to the paralleled inverters with interleaving.

FFT analysis of torque ripple is shown in Fig. 6.32. Without carrier phase-shift, the main torque harmonic is the second-order harmonic of the switching frequency, which is 6% of the rated torque value. With carrier phase-shift, the main harmonic of torque ripple, which was concentrated around the second order of switching frequency, is nearly eliminated and the whole torque ripple is decreased by 6% of the rated torque. To further validate the benefit for vibration and noise, motor vibration

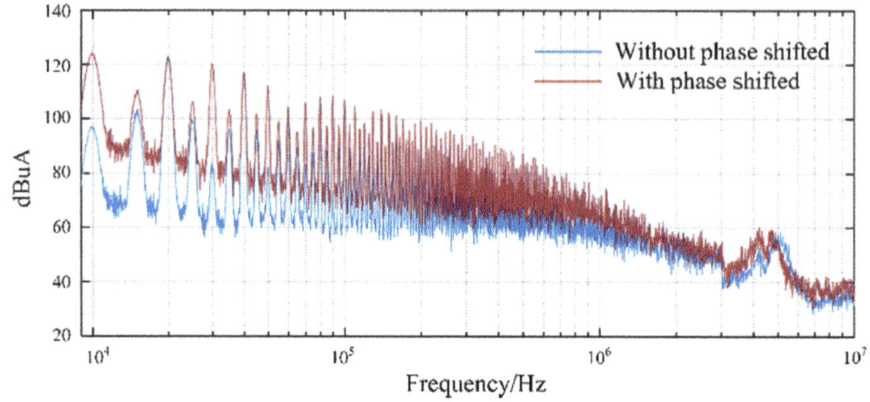

Fig. 6.31 Comparison of electromagnetic interference in frequency domain

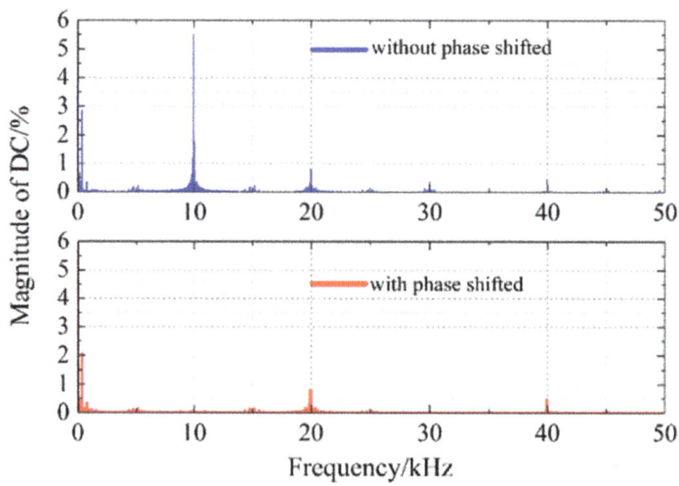

Fig. 6.32 FFT analysis of torque ripple

test has been done by vibration facility of Pulse Labshop from Bruel & Kjar. Experimental results are shown in Fig. 6.33. The vibration parameters are also dominated by the components near 10 kHz when there is no carrier phase-shift. With carrier phase-shift, the vibration amplitude near 10 kHz has been significantly reduced and the noise of the drive system has also been improved obviously.

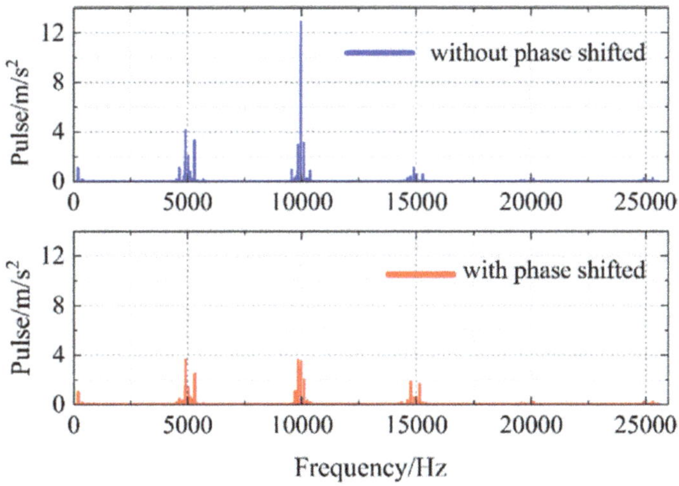

Fig. 6.33 FFT analysis of vibration

6.2.4 VSFPWM for Circulating Current of Paralleled Inverters [30]

As discussed in Sect. 6.2.1, for paralleled inverters with three-phase load, with interleaved carrier signals, the circulating current is introduced due to the instantaneous voltage difference of the paralleled-legs. Circulating current is a serious concern related to the losses and stresses in both active and passive components and should be suppressed. The implementation of coupled inductors provides a high impedance path for circulating current suppression, but a low impedance path for the output current. The magnetic coupling between the paralleled phase-legs, by the way of coupled inductors, is a preferred solution to achieve higher power density. If a coupled inductor has been already designed and capable of maintaining a certain effective inductance with the required bias current, the maximum value of the flux density in the core is determined by the circulating current peak value. Usually, to achieve a small size of coupled inductor, the maximum flux density is designed close to the saturation flux density of the selected cores. In order to make full use of the maximum flux density of coupled inductors of the paralleled VSIs system, the VSFPWM for circulating current control of paralleled inverters will be introduced in this section.

The topology of two paralleled inverters with coupled inductors is presented in Fig. 6.34, where three coupled inductors are added between the paralleled VSIs and the three-phase load. The line impedance stabilization network (LISN) is used here to prevent the external conductive noise of the DC source, and the common DC bus is shared.

To model the system, phase-a is taken as an example to analyze the circulating current, as shown in Fig. 6.35. The resultant current (i_a) is the sum of phase-a currents

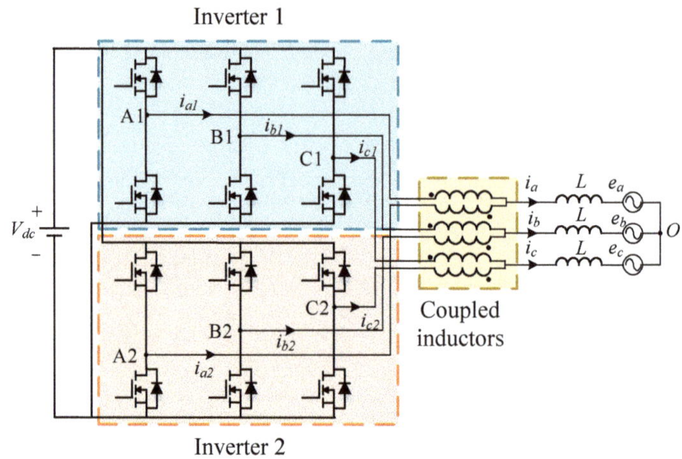

Fig. 6.34 Topology of two paralleled inverters

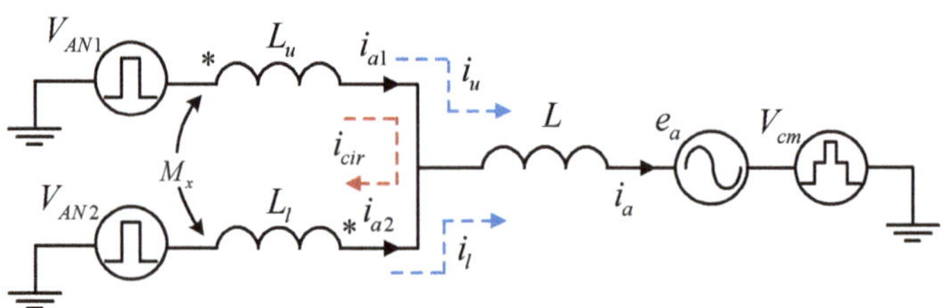

Fig. 6.35 Equivalent circuit for circulating current

of inverter 1 and inverter 2. L is the line filter to limit the current ripple and e_a represents the load voltage or grid voltage.

The phase current of each inverter (i_{a1}, i_{a2}) can be decomposed into two parts, where both i_u and i_l are the components of phase currents contributing to the resultant current, and the i_{cir} is defined as circulating current component. Ignoring the asymmetry in hardware or control, i_u and i_l are considered equal here. Their relationships are written as:

$$\begin{cases} i_{a1} = i_u + i_{cir} \\ i_{a2} = i_l - i_{cir} \end{cases} \tag{6.12}$$

$$i_{cir} = \frac{i_{a1} - i_{a2}}{2} \tag{6.13}$$

According to the Kirchhoff's law, the dynamic equation for circulating current is given as:

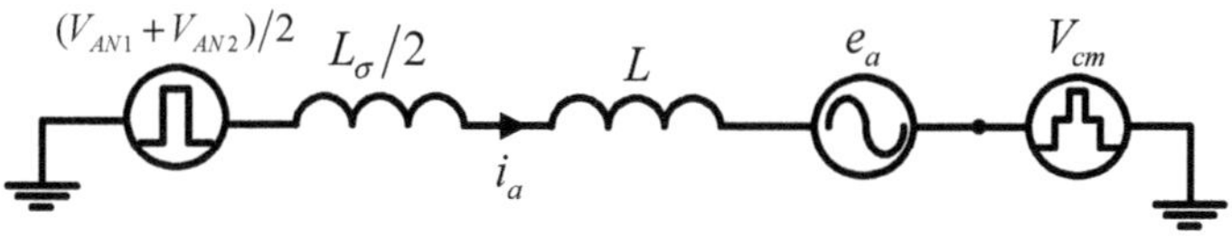

Fig. 6.36 Equivalent circuit for output current

$$V_{AN1} - V_{AN2} = (L_u + L_l + 2M_x)\frac{di_{cir}}{dt} \tag{6.14}$$

where L_u and L_l represent the self-inductance, and L_u is treated as equal to L_l here. M_x is the mutual inductance for the coupled inductor. V_{AN1} and V_{AN2} are the phase-a AC terminal voltages with respect to the DC-link middle point (N). From (6.14), it can be seen that the circulating current is introduced by the instantaneous voltage difference between paralleled phase-legs.

As shown in Fig. 6.36, the voltage equation of phase-a analysis model is written as:

$$\frac{V_{AN1} + V_{AN2}}{2} = \left(\frac{L_\sigma}{2} + L\right)\frac{di_a}{dt} + e_a + V_{cm} \tag{6.15}$$

In Eq. (6.15), L_σ represents the leakage inductance of the coupled inductor. From (6.15), if the two paralleled inverters are driven by the same PWM signals ($V_{AN1} = V_{AN2}$), the current will be equally shared by each inverter and the harmonics and EMI characteristics will be the same to a regular three-phase inverter. By utilizing the interleaved PWM signals (instantaneous voltage difference between V_{AN1} and V_{AN2} is introduced intentionally), the quality of output current and EMI characteristics will be improved, but the circulating current will be generated simultaneously.

As discussed in Sect. 6.2, the flux path for the circulating current is the main path in the coupled inductor and the leakage inductance is contributing to the AC line inductance. The flux density in the ferrite core can be expressed as:

$$B(t) = \frac{1}{2N_0 A}\int (V_{AN1} - V_{AN2})dt \tag{6.16}$$

In Eq. (6.16), A is the core area and N_0 is the number of turns. The maximum flux density is an important parameter for the coupled inductor design. Combining (6.14), it is clear that the maximum flux density is determined by the circulating current peak value.

For phase-a, Fig. 6.37 shows a typical case of 180° interleaving for two paralleled inverters in one switching cycle, where the reference signal is compared with the carriers to generate the PWM signals. It can be seen that the two inverters share the same reference voltage, but the carrier of inverter 2 is shifted 180° with respect to that of the inverter 1. Since the voltage difference between two inverters is balanced in each switching cycle, the average value of circulating current is 0. The circulating

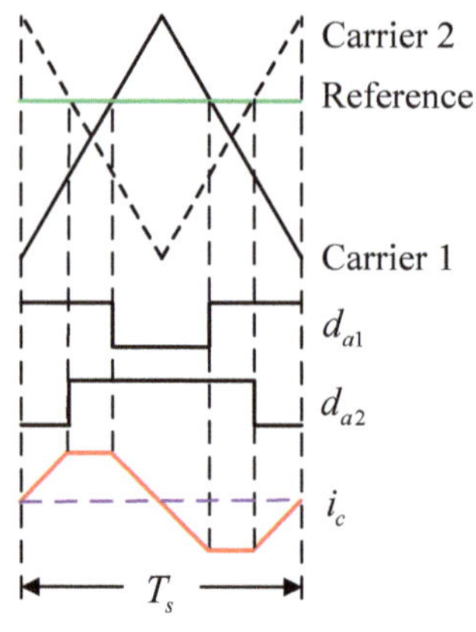

Fig. 6.37 Circulating current in one switching cycle

current is divided into five sectors by the four edges of duty cycles, and it is symmetric about the midpoint. In each sector, the slope of circulating current can be calculated by (6.14), and the action time of each sector is determined by the duty cycles. Therefore, the circulating current can be predicted based on the single-phase model. Having $V_{dc} = 200$ V, $m = 0.8$ and $f_s = 20$ kHz, the predicted peak values and the simulated results are plotted together as shown in Fig. 6.38. It can be observed that the predicted results match well with the simulated and the circulating current is reset in each switching cycle.

Based on the time-domain prediction model of circulating current, variable switching frequency PWM can be developed for circulating current control. Different from the traditional CSFPWM, the VSFPWM is with the circulating current prediction module and switching period updating module as shown in Fig. 6.39. When the circulating current prediction module receives a sampling signal, it reads three-phase duty cycles from the controller, and predicts the circulating current in real

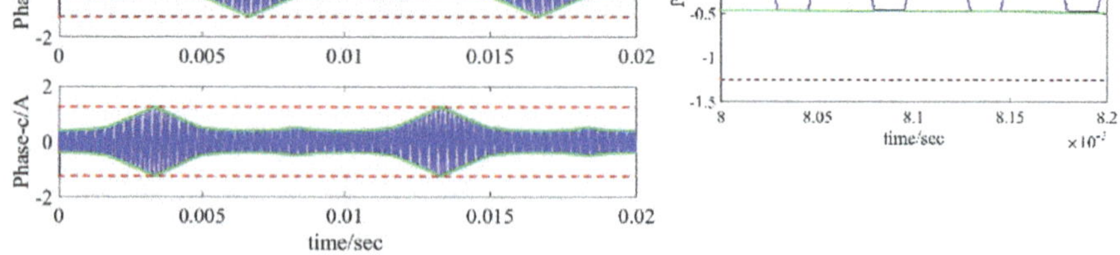

Fig. 6.38 Circulating current comparison between simulation and prediction

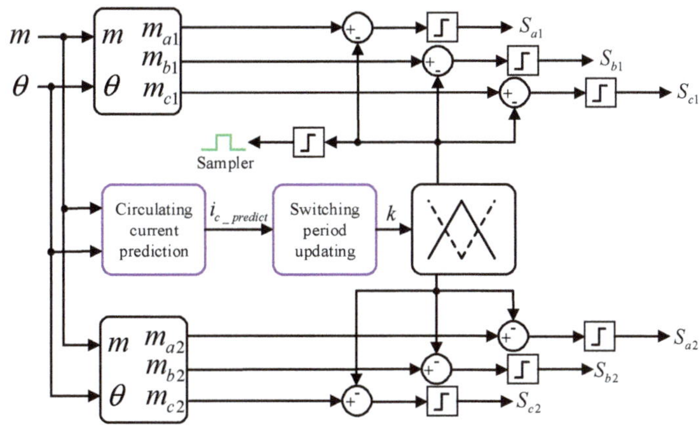

Fig. 6.39 VSFPWM control block diagram

time. The maximum peak value of three-phase circulating current is developed to switching frequency updating module. To assure all coupled inductors meeting the circulating current requirement, the maximum peak value is selected and the updated switching period is obtained by (6.17). If a complete carrier wave is generated, the sampler produces a sampling signal and the next cycle is coming. Here, the circulating current requirement is set to 1.25 A for VSFPWM, which is defined as the maximum peak value for CSFPWM in a fundamental period.

$$T_N = T_s \frac{i_{cir\_require}}{\max(i_{cir\_peak\_a}, i_{cir\_peak\_b}, i_{cir\_peak\_c})} \tag{6.17}$$

The PWM generation module should possess the ability of controlling the pulses position and switching frequency of switching pulses. The switching pulses are produced based on the modulation using synchronous saw-tooth carrier. In order to achieve the pulse position control, two comparison registers are utilized to control the rising-edge (*R*) and falling-edge (*F*) of each switching pulse, respectively, and there are two cases as shown in Fig. 6.40. Figure 6.41 gives the update method of the switching period, where the saw-tooth carrier wave is adopted with an up-counter. As shown in Fig. 6.41a, the peak value of the carrier wave is always set as a constant, and the counter slope is varied cycle by cycle to change the carrier/switching period. On the other hand, the peak value of the carrier wave can be varied cycle by cycle if the counter slope is always programmed as a constant, as shown in Fig. 6.41b. It is worth pointing out that the reference compared with the saw-tooth carrier should be in proportion to the peak value of the carrier wave to ensure volt-seconds balance. In the digital signal processor-based system, it is more convenient and efficient to change the peak value (period register) of the carrier wave.

In order to verify the proposed VSFPWM, the experiment has been carried out in paralleled VSIs connected to the *L-R* load. The experimental setup is shown in Fig. 6.42.

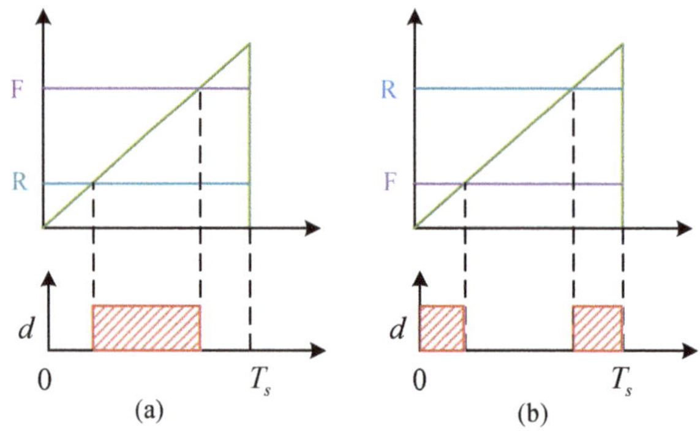

Fig. 6.40 Switching pulses generation: **a** F > R, **b** F < R

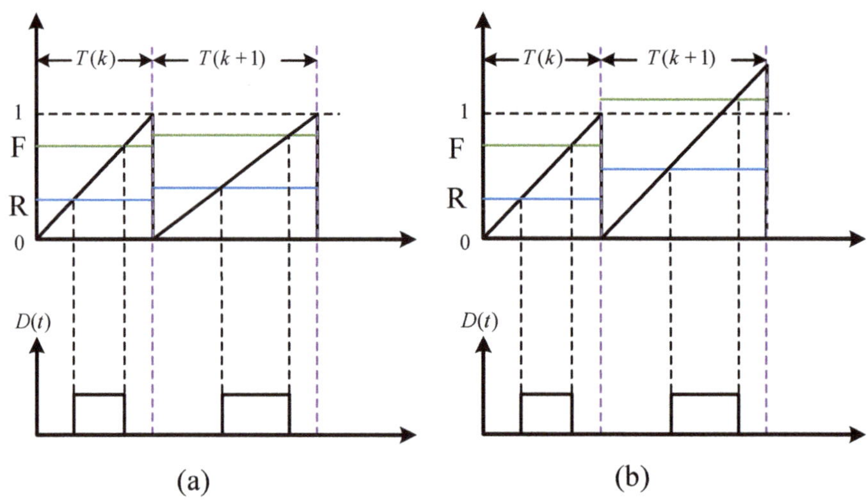

Fig. 6.41 Carrier-wave generator: **a** varying the counter slope with the constant peak value, **b** varying the peak value with the constant counter slope

Figure 6.43 shows the switching frequency comparison stored in DSP. It can be seen that the switching frequency of VSFPWM varies below the CSFPWM, from 8 to 20 kHz. A parameter describing the switching losses is defined as (6.18), and then loss saving can be obtained by (6.19). Focusing on Table 6.1, the VSFPWM can save up to 30% in switching losses for 180° interleaved PWM.

$$E_{sw} = \sum_{k=1}^{N} |i(t_k)| \tag{6.18}$$

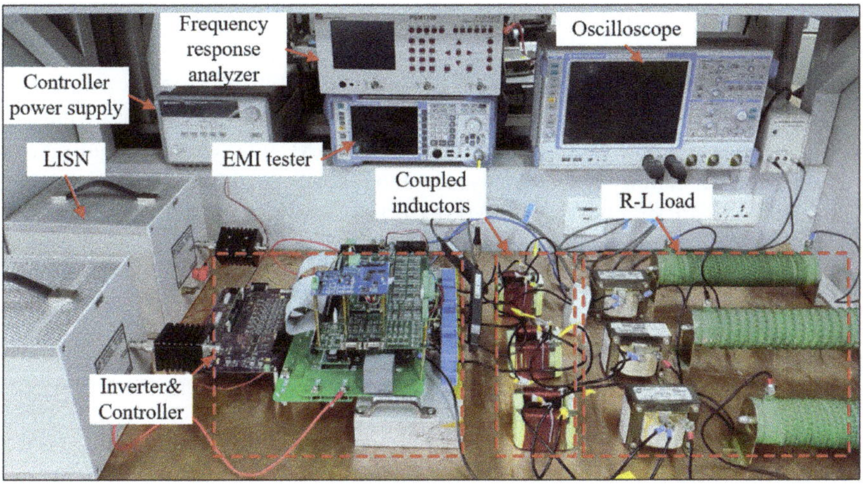

Fig. 6.42 Picture of experimental set-up

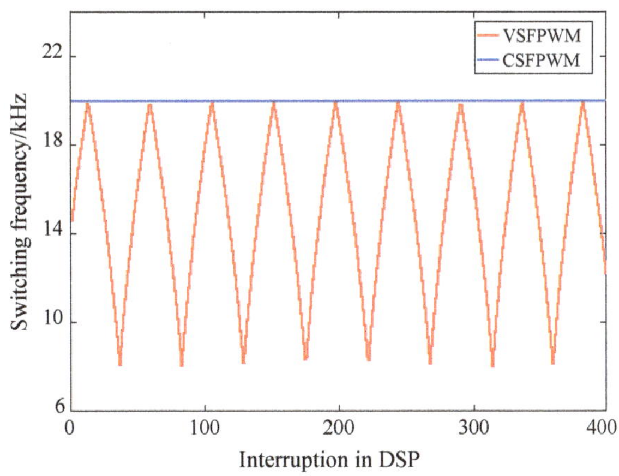

Fig. 6.43 Switching frequency comparison (experimental result)

Table 6.1 Comparison of average switching frequency and switching losses

PWM method	Peak value of circulating current	Number of commutations	E_{sw}
CSFPWM	1.25 A	200	2037.1 A
VSFPWM	1.25 A	141	1422.2 A

$$\text{Loss saving} = \left(1 - \frac{E_{sw,VSFPWM}}{E_{sw,CSFPWM}}\right) \cdot 100\% = \left(1 - \frac{\sum_{k=1}^{N1} |i(t_k)|}{\sum_{k=1}^{N2} |i(t_k)|}\right) \cdot 100\% \tag{6.19}$$

In the above formula, $i(t_k)$ is the instant current value in the kth commutation and N represents the number of commutations per fundamental period. Obviously, the number of commutations in VSFPWM is less than the CSFPWM ($N_1 < N_2$).

From Figs. 6.44 and 6.45, it can be clearly observed that the circulating current peak value of VSFPWM is the same with the CSFPWM, satisfying the requirement, which verifies the validity of the proposed method. Thanks to the switching frequency variation, the VSFPWM has better EMI performance with respect to the CSFPWM, where the EMI standard of DO-160E is adopted. Compared with CSFPWM, there is an approximate reduction of 20 dB in 150–800 kHz range with VSFPWM as shown in Fig. 6.46.

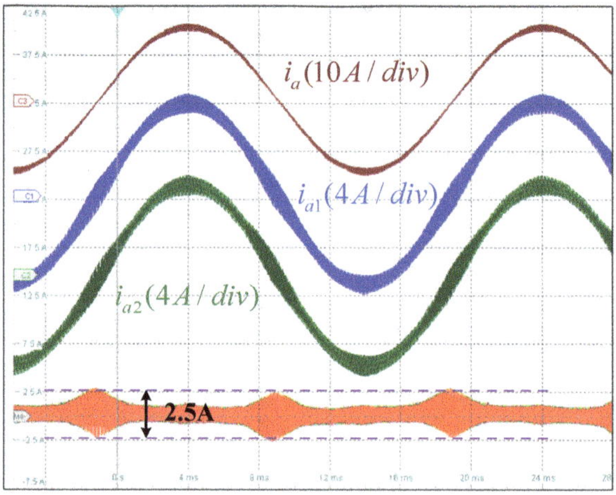

Fig. 6.44 Circulating current with CSFPWM (experimental result)

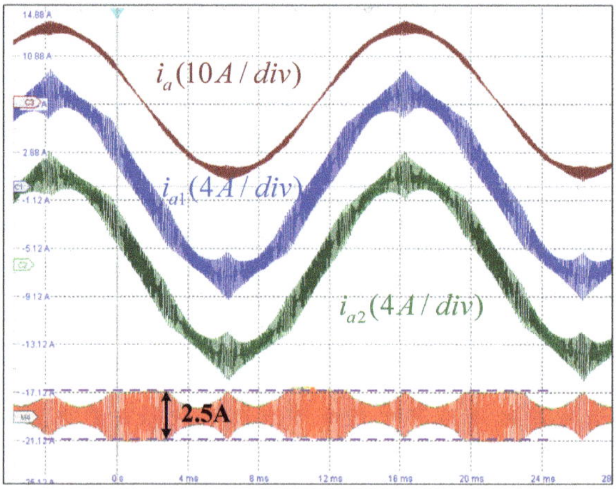

Fig. 6.45 Circulating current with VSFPWM (experimental result)

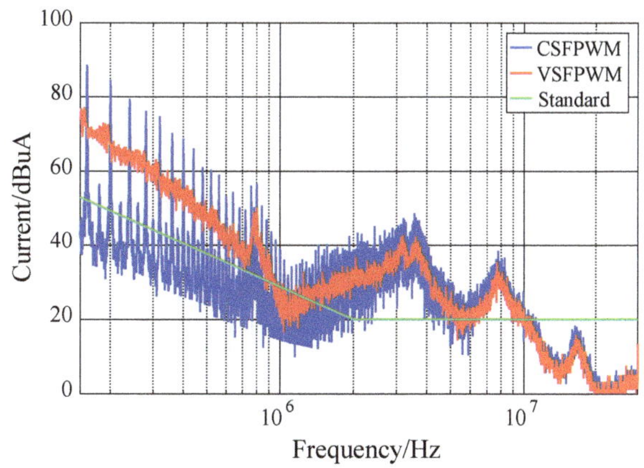

Fig. 6.46 Conducted EMI comparison

6.3 PWM Strategies for Multilevel Converters

The twin structure of a paralleled converter is a multilevel converter. As described in Sect. 6.1, the multilevel converter can implement PWM by space vector synthesis or multiple carrier comparison. Since the multilevel converter has more degrees of freedom for switching combination, there are more PWM schemes for multilevel converters than that of the ordinary two-level converter, and the space where the performance can be improved is larger. The improved PWM strategy to reduce common-mode noise for multilevel converter is introduced in the next chapter. This chapter introduces the VSFPWM technology based on CBPWM for multilevel converter. In addition, the neutral-point potential balance problem is also discussed, and VSFPWM is also applied to neutral-point potential-balance issues for better output performance [31].

6.3.1 VSFPWM for Three-Level Inverter

Figure 6.47 shows the topology of a three-level NPC inverter. The three terminals (*A*, *B*, and *C*) are switched with voltage of $V_{dc}/2$, 0 and $-V_{dc}/2$ with the DC neutral point *O* for reference. The load is representing general AC load, including inductors and balanced three-phase voltage sources. The AC voltage source can represent motor back-EMF for motor drive application or grid voltage for grid application. The voltage produced by inverter is rectangular pulse series whereas the terminal voltage of load is sinusoidal waveform, so their voltage difference imposed on inductance will result in plenty of high-frequency current ripple.

Figure 6.48 shows the phase currents and their 50 Hz fundamental component achieved by FFT analysis, which are the experimental results of the modulation with constant switching frequency called normal SVPWM. The experimental condition

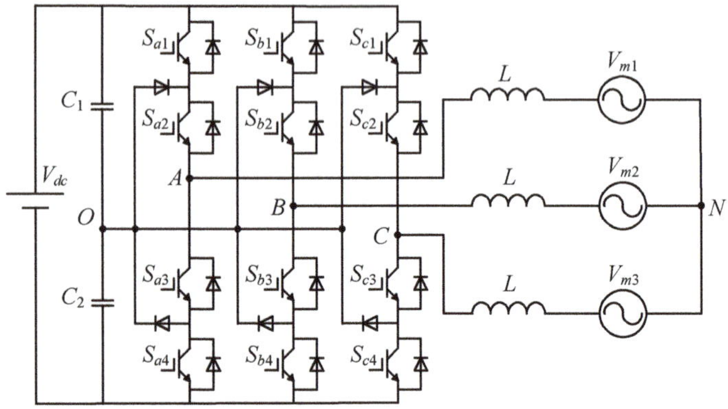

Fig. 6.47 Topology of three-level NPC inverter

Fig. 6.48 Phase current and its fundamental value of SVPWM for NPC inverter

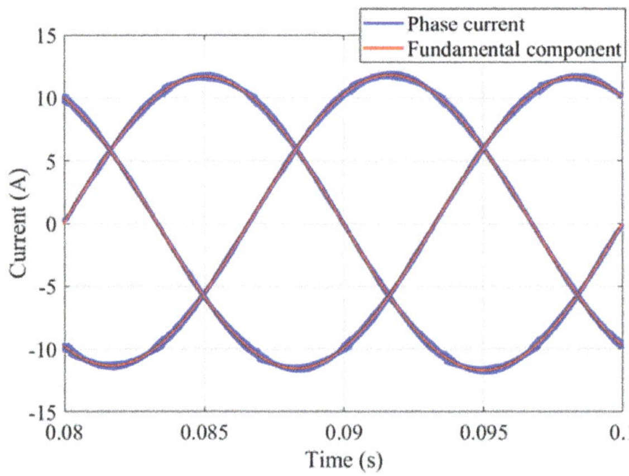

is with 270 V DC bus voltage, 20 kHz switching frequency and modulation index of 0.85.

Obvious ripple can be found in the phase current, which is caused by switching events and is synchronous with PWM signals. Figure 6.49 shows the three-phase output voltage together with phase-a current ripple during one switching cycle. The current ripple is following the switching actions of the three-phase rectangular pulse voltage. In addition, the variation of current ripple between each two adjacent switching actions is linear, which is the basic precondition for current ripple prediction. Nevertheless, three-level inverter's output voltage is switched between $V_{dc}/2$ and 0, or 0 and $-V_{dc}/2$, not between $V_{dc}/2$ and $-V_{dc}/2$, which is different from the current ripple generation in two-level VSI in Chap. 5. Then, the switching voltage model should be reconstructed to predict current ripple.

SVPWM, a rather complicated modulation for multilevel converters, is still considered as the most useful technique for power electronics converters. Owing

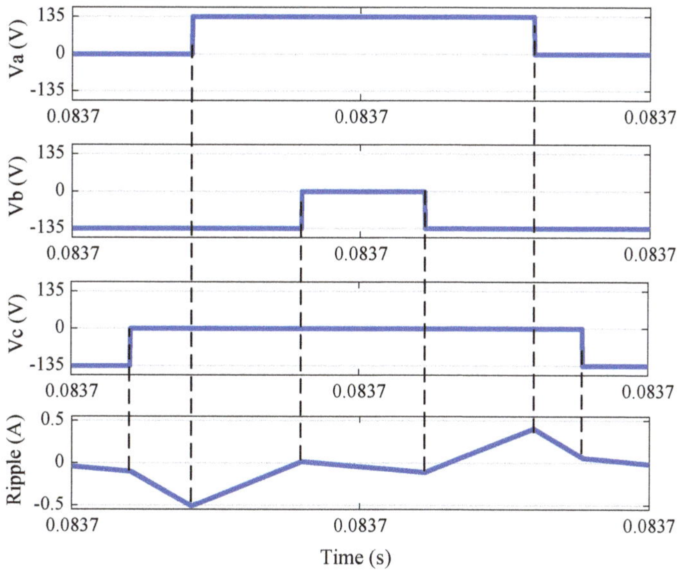

Fig. 6.49 Output voltage and current ripple during one switching cycle (simulation results)

to equivalent theory, carrier-based PWM could replace SVPWM to simplify the modulation and obtain the same performance for multilevel converters. Therefore, this paper is implementing the modulation SVPWM with carrier based PWM equivalently, rather than the complicated process of the vector synthesis with 27 space vectors for three-level inverters.

For the carrier-based PWM analysis of three-level inverters, there are two carriers in the modulator to compare with the reference value. Top carrier is used to generate output voltage of $V_{dc}/2$ and 0, and the bottom carrier is for 0 and $-V_{dc}/2$. Then, the output voltage and action time in each switching cycle can be determined. With the output voltage model, the current ripple can be derived through the equivalent circuit in Fig. 6.50, which is the equivalent model of Fig. 6.47. In each switching cycle, the pulse voltage sources (V_a, V_b, V_c) are determined by comparison between the duty

Fig. 6.50 Equivalent circuit for ripple prediction

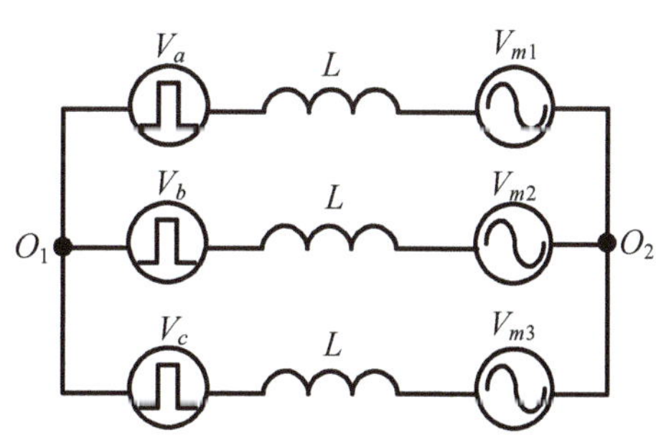

cycles and carrier. The balanced three-phase load average voltages (V_{m1}, V_{m2}, V_{m3} supposed to be constant value in a switching cycle) are directly determined by duty cycles. With the output voltage and action time, the current ripple can be predicted.

The current ripple prediction diagram is shown in Fig. 6.51. The duty cycles from the controller are the inputs. In each switching cycle, by comparing the duty cycles with the top and bottom carrier in the three-level inverter modulator, the action time for three phases will be determined as well as the terminal voltage. Then, by listing the six edges (three rising edges and three falling edges) in the switching cycle, the action time and output voltage between each two adjacent edges can be determined. With the equivalent model in Fig. 6.50, the di/dt between each two adjacent edges can also be calculated. With the di/dt and action time Δt between each two adjacent edges, the current ripple in the full switching cycle can be predicted.

Figure 6.52 shows the comparison of current ripple between experiment and prediction for all three phases. The predicted current of maximum and minimum value in each switching cycle are matching well with the experimental current ripple. The experimental results indicate that the prediction model is effective and possesses real-time prediction capability.

The VSFPWM for three-level inverter is similar to the structure of VSFPWM for two-level VSC in Chap. 5. The VSFPWM diagram is shown in Fig. 6.53. Duty cycle references (d_a, d_b, d_c) are generated from controller. They are sent to the ripple prediction module, the predicted current ripple is calculated and sent to the switching period generation block. The generated triangle carriers with variable frequency (top and bottom) are used to compare with duty cycles and generate the gate drive signals for the power devices in the three-level VSI. At the same time, the variable switching frequency sampler is generated to the controller for control.

The core for the VSFPWM algorithm is the switching period calculation module. With the real-time prediction of current ripple in each switching cycle, the maximum ripple current in three phases with nominal switching cycle T_{sN} is calculated. The updated switching cycle T_s is calculated in Eq. (6.20). The updated switching period

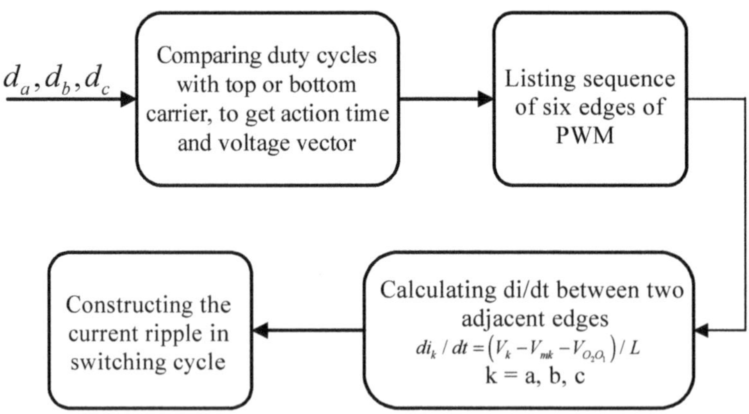

Fig. 6.51 Current ripple prediction diagram

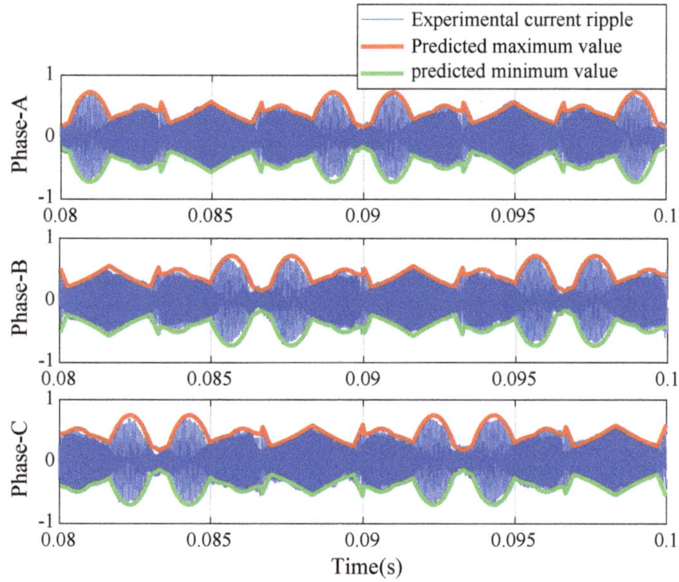

Fig. 6.52 Current ripple comparison between experiment and prediction (normal SVPWM)

Fig. 6.53 Structure of VSFPWM generation

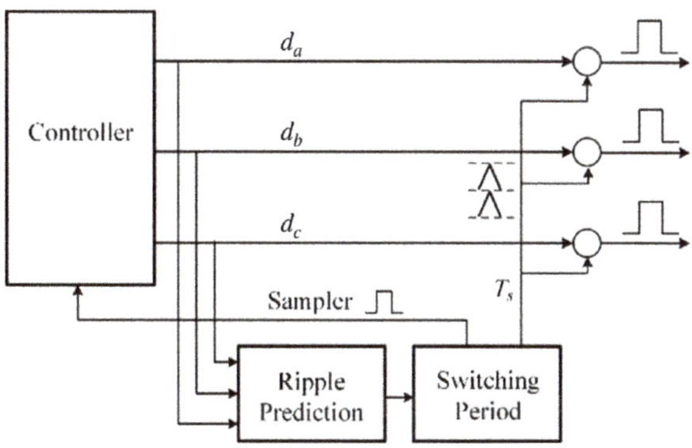

will control the maximum current ripple to be equal to the ripple requirement in each switching cycle.

$$T_s = T_{sN} \times \frac{I_{ripple\_require}}{I_{ripple\_\max}} \tag{6.20}$$

Compared with the model predictive VSFPWM in two-level inverter discussed in Chap. 5, the basic principle of VSFPWM in three-level inverter is similar: actively adjusting the switching frequency in each cycle to control the maximum current ripple clamped to the required value. The major difference for three-level inverter

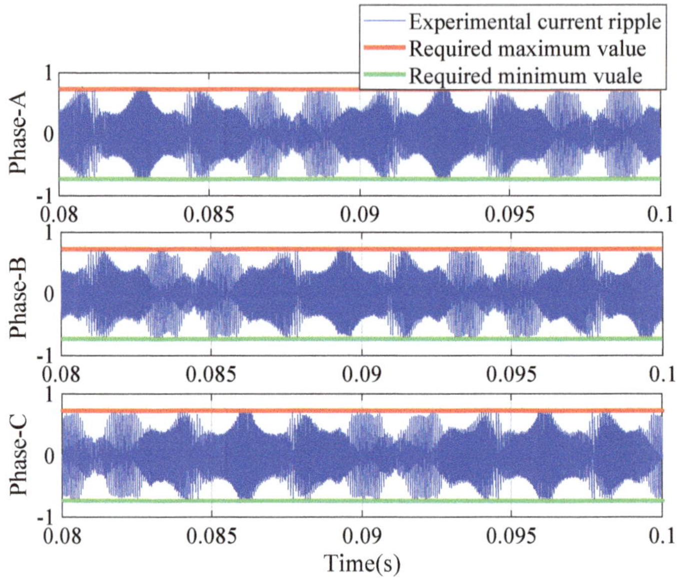

Fig. 6.54 Three-phase current ripple of VSFPWM

is double carrier, which makes prediction model more complicated than two-level. Because the variation range of current ripple in three-level inverter is big, there is a big range for switching frequency variation.

In order to verify the variable switching frequency methods proposed in previous parts, the experiments have been implemented in a three-phase PWM three-level NPC inverter. Three-phase currents are measured by high-precision current probes and recorded as data. With FFT analysis of the current data, fundamental current and plenty of harmonics components can be reconstructed. By subtracting the fundamental component from three-phase full current, current ripple of VSFPWM is obtained in Fig. 6.54. Compared with ripple in Fig. 6.52 for normal SVPWM, the VSFPWM has better effective utilization rate of space limited by the identical required current peak value ($\pm$0.73 A) apparently.

According to (6.20), the maximum current ripple of three-phase in carrier period and the required limitation of ripple determine the variation range of frequency. In Fig. 6.52, it is clear that the maximum ripple is not always clamped to required value (0.73 A) when normal SVPWM is applied. Nevertheless, the VSFPWM methods can make the maximum ripple equal to required value in every switching cycle as shown in Fig. 6.54. Figure 6.55 shows the switching frequency of VSFPWM, which is acquired from the register of DSP. Its switching frequency varies in a wide range from 10 to 20 kHz other than being fixed at a constant switching frequency (20 kHz), leading to an obvious reduction in switching loss. The reason for reducing switching loss is that the average switching frequency related to switching loss and efficiency has been reduced significantly with respect to the normal SVPWM (20 kHz).

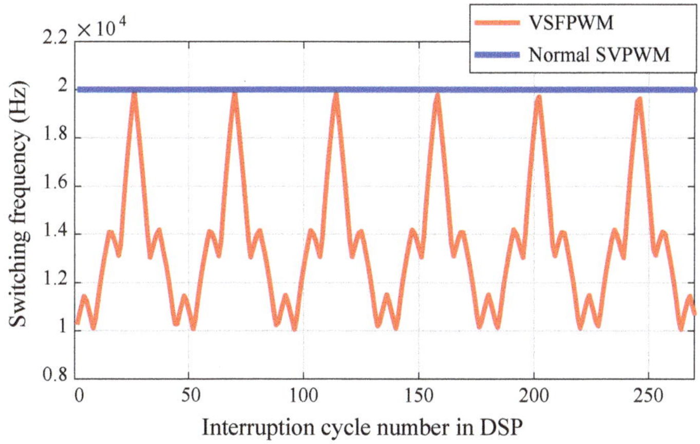

Fig. 6.55 Switching frequency variation of VSFPWM

DM-conducted EMI of two cases are compared in Fig. 6.56 with frequency range from 150 kHz to 30 MHz measured by EMI test receiver, which adheres to the narrowband standard DO-160E. EMI noise energy spread into a broad range, which makes the peak value obviously smaller than normal SVPWM. The EMI peak value could nearly reduce 10 dB around the carrier harmonics. The limit of EMC standard is also plot in Fig. 6.56. Although the advanced modulation can make conducted EMI achieve great attenuation, the EMI filter is also needed to satisfy the EMI requirement. However, reduction of EMI noise can increase cut-off frequency and decrease mitigation requirement for EMI filter, which could reduce the volume and weight of the passive components, and finally increase the power density of the whole system.

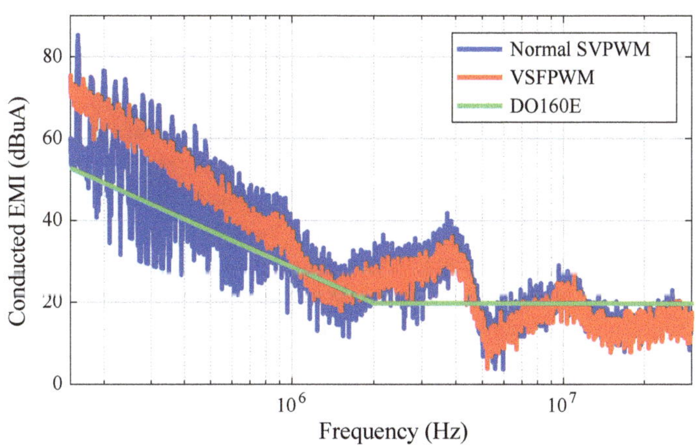

Fig. 6.56 Comparison of conducted EMI: VSFPWM and normal SVPWM

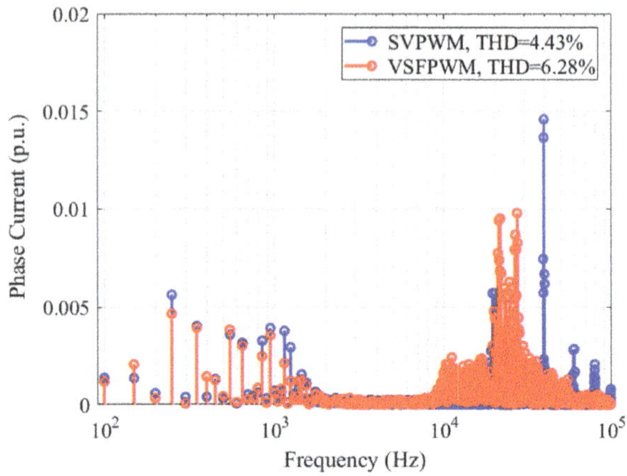

Fig. 6.57 Comparison of phase current spectrum: VSFPWM and normal SVPWM

The variable switching frequency methods will have impact on the low-frequency harmonics and the THD of the output currents. The THD increases from 4.43 to 6.28% after applying variable switching frequency method into normal SVPWM. The reason for the increase of THD is obvious that the switching frequency is kept smaller than 20 kHz, which results in the bigger average current ripple. Figure 6.57 shows the current spectrum comparison between VSFPWM and normal SVPWM. Compared with normal SVPWM, VSFPWM not only spreads the spectrum around carrier harmonics but also has little impact on low-frequency harmonics.

6.3.2 Neutral Point Potential Balance for Three-Level Inverter

In Sect. 6.3.1, VSFPWM has been developed for three-level inverter with a similar manner of that of two-level inverter. Actually, three-level inverter has its own requirement compared with two-level inverter, which gives new opportunities for VSFPWM. The neutral point potential imbalance is an inherent problem of the three-level NPC inverter, which refers to the fluctuation of neutral point potential caused by charging and discharging of the two capacitors at the DC side. The imbalance issues will distort the output voltage of the inverter and introduce harmonics to the load. In addition, voltage fluctuation of DC link capacitors will reduce the lifetime of capacitors. The NP potential balance is essential to maintain a good output waveform and guarantee the proper function of the converter. Therefore, it is treated as the control priority by most of the modulation schemes for three-level NPC converter.

For a three-level NPC inverter, each phase leg can be equal to a three-terminal switch connected to the positive, negative or neutral point, as shown in Fig. 6.58 [11]. There are totally 27 switching vectors, corresponding to 19 space vectors (6

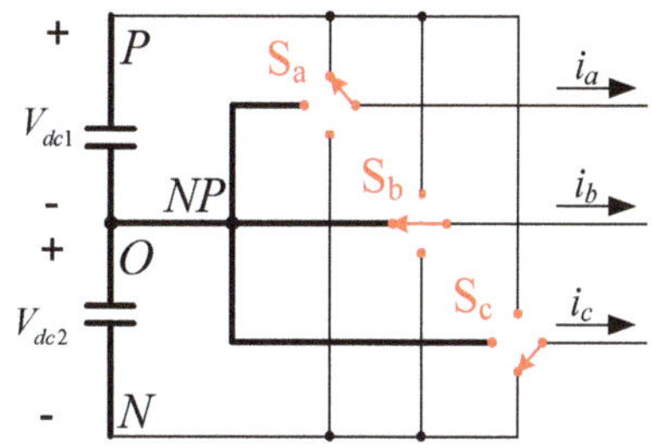

Fig. 6.58 Equal circuit for three-level NPC inverter

long vectors, 6 medium vectors, 6 small vectors and 3 zero vectors) in the line-to-line voltage space as shown in Fig. 6.59. The small and zero vectors have redundant switching states to choose, which is the premise of achieving neutral point potential balance through modulation.

The influence of different voltage vectors on the neutral point is different and it can be analyzed in Fig. 6.60. The analysis takes the first 60° triangle sector in Fig. 6.59 as an example. For the long vector in Fig. 6.60a, the three phases (a, b, c) have no loop with capacitor neutral point, so the long vector has no effect on capacitor neutral point voltage and does not create the neutral point potential imbalance. For the medium vector in Fig. 6.60b, three phases are connected with positive and negative port of DC bus and neutral point of capacitors, respectively. For neutral point, there is charging

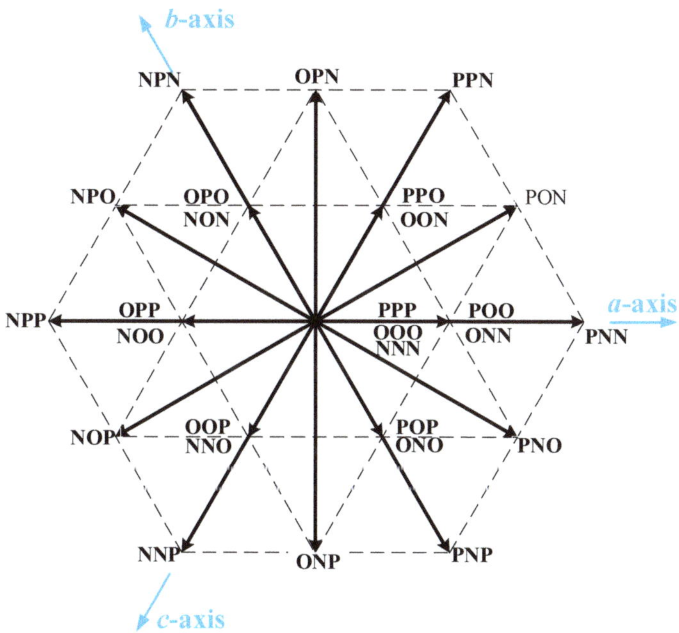

Fig. 6.59 Space vector diagram for three-level NPC inverter

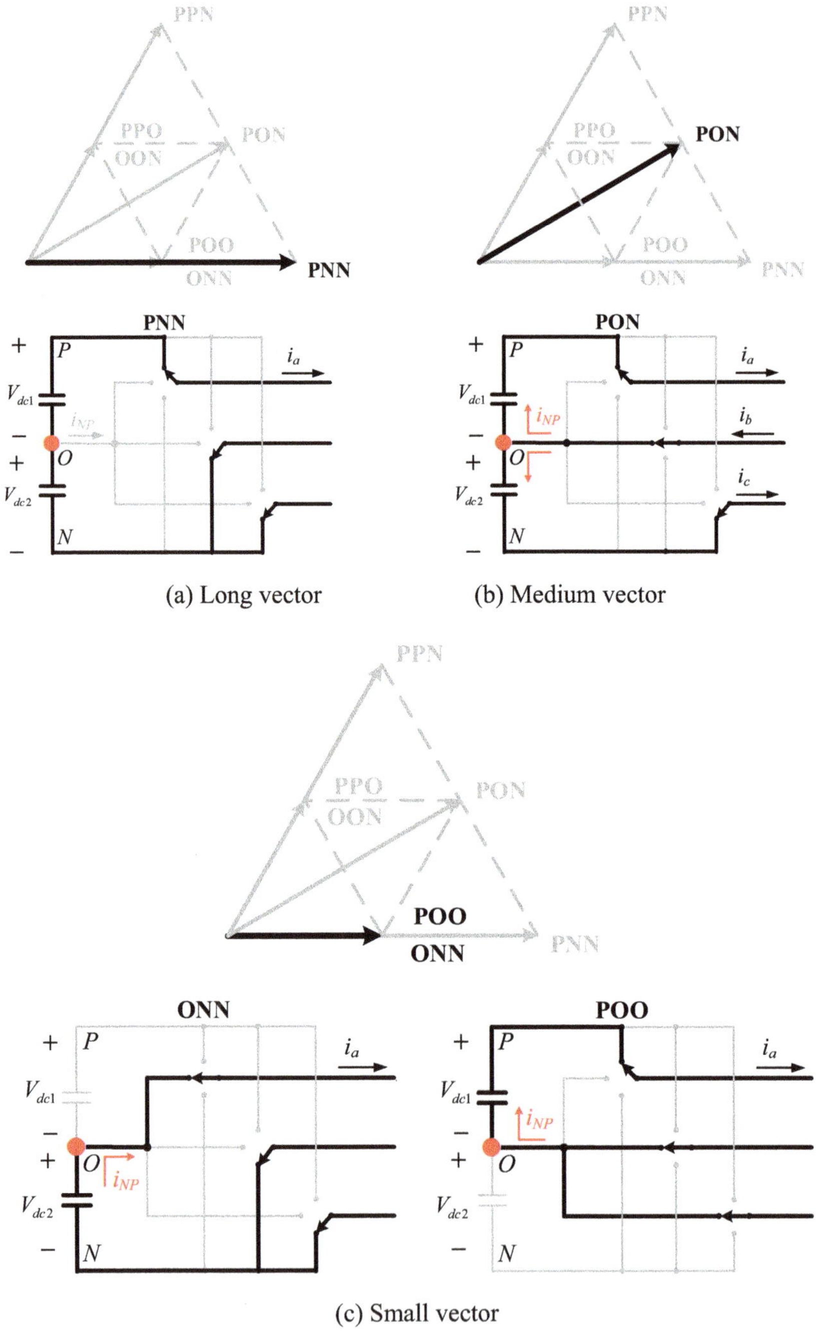

Fig. 6.60 Circuit diagram for vectors and the influence on NP voltage

or discharging issue by the neutral current and the effect on potential depends on the working conditions of the inverters. For the small vector in Fig. 6.60c, there are two switching states for one small vector. When the positive small vector (POO) acts on the inverter, the neutral point is charged through the load with phase-b and phase-c current, which makes the neutral point potential rise. Conversely, when the negative small vector (ONN) acts on the inverter, the neutral point of the capacitor is discharged through the load with phase-a current, causing the potential decreasing at the neutral point. Therefore, the two switching states for one small vector have an opposite influence on the neutral point potential and it is the basis for balancing the neutral point voltage.

The neutral point voltage balancing method of the NPC three-level inverter can be roughly classified into the modulation schemes and the non-modulation schemes. The modulation schemes include the SVPWM scheme and carrier-based PWM scheme, whose essence is the selection of suitable small vectors to achieve the NP balance. There are three approaches to select the small vectors for NP balance. The passive balancing method alternately uses two paired switching states in each switching cycle, where the neutral point voltage can be naturally balanced. (The two switching states for one small vector have an opposite influence on the NP voltage.) This method does not cause additional switching events and related losses in the switching cycle, but the balancing result largely relies on the power factor and it can only be operated in a balanced three-phase system.

The active balancing scheme uses two switching states for small vectors in each switching cycle to realize zero neutral point potential in one switching cycle, and the duty cycles of two switching states of each small vector are different. This method has a good balance effect, but since each switching cycle uses two switching states for each small vector, it inevitably introduces additional switching events, which will lead to the increase of switching losses. In addition, since the duty cycles of paired switching states for each small vector need to be calculated in each cycle, the implementation of this method requires more computation.

Unlike the active method, which uses two switching states of each small vector to balance the neutral point voltage in each switching cycle, hysteresis method uses only one switching state in each switching cycle and achieves neutral point voltage balance in two or more switching cycles. In each switching cycle, the switching state for small vector is selected based on the ΔV ($V_{\mathrm{ON}}-V_{\mathrm{PO}}$) and the phase current direction. This method controls the neutral point voltage to fluctuate around zero, so the NP balance effect is not as good as the active method. Nevertheless, since it uses only one switching state in each switching cycle, it has the advantage of the least switching events.

In addition to the modulation method, the method based on hardware circuit can also be used to balance the neutral point potential. The method parallels a voltage equalization circuit between the DC side power supply and the DC bus capacitor. The circuit adopts the buck-boost topology to absorb or store electrical energy through the inductor, and compensates neutral point potential fluctuation by charging and discharging the capacitor. This method uses extra hardware circuit, which is not as convenient as the modulation scheme.

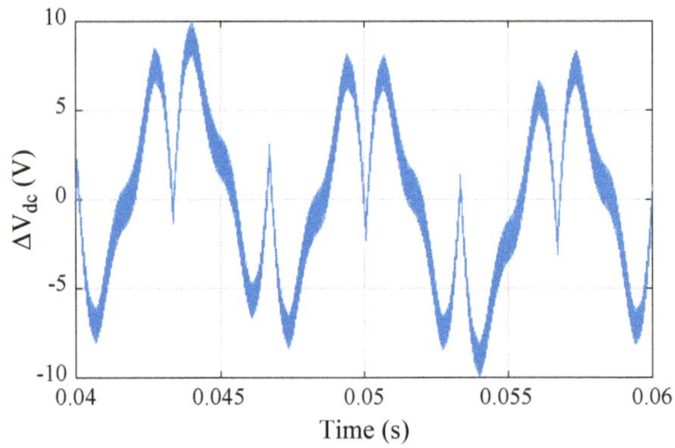

Fig. 6.61 The neutral point voltage imbalance under normal SVPWM

In Sect. 6.3.1, the VSFPWM for three-level inverters based on current ripple prediction has been introduced. This section will give theoretical analysis of neutral point voltage prediction based on neutral point voltage balance and introduce the variable switching frequency PWM to control the neutral point voltage. To distinguish the two methods, the VSFPWM based on current ripple prediction is marked as VSFPWM and VSFPWM based on neutral point voltage balance is marked as VSFPWM-NP.

It is assumed that the top cell capacitor voltage is V_{dc1} and the bottom cell capacitor voltage is V_{dc2}. The neutral point potential imbalance is defined as the difference between the top and bottom cell capacitor voltage $\Delta V_{dc} = V_{dc1} - V_{dc2}$. Figure 6.61 shows the ΔV_{dc} under normal SVPWM with the condition of 200 V DC-bus voltage, 68 uF DC-bus capacitor, 50 Hz fundamental frequency, 20 kHz switching frequency and modulation index of 0.8. It is obvious that ΔV_{dc} contains a large number of low-frequency components (mainly third-harmonic components of fundamental frequency) and these components are useless for VSFPWM-NP. Therefore, before the variable switching frequency operation, using the equivalent carrier-based PWM and the injected zero-sequence method can eliminate the low-frequency components of the neutral point potential fluctuation.

Assuming that the injected zero-sequence component is v_{zs} and the three-phase reference voltage can be expressed as (6.21). The v_{a,b,c_ref} is the reference of normal SVPWM. Dividing the space vector diagram into six sectors as shown in Fig. 6.62, then the symbols (positive or negative) of v_{a,b,c_ref} remain unchanged in each sector. For example, the $v_{a\_ref} \geq 0$, $v_{b\_ref} \leq 0$ and $v_{c\_ref} \leq 0$ in Sector I, $v_{a\_ref} \geq 0$, $v_{b\_ref} \geq 0$ and $v_{c\_ref} \leq 0$ in Sector II, and so on. When the carrier ratio is large, it can be assumed that the three-phase output current and reference voltage of the inverter are constant in every switching period. Then the average current of neutral point can be written as Eq. (6.22), where d_{NPx} (d_{Px}, d_{Nx}) is the duty cycle of x-phase connected with the neutral point (positive pole, negative pole).

$$v^*_{a,b,c} = v_{a,b,c\_ref} + v_{zs} \tag{6.21}$$

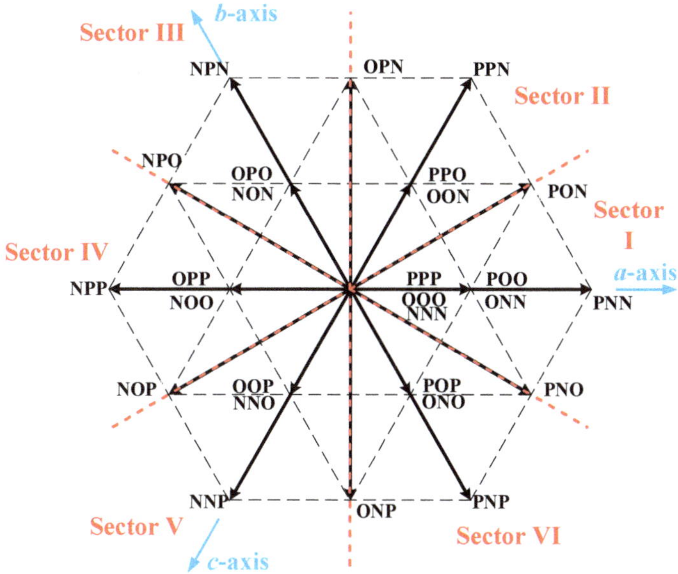

Fig. 6.62 Sector diagram for three-level NPC inverter

$$
i_{NP} = d_{NPa}i_a + d_{NPb}i_b + d_{NPc}i_c
$$
$$
\begin{cases} d_{Px} = v_x^* \\ d_{NPx} = 1 - v_x^* \\ d_{Nx} = 0 \end{cases}, when\, v_x^* \geq 0 \quad \begin{cases} d_{Px} = 0 \\ d_{NPx} = 1 + v_x^* \\ d_{Nx} = -v_x^* \end{cases}, when\, v_x^* < 0 \tag{6.22}
$$

Sector I is used as an example to illustrate the principle. In Sector I, the average current can be expressed as follows:

$$
\begin{aligned}
i_{NP} &= \left(1 - v_a^*\right)i_a + \left(1 + v_b^*\right)i_b + \left(1 + v_c^*\right)i_c \\
&= \left(v_a^* + v_b^*\right)i_b + \left(v_a^* + v_c^*\right)i_c \\
&= -2v_{zs}i_a - \left(v_{b\_ref}i_c + v_{c\_ref}i_b\right)
\end{aligned} \tag{6.23}
$$

In order to maintain the neutral point voltage balance, the average current flowing into or out of the neutral point during one switching cycle should be zero. In other words, the formula (6.24) should be satisfied. For Sector I, utilizing (6.23) and (6.24), the injected zero-sequence component can be obtained as (6.25). Similarly, the zero-sequence component expressions of the remaining five sectors can be deduced, as shown in Table 6.2.

$$
\frac{C}{T_s}\Delta v_{dc} + i_{NP} = 0 \tag{6.24}
$$

Table 6.2 Zero-sequence components of six sectors

Sector	v_{zs}	Sector	v_{zs}
I	$\frac{1}{2i_a}\left(\frac{C}{T_s}\Delta v_{dc} - \left(v_{b\_ref}i_c + v_{c\_ref}i_b\right)\right)$	IV	$-\frac{1}{2i_a}\left(\frac{C}{T_s}\Delta v_{dc} + \left(v_{b\_ref}i_c + v_{c\_ref}i_b\right)\right)$
II	$-\frac{1}{2i_c}\left(\frac{C}{T_s}\Delta v_{dc} + \left(v_{a\_ref}i_b + v_{b\_ref}i_a\right)\right)$	V	$\frac{1}{2i_c}\left(\frac{C}{T_s}\Delta v_{dc} - \left(v_{a\_ref}i_b + v_{b\_ref}i_a\right)\right)$
III	$\frac{1}{2i_b}\left(\frac{C}{T_s}\Delta v_{dc} - \left(v_{a\_ref}i_c + v_{c\_ref}i_a\right)\right)$	VI	$-\frac{1}{2i_b}\left(\frac{C}{T_s}\Delta v_{dc} + \left(v_{a\_ref}i_c + v_{c\_ref}i_a\right)\right)$

$$v_{zs} = \frac{1}{2i_a}\left(\frac{C}{T_s}\Delta v_{dc} - \left(v_{b\_ref}i_c + v_{c\_ref}i_b\right)\right) \tag{6.25}$$

The neutral point voltage under zero-sequence injection is presented in Fig. 6.63a and it can be seen that the neutral point voltage has substantially no low-frequency components. From the frequency spectrum showed in Fig. 6.63b, the neutral point voltage mainly consists of switching frequency components (20 kHz), while the low-frequency components are basically eliminated. Then the neutral point voltage can be easily predicted.

In a three-phase inverter, the switching cycle can be divided into seven zones which are symmetrical around the center. Each zone corresponds to a certain voltage vector. Thus, in each zone, the neutral point of capacitors is only connected to fixed phases and the neutral point current i_{NP} (inflow or outflow) is the sum of the currents of the connected phases from the equivalent circuit shown in Fig. 6.58. Neutral point current can be regarded as constant in each zone when the carrier ratio is large. Figure 6.64 shows the high-frequency neutral point voltage after balancing operation in one switching cycle. From Fig. 6.64, the neutral point voltage is divided into seven zones which are symmetrical around the center. In each zone, the neutral point current is constant, so the slope of neutral point voltage dv/dt is fixed and is equal to i_{NP}/C. As long as we know the active time (determined by the duty cycles) of each zone, the neutral point voltage can be predicted before the pulses are generated.

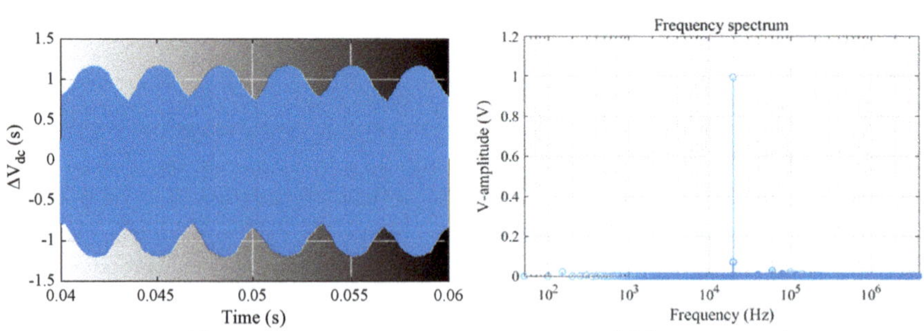

(a) waveform (b) frequency spectrum

Fig. 6.63 Neutral point voltage under zero-sequence injection of CSFPWM

Fig. 6.64 Neutral point voltage in one switching cycle

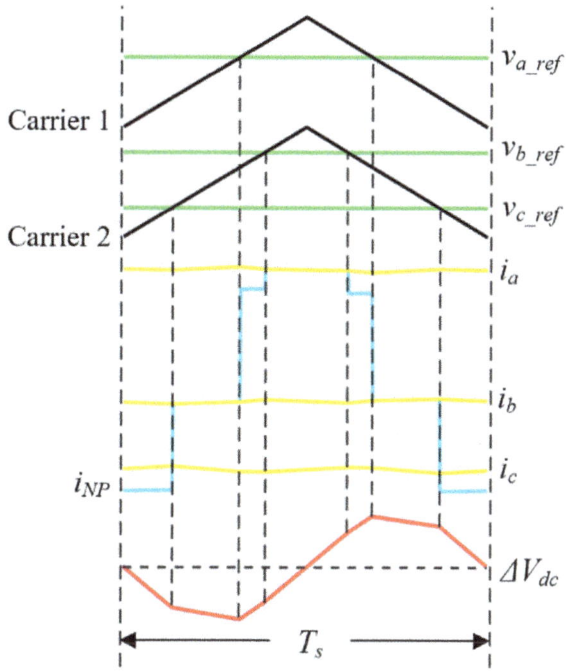

Figure 6.65 shows the comparison of neutral point voltage between simulation and prediction. The predicted voltages of maximum and minimum value in each switching cycle are matching well with the simulation NP voltage. The simulation results indicate that the prediction is effective and can possess real-time prediction capability.

Fig. 6.65 NP voltage comparison between simulation and prediction (CSFPWM)

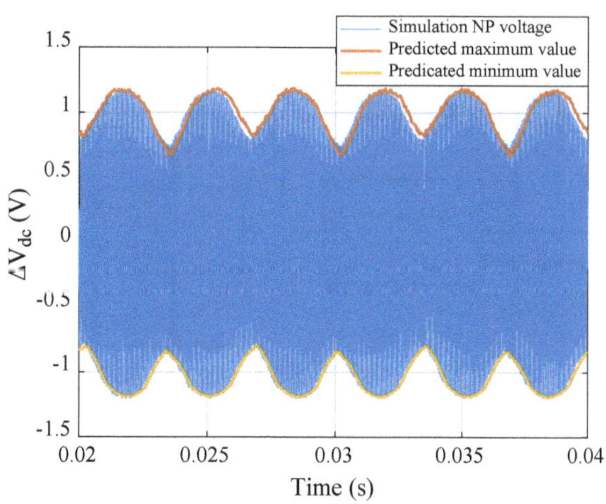

The high-frequency component of the neutral point voltage is proportional to the switching period T_s with a certain modulation method. Under zero-sequence injection, the low-frequency components are eliminated and neutral point voltage can be predicted with nominal switching period T_{sN}. Then the updated switching period will follow (6.26) with the neutral point voltage requirement.

$$T_s = T_{sN} \cdot \frac{V_{NP\_require}}{V_{NP\_\max}} \tag{6.26}$$

The updated switching period will control the maximum neutral point voltage to be equal to the requirement of voltage in each switching cycle. The voltage requirement is usually defined as the maximum neutral point voltage peak (the maximum value of the entire fundamental cycle), so the switching frequency f_s will vary below the nominal switching frequency f_{sN} ($f_{sN} = 1/T_{sN}$). Figure 6.66 shows the structure of VSFPWM-NP.

Comparing the neutral point voltage for CSFPWM (constant switching frequency PWM) shown in Fig. 6.63a with VSFPWM-NP shown in Fig. 6.67a, both methods can generate the neutral point voltage within the range ($\pm$1.2 V) and the neutral point voltage of VSFPWM-NP could satisfy the edge of range in more time than voltage of CSFPWM. Figures 6.63b and 6.67b show the spectrum of CSFPWM and VSFPWM-NP. Compared with normal CSFPWM, VSFPWM-NP spreads the spectrum around carrier harmonics.

Figure 6.68 shows the simulation results of the switching frequency variation of VSFPWM-NP method. Compared with f_s = 20 kHz of CSFPWM, the VSFPWM-NP's switching frequency varies from 14 to 20 kHz and the average switching frequency is reduced to 17.9 kHz. When the conditions of voltages and currents are similar, the switching losses will be reduced with the reduction of switching frequency. Thus, the VSFPWM-NP will obtain better efficiency.

The conducted EMI of AC and DC side for two cases are compared in Fig. 6.69 with frequency range from 150 kHz to 30 MHz measured by EMI test receiver.

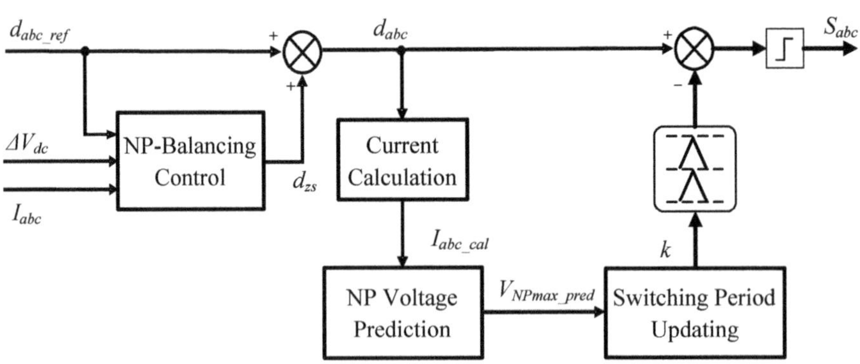

Fig. 6.66 The structure of VSFPWM-NP

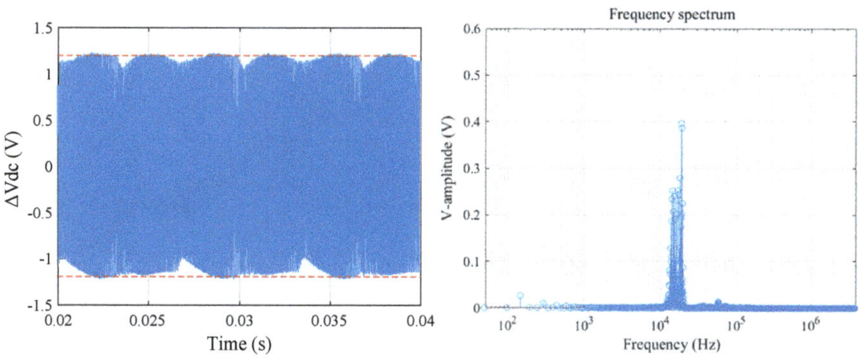

(a) waveform (b) frequency spectrum

Fig. 6.67 Neutral point voltage under zero-sequence injection of VSFPWM-NP

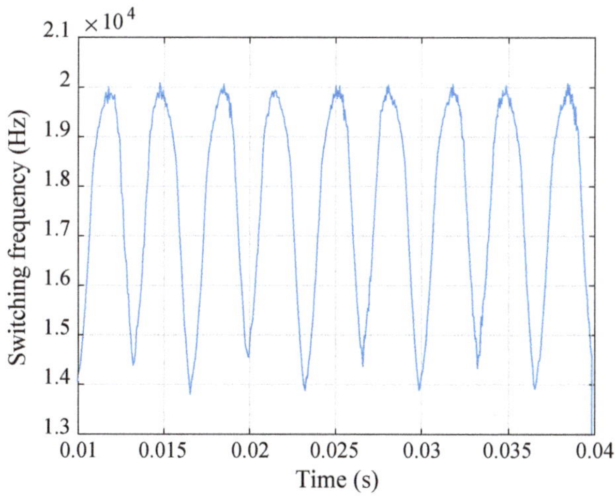

Fig. 6.68 Switching frequency variation for VSFPWM-NP

Compared with the CSFPWM, the conducted EMI of the VSFPWM-NP reduces approximately 10 dB in both AC and DC side, which proves that the EMI noise can be effectively attenuated by VSF operation.

Compared with the VSFPWM discussed in Sect. 6.3.1, the VSFPWM-NP has different control object. VSFPWM-NP controls the maximum neutral point voltage to reach the required value, while VSFPWM controls the maximum current ripple in AC side. But they both achieve the control goal by actively adjusting the switching frequency. Since the variation range of current ripple is larger than neutral point voltage, the switching frequency variation range of VSFPWM is larger than VSFPWM-NP.

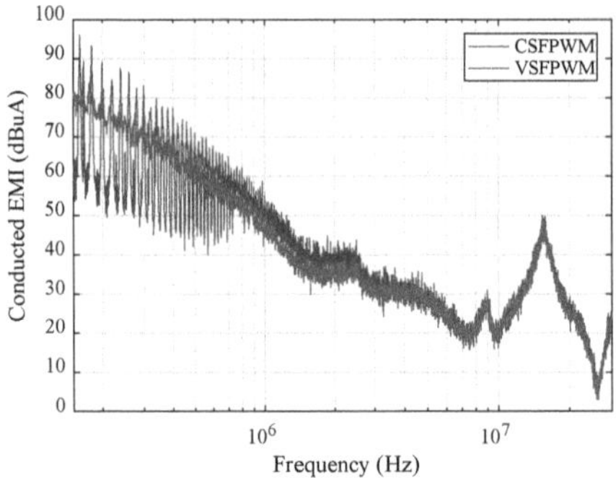

(a) DC-side

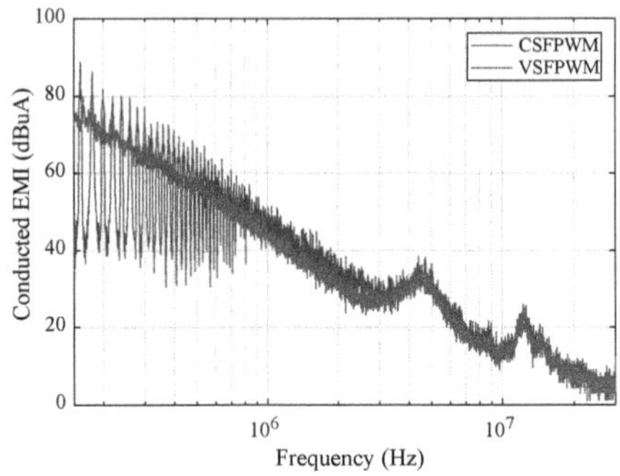

(b) AC-side

Fig. 6.69 Comparison of conducted EMI between CSFPWM and VSFPWM-NP in three-level inverter

6.3.3 VSFPWM for Voltage Ripple Control in Three-Level Flying Capacitor Inverter

Flying capacitor (FC) inverter is another member in the family of multilevel topologies, with the ability to self-balance the flying capacitor voltages. Well-known and most popular PWM method for the FC inverter is phase-shifted PWM, which has the FC voltage balancing function in every switching cycle. Considering the distortion of output phase voltage, the FC voltage ripple is often confined to be a certain limit. That is, the peak-to-peak value of voltage ripple is an important concern for the design and selection of flying capacitors. Actually, FC voltage ripple is determined by the PWM method and three-phase output voltage, which can be easily calculated and analyzed in time domain. Thus, the VSFPWM can also be extended to the voltage ripple control in FC multilevel inverters.

Figure 6.70 shows the topology of three-level FC inverter. For each phase, four power switching devices are connected in series, and a flying capacitor is connected between the series power switching devices. To achieve a three-level output voltage, the FC voltage should be charged to half of the DC-link voltage. Usually, the four power switching devices in one phase can be divided into two groups, where the same group is driven in a complementary way. Without losing generality, phase-a is taken as an example for FC voltage ripple analysis. There are four switching states for one phase, as shown in Fig. 6.71. If $i_a > 0$, it can be seen that the flying capacitor is charged in "1010" and discharged in "0101". For "0011" and "1100", the flying capacitor is bypassed.

If the phase-shifted PWM is used, the two groups in phase-a are driven with an interleaved delay equal to $T_s/2$, as shown in Fig. 6.72. The power switches S_1 is driven by the signal S_{a1}, and the power switches S_2 is driven by the signal S_{a2}. S_4 and S_3 are respectively driven in complementary with S_1 and S_2. There are five zones divided by two rising and two falling edges. In each zone, the flying capacitor is charged, discharged or bypassed. It can be seen that the voltage ripple is symmetry

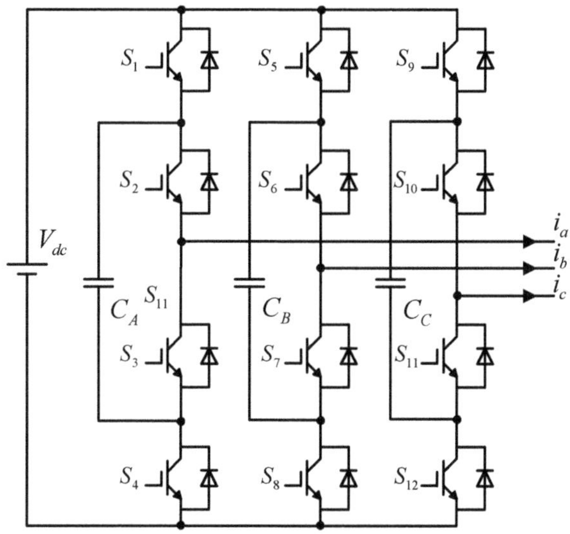

Fig. 6.70 Three-level flying capacitor inverter

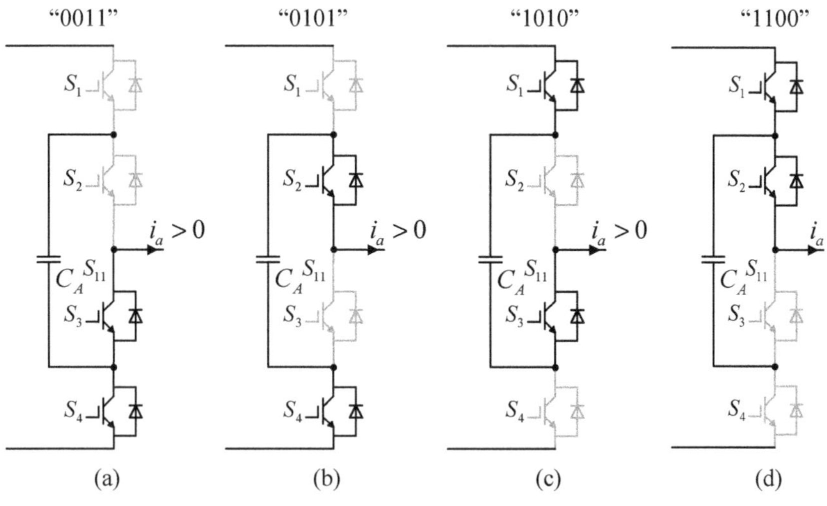

Fig. 6.71 Four switching states for phase-a

about the midpoint and reset in each switching cycle. Similarly, the voltage ripple can be predicted in each switching cycle with duty cycles and output currents from the controller.

Figure 6.73 shows the prediction diagram of phase-a FC voltage ripple in one switching cycle. There are two rising and two falling edges, and their locations can be calculated with the duty cycles. With the four edges, the sequence can be listed and

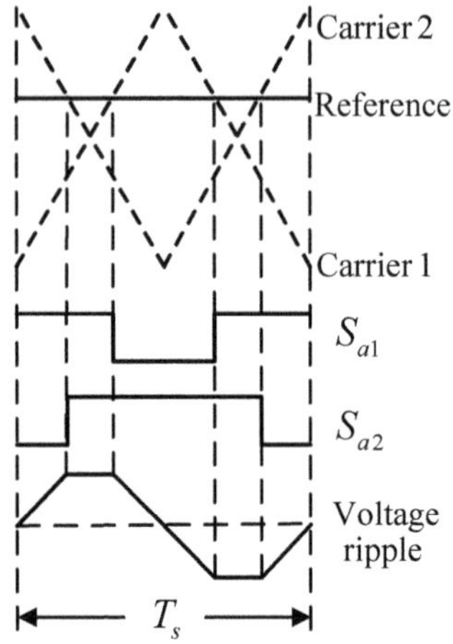

Fig. 6.72 FC voltage ripple in one switching cycle

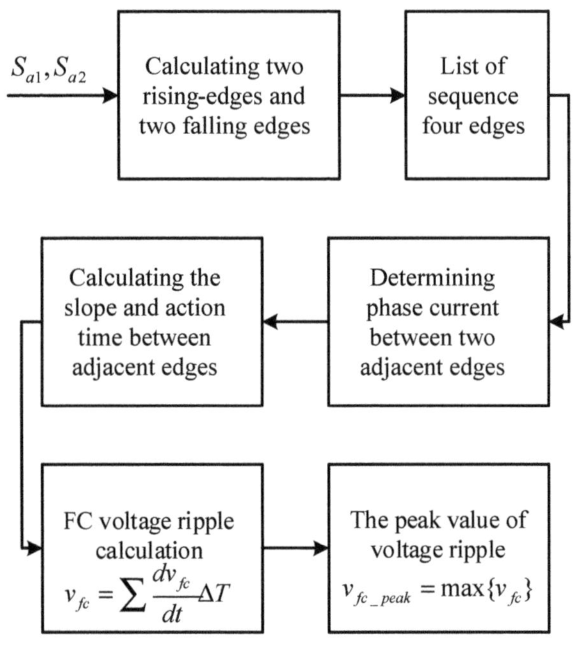

Fig. 6.73 FC voltage ripple prediction diagram

the phase current between two adjacent edges can be derived from the controller. The FC voltage ripple between two adjacent edges is linear and its slope is determined by the phase current and the capacitance. Combining the slope and the action time, the FC voltage ripple can be predicted in each switching cycle.

To assure three-phase flying capacitors meet the voltage ripple requirement, the maximum peak value is selected and the updated switching period is obtained by (6.27). Here, the voltage ripple requirement for VSFPWM is defined as the maximum peak value for CSFPWM in a fundamental period. When the newly triangular carriers are completed, the sampling signal occurs again and next cycle is coming. As a result, the switching frequency of the three-level FC inverter is changing cycle by cycle to control the FC voltage ripples within the requirement.

$$T_N = T_s \frac{v_{fc\_require}}{\max(v_{fc\_peak\_a}, v_{fc\_peak\_b}, v_{fc\_peak\_c})} \tag{6.27}$$

where T_s is the constant switching period and T_N is the updated switching period. $V_{fc\_require}$ is the voltage ripple requirement.

The three-level FC inverter has been implemented to verify the effectiveness of the proposed VSFPWM in voltage ripple control. Experiment has been done to further validate the effect.

Considering the symmetry of three-phase system, phase-a is taken as an example for further analysis. Figures 6.74 and 6.75, respectively, show the FC voltage ripple for CSFPWM and VSFPWM. Although there are nonideal characteristics in the experiment, both the CSFPWM and the VSFPWM satisfy the requirement of voltage ripple (6 V), which illustrates the availability of the proposed VSFPWM strategy.

The switching frequency is stored in each interruption of DSP and used for comparison of the PWM schemes. In the experiment, the real-time variation of switching frequency is realized by changing the periodic register of PWM module in each interruption. From Fig. 6.76, it can be seen that the switching frequency of VSFPWM is varying below 10 kHz used in CSFPWM, and a reduction of 15%

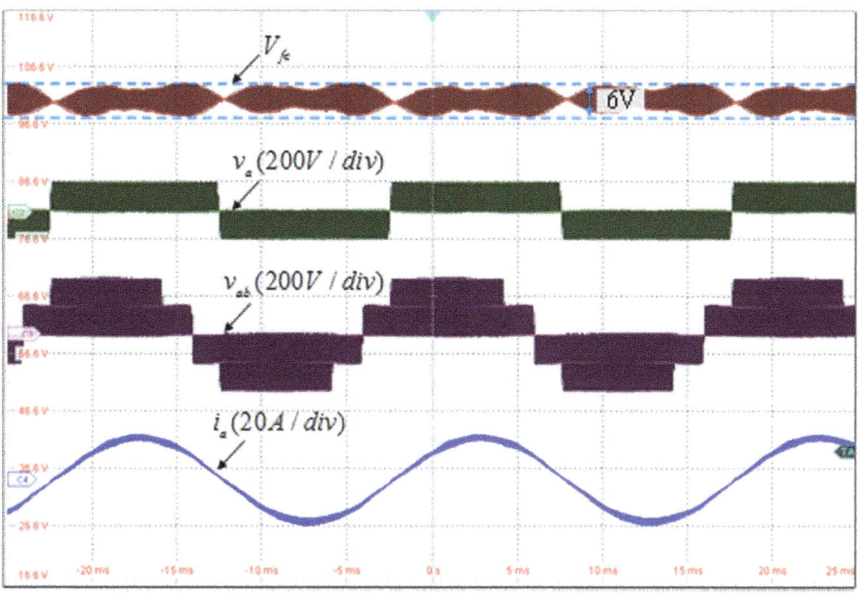

Fig. 6.74 Steady-state performance with CSFPWM

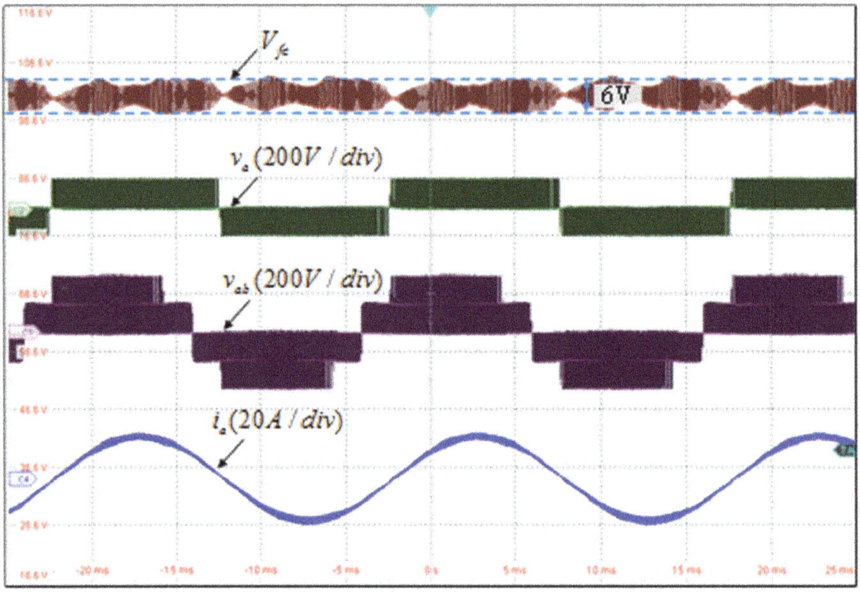

Fig. 6.75 Steady-state performance with VSFPWM

Fig. 6.76 Switching frequency comparison

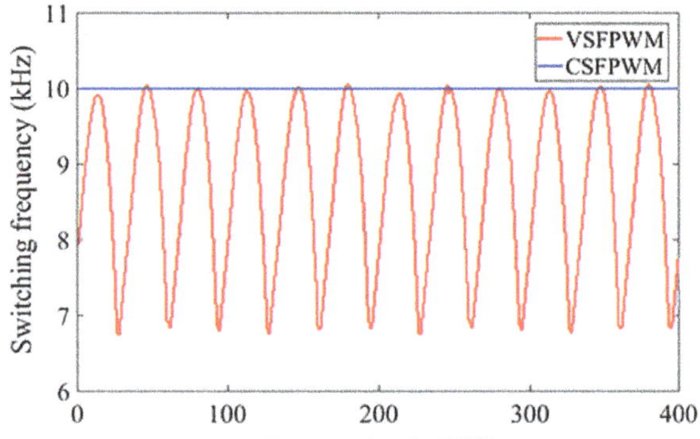

in average switching frequency have been obtained with VSFPWM. Thanks to the switching frequency variation, the VSFPWM shows a better EMI performance than the CSFPWM as shown in Fig. 6.77.

In order to investigate the dynamic performance, a step change of 0.4 in modulation ratio is applied here. As shown in Figs. 6.78 and 6.79, the dynamic response of VSFPWM is similar to the CSFPWM in the time domain. Under the VSFPWM control, when the modulation ratio is step-up changed, the switching frequency of three-level FC inverter is adjusted simultaneously to control the voltage ripple within the requirement.

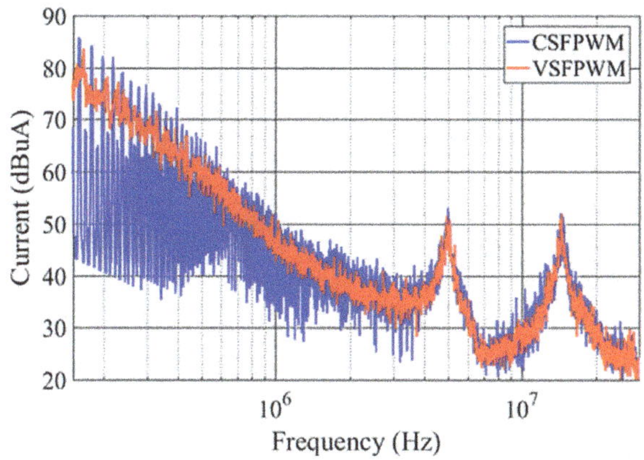

Fig. 6.77 EMI comparison

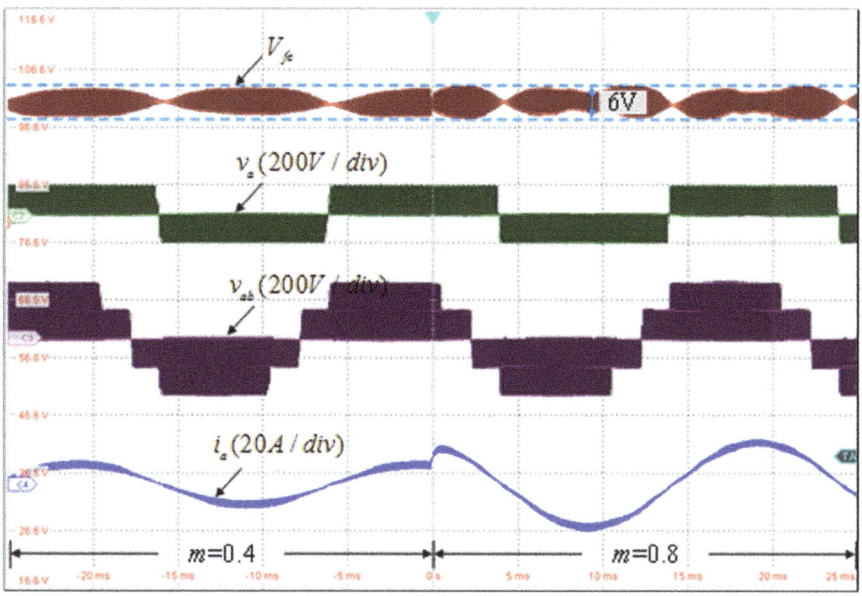

Fig. 6.78 Dynamic performance with CSFPWM

6.4 PWM Strategies for Current-Source Converters

Most of the work in this book is focused on power electronics VSC. The characteristics of VSCs are that the DC side is the voltage source and the AC side is the current source. At present, since the DC power supply of the actual application is with voltage source characteristics mostly, and the AC load represented by the motor is with current-source characteristic, the topology of the VSCs is adopted in most

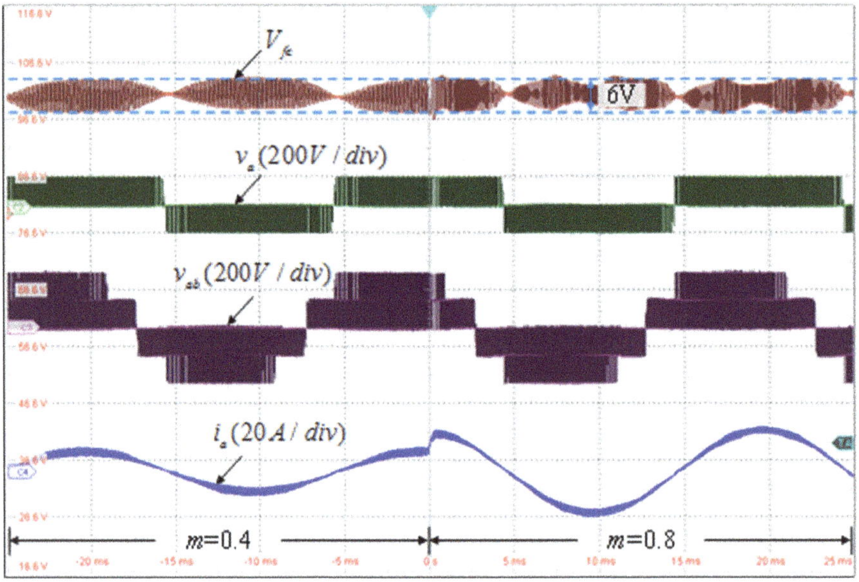

Fig. 6.79 Dynamic performance with VSFPWM

industrial applications. The paralleled converters and multilevel converters described in Sects. 6.2 and 6.3 of this chapter are the extensions of the regular two-level VSCs, which also conform to the characteristics that the AC side is current source and the DC side is voltage source. In this section, the twin topology of VSC:CSC is used for analysis.

6.4.1 General Control Method for Current Source Converters

As described in Chap. 3, the other class of topologies corresponding to VSC is CSC. Figure 6.80 shows a schematic diagram of the three-phase load with a typical three-phase current source inverter supplying. The CSC is characterized by current source characteristics on the DC side and voltage source characteristics on the AC side. Since most of the DC power supplies are with voltage source characteristics, the current source inverter adopts a series DC reactor to realize the current source characteristics on the DC side. On the AC side, three-phase capacitors are added to achieve the voltage source characteristics output and then connected to the load.

The current on the DC side flows into the load of the AC side through the switch of the three-phase bridge arm. Due to the characteristics of the current source on DC side, the switching device of each bridge arm is the structure with an active switching device and a diode in series that can withstand bidirectional voltage and only conduct unidirectional current, instead of the structure with an active switching device and an

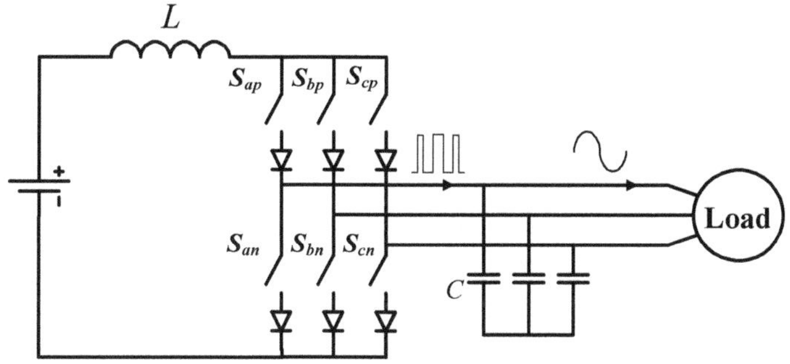

Fig. 6.80 Typical current source converter and load

anti-paralleled diode that can realize the bidirectional flow of current. The chopping target for CSCs is current, and the output pole is the pulsed current instead of the pulsed voltage.

Although the structure is more challenging than the widely used VSCs, CSCs also have their place in industrial applications. CSC has unique advantages: the presence of large inductors makes it possible to cope with the problem of DC bus shoot-through, and the reliability of the inductor is higher than the VSCs with chemical capacitors, especially for harsh environments. The Z-source converter, which has been a research area in the past decade, is also an extension of the CSC [12]. This chapter will introduce the PWM method for a typical CSC.

Structural changes make the control of CSCs more complicated. Figure 6.81 shows the control structure in *d–q* coordinates. Figure 6.81a shows the control structure in the *d–q* coordinates for the VSCs: since the output is with the characteristic of current source, the reference control variables and feedback are the currents in *d–q* coordinates. After the current regulators H_{id} and H_{iq}, the cross-decoupling variables ωLi_q and ωLi_d are added or subtracted, and the DC bus voltage is divided to obtain

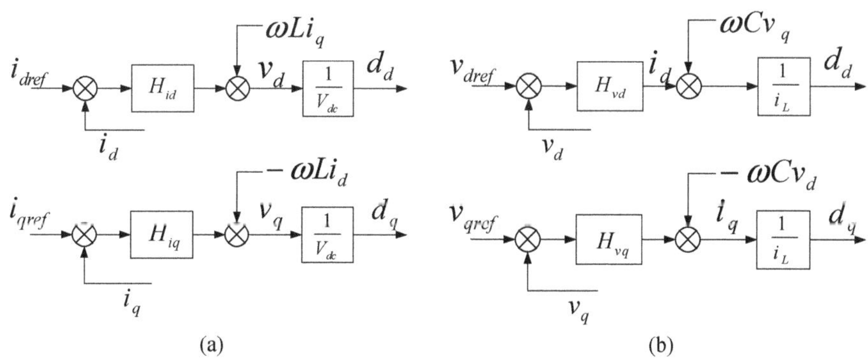

Fig. 6.81 Controller structure in *d–q* coordinates: **a** voltage source converter, **b** current source converter

the duty cycles d_d and d_q. Correspondingly, Fig. 6.81b shows the control structure in the d–q coordinates of the CSCs: since the output is with the characteristic of voltage source, the reference control variables and feedback are the voltages in d–q coordinates. After the voltage regulators H_{vd} and H_{vq}, the cross-decoupling variables ωCv_q and ωCv_d are added or subtracted, and the DC current is divided to obtain the duty cycles d_d and d_q.

The output duty cycles d_d and d_q for the controller are implemented through PWM. The PWM of CSC is different from that of VSC because its corresponding basic space vector is different. Figure 6.82 shows the switch combination principle of a typical three-phase CSC: the positive bus p and the negative bus n are connected to the same phase or different phase through the three-phase switches S_a, S_b and S_c at any time. Both positive bus p and the negative bus n have three connection methods with the three-phase bridge arms. Different from the eight switch vectors for VSC, the CSC has nine switch vectors, including three zero vectors, that is, three straight-through states. Table 6.3 shows the nine vectors and their corresponding current modes.

According to the current vector combination mode shown in Table 6.3, nine sets of current vectors can be obtained. Similar to the vector synthesis of VSC, three-phase currents of CSC are converted from a-b-c coordinates to α-β coordinates, as shown in (6.28).

$$I=\begin{pmatrix} i_\alpha \\ i_\beta \end{pmatrix}=T_{abc/\alpha\beta}\begin{pmatrix} i_a \\ i_b \\ i_c \end{pmatrix}=\sqrt{\frac{2}{3}}\begin{pmatrix} 1 & -\frac{1}{2} & -\frac{1}{2} \\ 0 & \frac{\sqrt{3}}{2} & -\frac{\sqrt{3}}{2} \end{pmatrix}\begin{pmatrix} i_a \\ i_b \\ i_c \end{pmatrix} \tag{6.28}$$

The length of the vector calculated by the formula (6.28) is the current vector length, and the phase angle is the angle of the current vector. Taking the ab vector in

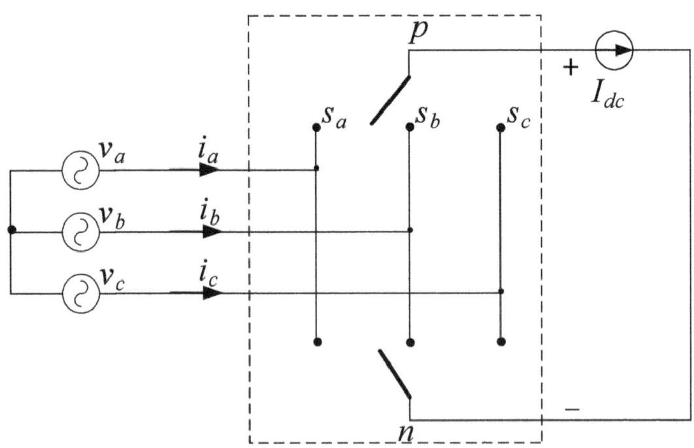

Fig. 6.82 Switching combination of three-phase current source converter [13]

Table 6.3 Switching vector and corresponding voltage and current distribution of three-phase current source converter [13]

P	n	Switching state	i_a	i_b	i_c	v_{pn}
A	a	aa	0	0	0	0
A	b	ab	I_{dc}	$-I_{dc}$	0	v_{ab}
A	c	ac	I_{dc}	0	$-I_{dc}$	$-v_{ca}$
B	a	ba	$-I_{dc}$	I_{dc}	0	$-v_{ab}$
B	b	bb	0	0	0	0
B	c	bc	0	I_{dc}	$-I_{dc}$	v_{bc}
C	a	ca	$-I_{dc}$	0	I_{dc}	v_{ca}
C	b	cb	0	$-I_{dc}$	I_{dc}	$-v_{bc}$
C	c	cc	0	0	0	0

Table 6.3 as an example, the calculation result is as shown in (6.29):

$$I_{ab} = \begin{pmatrix} i_\alpha \\ i_\beta \end{pmatrix} = \sqrt{\frac{2}{3}} \begin{pmatrix} 1 & -\frac{1}{2} & -\frac{1}{2} \\ 0 & \frac{\sqrt{3}}{2} & -\frac{\sqrt{3}}{2} \end{pmatrix} \begin{pmatrix} I_{dc} \\ -I_{dc} \\ 0 \end{pmatrix} = \begin{pmatrix} \sqrt{\frac{3}{2}} I_{dc} \\ -\sqrt{\frac{1}{2}} I_{dc} \end{pmatrix} \tag{6.29}$$

In this way, six nonzero current vectors are obtained as shown in Fig. 6.83. The six nonzero vectors are all 60° shift. The lengths of the three zero vectors are all zero. For the synthesis of the reference current, similar to VSC, two adjacent standard current vectors I_N and I_{N+1} are applied, and the corresponding action time is obtained by the sine theorem. The time remained for the switching cycle is assigned to the zero

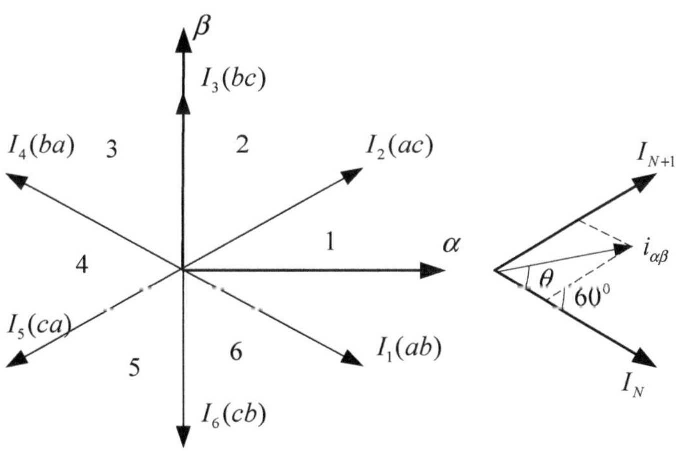

Fig. 6.83 Space vector and reference vector synthesis of current source converters

vectors. As shown in (6.30), the duty cycles d_1, d_2 and d_0 of the switching vector I_N, I_{N+1} and the zero vector in one switching cycle are finally obtained, and ρ is the amplitude of the reference vector.

$$\begin{cases} d_1 = \frac{T_N}{T_s} = \frac{2\rho}{\sqrt{6}I_{dc}}\sin(60^\circ - \theta) \\ d_2 = \frac{T_{N+1}}{T_s} = \frac{2\rho}{\sqrt{6}I_{dc}}\sin(\theta) \\ d_0 = 1 - d_1 - d_2 \end{cases} \tag{6.30}$$

Taking the first sector as an example, the two standard vectors are $\boldsymbol{I_1}(ab)$ and $\boldsymbol{I_2}(ac)$, and the zero vector is $\boldsymbol{I_0}(aa)$. According to the principle of vector synthesis (6.30), the three-segment duty cycles d_1, d_2 and d_0 are obtained, and the PWM is generated according to the timing of Fig. 6.84.

In order to achieve the timing of Fig. 6.84, the carrier comparison method shown in Fig. 6.85 is required. A saw-tooth-shaped carrier whose amplitude varies from 0 to 1 is compared with two sets of comparison values *h1 and h2*. The first set of

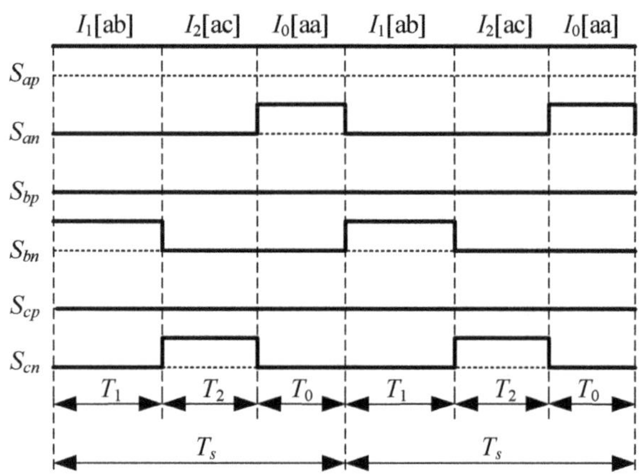

Fig. 6.84 PWM switching sequence example [13]

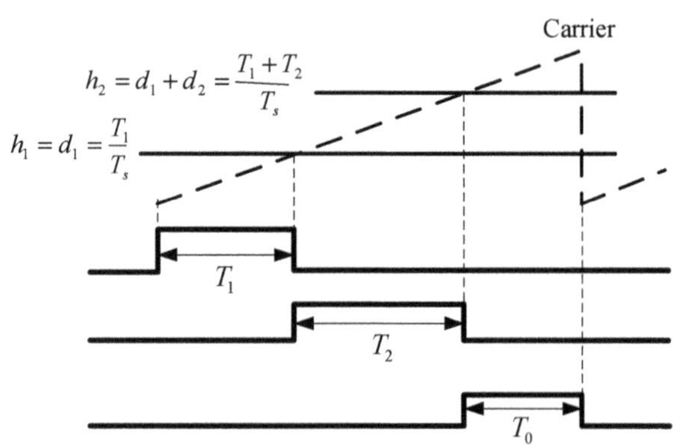

Fig. 6.85 Example of principle of implementing carrier comparison PWM to current source converter

comparison is used to generate the pulse of width T_1, the pulse between the first and second sets is used to generate the pulse of width T_2, and the last segment is the pulse of width T_0. Corresponding PWM is implemented by assigning these three sets of pulses to the space vector shown in Fig. 6.86.

According to the comparator of Fig. 6.85, the vector synthesis method of Fig. 6.86 is written into the interrupt service routine of the controller, and each interrupt cycle is executed once to complete the PWM synthesis of the switching cycle, as shown in Fig. 6.86. After the execution of the interrupt is started, the amplitude and phase of the reference current vector are first calculated to determine the corresponding sector and the angle in the sector; then, according to the vector synthesis equation, the T_1, T_2 and T_0 are calculated. According to the comparison method of Fig. 6.85, the comparison value and the carrier are determined, PWM is generated and assigned to the corresponding gate drive signal, finally the interruption is completed.

Figure 6.87 shows the gate drive signals of the six switching devices from simulation after applying the flowchart of Fig. 6.86. Taking the three segments of a cycle shown in the figure as an example: in the first segment, the upper switch of the phase-a (G_{ap}) is turned on, the lower switch of the phase-b (G_{bn}) is turned on and the two switches of the phase-c are turned off, thus realizing the output of the $\boldsymbol{I}_{ab}$ vector. In the second segment, both the upper switch and lower switch for phase-b are simultaneously off. The upper switch remains off but the lower switch (G_{cn}) is on for phase-c, thus realizing the output of the $\boldsymbol{I}_{ac}$ vector. In the third segment, the upper switch remains on and the lower switch (G_{an}) is also turned on for phase-a to form shoot-through, while all the switches for phase-b and phase-c are turned off, thus realizing the output of the $\boldsymbol{I}_{aa}$ vector. The above is the current PWM of a typical switching cycle of VSC.

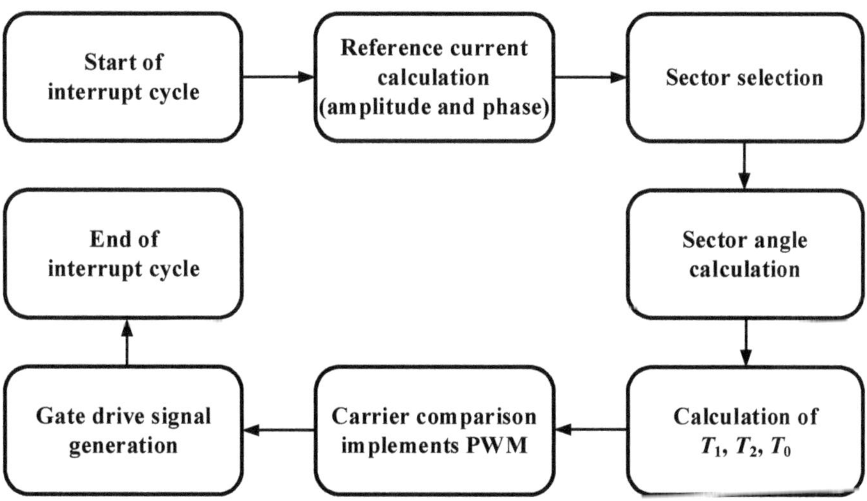

Fig. 6.86 Interrupt flowchart of real-time converter PWM implemented in software

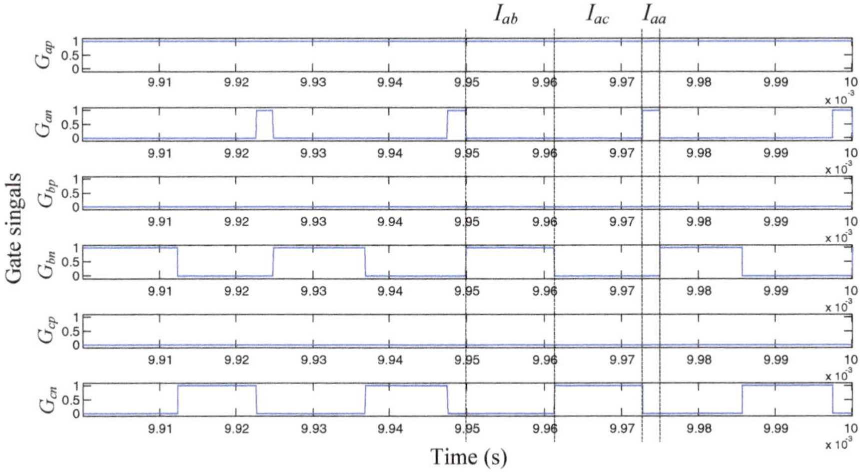

Fig. 6.87 Simulation result of PWM signal for CSC

Figures 6.88 and 6.89 compare the output of a VSC and a CSC. Figure 6.88 compares the output line voltage of the VSC and the output line current of the CSC. It can be seen that the CSC realizes the three-level pulse characteristic of the output line current through PWM, which is consistent with the output line voltage of the VSC. On this basis, through the filtering of the three-phase inductor, the output phase current of the VSC exhibits a sinusoidal characteristic, that is, the current ripple is added to the average current, as shown in Fig. 6.89a. Similarly, through the filtering of the three-phase capacitor, the output phase voltage of CSC exhibits a sinusoidal characteristic, that is, the voltage ripple is added to the average voltage, as shown in Fig. 6.89b. According to the current ripple prediction in Fig. 6.89a, VSFPWM can

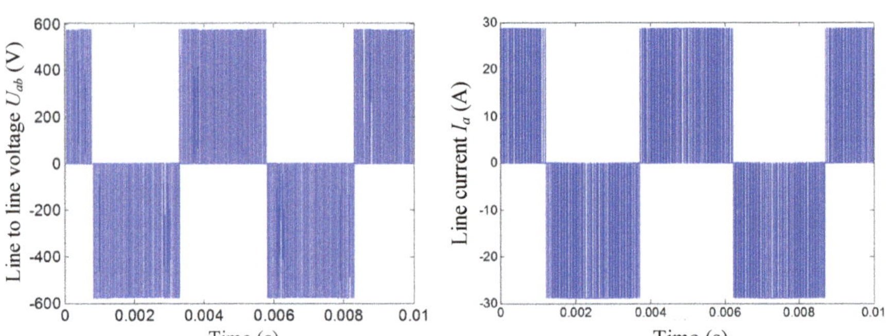

Fig. 6.88 Simulation comparison: **a** line voltage for voltage source converter, **b** line current for current source converter

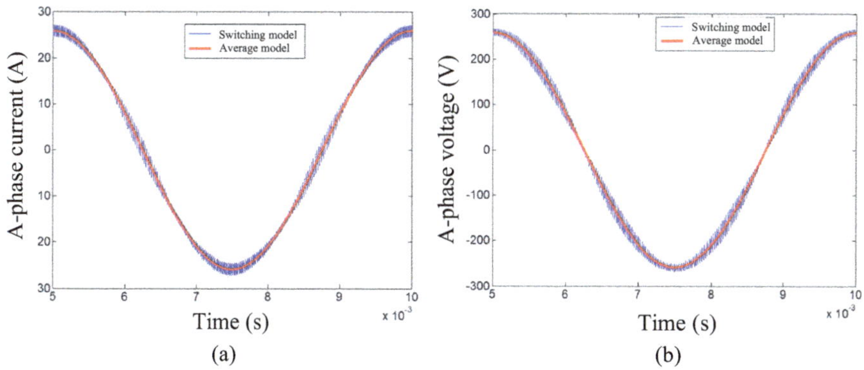

Fig. 6.89 Simulation comparison: **a** phase current for voltage source converter, **b** phase voltage for current source converter

be designed to control the current ripple. Similarly, the voltage ripple of the CSC can be controlled according to the voltage ripple prediction in Fig. 6.89b.

6.4.2 VSFPWM for DC-Link Current Ripple Control in CSI

The basic control and PWM for CSC have been introduced in Sect. 6.4.1. Similar to VSC, the VSFPWM can also be developed for CSC. For the direct thinking, the VSFPWM in CSC can be used to control the voltage ripple in AC side instead of the current ripple of VSC. However, the current ripple in the DC-link inductor of CSC is with more concerns in application since it is associated with the inductor performance. In this part, the VSFPWM of CSC is aiming at the DC-link current ripple.

The DC-link of current source inverter (CSI) needs a series inductor to ensure the stability and smoothness of DC current. As shown in Fig. 6.90, DC current $\boldsymbol{i_{dc}}$ is turned into three-phase PWM currents i_{pa}, i_{pb} and i_{pc} after the action of six switches Q_1–Q_6, and there are filter capacitors connected in parallel on the load side to filter out high-frequency harmonics in PWM currents and the load currents i_a, i_b and i_c are obtained.

In CSI, there are only two switches opened at any time. When two switches of the same phase are opened, V_{dc} is directly connected in parallel with inductor L_{dc}, and DC current $\boldsymbol{i_{dc}}$ increases. When two switches of different phases are opened at the same time and the voltage V_{pn} is higher than V_{dc}, part of energy stored in L_{dc} is transferred to the load side and $\boldsymbol{i_{dc}}$ decreases. In steady state, the increase and decrease of $\boldsymbol{i_{dc}}$ should be equal in one switching cycle. Therefore, there are nine switch combinations and corresponding standard current space vectors, including three zero vectors and six nonzero vectors which divide space into six sectors. The distribution of nine space vectors in α-β coordinate system is shown in Fig. 6.83. The

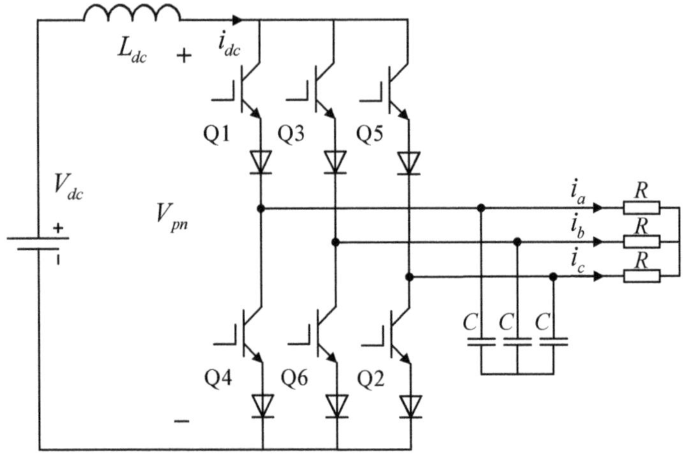

Fig. 6.90 The equivalent circuit of current source inverter

PWM scheme of CSI on which all the work in this section is based is five-segment SVPWM. As shown in Fig. 6.83, $\boldsymbol{i_{ref}}$ is supposed to be synthesized by adjacent nonzero vectors $\boldsymbol{i_a}$, $\boldsymbol{i_b}$ and zero vector $\boldsymbol{i_0}$. The action time of each current vector is expressed in (6.30).

From (6.30), the action time T_x, T_y and T_0 of $\boldsymbol{i_N}$, $\boldsymbol{i_{N+1}}$ and zero vector $\boldsymbol{i_0}$ can be calculated. With five-segment SVM, the switching sequence when $\boldsymbol{i_{ref}}$ is in Sector I is shown in Fig. 6.91. The selection of zero vector in each sector should minimize the total number of switching actions in a switching cycle.

For DC inductor, $u = L\mathrm{d}i/\mathrm{d}t$. As long as the voltage of inductor is obtained, the slope of DC current ripple $u = L\mathrm{d}i/\mathrm{d}t$ can be calculated. Table 6.4 shows the slopes of $i_L = i_{dc}$ corresponding to nine current vectors. The variation of DC current ripple can be predicted, as shown in Fig. 6.91. Taking $\boldsymbol{i_{ref}}$ in Sector I as an example: the first segment of $t < T_x/2$ corresponds to vector $\boldsymbol{i_{ab}}$ and $k_1 = \mathrm{d}i_L/\mathrm{d}t$, and the second segment corresponds to vector $\boldsymbol{i_{ac}}$ and $k_2 = \mathrm{d}i_L/\mathrm{d}t$. The values of current ripple at the first 2 turning points h_1 and h_2 are given in (6.31). Considering about the symmetry of ripple, current ripple peak value ΔI_{peak} equals to max($h1$, $h2$). When $\boldsymbol{i_{ref}}$ is located in other sectors, the derivation method is similar. The simulation results and prediction results are given in Fig. 6.92. The average value of DC current $\boldsymbol{i_{dc}} = 40$ A, and the predicted ripple peak value is almost exactly the same as the actual value when CSFPWM is applied in simulation model.

$$\begin{cases} h_1 = k_1 \dfrac{T_x}{2} \\ h_2 = k_1 \dfrac{T_x}{2} + k_2 \dfrac{T_y}{2} \end{cases} \tag{6.31}$$

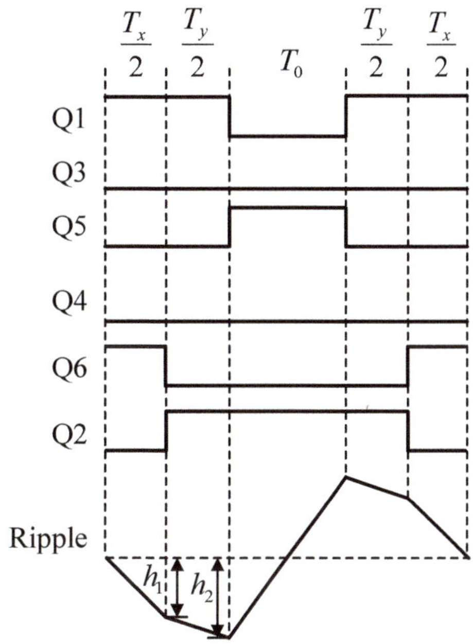

Fig. 6.91 Switching sequence with five-segment SVM

Table 6.4 Current vectors and corresponding current ripple slopes

Vector	$\mathrm{d}i_L/\mathrm{d}t$	Vector	$\mathrm{d}i_L/\mathrm{d}t$	Vector	$\mathrm{d}i_L/\mathrm{d}t$
$\overrightarrow{i_{61}}$	$(V_{dc}- V_a + V_b)/L$	$\overrightarrow{i_{34}}$	$(V_{dc}- V_b + V_a)/L$	$\overrightarrow{i_{14}}, \overrightarrow{i_{36}}, \overrightarrow{i_{25}}$	V_{dc}/L
$\overrightarrow{i_{12}}$	$(V_{dc}- V_a + V_c)/L$	$\overrightarrow{i_{45}}$	$(V_{dc}- V_c + V_a)/L$		
$\overrightarrow{i_{23}}$	$(V_{dc}- V_b + V_c)/L$	$\overrightarrow{i_{56}}$	$(V_{dc}- V_c + V_b)/L$		

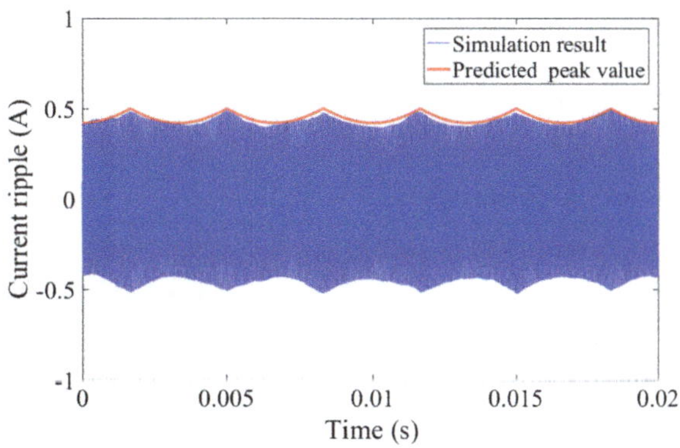

Fig. 6.92 Simulation result and peak prediction result of DC current ripple

The ripple peak of DC current is an important parameter to CSI which changes periodically in steady state, and the selection of DC inductance is mostly based on its maximum value which has been estimated in many papers. However, when CSFPWM is applied, current ripple peak is smaller than its maximum value for most of the period of a switching cycle. If DC current ripple peak can be predicted, the switching frequency can be reduced to keep ripple peak equal to its maximum value when it should decrease under CSFPWM scheme. Then the switching loss of CSI system can be reduced while all the requirements for CSI are still met, and EMI may also be suppressed just like when VSFPWM is applied to VSI.

Assuming V_{dc} and $\boldsymbol{i}_{dc}$ of CSI are known quantities. In microprocessor, $\boldsymbol{i}_{dc}$ amplitude ρ and output vector phase angle θ are known, which means the duty cycle of each current vector T_x/T_s, T_y/T_s and T_0/T_s can be calculated by (6.30) directly. Three-phase voltages on the load side can also be measured, so k_1 and k_2 can be calculated according to Table 6.4. Therefore, ΔI_{peak} can be derived from (6.31) as long as T_s is given.

To achieve VSFPWM of CSI, a standard switching period T_s is first specified and the corresponding current ripple peak $\Delta I_{peak\_predicted}$ can be calculated using the ripple prediction model. Therefore, if ΔI_{peak} is to be fixed at $\Delta I_{peak\_required}$, the required switching period T_{sN} can be obtained by (6.32):

$$T_{sN} = \frac{\Delta I_{peak\_required}}{\Delta I_{peak\_predicted}} T_s \tag{6.32}$$

Corresponding simulation is performed. CSI model has been presented in Fig. 6.90, with $V_{dc} = 50$ V, $L = 1$ mH, $R = 5\ \Omega$ and $C = 50\ \mu$F in steady state. When the maximum allowable ΔI_{peak} is designed to be 1 A, the result in Fig. 6.93 suggests that this requirement is exactly met when $f_s = 24.9$ kHz if using CSFPWM. When using VSFPWM scheme, the switching frequency is given in Fig. 6.94, which is periodically changed in the range of 21–24.9 kHz. The reduction of average switching frequency also means the improvement of efficiency. At the same time, DC current ripple peak is also controlled very well, as shown in Fig. 6.95. In addition, a reduction

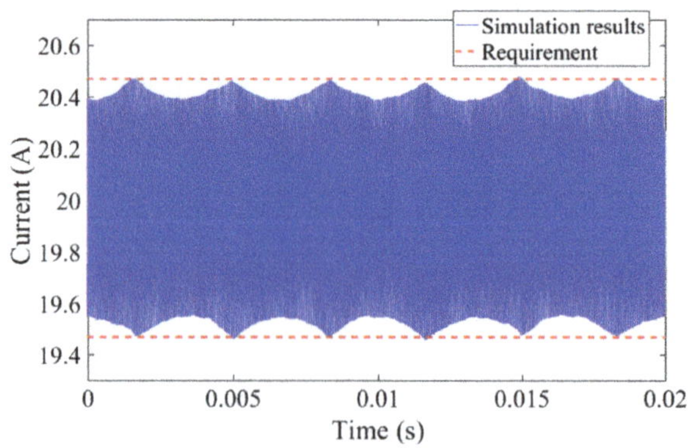

Fig. 6.93 DC current ripple with CSFPWM

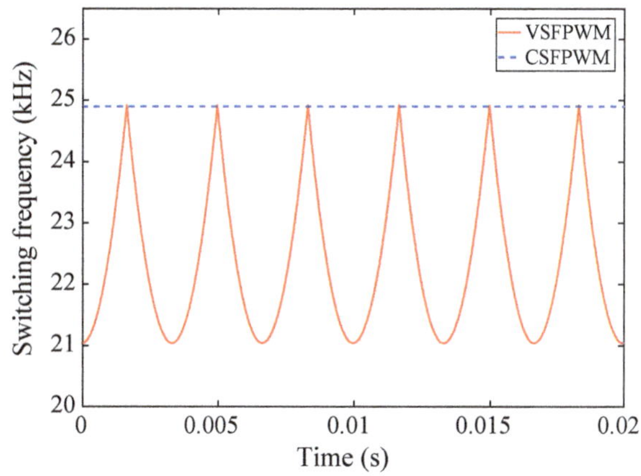

Fig. 6.94 Switching frequency comparison

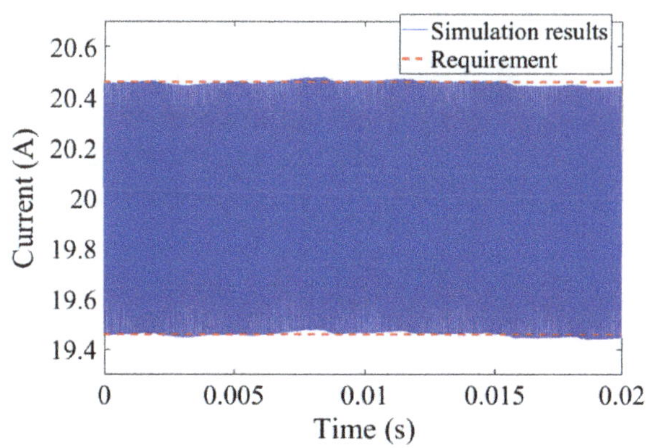

Fig. 6.95 DC current ripple with VSFPWM

of ~15 dB can be seen in the 150–800 kHz range in Fig. 6.96, which illustrated a better EMI performance for proposed VSFPWM.

6.5 PWM Strategies for OWMD with Novel Topology [25–27, 29]

While the widely used three-phase or multiphase loads are with star-connection for neutral point, there is another connection strategy for the load, which is open-winding connection. As shown in Fig. 6.97a, the open-winding motor drive (OWMD) is formed by opening the neutral point of the traditional Y-connected motor and use two converters to drive the open-winding motor. For OWMD with traditional three-phase H-bridge converter shown in Fig. 6.97b, the two types of PWM strategy have been introduced in detail in Sect. 6.1. It can achieve a multilevel power supply effect, reducing the harmonic components of the output voltage and current and improve the

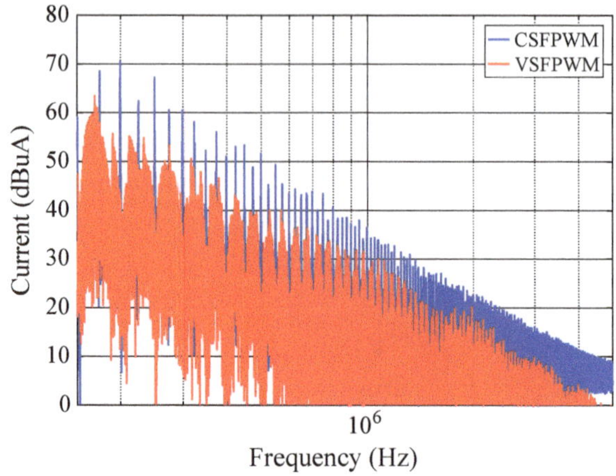

Fig. 6.96 EMI comparison

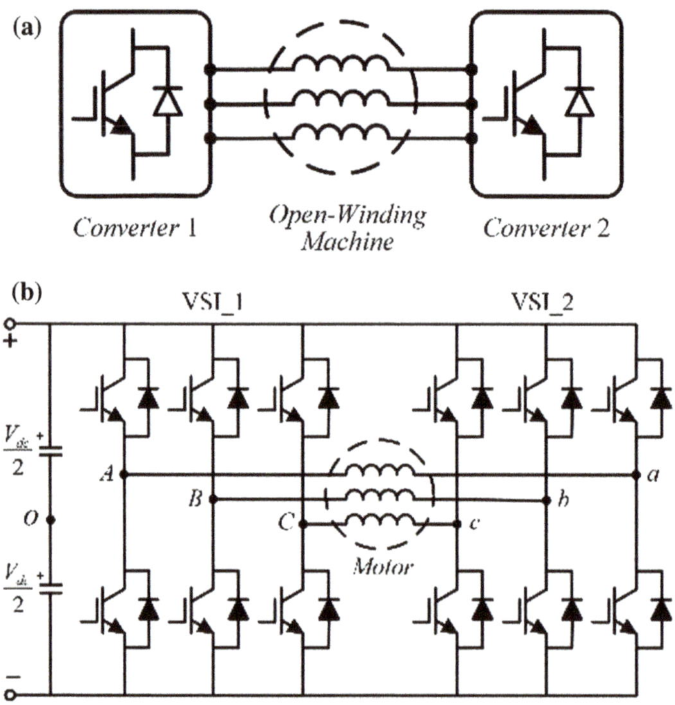

Fig. 6.97 **a** Open-winding motor drive. **b** Traditional three-phase H-bridge converter

DC voltage utilization. At the same time, due to the increase in the control freedom degree, the control of OWMD is flexible and can drive different types of motors.

However, the problem of the OWMD with traditional converter lies in the need for two general two-level VSCs, which leads to high cost, and low power density. Therefore, this section introduces a novel OWMD converter with multiplexed bridge leg and its corresponding PWM modulation method, which can reduce the cost, size and power loss while ensuring the modulation performance.

6.5.1 SVPWM for OWMD with Novel Three-Phase Four-Leg Converter

To overcome the shortcoming in the traditional converter, a novel three-phase four-leg converter has been proposed for OWMD, which is shown in Fig. 6.98. Through the bridge leg multiplexing technology, the bridge legs on both sides of phase-b are shared with the bridge legs of phase-a and phase-c, respectively. It has been proved that the advantages of traditional three-phase H-bridge converter in voltage utilization and zero-axis current control capability can be similarly realized through the three-phase four-leg converter. Compared with three-phase H-bridge converter, two bridge legs have been saved by the three-phase four-leg converter.

The SVPWM method for three-phase four-leg converter is different from general two-level VSCs. Because of the freedom of zero-axis voltage control, this topology requires three-dimensional voltage modulation at the same time.

According to Kirchhoff's voltage law, the relationship between the three-phase voltage and the leg voltage is shown in (6.33), where S_i is the switch variable of leg

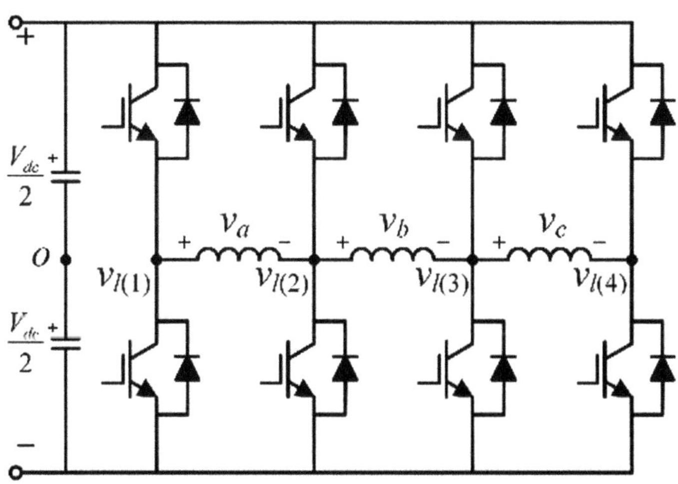

Fig. 6.98 Novel three-phase four-leg converter for OWMD with common DC bus

i, $i = 1, 2, 3, 4$. When $S_i = 0$, the upper switch is turned off and the below switch is turned on. When $S_i = 1$, the upper switch is turned on and the below switch is turned off.

$$\begin{bmatrix} V_a \\ V_b \\ V_c \end{bmatrix} = \begin{bmatrix} V_1 - V_2 \\ V_2 - V_3 \\ V_3 - V_4 \end{bmatrix} = \begin{bmatrix} S_1 - S_2 \\ S_2 - S_3 \\ S_3 - S_4 \end{bmatrix} \times V_{dc} \tag{6.33}$$

According to Clark transformation, the three-phase voltage in *abc*-axis frame can be transformed into the V_α, V_β, V_O in $\alpha\beta 0$-axis frame, and the voltage in $\alpha\beta 0$-axis frame can be expressed by switch variable as shown in (6.34).

$$\begin{cases} \dfrac{V_\alpha}{V_{dc}} = \dfrac{2}{3}S_1 - S_2 + \dfrac{1}{3}S_4 \\ \dfrac{V_\beta}{V_{dc}} = \dfrac{\sqrt{3}}{3}(S_2 - 2S_3 + S_4) \\ \dfrac{V_0}{V_{dc}} = \dfrac{1}{3}(S_1 - S_4) \end{cases} \tag{6.34}$$

According to the state of switch variable S_i in each leg, 16 voltage vectors of the three-phase four-leg converter in $\alpha\beta 0$-axis frame can be obtained, and Fig. 6.99 gives the space presentation of these voltage vectors in $\alpha\beta 0$-axis frame. Voltage vectors are labeled with decimal numbers, which, when converted into binary representation, reveal the values of the switch variable of leg 1 to leg 4, for example $V_6 = 0101$ means $S_1 = 0$, $S_2 = 1$, $S_3 = 0$, $S_4 = 1$.

It can be seen that although the voltage modulation area is three-dimensional, in the $\alpha\beta$-plane, the SVPWM for $V_{\alpha\beta}$ is basically the same as the traditional two-level VSC: six effective vectors divide the $\alpha\beta$-plane into six sectors. In each sector,

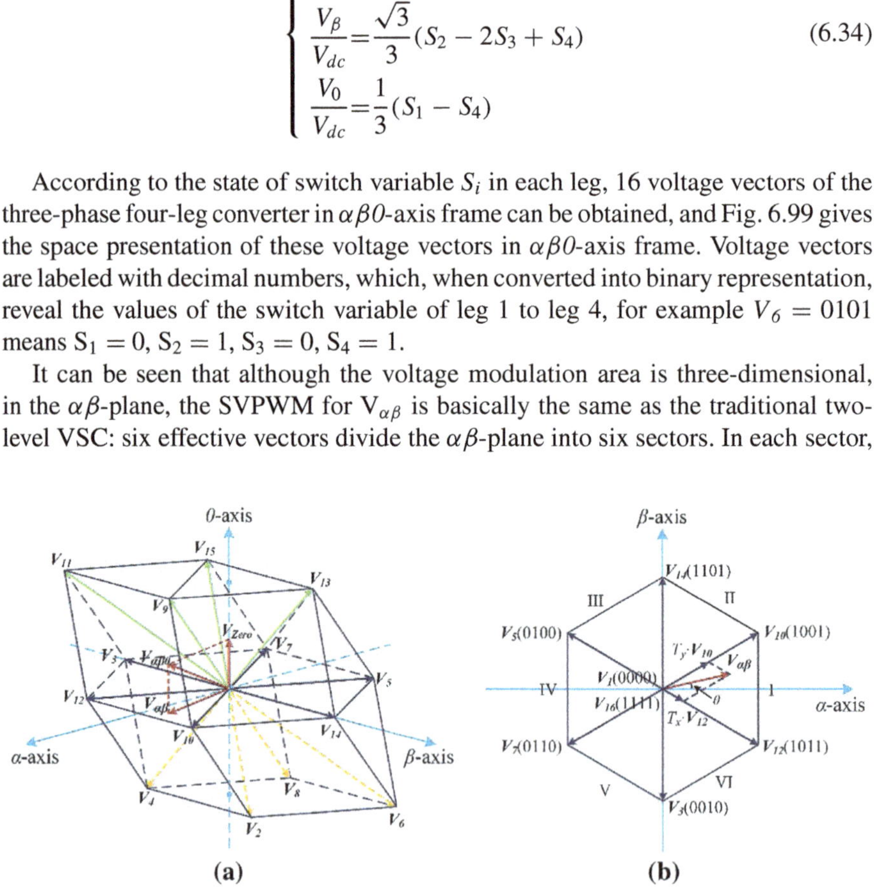

Fig. 6.99 **a** Spatial location of 16 voltage vectors in $\alpha\beta 0$-axis frame. **b** Eight voltage vectors in the $\alpha\beta$-plane

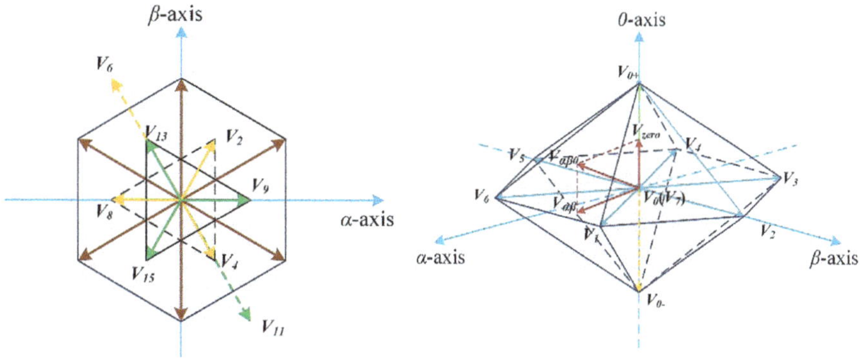

Fig. 6.100 **a** The voltage vectors projected onto the $\alpha\beta$-plane. **b** The motion space of reference vector $\boldsymbol{V}_{\alpha\beta\gamma}$

the effective vector on both sides and two zero vectors are used to synthesize the reference vector.

However, this converter also includes the modulation of the 0-axis voltage, which is different from traditional two-level VSC. Figure 6.100a shows the voltage vectors projected onto the $\alpha\beta$-plane. It can be seen that all voltage vectors will affect the reference voltage modulation in the $\alpha\beta$-plane and produce undesired harmonic components. Therefore, in order to eliminate this effect, when modulating the 0-axis voltage, V_9, V_{13}, V_{15} (or V_2, V_4, V_8) should be used at the same time and ensure that the effective time of the three vectors is equal. Thereby, the projected components in the $\alpha\beta$-plane will cancel each other out. Because the zero-axis voltage modulation is synthesized by three voltage vectors, the equivalent zero-axis voltage action time is reduced to one-third of the original. The final modulation range is shown in Fig. 6.100b, where V_{0+} and V_{0-} are synthesized from V_9, V_{13}, V_{15} and V_2, V_4, V_8, respectively.

Figure 6.101 shows the maximum pulse A-phase voltage generated by the three-phase four-leg converter and the corresponding FFT analysis. FFT analysis shows that the maximum amplitude of the fundamental phase voltage can reach DC bus voltage (V_{dc}). In addition, the pulse phase voltage does not include additional zero-sequence alternating voltages, such as the third-harmonic voltage.

6.5.2 *CBPWM for OWMD with Novel* N*-phase (*N + *1)-Leg Converter*

As shown in Fig. 6.102, the three-phase four-leg converter can be further extended to the N-phase (N + 1)-bridge-leg converter for N-phase OWMD. This type of structure has a distinct feature: all phase windings are connected in series, and then

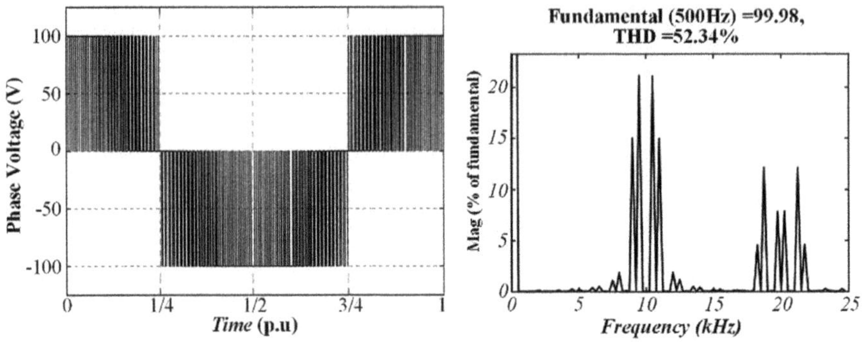

Fig. 6.101 The pulse A-phase voltage generated by the three-phase four-leg converter and the corresponding FFT analysis

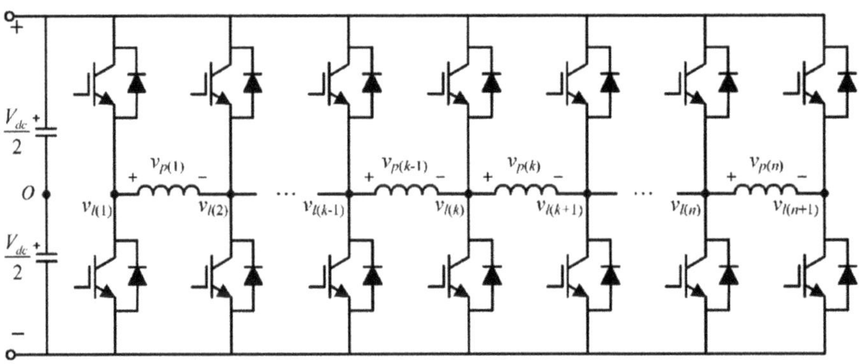

Fig. 6.102 Novel N-phase ($N + 1$)-leg converter for OWMD with common DC bus

$N + 1$ nodes are led out and connected to $N + 1$ converter bridge legs, respectively. Therefore, the N-phase OWMD with this type of converters is also called as series-end winding motor drive (SWMD). SWMD can be a potential alternative for OWMD with traditional H-bridge converter for multiphase AC motor. However, the SVPWM for N-phase SWMD with $N + 1$ legs requires the calculation and synthesis of 2^{N+1} voltage vectors, which is a very complex issue and make the application to be impossible.

In comparison, carrier-based PWM can be an easier way for multiphase SWMD. Carrier-based PWM for SWMD should be developed on the basis of understanding the relationship between phase and leg voltage. However, different from the N-phase VSC whose phase-leg voltage is determining the phase voltage directly, due to the special winding connection, it is difficult to find a direct formula between the phase-leg voltage and phase voltage in the stationary frame in SWMD.

Therefore, to realize the instantaneous modulation with all freedoms (including fundamental voltage, harmonic voltage and zero-axis voltage modulation), this section introduces a generalized modulation method combined with synchronous coordinate transformation for SWMD.

Assuming the leg voltage $v_{l(k)}$ is the midpoint output voltage of leg k, as shown in Fig. 6.102. The relationship between leg voltage and phase voltage is shown in (6.35).

$$v_{p(k)} = v_{l(k)} - v_{l(k+1)} \tag{6.35}$$

where $k = 1, 2, \ldots, n$.

According to the generalized Park inverse transformation, the N-dimensional phase voltage vector can be decomposed into different components as follows.

$$\begin{bmatrix} v_{p(1)} \\ \vdots \\ v_{p(k)} \\ \vdots \\ v_{p(n)} \end{bmatrix} = \sum_{m} \underbrace{\begin{bmatrix} \cos m\theta & -\sin m\theta \\ \cos m(\theta-\alpha) & -\sin m(\theta-\alpha) \\ \vdots & \vdots \\ \cos m(\theta-(n-1)\alpha) & -\sin m(\theta-(n-1)\alpha) \end{bmatrix} \begin{bmatrix} V_{d(m)} \\ V_{q(m)} \end{bmatrix}}_{subspace\,m} + \underbrace{\begin{bmatrix} 1 \\ 1 \\ \vdots \\ 1 \end{bmatrix} V_0}_{0axis} + \underbrace{\begin{bmatrix} 1 \\ -1 \\ \vdots \\ -1 \end{bmatrix} V_{0-}}_{0^{-}axis} \tag{6.36}$$

where $V_{d(m)}$ and $V_{q(m)}$ are the d–q-axis phase voltage components in subspace m. α is the angle difference between two adjacent phases with $\alpha = 2\pi/n$. V_0 is the zero-axis phase voltage component and $V_{0\text{-}}$ is the negative zero-axis phase voltage component. Component $V_{0\text{-}}$ in negative zero-axis only exists in even-phase motor. θ is taken as the corresponding electrical angle when the rotor rotates. $m \in (1, 3, 5, \ldots, n\text{–}2)$ in odd-phase motor, $m \in (1, 2, 3, \ldots, 0.5n\text{–}1)$ in even-phase motor.

From (6.36), it can be seen that the N-dimensional phase voltages are decomposed into many two-dimensional subplanes and axes. According to the superposition principle, the phase voltage components in each subplane and axis can be separately modulated. The process is to obtain the corresponding reference leg voltage components in each subplane and axis and then superimpose all of them to obtain the total reference leg voltage.

A. *Modulation in two-dimensional subplane*

From Fig. 6.102, taking subspace m as an example, the k-phase voltage component in subspace m can be expressed as

$$v_{p(k)}^{(m)} = [\cos m(\theta-(k-1)\alpha) \; -\sin m(\theta-(k-1)\alpha)] \begin{bmatrix} V_{d(m)} \\ V_{q(m)} \end{bmatrix} \tag{6.37}$$

According to (6.37), the phase voltage components in subspace m can be expressed by the corresponding voltage phasors. The magnitude and phase relationship between the k-leg and $(k + 1)$-leg voltage phasors and the k-phase voltage phasor is shown in the phasor diagram Fig. 6.103a.

According to the geometric relationship between phase voltage phasors and leg voltage phasors in Fig. 6.103a, the magnitude of the leg voltage and the value of λ

can be obtained as follows:

$$\begin{cases} \lambda = \dfrac{\pi - m\alpha}{2} \\ \left|V_{l(k)}\right| = V^{(m)}/2\sin\dfrac{m\alpha}{2} \end{cases} \tag{6.38}$$

where $k = 1, 2, \ldots, n + 1$.

Therefore, according to (6.38), both the phase voltage component and the leg voltage component can be drawn in the subplane m, as shown in Fig. 6.103b.

Define $V_{ld(m)}$ as the d-axis leg voltage in subplane m, and $V_{lq(m)}$ as the q-axis leg voltage in subplane m. According to the geometric relationship, $V_{ld(m)}$ and $V_{lq(m)}$ can be derived as follows:

$$\begin{cases} V_{ld(m)} = 0.5\left(V_{d(m)} + V_{q(m)}\cot\dfrac{m\alpha}{2}\right) \\ V_{lq(m)} = 0.5\left(V_{q(m)} - V_{d(m)}\cot\dfrac{m\alpha}{2}\right) \end{cases} \tag{6.39}$$

Furthermore, the k-leg voltage component in subspace m can be obtained by inverse Park transformation of $V_{ld(m)}$ and $V_{lq(m)}$.

B. *Modulation in zero-axis and in negative zero-axis*

According to (6.35) and (6.36), the relationship between the k-phase voltage component and leg voltage component in zero-axis is as follows:

$$V_{l(k)}^{(0)} - V_{l(k+1)}^{(0)} = V_{p(k)}^{(0)} = V_0 \tag{6.40}$$

where $k = 1, 2, \ldots, n$.

Accumulating N equations in (6.40),

$$V_{l(1)}^{(0)} - V_{l(n+1)}^{(0)} = nV_0 \tag{6.41}$$

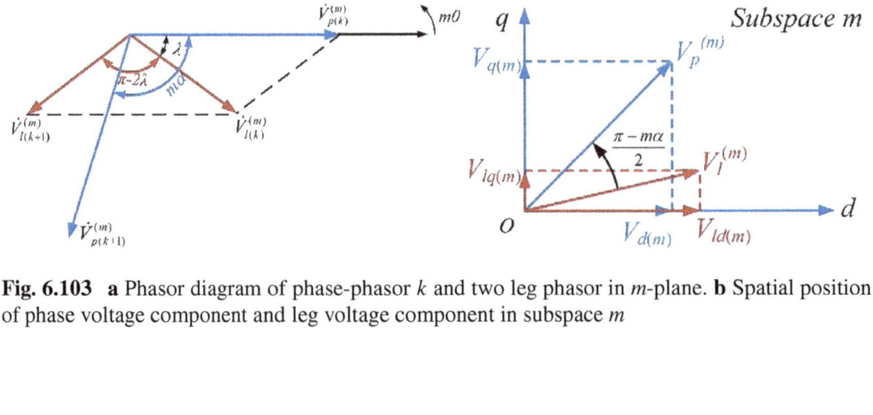

Fig. 6.103 **a** Phasor diagram of phase-phasor k and two leg phasor in m-plane. **b** Spatial position of phase voltage component and leg voltage component in subspace m

To have the minimum amplitude of leg voltage, the amplitude of $V_{l(1)}^{(0)}$ and $V_{l(n+1)}^{(0)}$ should be set to be equal. Therefore, the leg voltage component in zero-axis is calculated as follows:

$$V_{l(k)}^{(0)} = (0.5n - k + 1)V_0 \tag{6.42}$$

In even-numbered motors, there may be negative zero-axis components in the phase voltage. By a similar method to obtain the zero-axis leg voltage, the negative zero-axis leg voltage is calculated as follows:

$$V_{l(k)}^{(0-)} = 0.5(-1)^{k-1}V_{0-} \tag{6.43}$$

According to the above equation, when the amplitude of fundamental reference phase voltage is set to maximum, Fig. 6.104 shows the actual chopping leg voltage and actual chopping phase voltage. Although the actual 1-leg voltage before filtering only includes two levels, the actual A-phase voltage before filtering includes three levels with the proposed method, which can reduce the THD of phase current. After filtering, 1-leg voltage is consistent with the reference 1-leg voltage modulation.

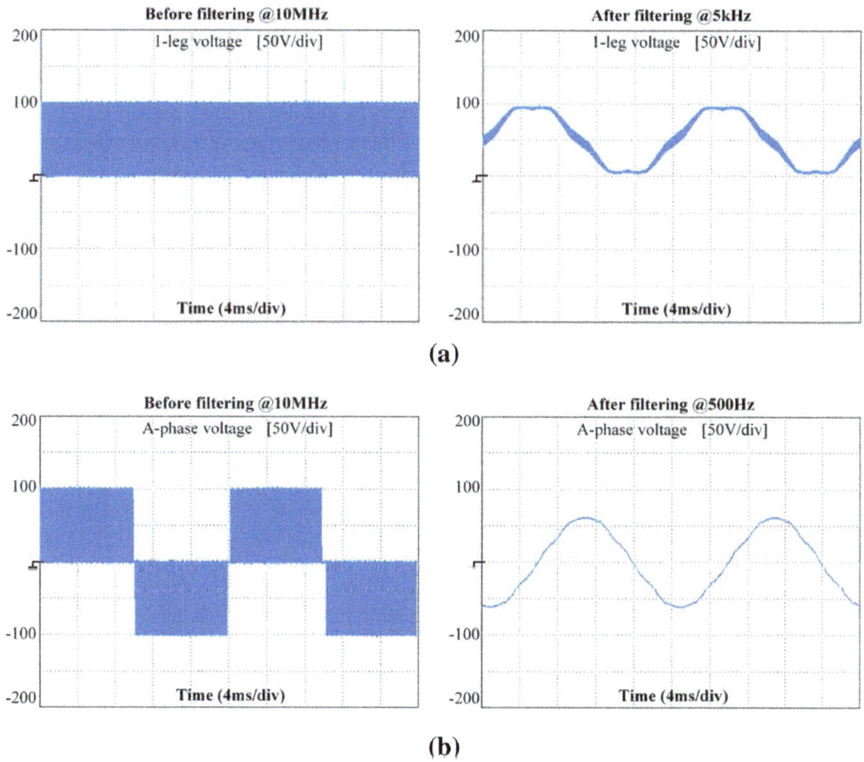

Fig. 6.104 **a** The actual chopping leg voltage before and after filtering. **b** The actual chopping phase voltage before and after filtering

6.6 Summary

The main research object in the previous chapters of this book is the two-level VSCs that are widely used in three-phase systems, but there are other topologies beyond two-level converters in terms of performance. Some structures are developed based on two-level VSCs: paralleled converters and multilevel converters, and some are the counterpart of VSCs, such as CSCs. These topologies have their own advantages and are irreplaceable in industrial applications, thus they are collectively referred to as complicated topologies in this chapter. This chapter begins with a comprehensive introduction to the characteristics of these topologies. The modulation methods of these complicated topologies are different from the ordinary two-level VSC: the paralleled converter can synthesize the target vector by performing space vector synthesis based on the vector synthesis for respective converters. Multilevel converters have more space vectors, and can also be obtained by comparing the multiple carriers with reference waves. The CSC synthesis target is a reference current vector rather than a reference voltage vector, and the standard current vector has its own special role. In this chapter, for the paralleled converter, the carrier phase-shift PWM method is introduced, which can realize the cancellation of the current harmonics related to the switching frequency and improve the power quality. This chapter introduces the VSFPWM strategy based on current ripple prediction for both paralleled and multilevel converters. The idea is similar to the VSFPWM of ordinary two-level converters, which can improve the system's performance on switching loss and electromagnetic interference. Besides, how to use the PWM methods to solve the neutral point potential balance problem of the clamp-type multilevel converter is another part for three-level converters. This chapter also introduces the space vector PWM technology of CSC and its VSFPWM method. Another family of topologies for motor drive is open-winding motor drive. Novel series-end winding motor drive can achieve the same modulation index of the full-bridge open winding motor drive, with much less power electronics devices. This chapter introduces the SVPWM and CBPWM for the novel series-end winding motor drives, too. It is worth noting that the advanced PWM methods for common-mode noise suppression of paralleled converters and multilevel converters are not described in detail in this section and are left to Chap. 7.

References

1. Baiju MR, Mohapatra KK, Kanchan RS, Gopakumar K (2004) A dual two-level inverter scheme with common mode voltage elimination for an induction motor drive. IEEE Trans Power Electron 19(3):794–805
2. Wang Y, Panda D, Lipo TA, Pan D (2013) Open-winding power conversion systems fed by half-controlled converters. IEEE Trans Power Electron 28(5):2427–2436
3. Zhang D, Wang F, Burgos R, Lai R, Boroyevich D (2010)Impact of interleaving on AC passive components of paralleled three-phase voltage-source converters. IEEE Trans Ind Appl 46(3):1042–1054

4. Zhang D, Wang F, Burgos R, Boroyevich D (2012) Total flux minimization control for integrated inter-phase inductors in paralleled, interleaved three-phase two-level voltage-source converters with discontinuous space-vector modulation. IEEE Trans Power Electron 27(4):1679–1688
5. Schweizer M, Kolar JW (2013) Design and implementation of a highly efficient three-level T-type converter for low-voltage applications. IEEE Trans Power Electron 28(2):899–907
6. Baumann M, Kolar JW (2005) A novel control concept for reliable operation of a three-phase three-switch buck-type unity-power-factor rectifier with integrated boost output stage under heavily unbalanced mains condition. IEEE Trans Ind Electrons 52(2):399–409
7. Hagiwara M, Akagi H (2009) Control and experiment of pulsewidth-modulated modular multilevel converters. IEEE Trans Power Electron 24(7):1737–1746
8. Ryu H-M, Kim J-H, Sul S-K (2005) Analysis of multiphase space vector pulse-width modulation based on multiple d-q spaces concept. IEEE TransPower Electron 20(6):1364–1371
9. Zhang D (2010) Analysis and design of paralleled three-phase voltage source converters with interleaving. Ph.D. Dissertation, Virginia Polytechnic Institute and State University, 2010
10. Zhang X, Mattavelli P, Boroyevich D, Wang F (2013)Impact of interleaving on EMI noise reduction of paralleled three phase voltage source converters. In: Proc Appl Power Electron Conf Expos (APEC), pp 2487–2492
11. Orfanoudakis GI, Youratich MA, Sharkh SM (2013) Hybrid modulation strategies for eliminating low-frequency neutral-point voltage oscillations in the neutral-point-clamped converter. IEEE Trans Power Electron 28(8):3653–3658
12. Wei Q, Wu B, Xu D, Zargari NR (2016) A natural-sampling-based SVM scheme for current source converter with superior low-order harmonics performance. IEEE Trans Power Electron 31(9):6144–6154
13. Peng FZ et al (2005) Z-source inverter for motor drives. IEEE Trans Power Electron 20(4):857–863
14. Franquelo LG, Rodriguez J, Leon JI, Kouro S, Portillo R, Prats MAM (2008) The age of multilevel converters arrives. IEEE Ind Electron Mag 2(2):28–39
15. Lai J-S, Peng FZ (1996) Multilevel converters-a new breed of power converters. IEEE Trans Ind Appl 32(3):509–517
16. Tolbert LM, Peng FZ, Habetler TG (1999) Multilevel converters for large electric drives. IEEE Trans Ind Appl 35(1):36–44
17. Rodriguez J, Lai JS, Peng FZ (2002)Multilevel inverters: a survey of topologies, controls, and applications. IEEE Trans Ind Electron 49(4):724–738
18. Rodriguez J, Bernet S, Wu B, Pontt JO, Kouro S (2007) Multilevel voltage-source-converter topologies for industrial medium-voltage drives. IEEE Trans Ind Electron 54(6):2930–2945
19. Han X, Jiang D, Zou T, Qu R, Yang K (2019) Two-segment three-phase PMSM drive with carrier phase-shift PWM for torque ripple and vibration reduction. IEEE Trans Power Electron 34(1):588–599
20. Jiao Y (2015) High power high frequency 3-level neutral point clamped power conversion system. Doctor Dissertation, Virginia Polytechnic Institute and State University
21. Celanovic N, Boroyevich D (2000) A comprehensive study of neutral-point voltage balancing problem in three-level neutral-point-clamped voltage source PWM inverters. IEEE Trans Power Electron 15(2):242–249
22. Lewicki A, Krzeminski Z, Abu-Rub H (2011) Space-vector pulse width modulation for three level NPC converter with the neutral point voltage control. IEEE Trans Ind Electron 58(11):5076–5086
23. Weidong J, Wang L, Wang J, Zhang X, Wang P (2018) A carrier-based virtual space vector modulation with active neutral point voltage control for neutral point clamped three-level inverter. IEEE Trans Ind Electron 65(11):8687–8696
24. Chen X, Huang S, Jiang D et al (2018) Fast voltage-balancing scheme for a carrier-based modulation in three-phase and single-phase NPC three-level inverters. J Electr Eng Technol 13
25. Li A, Jiang D, Kong W, Qu R (2019) Four leg converter for reluctance machine with DC biased sinusoidal winding current. IEEE Trans Power Electron 34(5):4569–4580

26. Li A, Jiang D, Zicheng L, Kong W (2019) Five-phase series-end winding motor controller: converter topology and modulation method. IEEE Energy Conversion Congress and Exposition (ECCE), Baltimore, MD, USA 2019:629–634
27. Li A, Jiang D, Liu Z, Sun X (2020) Generalized PWM method for series-end winding motor drive. IEEE Transactions on Power Electronics 36(4):4452–4462
28. Liu K, Liu Z, Jiang D, Wang Q, He Z (2020) Four-module three-phase permanent-magnet synchronous motor based PWM modulation strategy for suppressing vibration and common mode current. 2020 IEEE Energy Conversion Congress and Exposition (ECCE), Detroit, MI, USA, 2020:2328–2335
29. Li A, Sun X, Jiang D, Liu Z (2020) A generalized carrier-based PWM with zero-axis voltage elimination for open-end winding motor drive. 2020 IEEE Energy Conversion Congress and Exposition (ECCE), Detroit, MI, USA, 2020:5336–5340
30. Li Q, Jiang D, Shen Z, Zhang Y, Liu Z (2019) Variable switching frequency PWM strategy for high-frequency circulating current control in paralleled inverters with coupled inductors. IEEE Transactions on Power Electronics 35(5):5366–5380
31. Chen J, Jiang D, Li Q. (2018) Attenuation of conducted EMI for three-level inverters through PWM. CPSS Transactions on Power Electronics and Applications 3(2):134–145

Chapter 7
PWM Technology for Common-Mode Noise Reduction

For power electronics converters, common-mode (CM) noise is a relatively independent problem. This chapter focuses on PWM techniques to reduce the CM noise. First, the basic problem of CM noise is introduced; then, several improved PWM strategies are introduced to suppress the common-mode voltage (CMV) amplitude; in addition, to suppress the common-mode current, the PWM strategy needs to be coordinated with the CM loop; finally, when the control freedom increases in the topology, the freedom of suppressing CM noise also increases. In this chapter, three-level and paralleled inverters with zero-CM PWM technologies have been introduced, some common inverter schemes are also included. Especially, series work of paralleled inverters with zero-CM PWM has been discussed.

7.1 Common-Mode Noise Introduction

The concept of "common mode" is relative to "differential mode", which represents electrical variables in the same conducted direction. Common-mode voltage (CMV) denotes the codirectional voltage in the conducted direction, and common-mode current (CMC) denotes the codirectional current in the conducted direction. In the field of electromagnetic interference (EMI), common-mode (CM) noise represents the electromagnetic interference noise transmitted along the same conducted direction, usually in the form of high-frequency common-mode current.

In ideal linear circuits, the problem of CM noise does not exist. This is due to the voltage balance for each phase, which makes the sum of voltage in the same conducted direction zero. For example, in three-phase balancing circuit, the AC three-phase voltage sources always keep the sum zero, so the CMV is zero. At the same time, there is no CMC circulating path in the ideal circuit. Therefore, the problem of CM noise is caused by nonideal factors in practical application, namely CMV (noise source) and stray capacitance (conduction path).

D. Jiang et al., *Advanced Pulse-Width-Modulation: With Freedom to Optimize Power Electronics Converters*, CPSS Power Electronics Series,
https://doi.org/10.1007/978-981-33-4385-6_7

Because of the application of power electronics converter, high-frequency CM noise source: CMV is generated. Unlike the traditional linear AC voltage source, the pulse voltage sequence of each phase is formed by pulse width modulation (PWM) controlled power electronics converter. In this case, though the average sum of pulse voltage in each phase is zero in a fundamental period, the high-frequency CMV sequence exists in each switching cycle. Therefore, the existence of CMV source is determined by the PWM scheme. However, unlike previous chapters, the CM noise source must be determined by the interaction of all phase PWM. The different characteristics of noise sources and the dependence on nonideal conduction circuits determine the particularity of CM EMI.

Figure 7.1 shows the paths of DM and CM EMI for AC-fed motor drive. The EMI transmitting between the motor phases caused by the phase voltage difference of converter is the DM EMI, which also includes the conduction EMI between the front-end rectifiers; while the common-mode EMI which is generated by the CM component of the output voltage (CMV) of three-phase converter transmits to the ground through the stray capacitance of the motor windings, and then transmits back to the converter through the grounding terminal O_1 of the AC bus. The CMV of the whole converter system is very complex under AC power supply which is the synthetic product of the CM component for front-end rectifier and converter.

Figure 7.2 shows the DM and CM EMI paths of DC-fed motor drive. Unlike the AC power supply system shown in Fig. 7.1, DC power supply does not require a

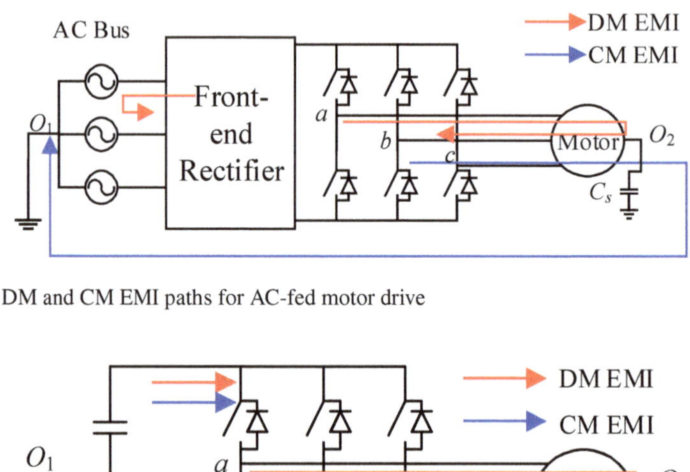

Fig. 7.1 DM and CM EMI paths for AC-fed motor drive

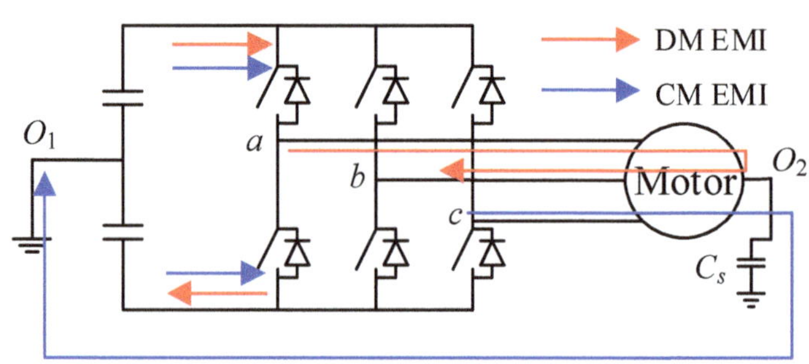

Fig. 7.2 DM and CM EMI paths for DC-fed motor drive

front-end rectifier, so the generation of CMV is more intuitive. In Fig. 7.2, the DC bus neutral point O_1 is grounded, and the CM loop is formed by the grounding neutral point O_2 of the motor winding through the stray capacitance C_s. Assuming that the DC bus voltage V_{dc} remains constant, the pole voltages at the output terminals *a*, *b*, *c* of the converter which are relative to the midpoint O_1 of the DC bus are pulse voltage sequences determined by PWM and switched between the positive and negative bus voltage. The expression of CMV is shown in (7.1). The CM loop includes the CM impedance of the motor and the stray capacitance to ground, and also includes the stray capacitance of the inverter to the heat sink and the transmission line to ground.

$$V_{cm} = \frac{1}{3}(V_{aO_1} + V_{bO_1} + V_{cO_1}) \tag{7.1}$$

Figures 7.1 and 7.2 are the simplified flow paths of the motor CMC, and the actual conduction path of the CMC inside the motor is very complicated. On one hand, the magnetic field asymmetry caused by the three-phase unbalanced voltage of the motor stator induces a voltage on the motor shaft, so there is a voltage difference between the motor shaft and the motor frame (ground), that is, the shaft voltage. The shaft voltage can form a shaft current loop through the shaft, bearing and stator frame, shown in Fig. 7.3 with solid line.

On the other hand, motor grounding and motor-to-ground stray parameters are also two important factors in the conduction path. Because of these two factors, the earth is introduced into the CM conduction loop, allowing the CMC to propagate. System grounding is owing to safety requirements, forcing a certain part of the electrical equipment to be connected to the ground to control the electrical potential. The grounding point is generally referred to the neutral point; and the ground stray parameter is referred to the coupling path between the physical element and the

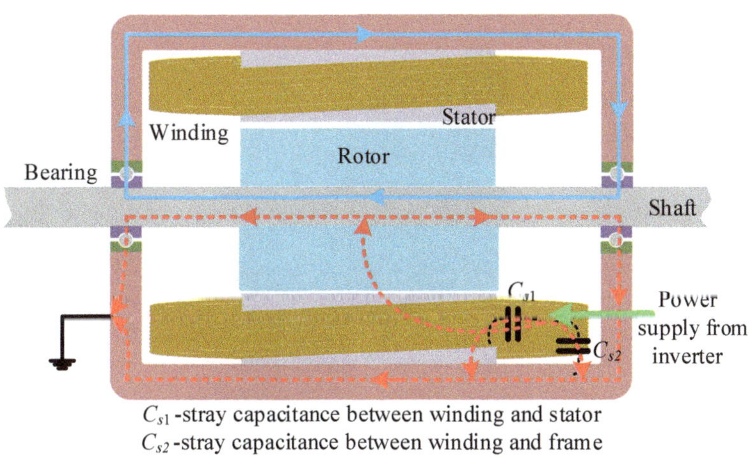

Fig. 7.3 Schematic diagram of CMC flow in AC motor

ground in the system, which is generally the stray capacitance to the ground. As shown in Fig. 7.3 with dotted line, taking the most important parasitic parameter C_{s1} as an example, C_{s1} represents the equivalent capacitance between the stator winding and the stator due to the small spacing and large contact area which connects to the ground through the motor frame. For the low-frequency CMV, it exhibits a large impedance and its induced CMC can be blocked. Unfortunately, due to the existence of high-frequency component of the CMV brought by the high-frequency switching action of the inverter, the inverter-fed motor makes CM-conducted EMI to become a non-negligible problem. With the application of the novel wide band-gap power electronics device whose switching speed and switching frequency are further increased, the CMC becomes more serious.

The main hazards caused by CM noise to power electronics and transmission systems include: further increase of EMI in the system on the basis of DM EMI, the motor long-term insulation damage due to continuous CMC flowing through the motor insulation, the motor bearing damage owing to the motor bearing current, and so on. Figure 7.4 shows the damage to the motor insulation due to the continuous flow of CMC through the stator winding insulation and motor frame. These problems make the suppression of CM noise and CMC to become an important issue [1].

The main common method to suppress the CM EMI noise and CMC for power electronics converters and motor drive systems is to add a CM filter which is represented by CM choke to increase the CM impedance in the CM loop [2]. The CM choke is essentially a large-value CM inductor whose volume and weight depend on the CM flux saturation of the core, which is the volt-second integral of the CMV. The larger the volt-second integral value, the easier is the core to saturate [3]. In industrial applications, the volume and weight of the CM choke are non-negligible parts in the entire power electronics converter system, especially in the case with strict EMI requirements. Therefore, how to reduce the weight and volume of the CM filter represented by the CM choke becomes an important problem in engineering.

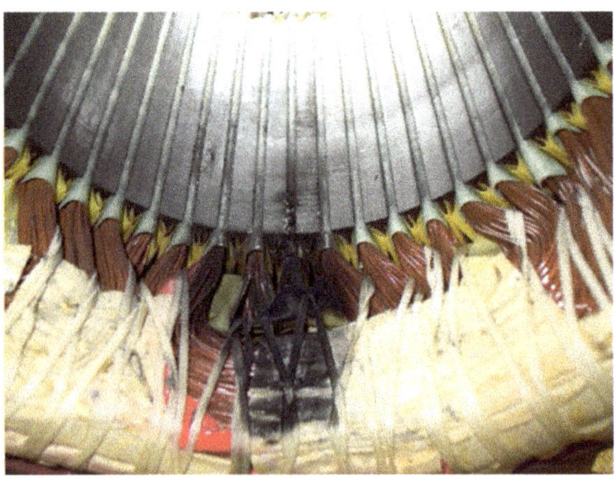

Fig. 7.4 The motor insulation damage due to CMC (picture from internet)

The essential approach for CM filter to solve the problem of CMC and CM noise is from the path. The other idea to solve the problem of the CM noise source is from the switching function of power electronics converter. Since the PWM mode determines the amplitude and frequency characteristics of the CM noise source, the CMC and CM noise can be suppressed from the source by improving the PWM method so that the requirement for the CM filter can be lower. Since the CMV is determined by the switching voltage pulses of all phases, the PWM improvement also needs to be implemented for all phases. This improvement utilizes the freedom of pulse distribution, and the goal is to suppress CM noise which is also in the category of advanced PWM.

Since the nature of CM noise is the CMC which is not only related to the CM noise source, but also related to the impedance of the conduction path, it is necessary to consider the characteristic of conduction path while designing the advanced PWM. In addition, with the change of power electronics converter topology, the available freedom and the space for improving the CMV can be increased. So advanced PWM can be further designed to achieve CM noise suppression. The main content of this chapter is centered around the work.

7.2 Common-Mode Voltage Suppression by Improved PWM Strategy

In order to design the advanced PWM strategy to suppress the CMV, the effect of conventional PWM on CMV should be first analyzed. Taking the most typical DC-fed inverter motor drive shown in Fig. 7.2 as an example, the most common space vector PWM (SVPWM) and discontinuous modulation PWM (DPWM) can be obtained in one switching cycle by the carrier-based PWM method. The switching function and the corresponding CMV are shown in Fig. 7.5a–c. For the SVPWM scheme shown in Fig. 7.5a, the three-phase switching functions are symmetric in one switching cycle, the $T_0/4$ period on both sides corresponds to the 000 vector, and the intermediate $T_0/2$ period corresponds to the 111 vector. With the definition of CMV shown in (7.1), 000 vector represents switching state of $-V_{dc}/2$ for each phase, so the CMV is $-V_{dc}/2$; 111 vector represents switching state of $V_{dc}/2$ for each phase, so the CMV is $V_{dc}/2$. The CMVs corresponding to the middle 100 and 110 vectors are $-V_{dc}/6$ and $V_{dc}/6$, respectively. Therefore, the CMV in a switching cycle presents four symmetrical steps with different values, with the peak amplitude of $V_{dc}/2$ in zero vector switching state. Figure 7.5b, c shows the switching functions and CMVs for the maximum and minimum clamped DPWM schemes. Similar to SVPWM, the maximum and minimum values for CMVs reach at zero vectors. The difference is that only 111 vector for maximum-clamped DPWM and 000 vector for minimum-clamped DPWM are utilized in a switching cycle. So it can be seen that the CMVs caused by zero vectors have the largest value for three-phase inverter. The main idea to reduce the CMV is to avoid using zero vectors.

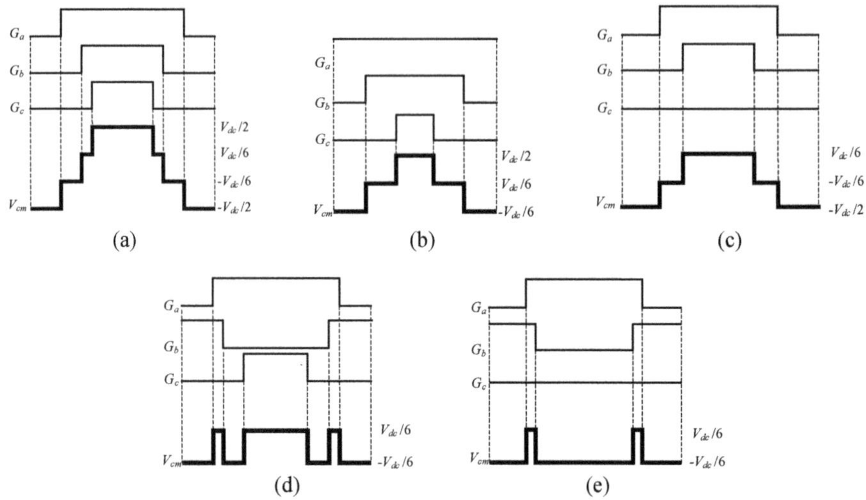

Fig. 7.5 Three-phase switching function and corresponding CMV in one switching cycle for different PWM schemes: **a** SVPWM, **b** DPWM_Max, **c** DPWM_Min, **d** AZSPWM, **e** NSPWM

In Fig. 7.5a–c, the reason for the zero vectors appearing is that the three-phase PWM utilizes a uniform alignment for PWM signals. If the switching function with the middle duty cycle (the phase-b in Fig. 7.5) is changed from centered alignment to double-side alignment by shifting 180°, the double-side 000 vector can be changed to 010 vector, and the centered 111 vector can be changed to 101 vector, which can avoid the use of 000 and 111 vectors throughout the switching cycle. Thus, the amplitude of CMV can be reduced from $V_{dc}/2$ to $V_{dc}/6$. For SVPWM scheme, the effect is shown in Fig. 7.5d, which is the active zero state PWM (AZSPWM). For the minimum clamped DPWM scheme, the effect is shown in Fig. 7.5e, which is the near state PWM (NSFPWM). It can be seen that this method can effectively reduce the amplitude of CMV to one-third of the previous value.

However, it should be pointed out that the improvement of DPWM cannot achieve the CMV reduction under every circumstance. Taking the DPWM shown in Fig. 7.5c as an example, when the modulation ratio m is small, even if the phase-a PWM is changed from the centered alignment to double-side alignment, there also exists period of time in which the zero vector cannot be avoided, as shown in Fig. 7.6. So, the suppression effect on the CMV for the improved DPWM scheme is limited by the modulation ratio.

The literatures [4–6] use the space vector method to analyze such improved methods in detail. In fact, the improved CMV PWM schemes implemented through the carrier comparison shown in Fig. 7.5 have the corresponding space vector interpretation. Figure 7.7 shows the space vector interpretation of the two typical improved PWMs in Fig. 7.5d, e. Figure 7.7a shows an improvement to SVPWM, named AZSPWM. This method does not use the two zero vectors of 000 and 111 in the synthesis of vector V_{ref}, but utilizes the opposite direction of two active vectors

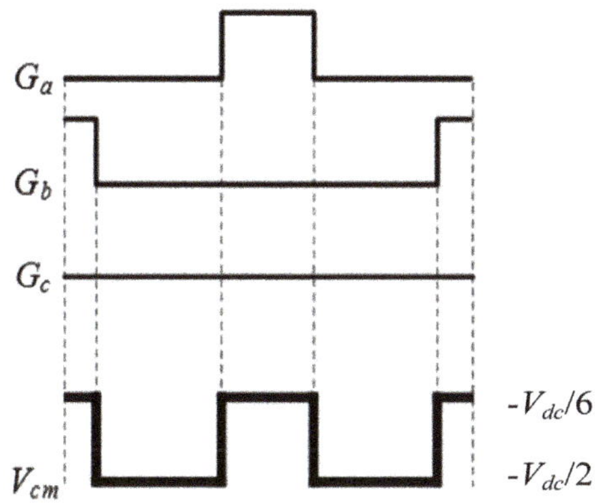

Fig. 7.6 Modified DPWM scheme losing the CMV effect at low modulation ratio

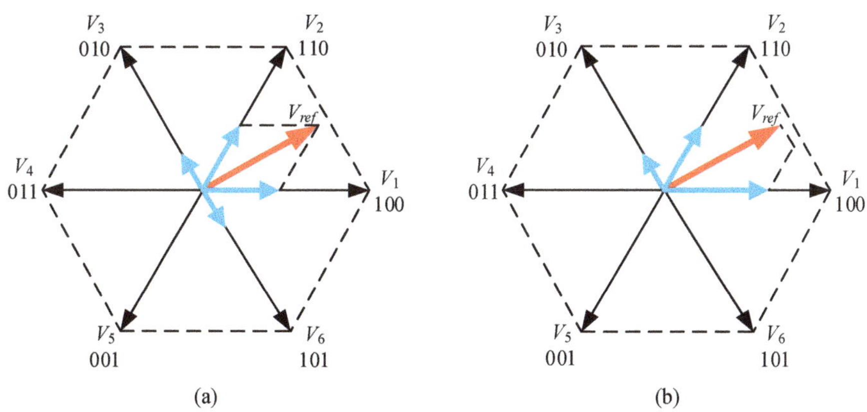

Fig. 7.7 Space vector analysis of reduced CMV PWM: **a** AZSPWM, **b** NSPWM [4]

which act at the same time to achieve the equivalent "dynamic zero vector". That is, the effect of zero vectors is realized by using two opposite vectors of V_3 (010) and V_6 (101) in Fig. 7.7a. This effect is consistent with the effect of Fig. 7.5d. Figure 7.7b shows an improvement to DPWM, named NSPWM. In this case, not only the two neighbor vectors (V_1 and V_2) are utilized to synthesize the V_{ref}, but also the third vector (V_3) of the adjacent sector is introduced to participate in the synthesis. In this way, the three action times can be solved, and the vector synthesis can be realized, without using the zero vectors. The effect is also consistent with Fig. 7.7e.

Considering the implementation equivalence of the space vector and carrier-based methods, and the ease of implementation for carrier-based method in the controller hardware and software, carrier-based method is used here as the implementation of CMV reduction. Figure 7.8 shows the principle of implementing two reduced CMV PWM strategies (AZSPWM and NSPWM) in microcontroller. After the controller

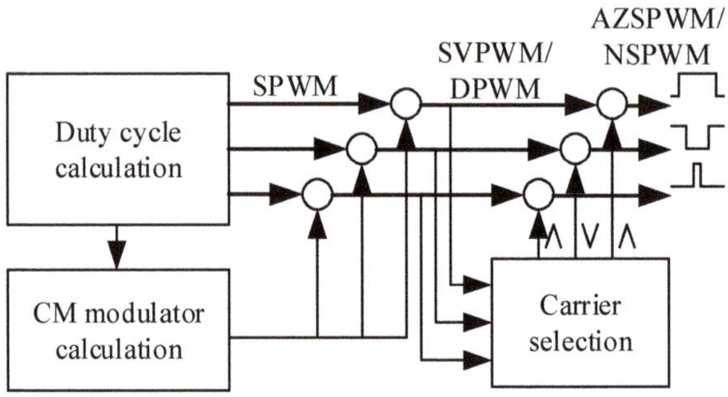

Fig. 7.8 The reduced CMV PWM generation method

calculates the basic duty cycles, the sinusoidal PWM (SPWM) is generated. As described in Chap. 2, the SPWM can be changed to SVPWM or DPWM by injecting different CM modulation functions. At this time, according to the length comparison of the three-phase duty ratios, the phase with middle length is phase-shifted by 180°, and SVPWM and DPWM can be changed to AZSPWM and NSPWM, respectively. Thus, the output CMV amplitude of the inverter can be reduced from $V_{dc}/2$ to $V_{dc}/6$.

Through this method, the output pole voltages and CMV corresponding to AZSPWM and NSPWM are partially amplified in experimental results, as shown in Fig. 7.9. It can be seen that after adopting AZSPWM and NSPWM, the pulse of the corresponding phase with middle duty cycle is shifted by 180°, and the peak value of the CMV is reduced to $V_{dc}/6$.

Comparing the test results of the CMV for the improved PWM and conventional PWM in a full fundamental period shown in Fig. 7.10, it can be seen that the CMV is reduced to one-third for AZSPWM compared with SVPWM; considering NSPWM versus DPWM, the CMV amplitude is also reduced to one-third.

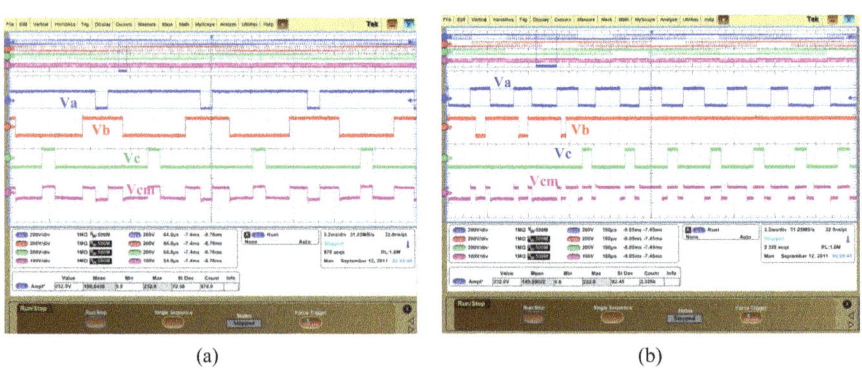

Fig. 7.9 Experimental results of partial amplification for each phase output pole voltage and CMV: **a** AZSPWM, **b** NSPWM

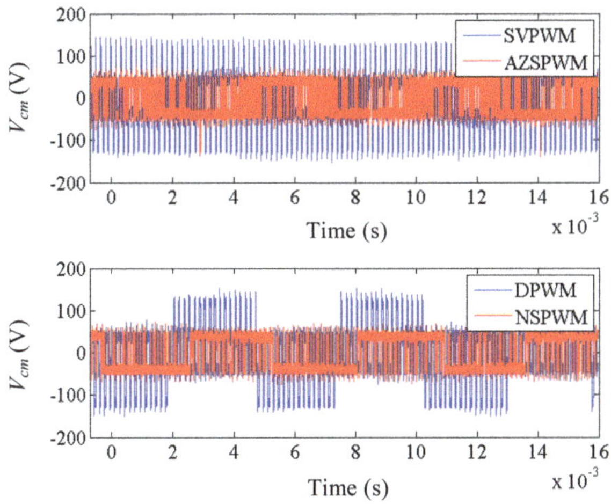

Fig. 7.10 Test result comparison of CMVs during the entire fundamental period

7.3 Common-Mode Loop Analysis and Common-Mode Current Suppression Method

The improved AZSPWM and NSPWM described in Sect. 7.2 are proved to effectively reduce the peak of CMV. But what really causes system problems is the conducted EMI, characterized by CMC which not only relates to the noise source but also closely relates to the impedance of the conduction loop. Since the CM conduction loop relies on parasitic parameters and has strong nonlinear characteristics, the amplitude reduction of CMV does not represent an effective suppression of the CMC. This section discusses this issue [7].

Considering the CM loop for the two typical inverter systems, that is, the photovoltaic (PV) inverter system and motor drive inverter system, the main parasitic parameters should be first identified. For PV system, the main parasitic parameters exist between the solar cell and the PV frame [8], as shown in Fig. 7.11a. For motor

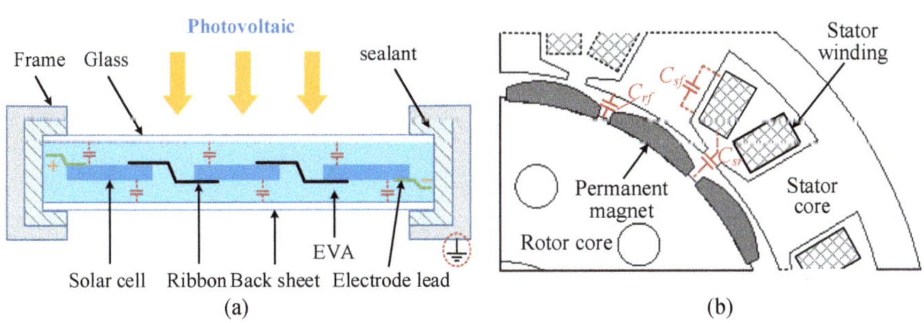

Fig. 7.11 Main parasitic parameters for typical inverter systems: **a** PV system, **b** motor drive system

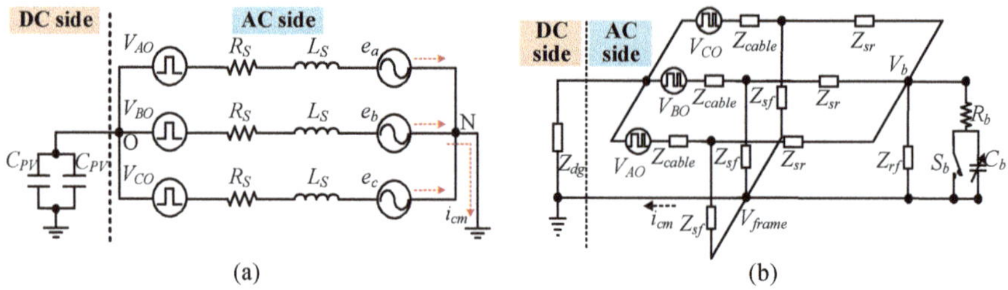

Fig. 7.12 The CM equivalent model for the typical inverter system: **a** PV system, **b** motor drive system

drive inverter system shown in Fig. 7.11b, the main parasitic parameters generate among the different parts of motor, such as the stray capacitance C_{sf}, C_{rf} and C_{sr} which indicate the capacitive coupling between stator winding and motor frame, rotor and motor frame stator winding and rotor, respectively.

According to the distribution of stray capacitance for the two typical inverter systems, the lumped CM equivalent model can be derived, as shown in Fig. 7.12. For PV inverter system, since the stray capacitors set up the coupling between the solar cell and PV frame, while the PV frame is grounded for safety, the stray capacitors can be equivalent to the two lumped grounding capacitor and provide the CM loop for CM current. Thus, the CM equivalent model for PV system is the DC side lumped grounding capacitor C_{PV} in series with the AC side three-phase load circuit which includes the three-phase pole voltage V_{XO} (X = A, B, C), grid-tied inductor and three-phase grid voltage e_x (x = a, b, c), as shown Fig. 7.12a. For motor drive inverter system, the CM equivalent model is shown Fig. 7.12b, where Z_{dg} is the DC-link midpoint grounding impedance, Z_{cable} is the transmission impedance of cable for each phase. Z_{sf} and Z_{sr} are the lumped impedances between stator winding and motor frame, stator winding and rotor for each phase winding, and Z_{rf} represents the entire impedance coupling between rotor and motor frame [9]. Moreover, the simplified model for the motor bearing is also considered which consists of the resistor R_b, the switch S_b and adjustable capacitor C_b [9].

Assuming the symmetry of AC side CM loop for the two typical inverter systems, the characteristic of CM loop can be deduced with further simplification, as shown in Fig. 7.13. It can be seen that the voltage source of the CM equivalent circuit is the CMV determined by PWM, and the CM loop can be simplified and regarded as L–R–C circuit, where L_{cm} and C_s are the total CM inductor and stray capacitor in CM loop. Thus, there exists the possibility of resonance between the CM inductor

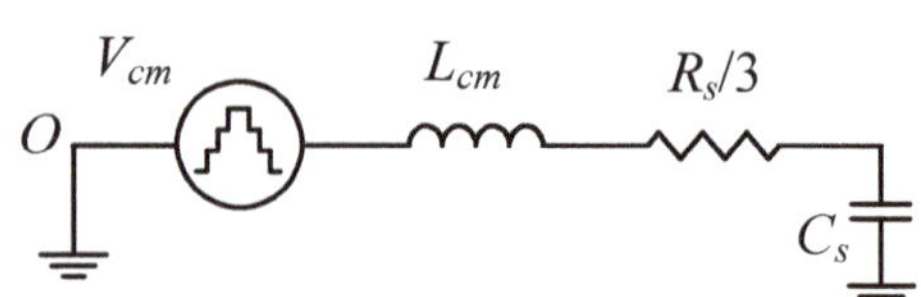

Fig. 7.13 The simplified CM equivalent circuit for the typical inverter system

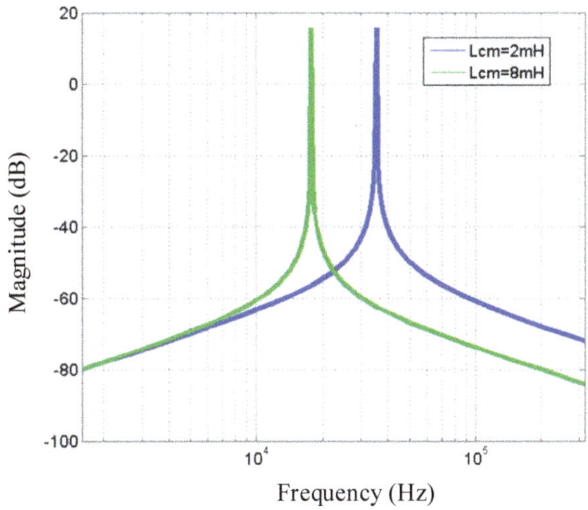

Fig. 7.14 Bode diagram of CM loop admittance

and the stray capacitor. The system impedance is shown in (7.2), and the resonant frequency is determined by both L_{cm} and C_s.

$$Z = \frac{R_s}{3} + j\omega L_{cm} + \frac{1}{j\omega C_s} = \frac{R_s}{3} + \frac{1 - \omega^2 L_{cm} C_s}{j\omega C_s} \tag{7.2}$$

Select a representative motor-to-ground stray capacitance value (10 nF), and select different CM inductors to get the admittance of the system, as shown in Fig. 7.14. It can be seen that at the resonance point of the CM inductor-stray capacitance, the system impedance is minimized. If the CMV has a major component near this frequency, the induced CMC will be significantly amplified, which reflects in the CMC such that the CMC near this frequency occupies the main component. The resonant frequency can be affected by the CM inductance.

Taking a fundamental period CMV of SVPWM and AZSPWM schemes for analysis, as shown in Fig. 7.15, it can be seen that though the CMV of AZSPWM is smaller than that of SVPWM in amplitude, the partial harmonic content in the spectrum is slightly increased, such as the harmonic of the fifth switching frequency. The CMVs corresponding to DPWM and NSPWM are more obvious, as shown in Fig. 7.16. Although the improved NSPWM suppresses the amplitude of CMV, its third switching frequency harmonic is significantly larger than that of DPWM. If the frequency corresponding to this harmonic is close to the CM resonant frequency of the system, the corresponding CMC will be significantly amplified.

The simulation effects of the two modulation strategies for CMC are shown in Figs. 7.17 and 7.18. In Fig. 7.17, the switching frequency is designed to be one-third of the CM resonant frequency, so the component of the third-order switching frequency harmonic coincides with the resonant frequency for the NSPWM scheme, causing the CMC of the NSPWM scheme significantly amplified and much higher

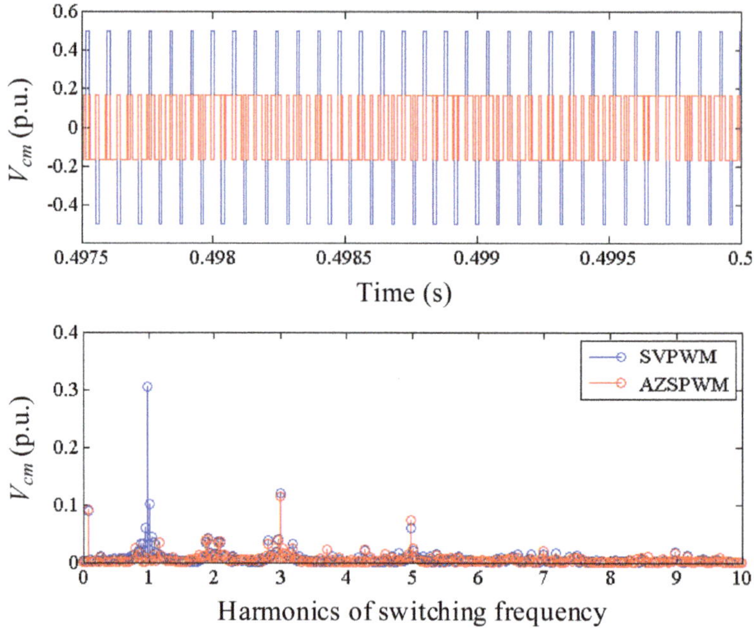

Fig. 7.15 CMVs and their spectrum comparison: SVPWM versus AZSPWM

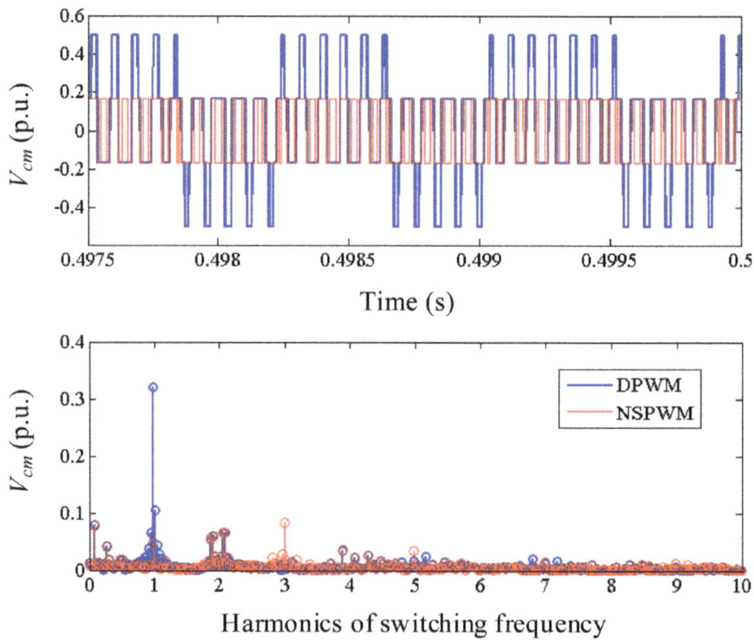

Fig. 7.16 CMVs and their spectrum comparison: DPWM versus NSPWM

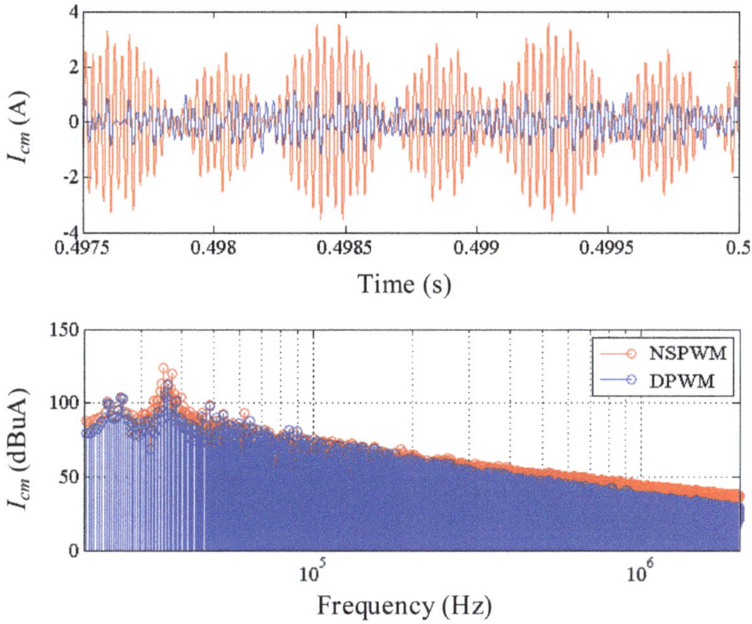

Fig. 7.17 CMCs and their spectrum comparison: DPWM versus NSPWM, when the resonant frequency is close to three times the switching frequency

than the DPWM scheme, which means the amplitude suppression of the CMV in turn increases CMC and CM EMI. Only if the design switching frequency is away from one-third of the CM resonant frequency, that is, the third switching frequency is far away from the CM resonant frequency, the reduction of the CMV leads to the suppression of CMC, as shown in Fig. 7.18.

Figure 7.19 shows the CMC comparison for the two different cases in the experiment. When the CM resonant frequency of system is close to three times of the switching frequency, the CMC of NSPWM is larger than DPWM due to the increase of the corresponding component. Only when the resonant frequency is far away from the three times of switching frequency, the CMC of NSPWM is less than DPWM, making the drop of the CMV play a positive role. The experimental results have good proof for this conclusion.

The research in this section shows that suppressing the CMV by the improved PWM does not necessarily represent the reduction of the CMC. The CMC is the combined action result of CMV and CM loop. Designing an improved PWM requires a comprehensive consideration of CM impedance characteristics and switching frequency to achieve the CMC suppression.

In addition, these improved PWMs change the order of output pulses and the output voltage stress during the switching period. A typical result is the change of output current ripple. Figure 7.20 shows the CMV and output phase current for DPWM and NSPWM schemes. It can be seen that the switching ripple of

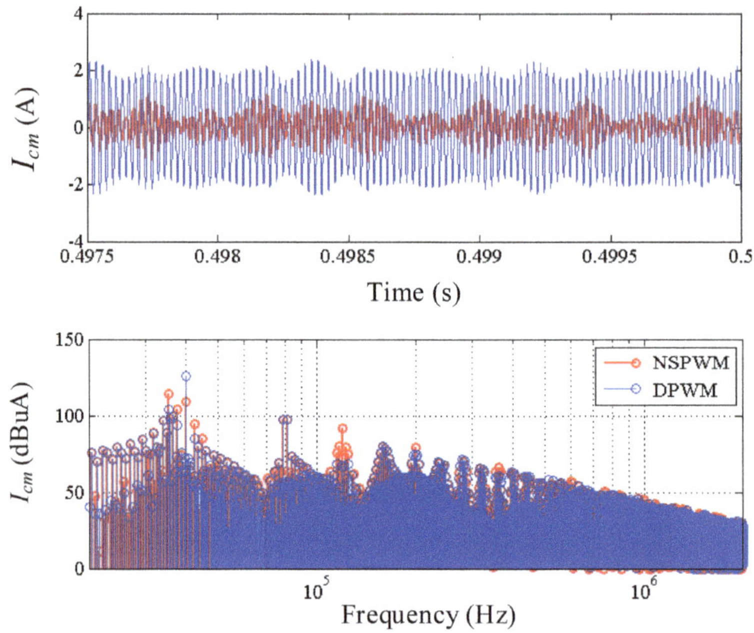

Fig. 7.18 CMCs and their spectrum comparison: DPWM versus NSPWM, when the resonant frequency is far from three times the switching frequency

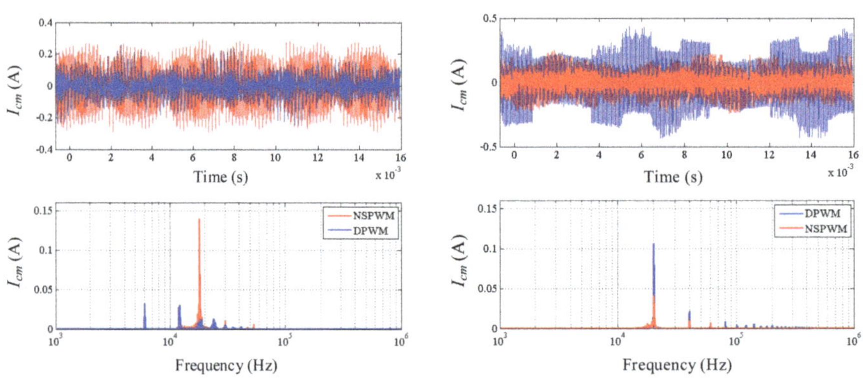

Fig. 7.19 CMC experimental results: **a** the resonant frequency is close to three times the switching frequency, **b** the resonant frequency is far from three times the switching frequency

output current for the NSPWM scheme is obviously different from DPWM because NSPWM changes the voltage distribution during the switching period. Therefore, the design of improved PWM for CMC reduction needs to take other factors into account, such as output current ripple.

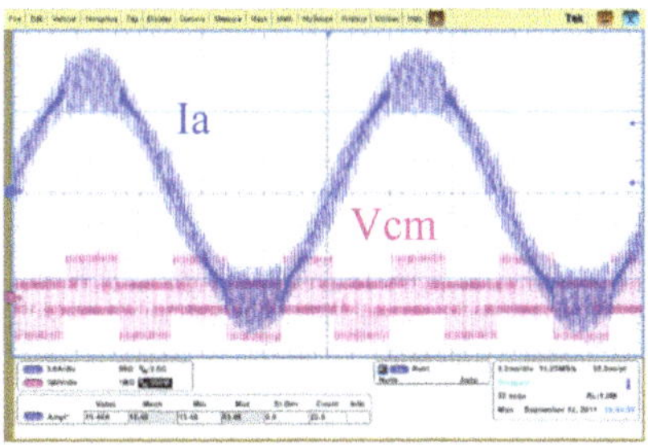

(a)

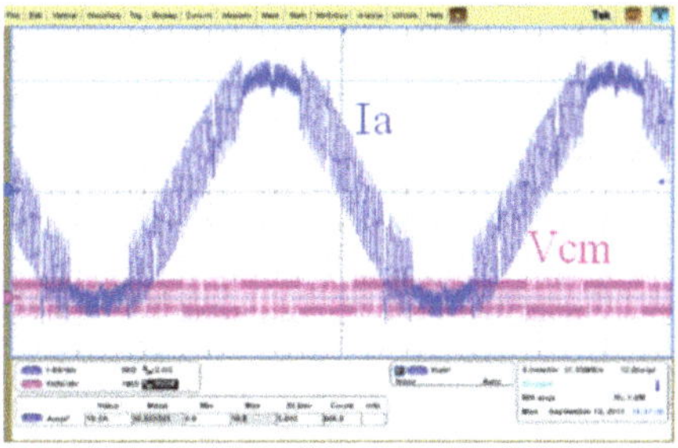

(b)

Fig. 7.20 CMV and output phase current experimental results: **a** DPWM, **b** NSPWM

7.4 Common-Mode Voltage Elimination by Three-Level Converter PWM Method

Though the improved PWM methods such as AZSPWM and NSPWM can effectively reduce the CMV amplitude by avoiding the zero vectors, the CMV cannot be eliminated because of the limitation of the general two-level three-phase inverter topology. Since the phase number of the two-level three-phase inverter is odd, and the output terminal voltage of each phase (relative to the midpoint of the DC-bus) is switched between the positive and negative DC bus voltages, so the summation of three-phase voltages cannot be zero, and it is doomed that the zero CMV cannot be realized by the ordinary two-level three-phase inverter.

On the other hand, through the modification of topology, more control freedoms can be added, and there is a possibility to achieve zero CMV in theory. A major improvement in topology is the application of multilevel converters, and zero CMV is achieved by increasing the probability of each phase output voltage; the other direction is to increase the quantity of bridge arms by paralleling the common two-level inverter. When the number of bridge arms is even, there is also the possibility of achieving zero CMV mathematically. Advanced PWMs for multilevel and parallel inverters have been introduced in Chap. 6, but is with little discussion of an improvement for CMV and CMC reduction. This section aims at CMV suppression, and introduces methods for achieving CMV cancellation in three-level topology.

7.4.1 *Three-Level Converter: Zero Common-Mode PWM [10, 11]*

Similar to Sect. 6.3, Fig. 7.21 shows a circuit diagram of a typical neutral point clamped (NPC) three-phase three-level inverter connected to the load. Each phase-leg can be connected to the positive DC bus, DC bus midpoint (O_1) and the negative DC bus by combination of the four switches, that is, the output voltage of each phase can have three states: positive bus voltage (p), zero voltage (o) and negative

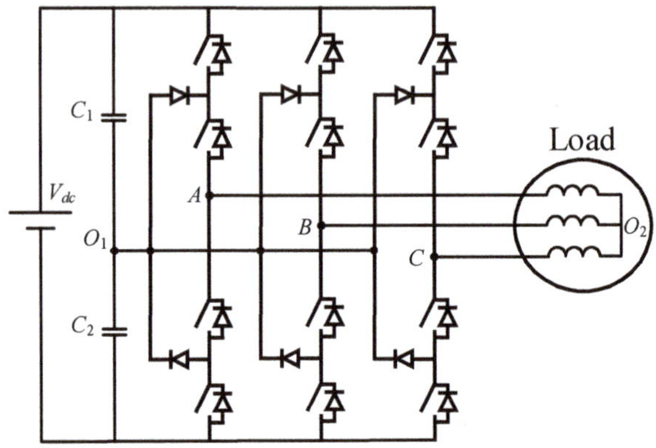

Fig. 7.21 Circuit diagram of typical three-phase three-level inverters connected to load

bus voltage (n) with respect to point O_1. Thus, even though it is still a three-phase inverter, the sum of the three-phase voltages can be zero in reference with O_1. The load is typically a three-phase AC motor with star-connected winding with neutral point of O_2.

Figure 7.22 shows the space vector distribution of the three-level converter. As discussed in Sect. 6.3, according to the length of vectors, they can be divided into large vectors (such as pnn), medium vectors (such as pon), small vectors (such as poo) and zero vectors (including ppp, ooo and nnn). Among them, the particular six medium vectors (pon, opn, npo, nop, onp and pno) marked by thick lines are composed of three states: p, n and o, and their corresponding CMVs are exactly zero. At the same time, the CMV corresponding to ooo vector is zero in the three zero vectors. In this way, the reference voltage can be synthesized by using these

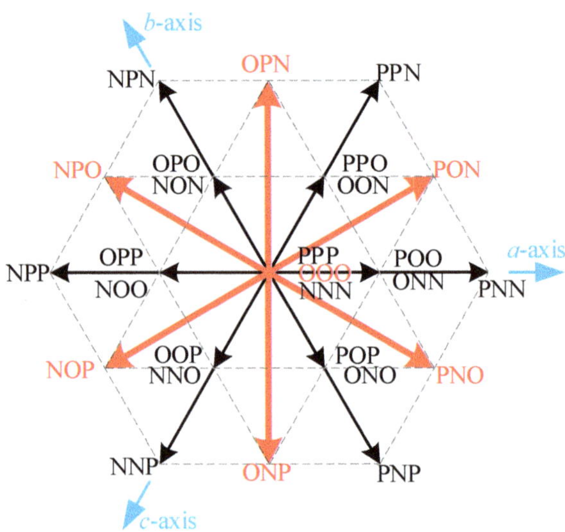

Fig. 7.22 Space vector distribution of a typical three-phase three-level inverter

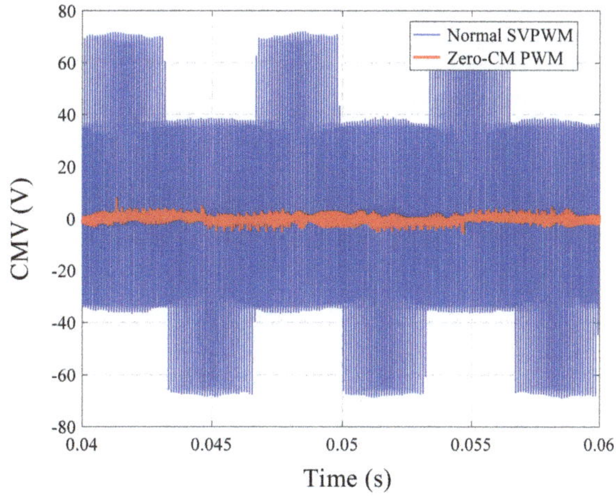

Fig. 7.23 Experimental results comparison of CMV for three-level converters: normal SVPWM versus zero-CM PWM

seven vectors, and the CMV can be kept at zero in the full switching period. The six medium vectors divide the voltage space into six sectors, and the reference voltage vector can be synthesized by two adjacent medium vectors and ooo vectors in the corresponding sector, which is similar to the two-level space vectors.

Figure 7.23 shows the experimental results of CMV comparison for three-level converter. It can be seen that there is a CMV pulse sequence for three-level converters with ordinary SVPWM, while the CMV under zero-CM PWM can be significantly reduced and is basically controlled to be near zero except for a few spikes. These spikes are caused by the nonideal characteristics of the system [10, 11].

The three-level zero-CM PWM algorithm can also be implemented by a carrier-based method represented by sinusoidal pulse width modulation (SPWM). The normal SVPWM for three-level converter mainly adopts the phase disposition method, while the zero-CM PWM scheme adopts the single-carrier method similar to two-level converter [10]. As shown in Fig. 7.24, the three-phase symmetric sinusoidal reference waves V_1, V_2 and V_3 are compared with the triangular carrier V_r that varies between ± 1 to obtain reference signals V_{g1}, V_{g2} and V_{g3}; then according to formula (7.3), the pole voltages V_a, V_b and V_c of three phase-legs can be determined. Obviously, the sum of three-phase voltages in (7.3) is zero, that is, the output CMV is zero.

$$
\begin{aligned}
V_a &= \frac{V_{g1} - V_{g2}}{2} \\
V_b &= \frac{V_{g2} - V_{g3}}{2} \\
V_c &= \frac{V_{g3} - V_{g1}}{2}
\end{aligned}
\tag{7.3}
$$

In practical applications, since the carrier-based PWM algorithm is simple and easy to implement, it is commonly used to achieve SVPWM equivalently. In general,

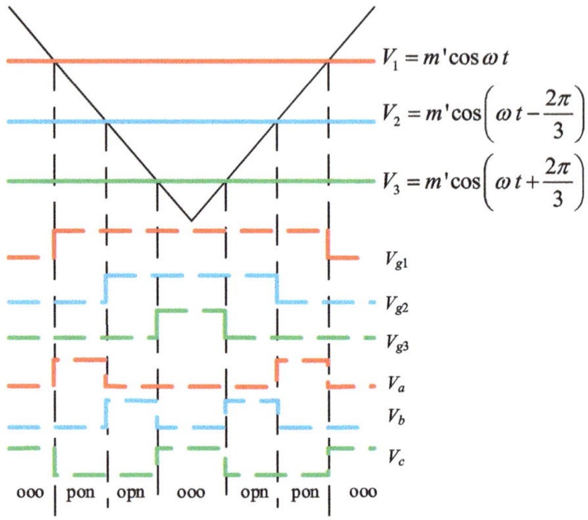

Fig. 7.24 Pole voltage generation of zero-CM PWM with carrier-based principle for three-level converter

SVPWM uses a seven-segment vector arrangement sequence (zero vector → effective vector 1 → effective vector 2 → zero vector → effective vector 2 → effective vector 1 → zero vector) in each carrier cycle, which is exactly the same as the vector sequence generated by carrier-based method shown in Fig. 7.24, so only the same action time of each segment needs proving. Space vector modulation can calculate the vector action time by the principle of vector synthesis, while the carrier-based method can also calculate the action time by using the theory of plane geometry. Regardless of the theoretical calculation or the volt-second balance, the effective vector of the two modulation methods and the action time of the zero vector are consistent. The only difference between the two PWMs is that the zero vector has different distribution ratios at the middle and double-side time segments, and the carrier-based method can change the action time of the zero vector in the middle and double-side by injecting the zero-sequence component without affecting the allocation of the active vector, which can achieve complete equivalence with zero-CM space vector modulation. The equivalent method is to inject the zero-sequence voltage component $v_z = -0.5(v_{\max} + v_{\min})$ into the three-phase reference waves V_1, V_2 and V_3, where $v_{\max}(v_{\min})$ represents the one phase reference wave with the maximum (minimum) instantaneous value in three-phase reference waves.

However, the equivalent theory of the two zero-CM PWMs is different from the equivalent principle of the two normal PWMs for three-level converter. First, the resulting phase-A drive waveform is the difference between V_1 and V_2 divided by 2, which is half of the line voltage, as shown in Fig. 7.24. So the fundamental voltage amplitude of phase-A is only 0.866 of the modulated wave V_1 and the phase-angle is advanced by 30°. The modulation ratio (m') of the three-phase symmetric reference wave in Fig. 7.24 indicates that it is not the true modulation ratio. The maximum modulation ratio is only 0.866 compared with normal SVPWM, which also explains the reason for the DC bus voltage utilization reduction. Second, six effective medium vectors are shown in Fig. 7.22. Though in vector synthesis, the send order of the two effective vectors has no effect on the fundamental characteristics. However, in order to achieve exactly the same effect of the two zero-CM PWMs, it is necessary to specify the sending order of effective vectors for the zero-CM SVPWM: in the first half switching cycle, the odd vector in the sector is always sent first, and then the even vector is sent. The sending order in the last half switching cycle is symmetrical to the first half of the cycle. According to the equivalent law of the two zero-CM PWMs, the CMV can be eliminated by carrier-based method, thereby simplifying the control algorithm.

7.4.2 Three-Level Converter: Evaluation of Zero-CM PWM

Either with the carrier-based PWM or space vector PWM, the three-level converter can theoretically eliminate the CMV, thereby suppress CMC and CM EMI at the source. But three-level zero-CM PWM also brings some disadvantages. First, the range of the modulation ratio is reduced. Since the length of the medium vector is only 0.866 times that of the large vector, the length of the reference vector synthesized by the medium vectors can only reach 0.866 of the normal SVPWM using the large vector. As shown in Fig. 7.22, the linear modulation area of the zero-CM PWM is the inscribed circle of the regular hexagon formed by six medium vectors. Second, the switching loss is nearly twice of normal SVPWM. As shown in Fig. 7.25, the zero-CM PWM switches four times per phase-leg in one carrier cycle, while the normal SVPWM has only two switching actions. This is because only one zero vector can make the CMV zero, and no redundant zero vector to maintain the same output state, so that the zero-CM PWM must return to zero in the middle time segment, causing extra switching actions. In fact, the true switching frequency in this modulation mode is twice the carrier frequency.

In addition, the current ripple and THD increase. Normal SVPWM has 27 switching vectors to realize voltage synthesis, making full use of the switch combination to optimize the output voltage stress of the converter. The difference between the reference voltage and the switching voltage vector is relatively small, so the current ripple is suppressed. The use of only six medium vectors and one zero vector to

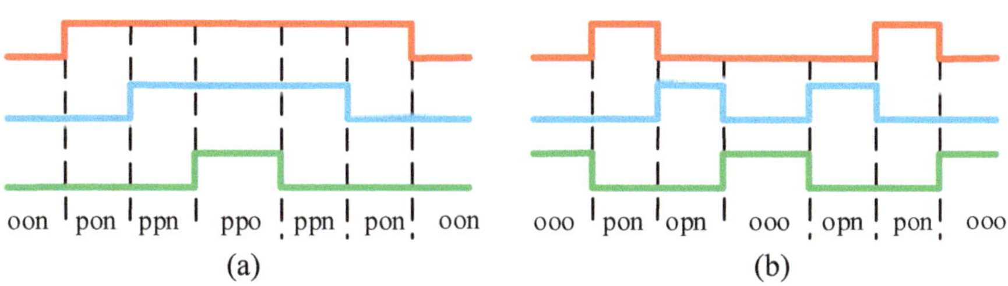

Fig. 7.25 Switching sequence of three-level converter: **a** normal SVPWM, **b** zero-CM PWM

achieve reference voltage synthesis is equivalent to sacrificing the role of other switch combinations in voltage synthesis. So the voltage stress is increased and the current ripple is also significantly increased for the zero-CM PWM. Figure 7.26 shows an example of a three-level space vector synthesis reference voltage. Figure 7.26a shows the reference voltage vector synthesis of normal SVPWM which uses the three closest voltage vectors (ppn, pon and ppo/oon) to synthesize the reference vector, while Fig. 7.26b shows the reference voltage vector synthesis of the zero-CM PWM, only the zero-CM vectors (pon, opn and ooo) can be used to synthesize the reference voltage vector. So the difference between the selected vectors and the reference vector is larger, which brings greater voltage stress and current ripple on the inductor. Figure 7.27 shows the test results of the current ripple comparison. The output current THD of the normal SVPWM is 7%, while that of the zero-CM PWM is increased to 13%.

Moreover, the only use of medium vectors brings a serious challenge to the voltage balance of DC bus midpoint for three-level converter. The fluctuation of the midpoint

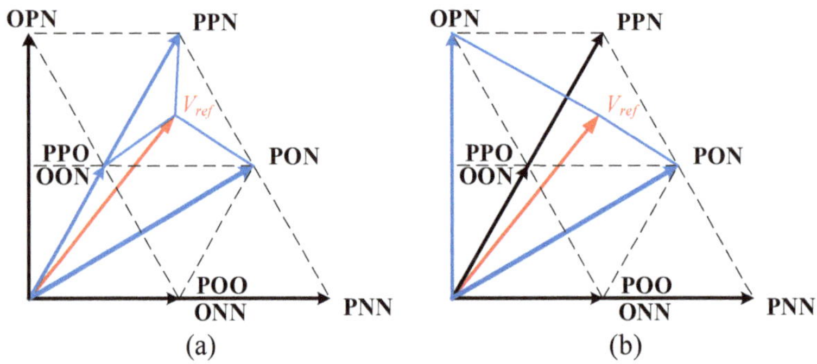

Fig. 7.26 Three-level converter space vector synthesis and voltage difference: **a** normal SVPWM, **b** zero-CM PWM

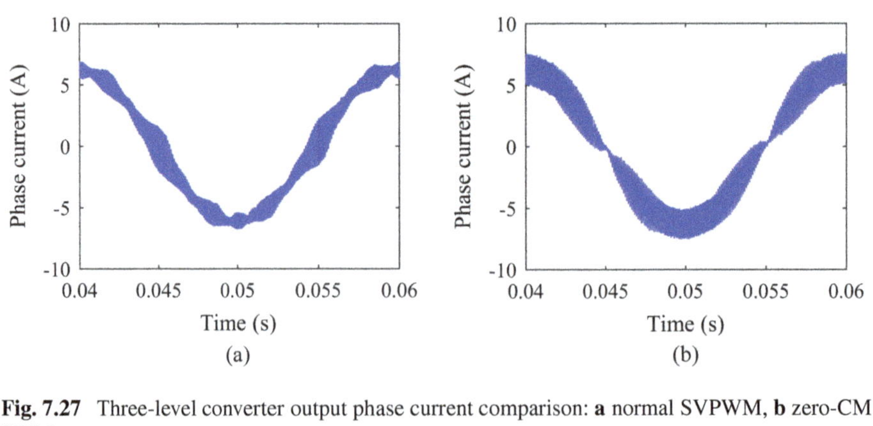

Fig. 7.27 Three-level converter output phase current comparison: **a** normal SVPWM, **b** zero-CM PWM

potential is an inherent problem of the NPC three-level converter. As discussed in Chap. 6, each type of vector has a different effect on the midpoint potential: the large vector and the zero vector do not inject current into the midpoint of the DC bus, so they do not affect the midpoint potential; the medium vector which has one phase connected to the midpoint affects the midpoint potential, especially in the case of high modulation ratio and low power factor; pairs of small vectors have the opposite effect on the midpoint potential, which is the key to balancing the midpoint voltage. The zero-CM PWM has no pair of small vectors to use, so the midpoint-voltage balance ability is greatly reduced. Figure 7.28 shows the midpoint potential fluctuations of the two PWMs ($V_{dc} = 200$ V, $m = 0.9$, $C_1 = C_2 = 450$ uF, PF ≈ 1). Obviously, the midpoint potential fluctuation caused by zero-CM PWM is much larger.

Table 7.1 summarizes the comparative results on various performances of two kinds of PWM. Obviously, three-level topology has the freedom to suppress CMV. By choosing appropriate space voltage vectors, CMV can be eliminated theoretically. However, the elimination of CMV is at the cost of the reduction of DC voltage utilization, the increase of switching loss, the increase of neutral-point voltage fluctuation and the deterioration of harmonics characteristic.

Fig. 7.28 Comparison of neutral potential fluctuations for three-level converter under normal SVPWM and zero-CM PWM

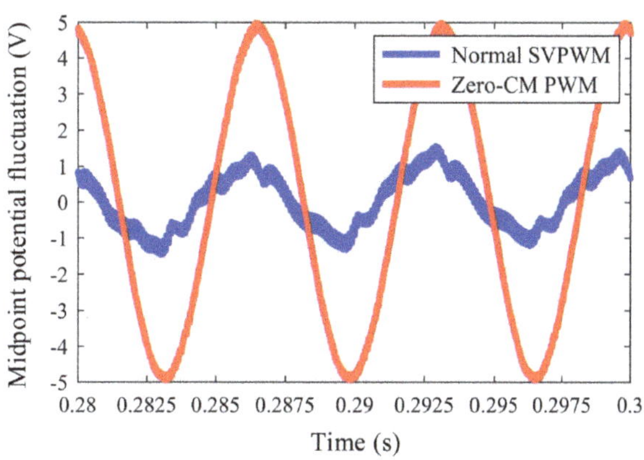

Table 7.1 Performance comparison of three-level converter under normal SVPWM and zero-CM PWM

Performance	Normal SVPWM	Zero-CM PWM
Phase current ripple	Small	Large
Neutral potential fluctuation	Small	Large
Maximum modulation index	1.15	1
Peak amplitude of CMV	One-third of bus-voltage	Nearly zero
Switching actions	Twice per phase-leg per switching cycle	Four times per phase-leg per switching cycle

7.4.3 Three-Level Converter: Zero-CM VSFPWM

Zero-CM PWM has the problem of double switching loss and large harmonic increase. These shortcomings can be mitigated by the freedom of switching frequency in a way. In view of the basic principle and implementation method of variable switching frequency have been introduced in detail in the previous sections, the same part is no longer described. This section focuses on the performance improvement of zero-CM PWM by adding the variable switching frequency algorithm.

The ripple prediction model of zero-CM PWM is based on the Thevenin equivalent circuit instead of the single-phase model. Because the modulation method has only seven vectors, unlike the normal three-level having 27 Thevenin equivalent circuits to be established. In addition, zero-CM PWM has two pulses per carrier cycle, increasing the complexity of the single-phase model. Figure 7.29 shows the circuits of four typical vectors and the equivalent circuit models of phase-A. The current slope of the four equivalent circuits can be calculated by the Eqs. (7.4)–(7.7), where V_{sum} represents the sum of the three-phase voltages, which is generally considered to be zero in three-phase balance condition.

$$\frac{di_a}{dt}(ooo) = \frac{1}{3L}(V_{sum} - 3V_a) \tag{7.4}$$

$$\frac{di_a}{dt}(pon) = \frac{1}{3L}(\frac{3}{2}V_{dc} + V_{sum} - 3V_a) \tag{7.5}$$

$$\frac{di_a}{dt}(opn) = \frac{1}{3L}(V_{sum} - 3V_a) \tag{7.6}$$

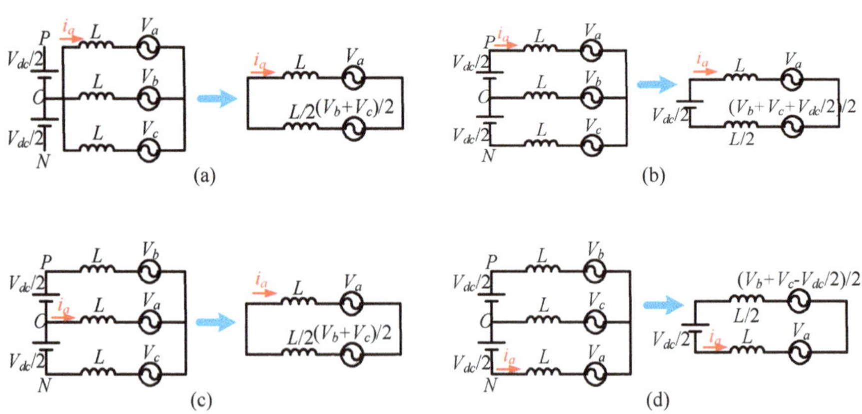

Fig. 7.29 The circuits of four typical vectors and the equivalent circuit models of phase-A: **a** ooo, **b** pon, **c** opn, **d** npo

$$\frac{di_a}{dt}(nop) = \frac{1}{3L}\left(-\frac{3}{2}V_{dc} + V_{sum} - 3V_a\right) \tag{7.7}$$

The current ripple slopes of other vectors and the current ripples of phase-B and phase-C are similar to the four cases in Fig. 7.29. Therefore, the expressions of the ripple slope in various cases are summarized in Table 7.2, where i_k represents the current of the phase-k, and V_k represents the voltage of the phase-k. It can be seen from Table 7.2 that the current slope is only related to the position of the phase clamped to the DC bus, regardless of the clamped state of the other two phases. This conclusion proves that it is very simple to predict the current ripple of the zero-CM PWM method by using the Thevenin equivalent circuit.

Figure 7.30 shows the comparison of the high-frequency current ripple extracted by MATLAB/Simulink simulation model with the current-ripple peak-value calculated based on the prediction model. The high agreement between the two results demonstrates the accuracy of the prediction model. Based on the accurate ripple prediction model of zero-CM PWM, the algorithm of zero-CM PWM with variable switching frequency (zero-CM VSFPWM) can be generated. The control process has been described in the previous chapter. It should be noted that in the zero-CM VSFPWM algorithm, the PWM generation is based on carrier-based comparison

Table 7.2 Current ripple slope in various situations

Position of output pole	Current slope
k-phase clamp to positive DC bus	$\frac{di_k}{dt} = \frac{1}{3L}(\frac{3}{2}V_{dc} + V_{sum} - 3V_k)$
k-phase clamp to the midpoint of the DC bus	$\frac{di_k}{dt} = \frac{1}{3L}(V_{sum} - 3V_k)$
k-phase clamp to negative DC bus	$\frac{di_k}{dt} = \frac{1}{3L}(-\frac{3}{2}V_{dc} + V_{sum} - 3V_k)$

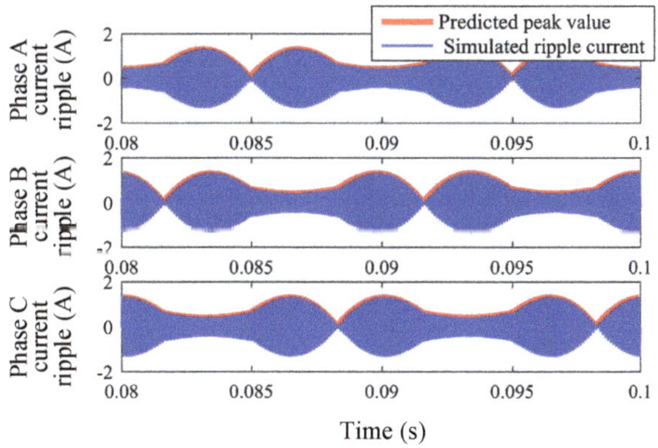

Fig. 7.30 Simulated current ripple versus predicted current ripple peak value

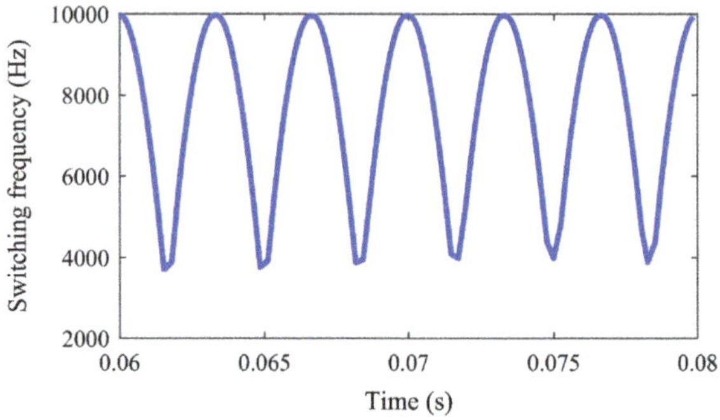

Fig. 7.31 Switching frequency change of zero-CM VSFPWM1

because it is simple to implement, but the ripple prediction is based on the space voltage vector because of the simplicity of the Thevenin equivalent model.

Two modes of zero-CM VSFPWM can be applied, one is VSFPWM1 based on ripple-peak control, and the other is VSFPWM2 based on ripple-RMS-value control. The zero-CM VSFPWM1 is mainly used to reduce the switching loss. As shown in Fig. 7.31, the switching frequency varies from 3.8 to 10 kHz. Compared with the original fixed 10 kHz, the switching loss can be reduced by about 30%.

Zero-CM VSFPWM2 is mainly used to reduce conducted DM EMI. Under the experimental conditions of 200 V DC voltage, modulation ratio of 0.9, load of 0.5 mH inductor and 9.7 Ω resistor, the harmonic content of zero-CM PWM is 11.99%. In order to obtain the same THD, the RMS value of each carrier cycle can be controlled to 0.85 A, and the THD is about 12.07% at this time. Figure 7.32 shows the variation range of the switching frequency. The switching loss is basically the same as the original. Figure 7.33 shows conducted DM EMI comparison, the zero-CM VSFPWM2 has 10 dB attenuation over a wide spectral range. Thus, the size of the EMI filter can be reduced which can increase the power density of the system.

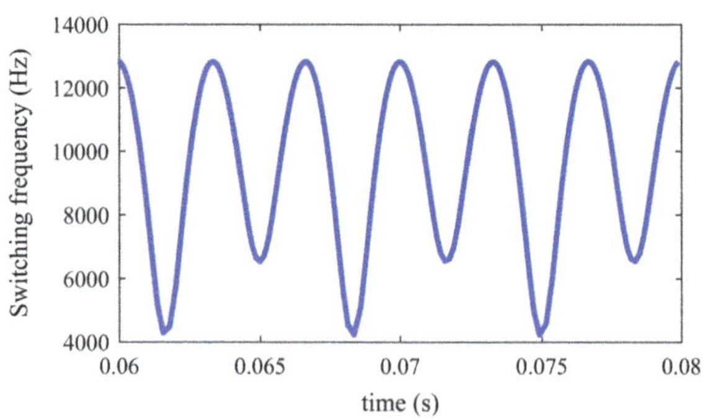

Fig. 7.32 Switching frequency change of zero-CM VSFPWM2

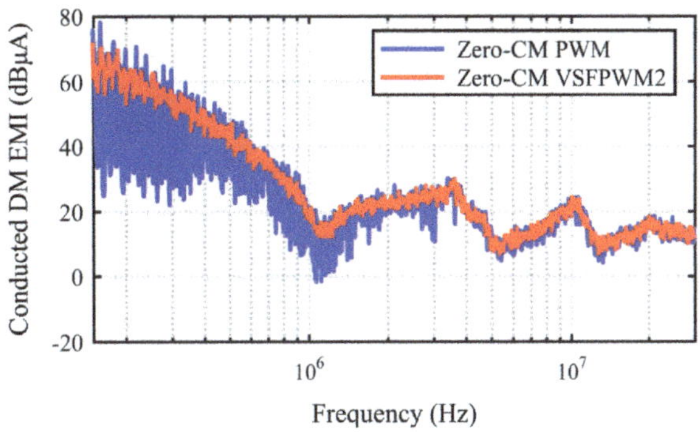

Fig. 7.33 Conducted DM EMI comparison between zero-CM PWM and zero-CM VSFPWM2

7.5 Common-Mode Voltage Reduction and Elimination by Paralleled Two-Level Converter PWM Methods

Paralleled two-level three-phase converters increase the number of phase-legs compared with conventional single three-phase converter, which has more freedoms for CMV reduction. This section takes CMV suppression as the goal, and introduces the PWM methods of CMV reduction and elimination under the topology of paralleled converters.

7.5.1 Paralleled Converters: Interleaved SVPWM

Similar to multilevel converters, another mainstream converter topology is the paralleled converters. In Chap. 6, the application of carrier phase shifting (interleaving) PWM in paralleled converters is introduced. Interleaved PWM can effectively suppress the input and output current ripple and DM EMI noise by shifting the switching function between two converters [12, 13]. This section will further introduce the role of paralleled converters and their PWM strategies in CM noise reduction.

Figure 7.34 shows a typical structure of three-phase paralleled inverters. Two three-phase inverters share the same DC bus and are connected by coupled inductors (CIs) on the AC side. The common output terminals of the CIs are connected to three-phase load. The function of the CI is to suppress the circulating current by the main inductor which is caused by the difference of the output pole voltages of the two paralleled phase-legs, and to average the output voltages of the two connected phase-legs to the load. The CMV of the paralleled inverter is shown in Eq. (7.8), that is, the average value of the output three-phase voltages relative to the DC bus voltage which is equivalent to the average value of all output pole voltages of the two inverters. The six pole voltages in Eq. (7.8) are all digital quantities switched between $V_{dc}/2$ and $-V_{dc}/2$. Equation (7.8) converts the CMV from an odd number of

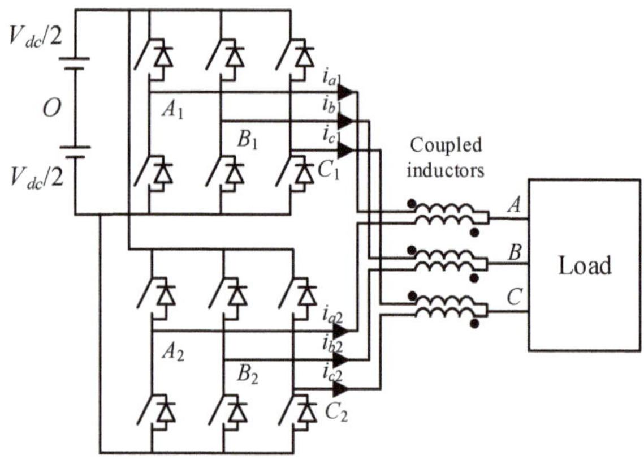

Fig. 7.34 Paralleled inverter topology

digital quantities summation for conventional three-phase inverter to an even number of digital quantities summation, theoretically making it possible to be zero for CMV. This is similar to the three-level inverter but has some difference in principle. The three-level inverter changes the two-dimensional digital quantity (+, −) to three-dimensional digital quantity (+, 0, −), thereby realizing the possibility of theoretical zero CMV. In the space vector of two paralleled inverters, if there are three "1"s and three "0"s in (7.8), the CMV remains zero.

$$V_{cm} = \frac{1}{3}(V_{AO} + V_{BO} + V_{CO}) = \frac{1}{6}(V_{A1O} + V_{A2O} + V_{B1O} + V_{B2O} + V_{C1O} + V_{C2O}) \tag{7.8}$$

First, the suppression effect of CMV by interleaved PWM is analyzed. Figure 7.35 is a schematic diagram of a 180° interleaved SVPWM, where the reference voltage and duty cycles keep the same, while the carriers of the two inverters are interleaved

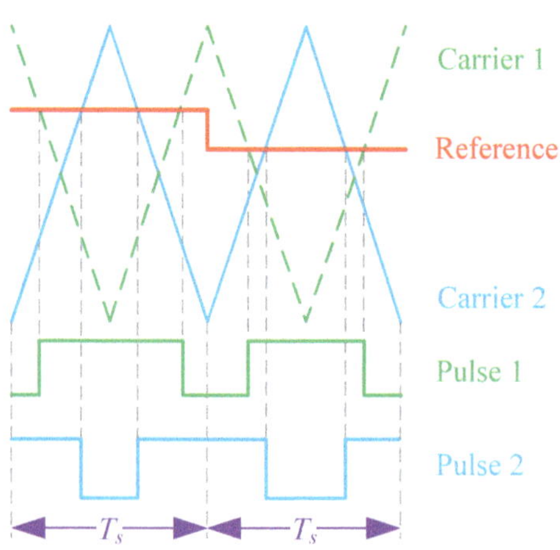

Fig. 7.35 Principle of interleaved PWM

by 180°. So the output pulse of the inverter 2 is changed from middle alignment to the double-side alignment, which forms a phase-shift effect compared with the output pulse of the inverter 1. In Chap. 6, it is analyzed that such a phase-shift method can achieve the effect of cancellation on the input and output switching ripples.

Figure 7.36 shows the effect of the 180° interleaving method on the CMV for SVPWM scheme. Without loss of generality, considering that $d_a \geq d_b \geq d_c$, the three-phase output voltage logic of inverters 1 and 2 is divided into five segments during a half switching cycle, shown in Fig. 7.36, and the voltage vectors and corresponding CMV state are shown in Table 7.3. In the first segment, the inverter 1 output voltage vector is 000 and the inverter 2 output voltage vector is 111, so the CMV is zero. In the second segment, the output voltage vector of the inverter 1 is 100 and the output voltage vector of the inverter 2 is 110, and the CMV also remains zero. In the third segment, the output voltage vector of the inverter 1 is 110 and the output voltage vector of the inverter 2 is 110, so the CMV changes to $V_{dc}/6$. In the fourth segment, the output voltage vector of the inverter 1 is 110 and the output voltage vector of the inverter 2 is 100, the CMV is zero. In the fifth segment, the output voltage vector of the inverter 1 is 111 and the output voltage vector of the inverter 2 is 000, so the

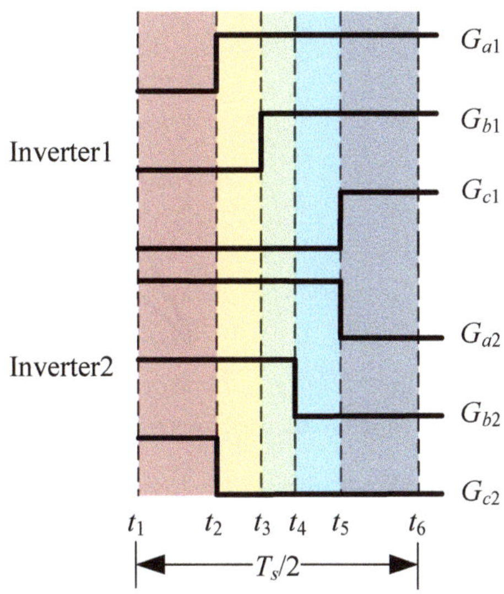

Fig. 7.36 Output voltage logic for 180° interleaved SVPWM in half switching cycle

Table 7.3 Voltage vectors and CMV in half switching cycles for interleaving

Time segment	Voltage vector	CMV
1 (t_1–t_2)	(000) + (111)	0
2 (t_2–t_3)	(100) + (110)	0
3 (t_3–t_4)	(110) + (110)	$V_{dc}/6$
4 (t_4–t_5)	(110) + (100)	0
5 (t_5–t_6)	(111) + (000)	0

CMV is zero. In the last half switching cycle, the switching sequence and CMV will reverse the process of the first half cycle.

It can be seen that there are only two narrow voltage pulses with an amplitude of $V_{dc}/6$ for the CMV in one switching cycle when utilizing the interleaving scheme for paralleled inverters. To analyze these two narrow pulses, Fig. 7.37 shows voltage vector synthesis principle of SVPWM. The action times of V_2 and V_1 vectors are d_2T_s and d_1T_s, respectively. Due to the difference between the two action times, the third period in Table 7.3 is generated, and the length is $(d_2 - d_1)T_s/2$. Therefore, there are nonzero pulses of CMV. If $d_2 < d_1$, the CMV will become $-V_{dc}/6$ in the third segment and the length of time is $(d_1 - d_2)T_s/2$. Figure 7.38 shows the CMV and CM volt-second in one switching cycle in the $d_2 > d_1$ and $d_2 < d_1$ conditions. Compared with the CMV of two-level three-phase inverter switched between $V_{dc}/2$ and $-V_{dc}/2$ under SVPWM scheme, the paralleled inverter with interleaved SVPWM can significantly improve both the CMV amplitude and duration. However, due to the existence of CMVs with amplitudes of $V_{dc}/6$, CM noise is still a major problem in inverter systems and needs to be suppressed by CM filters. The volt-second integral of the two CMV pulses affects the saturation of the CM inductor, which is a limitation of volume and weight in the filter design.

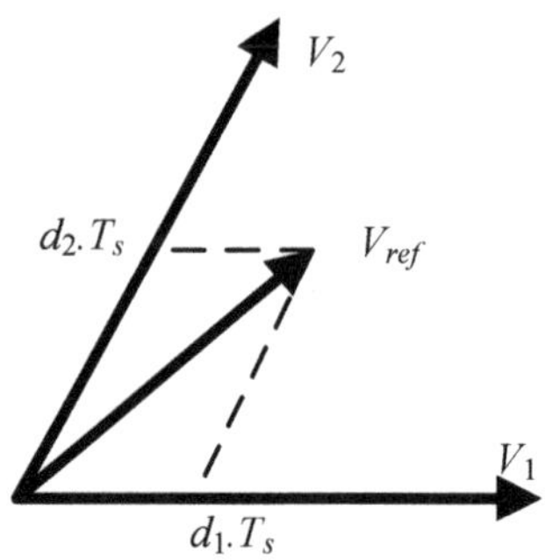

Fig. 7.37 Space vector combination in one sector

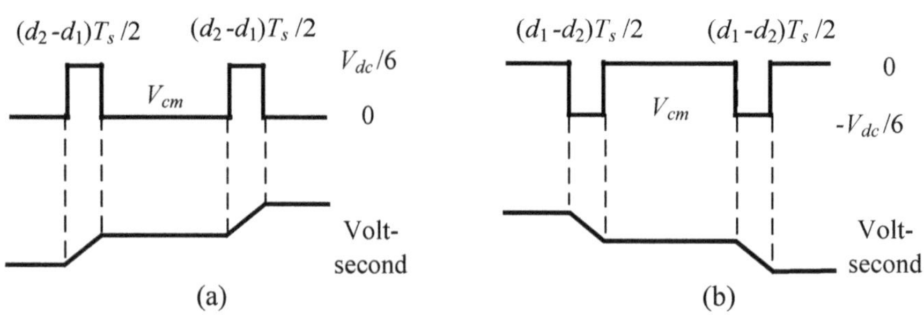

Fig. 7.38 CMV and volt-second for paralleled inverter with interleaved SVPWM: **a** $d_2 > d_1$, **b** $d_2 < d_1$

7.5.2 Paralleled Converters: Zero-CM PWM

Though interleaved SVPWM can reduce the CMV, it does not fully utilize the freedom of (7.8) to keep the CMV at zero theoretically. This section describes another modulation method for paralleled inverters that can reduce the CMV to zero theoretically [14–16].

Figure 7.39 shows paralleled inverter vector synthesis diagram based on the voltage vector of standard two-level three-phase inverter. According to the six effective vectors (100, 110, 010, 011, 001, 101) of the standard two-level three-phase inverter, six new paralleled voltage vectors (210, 120, 021, 012, 102, 201) can be synthesized with two adjacent voltage vectors. For example, the 210 vector is a paralleled voltage vector synthesized by two vectors of 100 and 110. When one inverter applies 100 vectors and the other inverter applies 110 vectors, the resultant effect is equivalent to application of 210 vector for paralleled inverters. In addition, there are also two paralleled zero vectors (111 + 000) and (000 + 111). Therefore, the reference voltage of the paralleled inverters can be driven by the six paralleled vectors and two paralleled zero vectors. More importantly, these eight paralleled voltage vectors satisfy the characteristics of three "1"s and three "0"s at the same time, so the output CMV is guaranteed to be zero.

Figure 7.40 shows a schematic diagram of reference voltage synthesis using the paralleled voltage vectors. The standard paralleled voltage vector is with amplitude of V_c. Similar to the conventional two-level inverter SVPWM, according to the

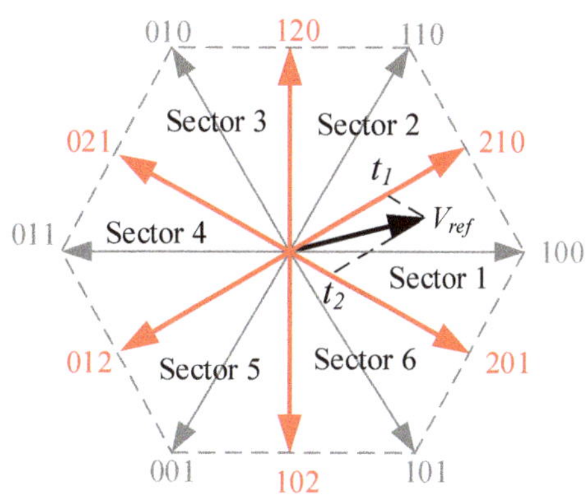

Fig. 7.39 Synthetic vector of paralleled inverters

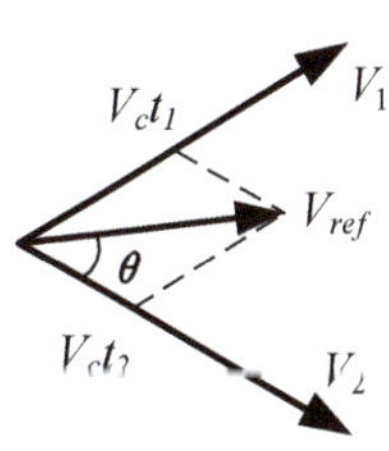

Fig. 7.40 Reference voltage combination by paralleled voltage vectors

position of the reference voltage vector, the sector and the corresponding two adjacent paralleled voltage vectors V_1 and V_2 can be determined first. Then, according to the amplitude and phase of reference voltage vector in the sector, two paralleled voltage vector action times t_1, t_2, and the remaining paralleled zero vector action time t_0 can be calculated according to Eq. (7.9).

$$\begin{cases} \dfrac{V_c \cdot t_1}{\sin\theta} = \dfrac{V_c \cdot t_2}{\sin(\frac{\pi}{3} - \theta)} = \dfrac{V_{ref} \cdot T_s}{\sin(\frac{2\pi}{3})} \\ t_0 = T_s - t_1 - t_2 \end{cases} \tag{7.9}$$

According to (7.9), the action time of each paralleled vector can be obtained. The more critical question is how to decompose the paralleled voltage vectors into two inverters and sort them to generate the PWM signals. If the traditional seven-segment SVPWM sequence is used, taking the first sector as an example, the PWM is divided into seven zones in one switching cycle, as shown in Table 7.4. In t_1 period, V_1 is 210 vector, and is assigned to inverter 1 with 110 vector, inverter 2 with 100 vector; in t_2 period, V_2 is 201 vector, and is assigned to inverter 1 with 100 vector, inverter 2 with 101 vector; the zero vectors are assigned with 111 + 000 in the first and last time period, and 000 + 111 in middle time period. This effect is similar to the normal seven-segment SVPWM. The main difference is that t_1 and t_2 are calculated by the paralleled vectors.

A fatal problem with this vector allocation scheme is that the output voltages of the two inverters are unbalanced. Since the two inverters use different voltages during the same period (t_1 and t_2), the duty ratios of the inverters 1 and 2 are unbalanced during the switching period, so there is a low-frequency voltage difference between the two inverters which brings the serious circulating current problem. The interleaved PWM scheme does a good job which keeps the duty ratios of the two inverters the same within one switching cycle, so the voltages of the two inverters are balanced in one switching cycle, and only a small coupled inductance is required to suppress the circulating current. In the manner of Table 7.4, the voltage difference between the two inverters cannot be balanced during one switching period, so the circulating current has low-frequency components. Thus, the CI should be designed very large and is unpractical.

Table 7.4 Zero-CM PWM regardless of voltage balance issue in one switching cycle (taking the first sector as an example)

	Time	Inverter 1	Inverter 2
Zone 1	$t_0/4$	1, 1, 1	0, 0, 0
Zone 2	$t_1/2$	1, 1, 0	1, 0, 0
Zone 3	$t_2/2$	1, 0, 0	1, 0, 1
Zone 4	$t_0/2$	0, 0, 0	1, 1, 1
Zone 5	$t_2/2$	1, 0, 0	1, 0, 1
Zone 6	$t_1/2$	1, 1, 0	1, 0, 0
Zone 7	$t_0/4$	1, 1, 1	0, 0, 0

Therefore, based on Table 7.4, further improvements are needed. During t_1 period, the allocation of vectors needs to be rebalanced: at the first $t_1/2$ period, the 110 vector is assigned to the inverter 1, and the 100 vector is assigned to the inverter 2; at the second $t_1/2$ period, the 110 vector is assigned to the inverter 2, and the 100 vector is assigned to the inverter 1. Similarly, at the first $t_2/2$ period, the 100 vector is assigned to the inverter 1, and the vector 101 is assigned to the inverter 2; at the second period $t_2/2$, the vector 100 is assigned to the inverter 2, the 101 vector is assigned to the inverter 1. In this way, the two inverters exchange vector distribution in the first and last half switching cycles, achieving a voltage balance in one switching cycle, so that the circulating current can be suppressed by the smaller CIs. The sequence of voltage vectors is shown in Table 7.5.

However, the resultant PWM of Table 7.5 still has significant problems. Due to the change of the vectors sending sequence, the switching actions of some phases are increased. For example, the phase-C of the inverter 1 has undergone the switching sequence of 1–0–0–0–1–0–1, and the phase-B of the inverter 2 has experienced 0–0–0–1–0–1–0. The switching state changes four times in one switching cycle which obviously increases the switching losses. Therefore, the vector sending sequence of Table 7.5 needs to be further modified: exchanging the order of the fifth and sixth period segment. That is, in the second half of the switching cycle, the vector of $t_1/2$ is executed first, and then the vector of $t_2/2$ is executed, as shown in Table 7.6. In this way, two inverters' voltage balance and zero CMV are achieved while maintaining the number of switching times.

Table 7.5 Zero-CM PWM considering voltage balance in one switching cycle (taking the first sector as an example)

	Time	Inverter 1	Inverter 2
Zone 1	$t_0/4$	1, 1, 1	0, 0, 0
Zone 2	$t_1/2$	1, 1, 0	1, 0, 0
Zone 3	$t_2/2$	1, 0, 0	1, 0, 1
Zone 4	$t_0/2$	0, 0, 0	1, 1, 1
Zone 5	$t_2/2$	1, 0, 1	1, 0, 0
Zone 6	$t_1/2$	1, 0, 0	1, 1, 0
Zone 7	$t_0/4$	1, 1, 1	0, 0, 0

Table 7.6 Zero-CM PWM considering voltage balance and switching times in one switching cycle (taking the first sector as an example)

	Time	Inverter 1	Inverter 2
Zone 1	$t_0/4$	1, 1, 1	0, 0, 0
Zone 2	$t_1/2$	1, 1, 0	1, 0, 0
Zone 3	$t_2/2$	1, 0, 0	1, 0, 1
Zone 4	$t_0/2$	0, 0, 0	1, 1, 1
Zone 5	$t_1/2$	1, 0, 0	1, 1, 0
Zone 6	$t_2/2$	1, 0, 1	1, 0, 0
Zone 7	$t_0/4$	1, 1, 1	0, 0, 0

According to the voltage vector sending sequence shown in Table 7.6, the corresponding carrier-based process of the two inverters can be obtained, as shown in Fig. 7.41. Since the symmetrical pulse distributions in Table 7.5 are adjusted to Table 7.6, the pulses of some phases are changed to be asymmetrical in the switching period. Reflected in the carrier-based principle, as shown in Fig. 7.41, the reference values are different in the first and last half switching period. In Fig. 7.41, for phase-B of the inverter 1, the reference value is $1 - (t_0 + 2t_2)/T_s$ in the first half cycle, and changes to $1 - (t_0 + 2t_1 + 2t_2)/T_s$ in the last half cycle; for the phase-C of the inverter 1, the reference value is $1 - (t_0 + 2t_1 + 2t_2)/T_s$ in the first half cycle, and changes to $1 - (t_0 + 2t_1)/T_s$ in the last half cycle. Only the phase-A remains the same reference value of $1 - t_0/T_s$ in the two half cycles. The carrier of the inverter 2 has 180° phase shift compared with the inverter 1, so the inverter 1 starts and ends from the 111 vector, and the inverter 2 starts and ends from 000 vector. The reference value of phase-A of inverter 2 is $1 - t_0/T_s$, the reference value of phase-B in the first half cycle is $1\text{-}(t_0 + 2t_1 + 2t_2)/T_s$ and changes to $1 - (t_0 + 2t_2)/T_s$ in the last half cycle; the reference value of phase-C in the first half cycle is $1 - (t_0 + 2t_1)/T_s$ and changes to $1 - (t_0 + 2t_1 + 2t_2)/T_s$ in the last half cycle. In this way, the PWM obtained by carrier-based process is exactly the same as the sequence in Table 7.6, and the zero CMV can be realized while satisfying the voltage difference balance between the two inverters in one switching cycle.

The analysis above is focused on the first sector which uses the paralleled vectors 201 and 210. The same analysis can be done for the other five sectors, and the resulting switch sequence table and corresponding carrier-based comparison method are listed in the appendix of this chapter. By utilizing the zero-CM PWM scheme in these six sectors, the CMV in the whole fundamental period can remain zero, and the voltage difference between the two inverters can be balanced in each switching cycle.

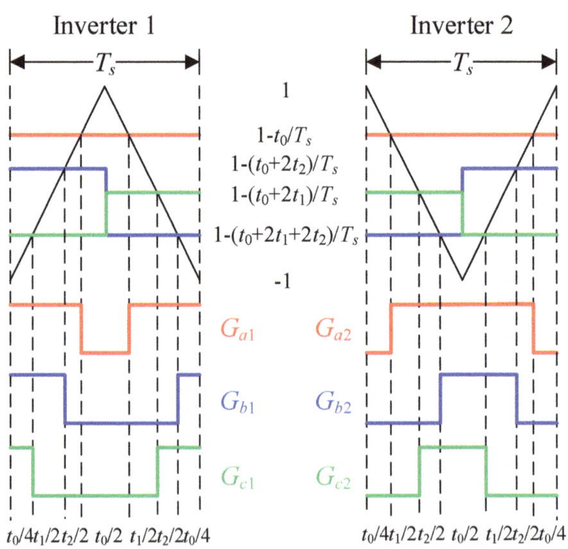

Fig. 7.41 Carrier-based method for zero-CM PWM (first sector)

Through the theoretical analysis above, the relevant simulation model is built for verification, and the simulation results are shown in Figs. 7.42, 7.43, 7.44 and 7.45. Figure 7.42 shows the CMV comparison of the three modulation algorithms for paralleled inverters described above. It can be seen that when the normal SVPWM is used, the CMV is switched between the positive and negative DC bus voltages (±300 V), and the interleaved SVPWM modulation algorithm can reduce the CMV by 1/3 (±100 V) compared with the normal SVPWM scheme. Finally, if zero-CM PWM modulation algorithm is implemented, the parallel inverters can theoretically achieve zero CMV output.

Figure 7.43 shows the comparison of output phase current. Similar to interleaved SVPWM, zero-CM PWM also achieves a certain switching voltage cancellation effect. In Fig. 7.43, the utilization of zero-CM PWM not only eliminates the CMV

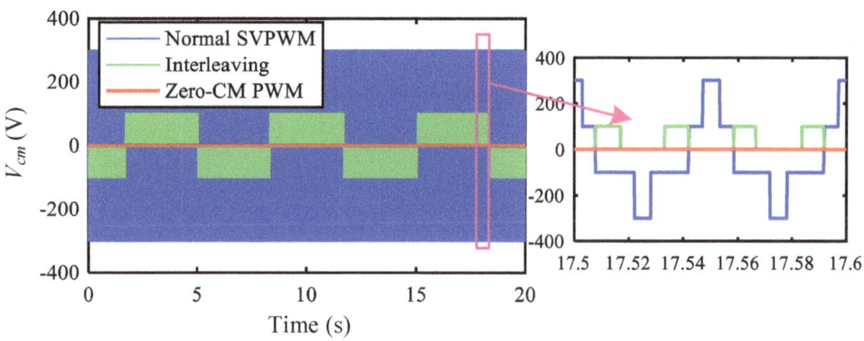

Fig. 7.42 Simulation result comparison: CMV

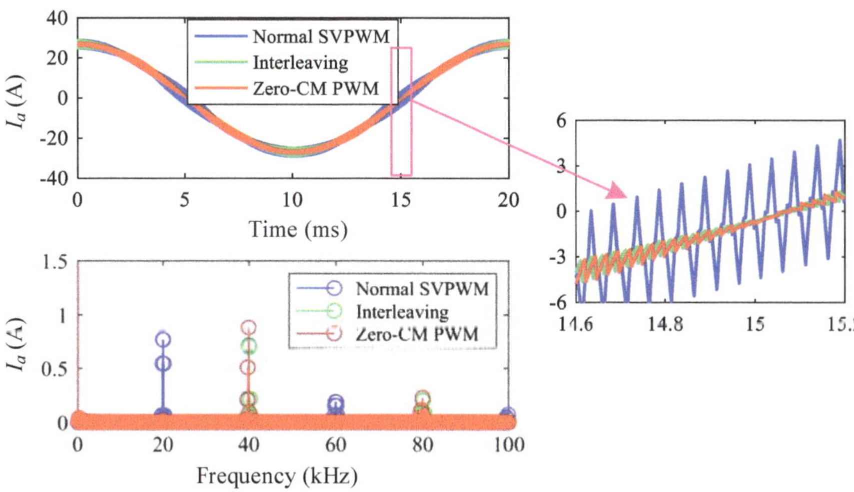

Fig. 7.43 Simulation result comparison: output phase current

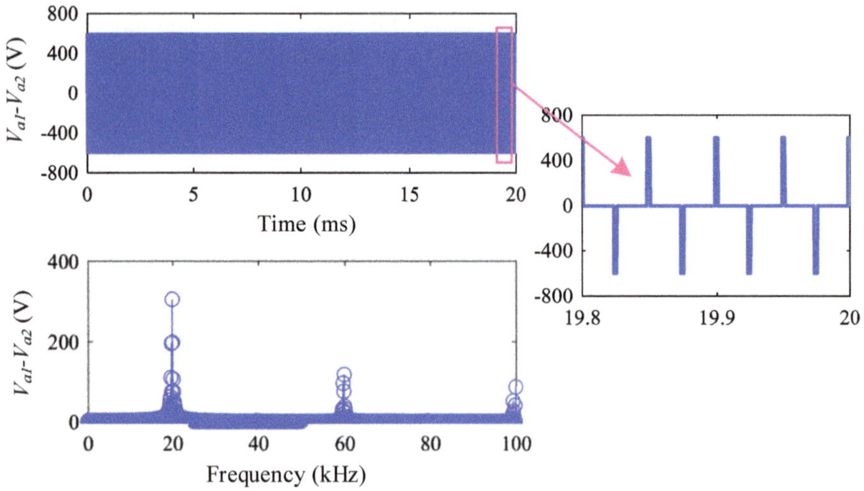

Fig. 7.44 Simulation results: voltage difference of paralleled phase-legs

Fig. 7.45 Simulation results: two paralleled phase currents and circulating current

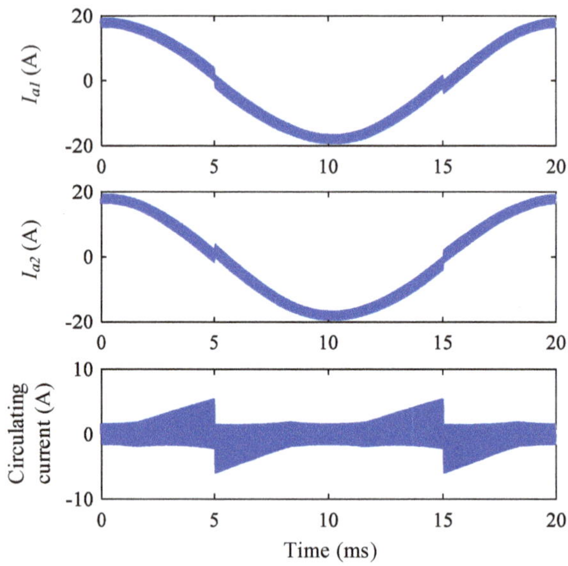

but also reduces the switching current of the output phase current, especially the odd switching frequency harmonics.

Considering the voltage difference of the paralleled inverter zero-CM PWM algorithm, the analyzing result is shown in Fig. 7.44. It can be seen that the voltage difference is balanced in each switching cycle from the time-domain waveform. In addition, the spectrum of the voltage difference also reflects that the main frequency components are near the switching frequency and without low-frequency components. The voltage difference of the interleaved SVPWM scheme also has similar properties to the zero-CM PWM algorithm.

Considering the circulating current suppression and current sharing effect of the zero-CM PWM algorithm, the result is as shown in Fig. 7.45. In the condition of 1 mH CI, the zero-CM scheme can suppress the circulating current within 10 A, while the amplitudes of the paralleled phase currents remain the same, indicating that the algorithm can achieve the current sharing effect, and the circulating current is also effectively suppressed. However, there is an obvious jump in circulating current near the zero-crossings. This issue will be further studied in the next section.

On the basis of simulation, the CMV suppression effect of zero-CM scheme of paralleled inverters is further studied through experiments. Figure 7.46 shows the PWM waveform of zero-CM scheme. In Fig. 7.46, the switch states of paralleled inverters for the first time are 111 and 000, then turn into 110 and 100, 100 and 101, 000 and 111 and so on, which always keeps the CMV zero.

Figure 7.47 shows the CMV in three different cases: normal SVPWM, interleaved SVPWM and zero-CM PWM. It can be seen that with normal SVPWM, the CMV

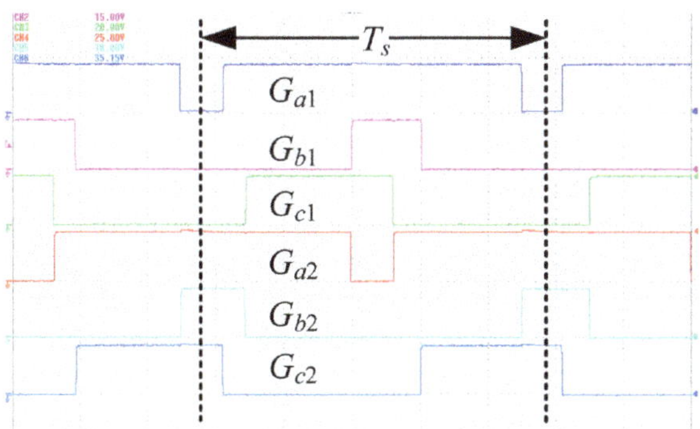

Fig. 7.46 Experimental results: PWM waveform of gate drivers using zero-CM PWM scheme

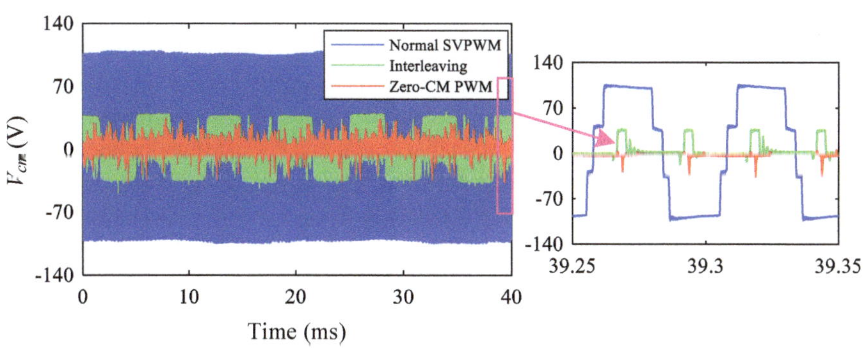

Fig. 7.47 Experimental result comparison: CMV

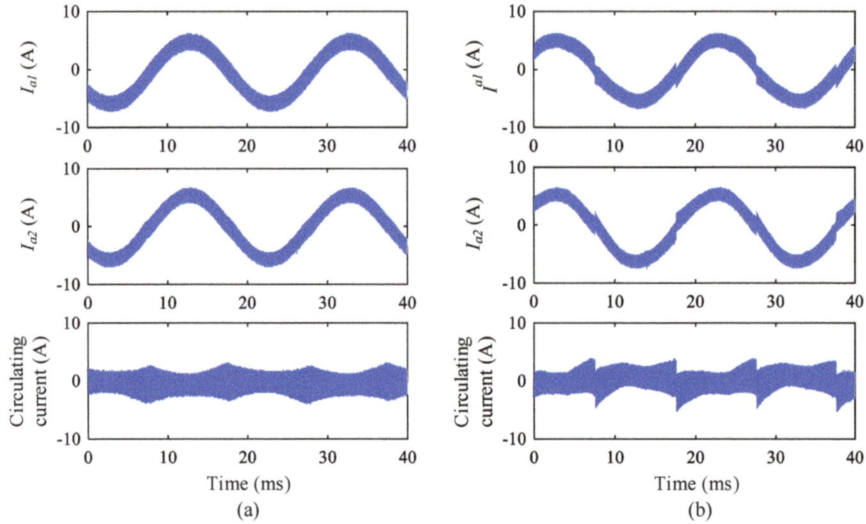

Fig. 7.48 Experimental results: phase currents and circulating current in paralleled inverters: **a** interleaved PWM, **b** zero-CM PWM

is switched between the positive and negative DC bus voltages (100, −100 V). With interleaved SVPWM, the CMV peak can be reduced to one-third of the previous value. With zero-CM PWM, the CMV can be further reduced. Due to the nonideal characteristics of the system, such as deadtime and commutation process, the CMV can theoretically be reduced to zero, but there still exist some voltage spikes.

The test results for circulating current are shown in Fig. 7.48. The 0.5 mH coupled inductors are used for the paralleled inverters. It can be seen that both interleaved PWM and zero-CM PWM can suppress the circulating current to less than 5 A, because both of them guarantee the voltage balance. The currents of two paralleled phase-legs are basically the same, and there is no current uneven or circulating current degradation phenomenon.

In addition, the CMCs of three different cases are measured under the condition of 5 nF grounding capacitance, as shown in Fig. 7.49. The CMC with interleaved SVPWM can be suppressed compared with the CMC of normal SVPWM. With zero-CM PWM, the CMC can be almost eliminated which shows the superiority. Conducted CM EMI is measured in three cases using an EMI test instrument, and the result is shown in Fig. 7.50. It can be seen that conducted CM EMI under interleaving manner mainly suppresses the components near the odd switching frequency, while zero-CM PWM suppresses EMI over the entire frequency domain. The experimental results effectively demonstrate the suppression effect of zero-CM PWM on CMC and CM EMI.

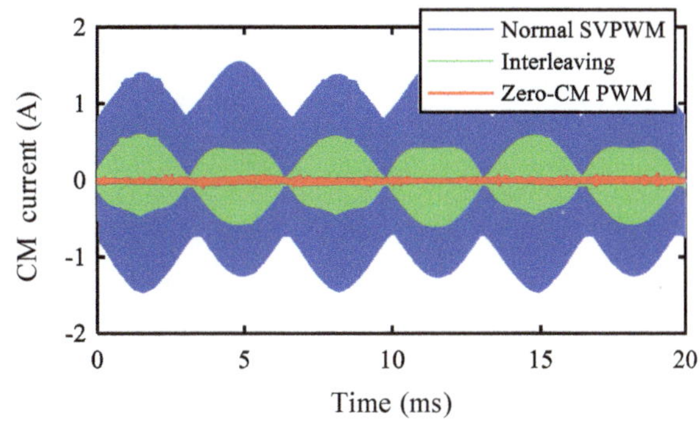

Fig. 7.49 Experimental result comparison: CMC

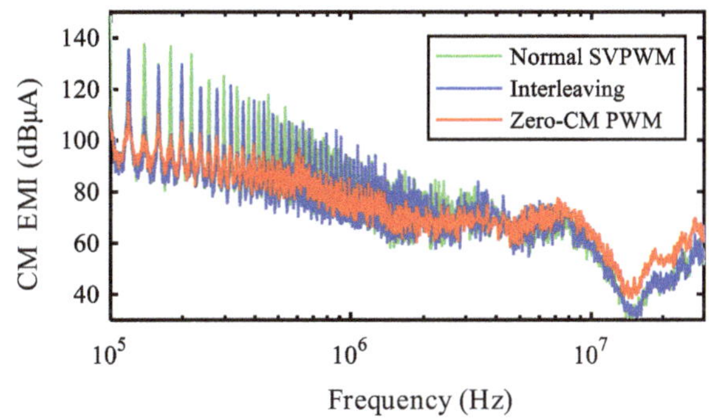

Fig. 7.50 Experimental result comparison: conducted CM EMI

7.5.3 Modified Zero-CM PWM for Paralleled Converters: Circulating Current Suppression

Though the paralleled inverter topology brings the benefit of CMV elimination, it also has the inherent problems, such as differential-mode circulating current (DMCC) and zero-sequence circulating current (ZSCC) [17]. The principle is shown in Fig. 7.51.

The DMCC is often caused by the instantaneous voltage difference between paralleled phase-legs, and the DMCC is defined as the difference of paralleled phase-leg currents which can be expressed as

$$I_{x\_DMCC} = I_{x1} - I_{x2} \tag{7.10}$$

In addition, the ZSCC is usually generated by the instantaneous CMV difference between each inverter. For inverter 1, the ZSCC can be given by

$$I_{ZSCC1} = I_{A1} + I_{B1} + I_{C1} = -I_{ZSCC2} \tag{7.11}$$

For the convenience of subsequent analysis, the magnitude relationship of the action time t_1 and t_2 is selected as a standard to further subdivide each sector, as

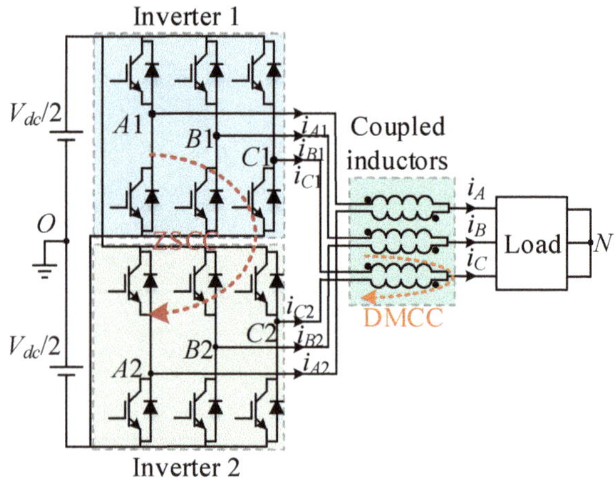

Fig. 7.51 Differential-mode and zero-sequence circulating current for paralleled inverter with CIs

shown in Fig. 7.52. Taking Sector 1 as an example, if $t_1 > t_2$, the synthetic reference voltage vector falls in the range of subsector 2; if $t_1 < t_2$, the synthetic reference voltage vector falls in the range of subsector 1.

The traditional modulation method keeps the PWM symmetric in one switching cycle, and the symmetry of PWM does not change after the 180° carrier phase-shift. However, for the parallel zero-CM PWM scheme, the PWM does not satisfy this relationship. Taking the two PWM signals of the phase-A paralleled phase-legs as an example, the PWM waveform in different subsectors of six sectors is drawn according to the switching state table, as shown in Fig. 7.53.

As can be seen from Fig. 7.53, the PWM signals of phase-A are asymmetric except for Sector 1 and Sector 4. The reason for the asymmetry is the time segment sequence asymmetry, that is, t_1 is always arranged ahead of t_2 in the first and last half switching cycles. This asymmetric arrangement produces a transient unbalance of output volt-second for the phase-legs during certain time slot, resulting in current jump in the

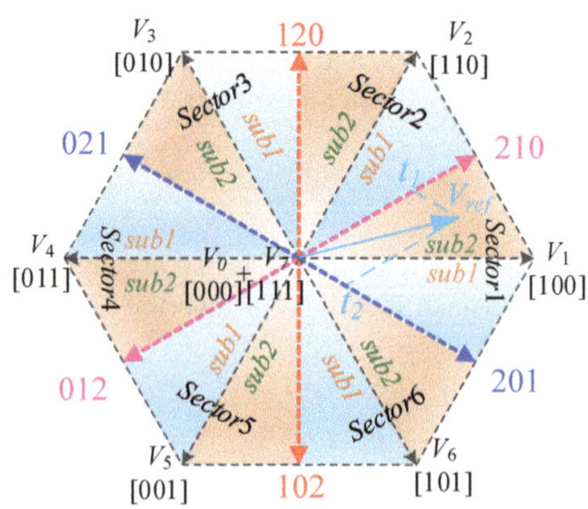

Fig. 7.52 Sector and subsector division of zero-CM PWM algorithm

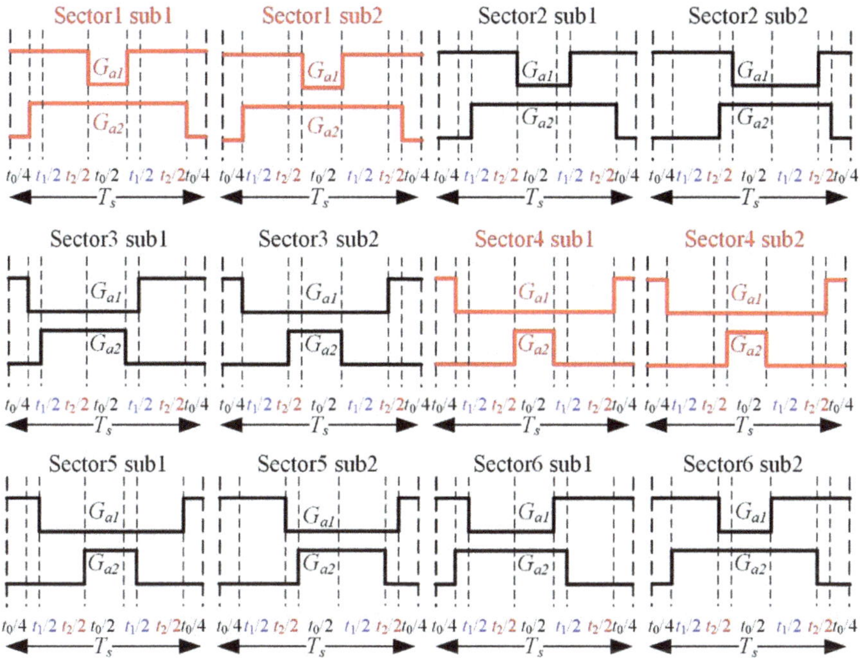

Fig. 7.53 PWM signals for phase-A in six sectors

phase-leg current, which also amplifies the peak value of the DMCC and ZSCC. For this special modulation algorithm, the sector switching process should be focused on. Considering the characteristic of PWM signals, the sector switching process can be divided into two types: (1) switching in asymmetric and symmetric sectors; and (2) switching only in asymmetric sectors. The first case can be subdivided into two states: the asymmetric sector enters the symmetric sector and the symmetric sector enters the asymmetric sector. In this case, selecting Sector 1 as a symmetric sector can introduce two sector switching processes in adjacent two switching cycles. The PWM waveform is shown in Fig. 7.54.

It can be found that the paralleled phase-legs do not have obvious volt-second unbalance in the above sector switching process, which does not cause current jump in phase-leg current. Considering all the asymmetric sectors switching process in one fundamental period, the transition process is shown in Fig. 7.55.

In the above sector switching process, there exists obvious volt-second unbalance between the paralleled phase-legs, and the difference among duty ratios becomes larger as the modulation index increases, resulting in the current jump of phase-leg currents and the peak value amplification of the DMCC in high modulation index condition.

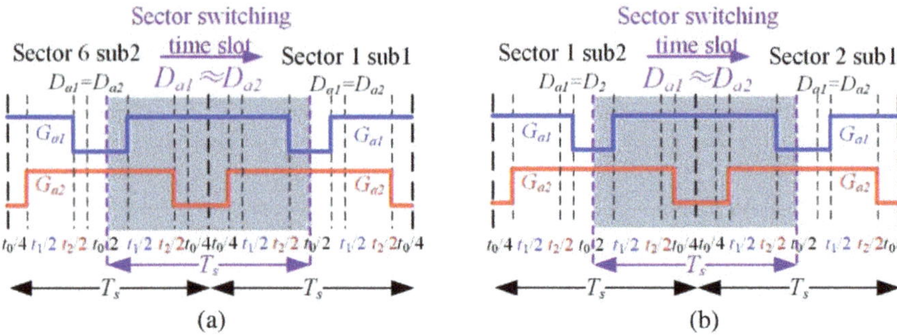

Fig. 7.54 Sector switching process of the phase-A PWM signals: **a** the asymmetric sector switching to the symmetric sector, **b** the symmetric sector switching to the asymmetric sector

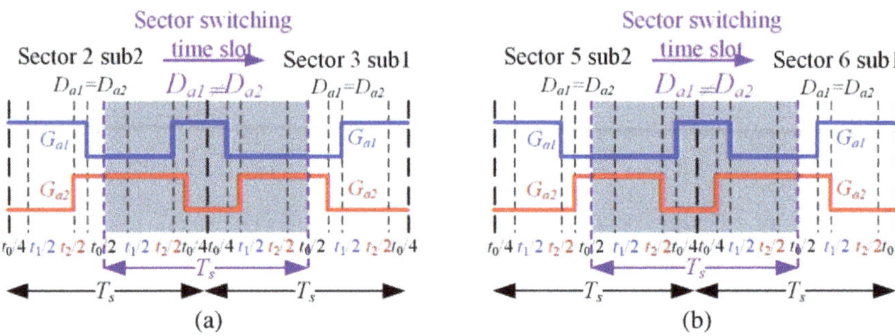

Fig. 7.55 Asymmetric sector switching process: **a** asymmetric Sector 2 switching to asymmetric Sector 3, **b** asymmetric Sector 5 switching to asymmetric Sector 6

To mitigate the current jump and reduce the circulating current, the freedom of zero-CM PWM scheme can be considered. The conventional SVPWM scheme exchanges the sequence of t_1 and t_2 in adjacent two sectors, while in the zero-CM PWM algorithm, the time segment sequence keeps the same in all sectors, that is, t_1 is always ahead of t_2 whether in the first or the last half switching cycles. So by utilizing the freedom of time segment sequence, the other different zero-CM PWM scheme can be deduced. Taking the subsector 2 of the Sector 1 as an example, the PWM signals of original and the other schemes in Sector 1 subsector 2 can be drawn and shown in Fig. 7.56a, b. In Fig. 7.56b, the time segment sequence of active voltage vectors is reversed compared with the original scheme by implementing the freedom of time segment sequence exchange. It can be found that the deduced PWM scheme only changes the sequence of voltage vectors in paralleled inverters, but the combination of voltage vectors is invariable. In this condition, the CMV can also keep zero in one switching cycle.

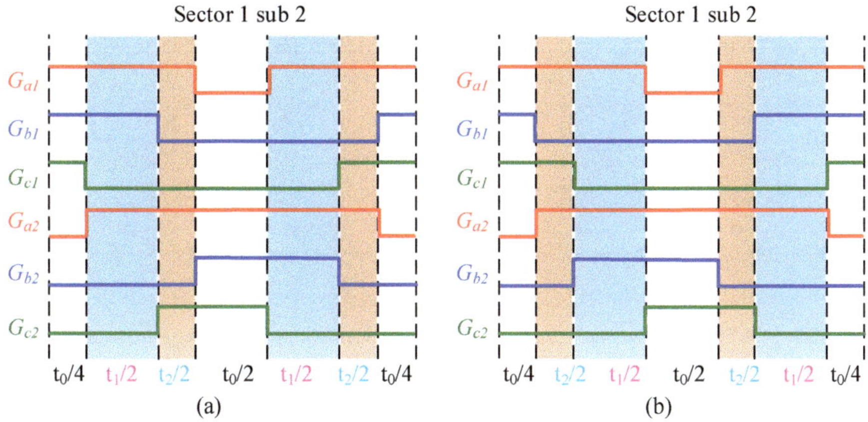

Fig. 7.56 The two different zero-CM modulation schemes in Sector 1 subsector 2: **a** the original modulation scheme, **b** the other modulation scheme

Comparing Fig. 7.56a, b, it can be found that the active vector reversion can change the duty cycles in the first and last half switching cycles, so the time segment sequence exchange can be implemented in the problematic sector switching process considering this special effect. Taking phase-A for example, when the sector is switching from 2 to 3 or 5 to 6, the principle of the operations is shown in Fig. 7.57a, b.

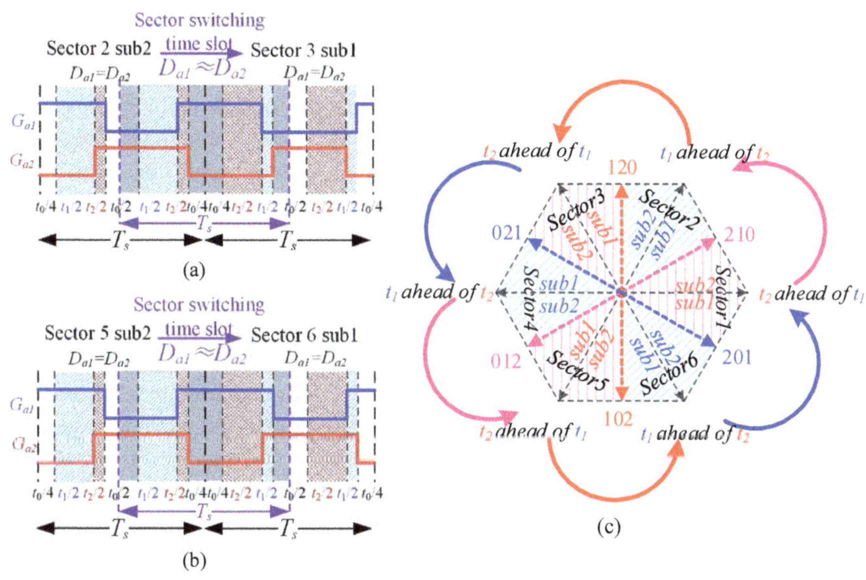

Fig. 7.57 Time segment sequence exchange for phase-A when relevant sector is switching from: **a** Sector 2 to Sector 3, **b** Sector 5 to Sector 6. **c** The principle of the operation considering three-phase voltage balance

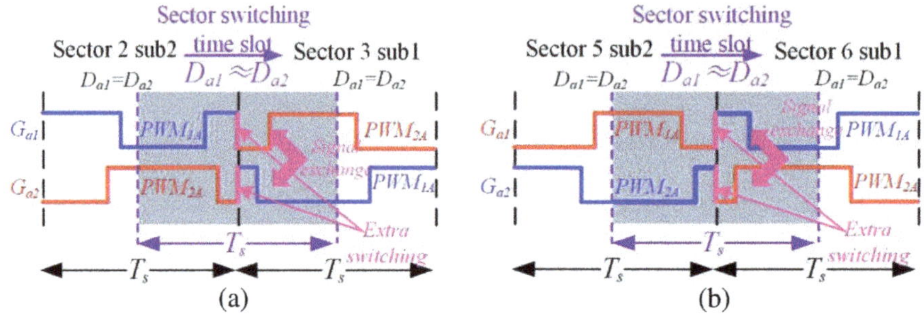

Fig. 7.58 PWM signal exchange for phase-A when relevant sector is switching from: **a** Sector 2 to Sector 3, **b** Sector 5 to Sector 6

By exchanging the time segment sequence in the corresponding sector switching process, the volt-second unbalance can be eliminated and the current jump can be mitigated. In addition, the freedom of time segment sequence demands to change all PWM signals. Thus, considering volt-second unbalance problem of other phases, the freedom should be utilized in every sector switching process and the principle of one kind of operation is shown in Fig. 7.57c.

In addition, the PWM signal exchange is another useful freedom to mitigate the current jump. Similar to time segment sequence exchange, the PWM signal exchange method can redistribute the duty cycles in the first and last half switching cycle. Extracting the problematic sector switching process for phase-A, the principle is shown in Fig. 7.58a, b. It can be clearly seen that the duty cycle in the problematic sector switching process can also be balanced after the PWM signal exchange being implemented, so the volt-second unbalance can be mitigated.

The high-frequency ZSCC is closely related to the output voltage vectors in each VSI, especially the duration time of zero voltage vectors, and the PWM signal exchange method can be used to avoid zero voltage vectors to limit the peak value of ZSCC which focuses on the subsector switching process. For instance, when synthesized voltage vector is sweeping from subsector 1 to subsector 2 in Sector 1 and Sector 4 for phase-A, based on the aforementioned signal exchange method, the PWM signal exchange scheme can be further utilized in subsector switching process shown in Fig. 7.59a, b, and the PWM scheme for phase-A in the whole fundamental period can be deduced and shown in Fig. 7.59c. By utilizing the modified scheme, the PWM signals should be exchanged four times in every fundamental period, and the time interval of the two extra switching actions is one-fourth of fundamental period.

First, the above several improved methods are compared with the original method in simulation. The phase-leg currents (I_{a1}, I_{a2}) and output phase current (I_{a_sw}) of original modulation scheme are shown in Fig. 7.60a. Though the output phase current can keep sinusoidal, both the two phase-leg currents generate obvious current jump. In addition, the DMCC and ZSCC of original scheme are shown in Fig. 7.60b. Accompanied with sector switching process, the DMCC and ZSCC also generate

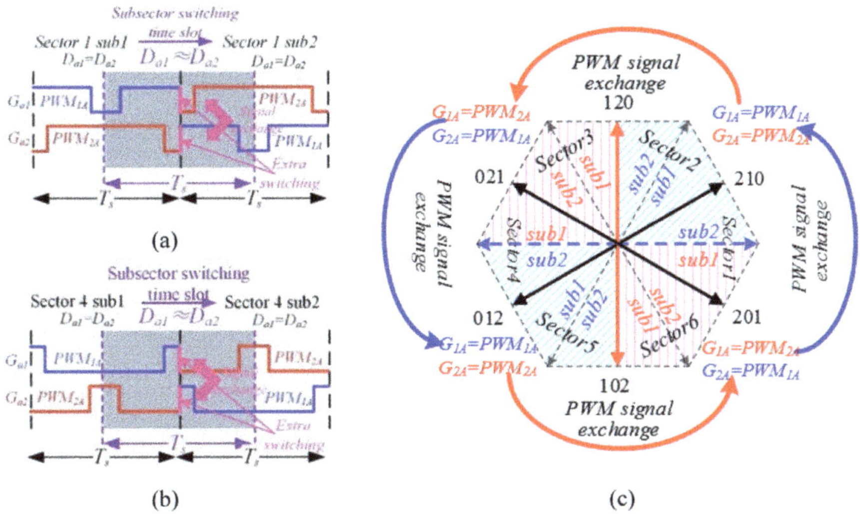

Fig. 7.59 PWM signal exchange for phase-A when relevant subsector is switching from: **a** subsector 1 to sector 2 in sector 1, **b** subsector 1 to sector 2 in sector 4. **c** The principle of operation

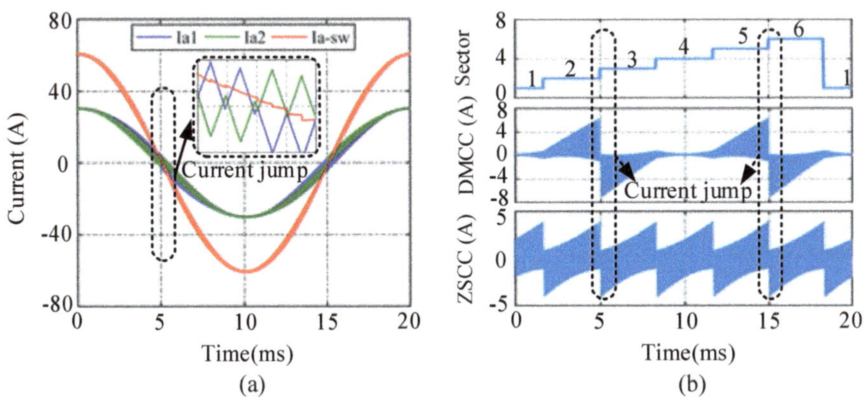

Fig. 7.60 The currents for original method: **a** phase-leg currents and output phase current, **b** DMCC and ZSCC

obvious current jump. The current jump in DMCC and ZSCC amplifies the peak values which is adverse for the CI and common mode inductor (CMI) design.

Figure 7.61 shows the modified method by utilizing the freedom of time segment sequence exchange. With the freedom implemented in every sector switching process, the current jump in paralleled phase-leg currents can be mitigated shown in Fig. 7.61a. Meanwhile, the DMCC and ZSCC change smoothly which means the current jump can also be mitigated shown in Fig. 7.61b. The peak value of DMCC

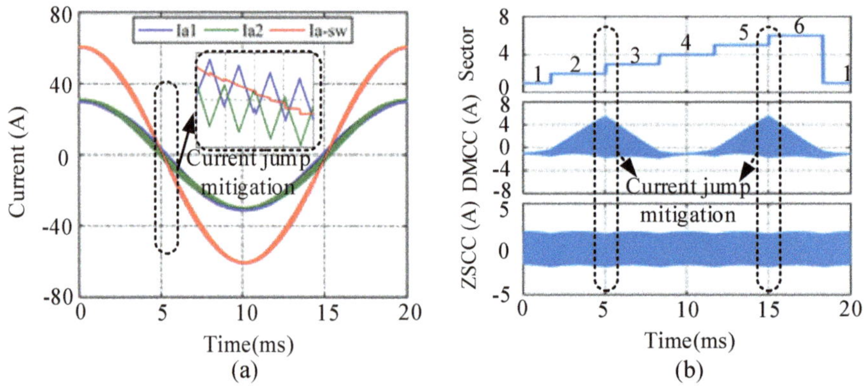

Fig. 7.61 The currents for time segment sequence exchange scheme: **a** phase-leg currents and output phase current, **b** DMCC and ZSCC

can be reduced which has an effect in size reduction of CI, the reduction effect of ZSCC is limited to some extent.

The modified method by utilizing the freedom of PWM signal exchange is shown in Fig. 7.62. By adding extra switching actions to corresponding subsector switching process, the signal exchange scheme can be implemented. The current jump mitigation in phase-leg currents can be realized shown in Fig. 7.62a, and the DMCC and ZSCC can also be modified shown in Fig. 7.62b. With this modified method, the positive and negative peak values of DMCC are different in one fundamental period and the maximum value can be reduced. The ZSCC also has obvious reduction performance compared with original scheme. So, by utilizing the freedom of PWM signal exchange, the current jump mitigation and circulating current reduction can be realized.

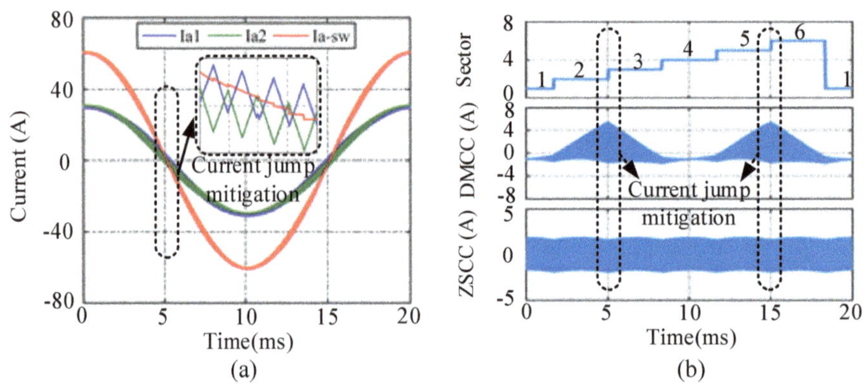

Fig. 7.62 The currents for one-fourth period signal exchange scheme: **a** phase-leg currents and output phase current, **b** DMCC and ZSCC

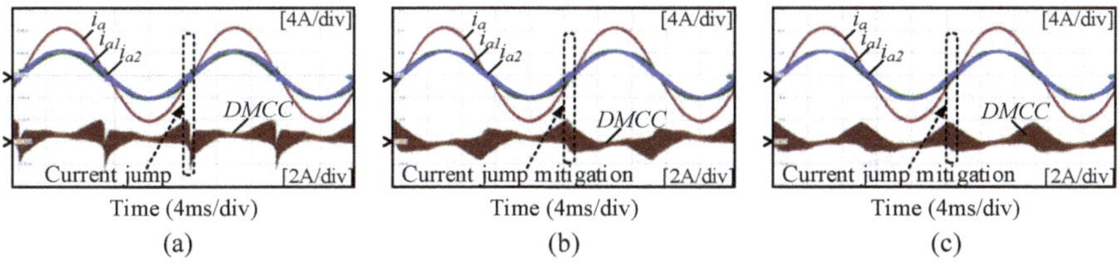

Fig. 7.63 The paralleled phase-leg currents, phase current and DMCC comparison for different schemes when $m = 0.9$: **a** the original scheme, **b** the time segment sequence exchange scheme, **c** the PWM signal exchange scheme

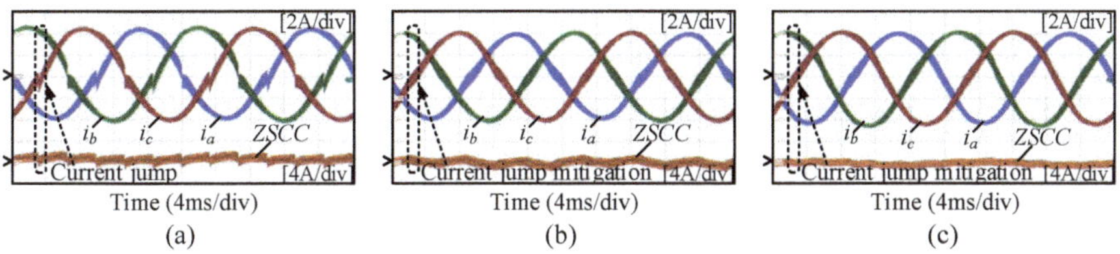

Fig. 7.64 The phase-leg currents and ZSCC comparison for different schemes when $m = 0.9$: **a** the original scheme, **b** the time segment sequence exchange scheme, **c** the PWM signal exchange scheme

In addition, all the above schemes are further verified in the experiment. In original scheme, the current jump between phase-leg currents is obvious which amplifies the peak value of DMCC, as shown in Fig. 7.63a. Through the two modified schemes, the volt-seconds balance can be realized in sector switching process and the current jump can be restrained in phase-leg currents. So the DMCC of two modified schemes can be reduced, as shown in Fig. 7.63b, c.

Meanwhile, the ZSCC comparison has been carried out shown in Fig. 7.64. In original scheme, the three phase-leg currents generate obvious current jump which leads to the ZSCC generating palpable current jump in every sector switching process shown in Fig. 7.64a. With the two modified schemes shown in Fig. 7.64b, c, the current jump in three phase-leg currents can be mitigated; moreover, the ZSCC can be more smooth and the maximum values of ZSCCs for the two modified schemes can be reduced, with especially the PWM signal exchange scheme.

7.5.4 General Pulse Delay Compensation Method

The zero-CM PWM algorithm for paralleled inverters can realize CMV elimination in ideal condition, but due to the deadtime of the inverter in reality, the performance of CMV elimination cannot be well achieved like ideal case. Usually, the deadtime is embedded in the PWM gate signals to prevent the simultaneous conduction of two switching devices in the same phase-leg [18, 19]. This operation brings the problem

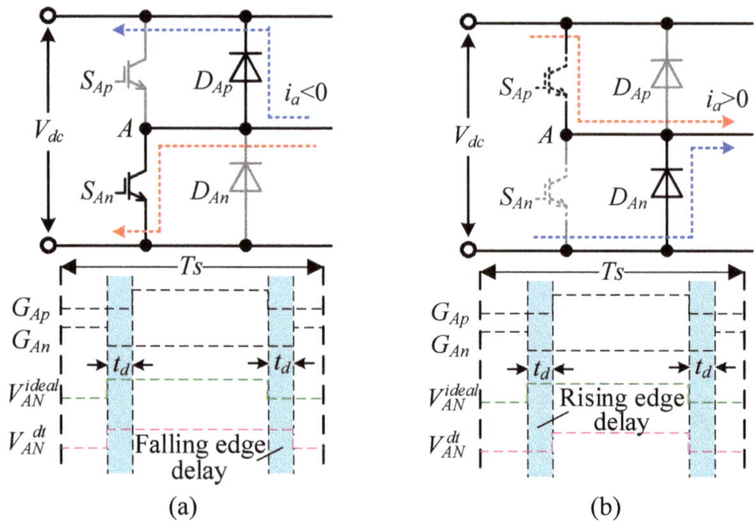

Fig. 7.65 Channel flow of phase current: **a** $i_a < 0$, **b** $i_a > 0$

of synchronization and generates the pulse delay between the pole voltage and corresponding PWM signal, especially the deadtime-induced freewheeling and the turn-on/turn-off process of phase-leg [20, 21]. Thus, to realize the CMV elimination in reality, the pulse delay problem should be solved.

First, the effect of pulse delay caused by deadtime freewheeling is analyzed, as shown in Fig. 7.65, where S_{Ap}, S_{An}, D_{Ap}, D_{An} are the upper and lower switching devices and corresponding anti-paralleled diodes, G_{Ap}, G_{An} are the actual gate signals with the inserted deadtime, V_{AN}^{ideal} is the ideal pole voltage, and V_{AN}^{dt} is the delayed pole voltage which considers the phase current freewheeling in deadtime period. Taking the phase-A as an example, Fig. 7.65a shows channel flow of phase current when $i_a < 0$. In the whole switching cycle, phase current passes through D_{Ap} and S_{An}, and the output voltage is clamped to the positive pole of DC bus during the deadtime interval, so the falling edge delay is generated for V_{AN}^{dt}. Similarly, Fig. 7.65b shows channel flow of phase current when $i_a > 0$, and rising edge delay is generated for V_{AN}^{dt}.

The above situation introduces the pulse-delay effect with normal diode freewheeling. If the phase-leg current is close to zero, there may be a zero-crossing point in one switching cycle, and it has the possibility to generate zero-current clamping (ZCC) phenomenon if phase-leg current changes the direction in one switching cycle. Figure 7.66 shows all kinds of ZCC situations and the corresponding output pole voltages for phase-leg A, where t_1 is the period of normal freewheeling in the deadtime interval, and t_2 is the period of ZCC in the deadtime interval.

The above content shows the different effects of deadtime freewheeling. Since the phase-leg current of inverter always accompanies with current ripple which leads to fundamental component enveloped by the peak values, the deadtime effect of rising

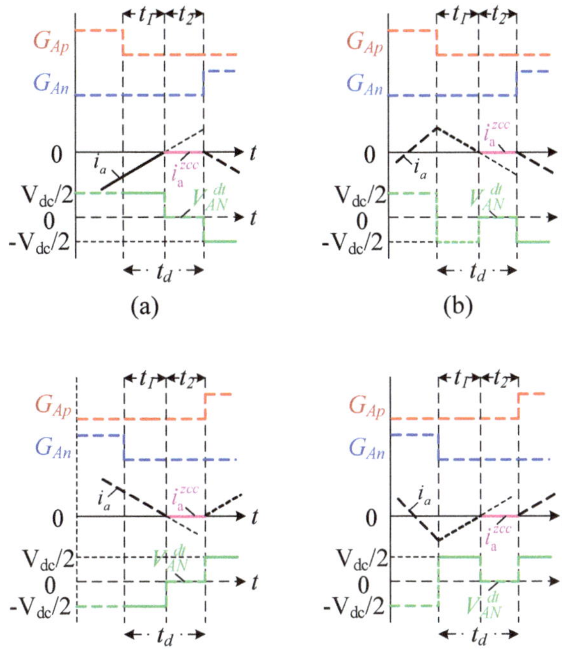

Fig. 7.66 All kinds of ZCC situations and the corresponding output pole voltages in ideal condition

and falling edges should be analyzed separately. To analyze the deadtime effect under various conditions, Fig. 7.67 gives a detailed description of the rising process of phase-leg current with simplified current ripple and corresponding pole voltage which only considers the deadtime freewheeling and neglects the effect of turn-on and turn-off time [20].

According to Fig. 7.67, the deadtime effect can be divided into four cases:

1. The switching phase current i_a^{sw} locates in P-zone (positive zone) which has no zero-crossing point: that is to say the minimum phase current meets the condition that $i_a^{rise} \geq i_a^{bd}$ which can keep the normal freewheeling of current in deadtime interval. i_a^{bd} is the defined boundary between P-zone and Z-zone (zero-zone). The freewheeling of phase current in deadtime interval leads the positive time of actual output pole voltage V_{AN}^{dt} to decrease compared with ideal output voltage V_{AN}^{ideal} ($\Delta d_a = -t_d / T_s$).
2. The switching phase current i_a^{sw} located in N-zone (negative zone) also has no zero-crossing point. Conversely, when the maximum phase current meets the condition that $i_a^{fall} \leq -i_a^{bd}$ in one switching period, the switching current can also keep the normal freewheeling in deadtime interval. $-i_a^{bd}$ is the defined boundary between N-zone and Z-zone. The freewheeling of current in deadtime

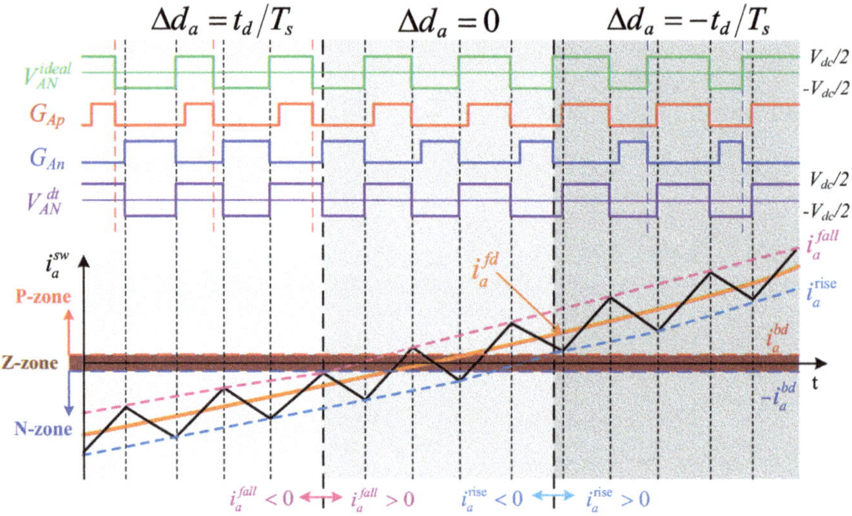

Fig. 7.67 The deadtime effect of rising process of phase-leg current

interval leads to actual output voltage's positive time increase compared with ideal output voltage ($\Delta d_a = t_d / T_s$).

3. The switching phase current i_a^{sw} has zero-crossing points but all the peak values are out of Z-zone: that is to say the minimum phase current meets the condition $i_a^{rise} \leq -i_a^{bd}$ and the maximum phase current meets the condition $i_a^{fall} \geq i_a^{bd}$ in one switching period. Although deadtime interval exists under this circumstance, there is no ZCC happening because of the margin of current freewheeling. At the same time, the deadtime effect can offset each other on duty cycle variation, and the main duty cycle variation is caused by the turn-on and turn-off time ($\Delta d_a = 0$).
4. The switching phase current i_a^{sw} has zero-crossing points and one of peak values is in Z-zone (zero zone): that is to say the minimum phase current $i_a^{bd} \geq i_a^{rise} \geq -i_a^{bd}$ or the maximum phase current $i_a^{bd} \geq i_a^{fall} \geq -i_a^{bd}$ in one switching period. As previously analyzed, phase current may generate ZCC because of the possibility of peak currents changing directions. Since the influence of ZCC is limited because of the shorter time (less than deadtime interval) and the lower occurrence possibility in one fundamental period, so the influence on the pulse delay is ignored.

Based on the above analysis, the pulse delay caused by the deadtime freewheeling can be clarified. In addition, the switching-edge delay caused by the turn-on and turn-off process should also be considered. Taking the parasitic capacitor into consideration [21], Fig. 7.68a illustrates the transition process of the phase-leg from on-state to off-state for $i_a > 0$. In the first on period stage, i_a flows through S_{Ap}. During the second deadtime stage, the upper parasitic capacitor C_{Ap} begins to charge and the

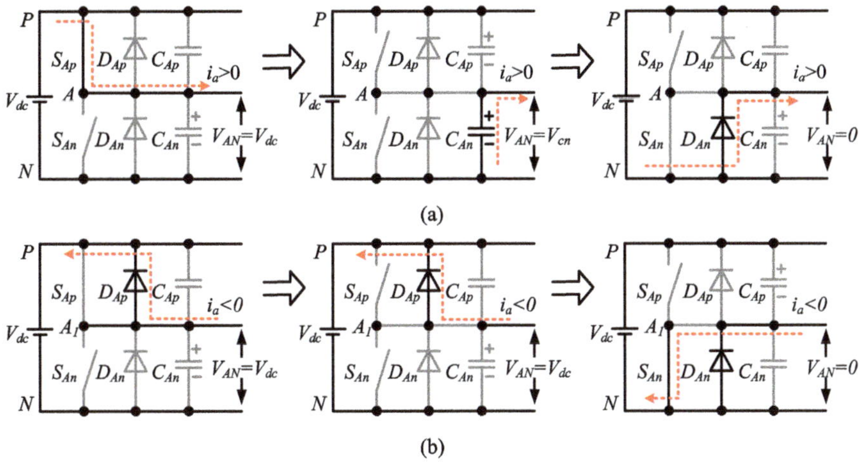

Fig. 7.68 Turn-off transition process of phase-leg: **a** $i_a > 0$, **b** $i_a < 0$

lower capacitor C_{An} discharges to make phase current flow. Since the discharge duration increases with the decrease of phase current magnitude, the turn-off time t_{off} increases with a decrease in the magnitude of phase current when $i_a > 0$. In the final off-period stage, i_a flows through the lower diode D_{An}.

Moreover, Fig. 7.68b shows the turn-off transition process for $i_a < 0$. In the first on period stage, i_a flows through D_{Ap}. During the second deadtime stage, i_a continues to flow through D_{Ap}. In the final off-period stage, the inverter output voltage becomes zero immediately after the S_{An} turns on. Thus, t_{off} is independent of the magnitude of phase current when $i_a < 0$. Similarly, the effect of parasitic capacitance from off-state to on-state can be derived in the same way. In this case, it can be concluded that the turn-on and turn-off processes are not only influenced by the direction of phase current, but also related with the amplitude of phase current.

Figure 7.69 shows the turn-off and turn-on time for the pole voltage when the conventional 2 μs deadtime is added between the upper and lower PWM signals. For the turn-off time shown in Fig. 7.69a, it can be seen that when $i_a < 0$, the transient time

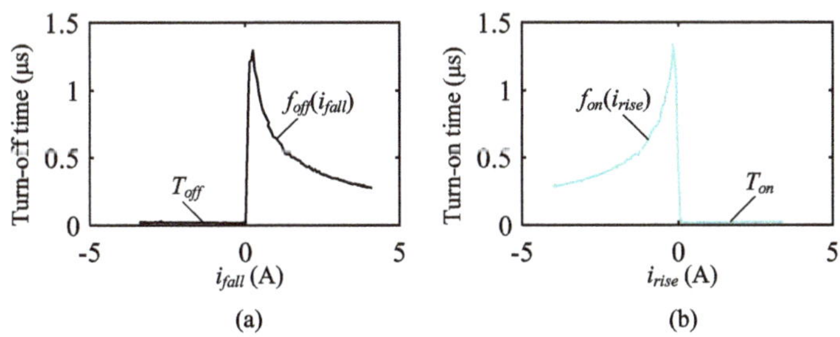

Fig. 7.69 The relationship of the phase-leg's switching edges delay time with the current: **a** turn-off time with current, **b** turn-on time with current

is short and almost maintaining unchanged which corresponds to the fast discharging of lower parasitic capacitance. When $i_a > 0$, the turn-off time is significantly increased and reduced with the phase current increase which corresponds to situation. As for the case of turn-on time, it is opposite to the turn-off time situation, as shown in Fig. 7.69b.

So the tested curves are consistent with the theoretical analysis. From the tested curves of switching times, the turn-on and turn-off time can be expressed as:

$$t_{off}(i) = \begin{cases} f_{off}(i) & (i > 0) \\ T_{off} & (i < 0) \end{cases} \quad t_{on}(i) = \begin{cases} T_{on} & (i > 0) \\ f_{on}(i) & (i < 0) \end{cases} \tag{7.12}$$

where T_{on} and T_{off} are the constant turn-on and turn-off time, and $f_{on}(i)$ and $f_{off}(i)$ are the function of turn-on and turn-off time. In this case, the total deadtime-effect-caused time-delay can be deduced as:

$$T_d^{on}(i) = \begin{cases} T_{on} + t_d & (i > 0) \\ f_{on}(i) & (i < 0) \end{cases} \quad T_d^{off}(i) = \begin{cases} f_{off}(i) & (i > 0) \\ T_{off} + t_d & (i < 0) \end{cases} \tag{7.13}$$

where T_d^{on} and T_d^{off} are the total time delay of the rising and falling edges. Thus, the total on-time and off-time delay effects can be defined and shown in Fig. 7.70a, b, where V_{AN}^{delay} is the actual pole voltages with the turn-on and turn-off time of switching devices. Figure 7.70c shows the final measured time delay of switching edges with the phase current which are consistent with the theoretical analysis. Thus, with the method of curve fitting, the time delay functions with the phase current as variable can be set up, and the pulse delay compensation can be precisely implemented.

Though the time delay can be precisely obtained with the tested time delay curves, the other challenging task is the real-time phase current estimation. Figure 7.71 shows the voltage source inverter's output phase current i_{sw} which is formed by fundamental current i_{fd} (low frequency) and ripple current i_{ripple} (high frequency). Generally, the control processor with sampling circuit can only extract the low-frequency fundamental current, while the high-frequency ripple current is often neglected. In this case, the time delay calculation of the rising and falling edges cannot be decoupled which leads to the compensation errors of pulse delay. On the contrary, if the high-frequency i_{ripple} can be predicted with the current ripple prediction method introduced in Chap. 4, i_{sw} can be reconstructed by adding the two components (i_{fd} and i_{ripple}) together. Under these circumstances, the phase current of the phase voltage's rising and falling edges can be independently calculated and the pulse delay compensation can be implemented more precisely without adding any hardware cost. With the current ripple prediction method in Chap. 4, this delay can be compensated.

To verify the above method, the zero-CM PWM scheme for paralleled inverter is selected to test the CMV and CMC for different schemes. For the zero-CM PWM scheme without pulse delay compensation shown in Fig. 7.72a, the obvious CMV still exists owing to the pulse delay effect. As for the zero-CM PWM scheme with the

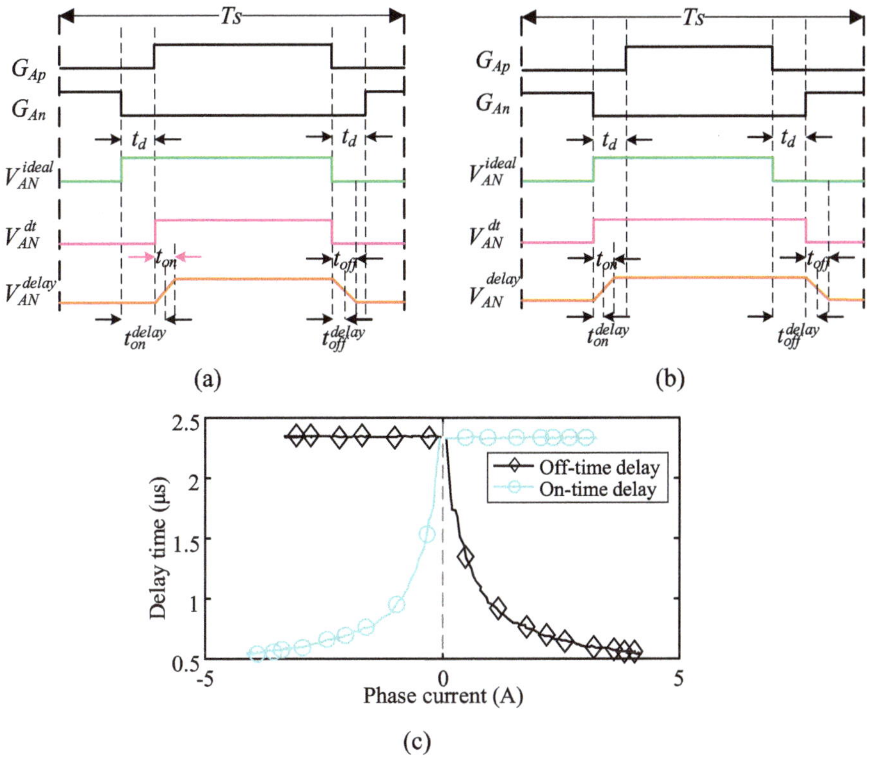

Fig. 7.70 The relationship of the phase-leg's switching edges delay time with the current: **a** turn-on time effect, **b** turn-off time effect, **c** turn-on time with current

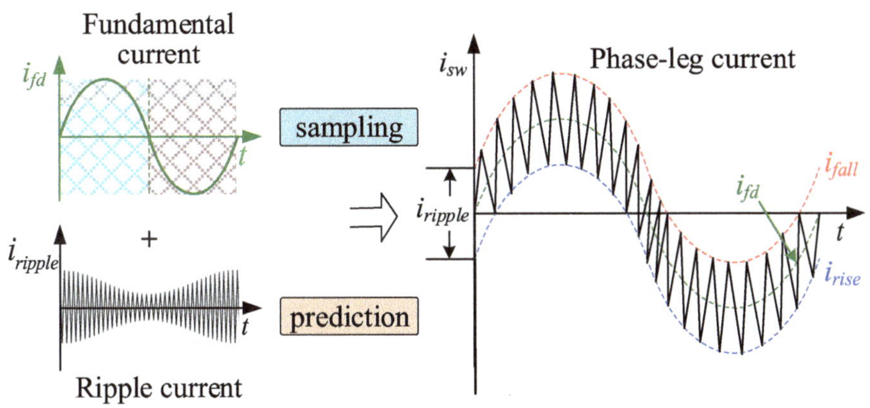

Fig. 7.71 The relationship among actual phase current, fundamental current and ripple current

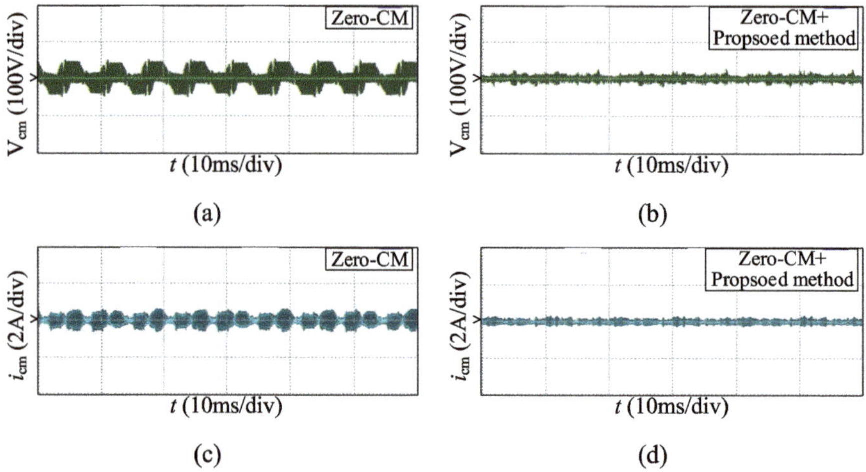

Fig. 7.72 The CMV and CM comparison for different schemes: **a** CMV for zero-CM without pulse delay compensation, **b** CMV for zero-CM with pulse delay compensation, **c** CMC for zero-CM without pulse delay compensation, **d** CMC for zero-CM with pulse delay compensation

proposed method shown in Fig. 7.72b, the CMV can be well suppressed which proves the effectiveness of the proposed method. The CMC comparison between Fig. 7.72c, d also proves the effect of the proposed pulse delay compensation method.

In addition, the three-phase currents are compared for different schemes, as shown in Fig. 7.73. For the zero-CM PWM scheme without deadtime compensation (DTC) method, the phase voltage distortion induced low-frequency current (mainly the fifth and seventh) is inevitable, as shown in Fig. 7.73a. For the zero-CM PWM scheme with the proposed pulse delay compensation method shown in Fig. 7.73b,

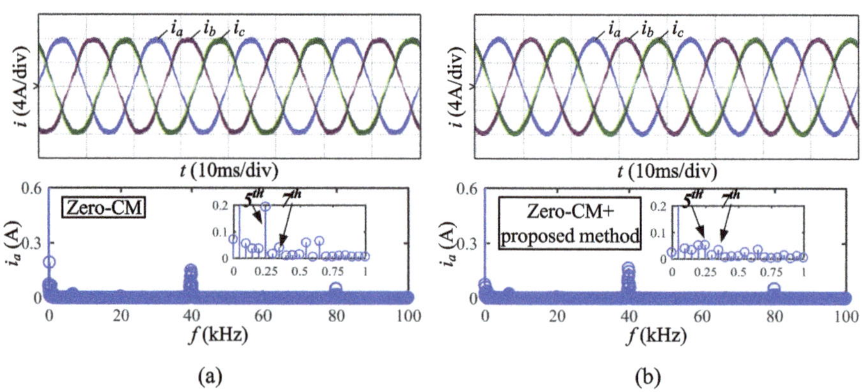

Fig. 7.73 The three-phase currents comparison for different schemes: **a** zero-CM PWM scheme without pulse delay compensation, **b** zero-CM PWM scheme with pulse delay compensation

the low-frequency harmonic currents can be well suppressed which also shows the compensation effect of the proposed method.

In summary, the freewheeling of phase-leg current in the commutation process may change the polarity of the output pole voltage, thereby adding the pulse delay and changing the pulse edge position for the pole voltage, which makes the pole voltage no longer synchronous with the PWM signal, and causes the short time CMV pulse. In the actual situation, the CMV elimination effect for parallel inverters needs to consider the influence of the pulse delay of inverter. The proposed pulse delay compensation method can be not only used to eliminate most CMV pulses for the zero-CM modulation algorithm, but also can be extended to different topologies with different modulation schemes if the current ripple can be estimated with general current ripple prediction method.

7.6 Common-Mode Voltage Elimination for Dual Inverter-Fed Dual Three-Phase PMSM

In the previous sections, the concept of paralleled inverters with zero-CM PWM and the main technical issues have been introduced. The control object is a three-phase load, mainly three-phase motor. Then, three CIs are needed. However, the load motor and CIs are both magnetic components. So there is a possibility to combine the function of CIs to the motor itself and save the CIs. Since the paralleled inverters are with six terminals, the load motor should also be with six terminals which corresponds to the conventional dual three-phase motor. The dual three-phase motor has two groups of three-phase windings. They can be with or without phase-shift, more or less coupling. The simplest case is the dual three-phase motor with zero phase-shift, which is similar to paralleled inverters with coupled-inductors and three-phase motor in the previous sections. Another typical dual three-phase motor is with 30° phase-shift. More generally, arbitrary phase-shift can be designed in the dual three-phase motor. In this section, dual inverters with dual three-phase motor for zero-CM PWM will be studied, with the case of phase-shift of 0°, 30° and arbitrary angle. Permanent magnet synchronous motor (PMSM) is used as an example. For induction motor, the situation is similar.

7.6.1 Common-Mode Voltage Elimination for Dual Three-Phase PMSM with 0° Angle Displacement [32]

For paralleled inverter-fed three-phase AC motor shown in Fig. 7.74a, the paralleled manner realization is contributed to the three extra CIs which limit the circulating current for paralleled phase-legs. Compared with other CMV elimination schemes' topologies [10, 11], though the zero-CM PWM scheme has the same potential to

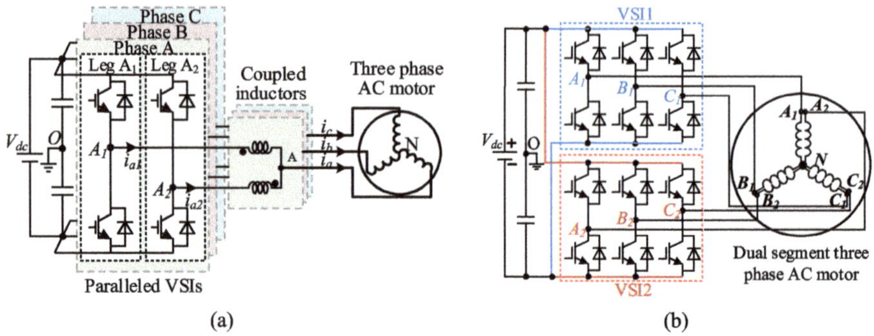

Fig. 7.74 The system configuration comparison: **a** paralleled inverters-fed three-phase AC motor with CIs, **b** dual inverter-fed dual segment three-phase motor

cancel out the big CM choke, the added three extra CIs not only increase the volume and weight for the system but also make the whole system complex. Actually, the pivotal role of CI is to provide the suitable inductance in the circulating path to restrain the circulating current, but there exists the other kind of inductor in the system, that is, the stator inductor of motor which has the potential to offer the so-called "suitable inductance". In order to make the stator inductor integrated with the function of circulating current suppression to cancel the CI, the special motor should be designed to work with dual inverter, for example, the dual segment three-phase AC motor [10], as shown in Fig. 7.74b.

To deduce the principle of CI elimination, two paralleled phase-legs with single CI and the corresponding phase loads are extracted to simplify the analysis. Taking phase-A for example, the single-phase model can be built and shown in Fig. 7.75a. It can be seen that both the CI and phase load have inductors. The main inductor of CI is to limit the circulating current between the paralleled phase-legs caused by the instantaneous pole voltage difference, while the inductor of phase load is the stator inductor of motor. If these two inductors can be integrated to one inductor, the simplified system which eliminates CIs can be realized.

By utilizing the circuit decoupling equivalence principle, Fig. 7.75b depicts the equivalent circuit with the CI decoupling, the external inductor and the stator inductor

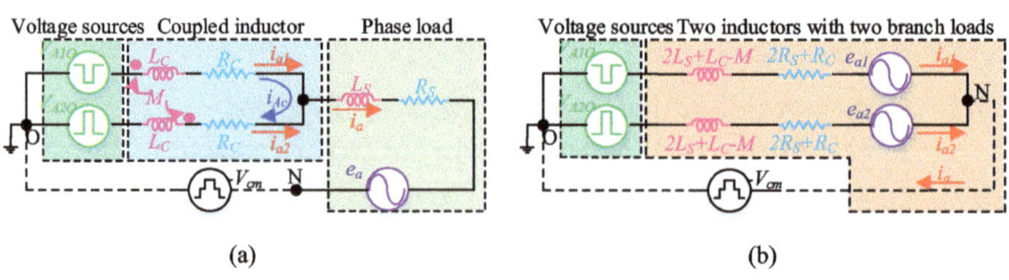

Fig. 7.75 The process of CI elimination: **a** the original equivalent circuit model of one phase, **b** the equivalent circuit model without external inductors of one phase

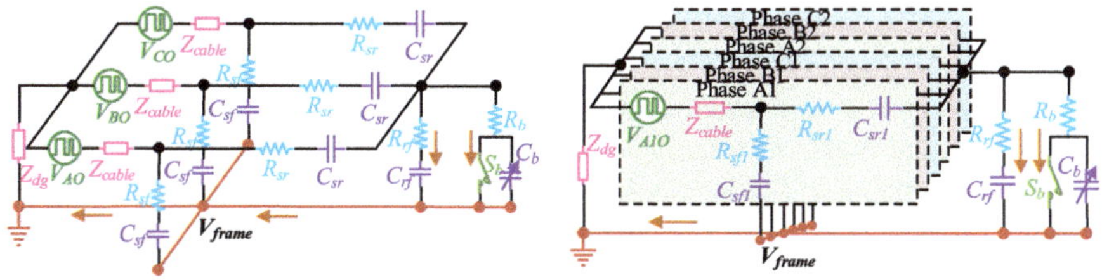

(a) (b)

Fig. 7.76 The CM equivalent model comparison for: **a** the typical three-phase motor drive system, **b** the dual-three-phase motor drive system

can be merged to one inductor with the inductance of $2L_S + L_C - M$, and the merged resistance follows the same law with the value of $2R_S + R_C$. In this case, the combined inductor not only provides the function of circulating current suppression but also can be regarded as the stator inductor, so the external inductor can be eliminated. Meanwhile, considering the mandatory rule that one stator inductor should have its own back-EMF, the back-EMF of one phase should be decomposed into two back-EMFs and the phase load can be regarded as two branches of loads. The increasing number of stator inductors means the physical changing requirement of motor that one set of stator windings should increase to two, so the conventional three-phase motor should be modified to dual three-phase winding motor. In addition, according to the circuit principle, the relationship $e_{a1} = e_{a2} = e_a$ should be satisfied, so each homologous two stator inductors should have identical back-EMFs.

In addition, the influence of stator winding modification for the CMV-related problem should be analyzed. In conventional three-phase motor with single set of stator winding, taking the main impedance of system CM path into consideration, the simplified CM equivalent model shown in Fig. 7.12b can be redrawn and shown in Fig. 7.76a, where Z_{dg} is the DC-link midpoint grounding impedance, Z_{cable} is the transmission impedance of cable for each phase. R_{sf}, R_{sr}, and R_{rf} are the resistances between stator and frame, stator and rotor, rotor and frame. C_{sf} and C_{sr} are the capacitances between stator winding and motor frame, stator winding and rotor for each phase, and C_{rf} represents the entire capacitive coupling between rotor and motor frame. The simplified model for the motor bearing consists of the resistor R_b, the switch S_b and the adjustable capacitor C_b. Owing to the symmetric characteristic of the CM equivalent branch circuit for each phase in ideal condition, it can be easily deduced that the induced motor frame voltage can be restrained to zero with the zero-CM PWM scheme.

Similarly, considering the stray coupling in dual winding motor, the corresponding CM equivalent model can also be set up and shown in Fig. 7.76b. Even when the number of stator windings increase and the parasitic coupling parameters change, the symmetric characteristic of CM circuit keeps unchanged, so the high-frequency CM leakage current can also be cancelled out regardless of the increase of parasitic coupling branch.

To verify the feasibility of the proposed method, a typical PMSM prototype with two sets of symmetrical three-phase windings has been designed. Design of dual three-phase motor is similar to that of dual-segment motor in Sect. 6.2.3. As shown in Fig. 7.77, the motor is constructed with a 12-slot stator wound with double-layer nonoverlapping coils. The 8-pole-pair permanent magnet (PM) rotor constitutes a slot/pole combination of 12/16 together with the stator. According to classical winding theory of AC machines, the periodicity number of the 12/16 motor is 4. Figure 7.78 gives the FEA-based flux contour plots of the motor under open circuit and rated load conditions, and it can be clearly seen that the flux distribution repeats itself four times along the periphery. Due to this periodicity, the coils can be divided

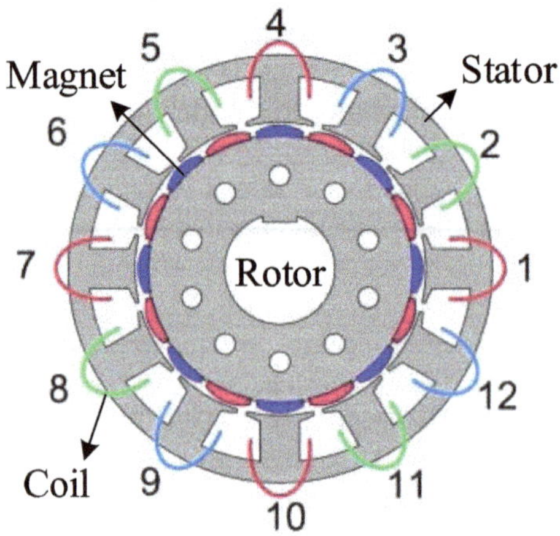

Fig. 7.77 Sketch of the motor topology

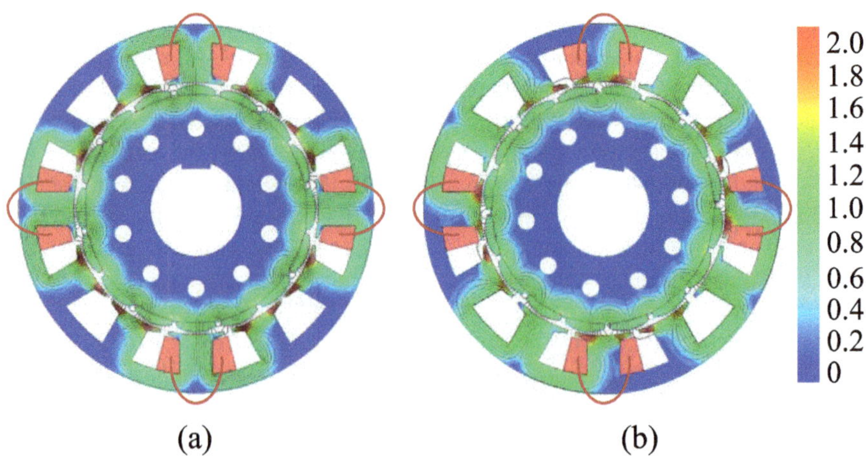

Fig. 7.78 Flux contour plots of the machine: **a** open circuit, **b** rated load

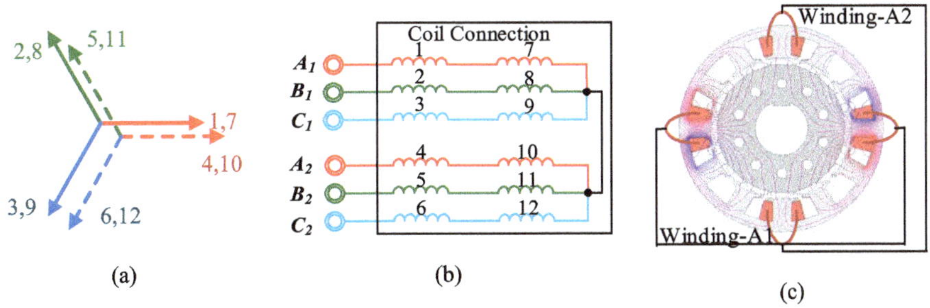

Fig. 7.79 Stator winding connection scheme: **a** star of slots' vector gram, **b** two-segment windings connection, **c** armature reaction of the winding A_1

into three groups according to their difference in space vector, which are shifted by 120° electrical angle. Meanwhile, the four coils in each group can induce back-EMF with exactly the same waveform.

Hence, one set of three-phase winding can be simply designed by selecting any two coils from each group. For example, the dual-segment three-phase winding is implemented according to the star of slots plot shown in Fig. 7.79a, and the winding is specifically configured and illustrated in Fig. 7.79b. In order to investigate the mutual coupling of corresponding phases from each set of winding with the same back-EMF, the armature reaction of the motor is further evaluated through finite element analysis (FEA). As illustrated in Fig. 7.79c, there is only weak coupling between phase A_1 and A_2. So this special designed motor can prevent the stator inductor from being overly weakened by the mutual-inductance of different windings, which is favorable to restrain the increase of current ripple in phase-leg current.

Though the designed dual-segment motor has multiset stator windings, the stator winding is only with three-phase back-EMFs. More importantly, under the FEA, the phase windings with the same back-EMF in different sets have little magnetic coupling, so the influence of different stator winding sets can be neglected to reduce the complexity of the system. In this case, the modeling of the designed motor can be regarded as the combination of two conventional three-phase motors, and the control method can follow the conventional three-phase AC motor's control strategy. So ignoring the small magnetic coupling between the two sets of windings, the mathematical model of each set of winding is identical with regular three-phase PM machine, and the flux and voltage equations in the *d-q* reference frame can be written as (7.14) and (7.15).

$$\begin{bmatrix} \psi_{d1} \\ \psi_{q1} \\ \psi_{d2} \\ \psi_{q2} \end{bmatrix} = \begin{bmatrix} L_{d1} & 0 & 0 & 0 \\ 0 & L_{q1} & 0 & 0 \\ 0 & 0 & L_{d2} & 0 \\ 0 & 0 & 0 & L_{q2} \end{bmatrix} \begin{bmatrix} i_{d1} \\ i_{q1} \\ i_{d2} \\ i_{q2} \end{bmatrix} + \begin{bmatrix} \psi_f \\ 0 \\ \psi_f \\ 0 \end{bmatrix} \tag{7.14}$$

$$\begin{bmatrix} U_{d1} \\ U_{q1} \\ U_{d2} \\ U_{q2} \end{bmatrix} = \begin{bmatrix} R_S & 0 & 0 & 0 \\ 0 & R_S & 0 & 0 \\ 0 & 0 & R_S & 0 \\ 0 & 0 & 0 & R_S \end{bmatrix} \begin{bmatrix} i_{d1} \\ i_{q1} \\ i_{d2} \\ i_{q2} \end{bmatrix} + \frac{d}{dt} \begin{bmatrix} \psi_{d1} \\ \psi_{q1} \\ \psi_{d2} \\ \psi_{q2} \end{bmatrix} + \omega_r \begin{bmatrix} -\psi_{q1} \\ \psi_{d1} \\ -\psi_{q2} \\ \psi_{d2} \end{bmatrix} \tag{7.15}$$

where ψ_{dk}, ψ_{qk}, U_{dk}, U_{qk}, i_{dk}, i_{qk} $(k = 1, 2)$ are the d-q axis flux linkages, voltages and currents, respectively. Ψ_f is the flux linkage produced by permanent magnets, R_S is the stator resistance per phase. ω_r stands for the electrical angular speed of the rotor. In this condition, the whole torque is superimposed by each segment torque of the motor. Considering the characteristic $L_{dk} = L_{qk}$ in this surface-mounted PMSM, the simplified equation for the total torque can be deduced and shown in (7.16).

$$\begin{aligned} T_e &= \frac{3}{2}p\left[\psi_f\left(i_{q1} + i_{q2}\right) + \left(L_{d1} - L_{q1}\right)i_{d1}i_{q1} + \left(L_{d2} - L_{q2}\right)i_{d2}i_{q2}\right] \\ &= \frac{3}{2}p\psi_f\left(i_{q1} + i_{q2}\right) \end{aligned} \tag{7.16}$$

Thus, for the designed dual-segment three-phase motor, the output torque is decided by the summation of two q-axis currents in each set of stator winding.

Based on the design and control principle introduction for the dual-segment three-phase motor, the CMV and motor frame voltage comparison for paralleled inverters with CI scheme and dual inverter with dual sets of windings scheme is tested to verify the feasibility of CI elimination. Three different modulation schemes are implemented, including the conventional SVPWM, the SVPWM with interleaving and the zero-CM PWM which has been introduced in Sect. 7.5. Figure 7.80 shows the test results of stator winding CMV, motor frame voltage and the phase-leg current. For conventional SVPWM utilizing the structure of paralleled inverters with CI and dual inverter with dual winding shown in Fig. 7.80a, b, the CMV is a symmetric six-step wave with the peak value of $\pm V_{dc}/2$ regardless of the instantaneous voltage overshoot shown in the enlarged two switching cycle views. More importantly, the motor frame voltage is directly related with the CMV which can be almost recognized as the replica, and only the amplitude is smaller than the CMV. In addition, for the SVPWM with interleaving shown in Fig. 7.80c, d, it can be found that the CMVs in different structures have been reduced to $\pm V_{dc}/6$ which coincide with the theoretical analysis, and the motor frame voltages also keep the replicating relationship with CMV. As for the zero common-mode voltage (ZCMV) scheme with two structures shown in Fig. 7.80e, f, the CMV can be further reduced in comparison with interleaving manner so that only small spikes are left. These spikes are owing to the short deadtime and switching cancellation transient process and so on. So even with some nonideal factors, the motor frame voltage can also be further reduced which is beneficial for the CM leakage current suppression.

Considering the experimental results above, it can be concluded that the dual inverter can directly drive the designed two-segment motor with similar CMV and motor frame voltage suppression.

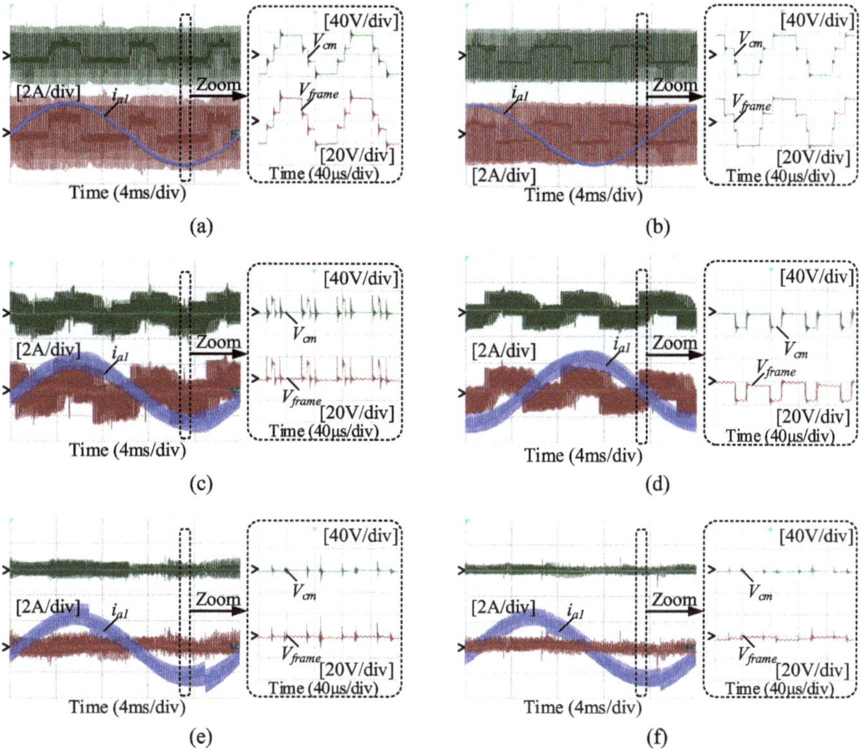

Fig. 7.80 The CMV, motor frame voltage and phase-leg current comparison for different schemes: **a** the conventional SVPWM for paralleled inverters with CI, **b** the conventional SVPWM for dual inverter with dual winding, **c** the SVPWM for paralleled inverters with CI utilizing interleaving manner, **d** the SVPWM for dual inverter with dual winding utilizing interleaving manner, **e** the ZCMV for paralleled inverters with CI, **f** the ZCMV for dual inverter with dual winding

Figure 7.81 shows the CMC and CM EMI comparison for the proposed dual inverter-fed dual winding motor with three modulation schemes. In Fig. 7.81a, it can be seen that the SVPWM scheme with the largest motor frame voltage leads to the largest CMC in the three schemes. The ZCMV scheme which can almost eliminate the motor frame voltage has the best CMC suppression effect which proves the feasibility of the proposed scheme. The left CMC is induced by the left frame voltage spikes.

In addition, the DC side CM EMIs for different schemes have been measured and compared with the EMI standard DO-16050 shown in Fig. 7.81b. It can be seen that in low EMI frequency range (150 kHz–1 MHz), the SVPWM with interleaving scheme can restrain the CM EMI near the odd switching frequency compared with conventional SVPWM scheme which leads to about 10 dB EMI peak reduction. Considering the ZCMV scheme, the proposed ZCMV scheme can significantly reduce CM EMI regardless of the odd or even switching frequency and the peak EMI can be further reduced about 10 dB compared with SVPWM with interleaving scheme from 150

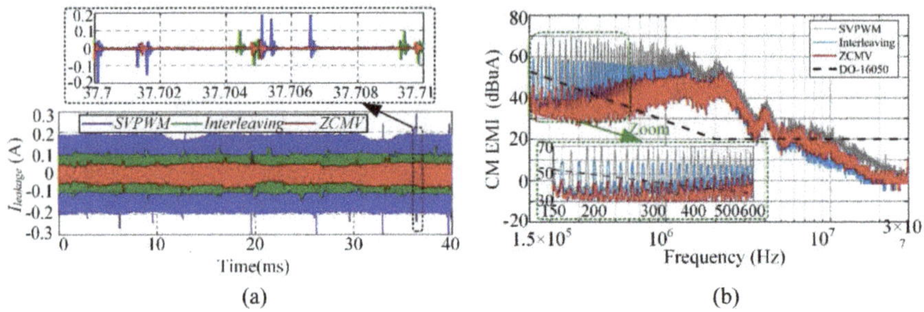

Fig. 7.81 The CMC and CM EMI comparison of proposed dual inverter with dual winding structure for different schemes: **a** the CMC comparison, **b** the CM EMI comparison

to 600 kHz. At high EMI frequency range (2–30 MHz), the influence of PWM is not dominated, so the three cases are close to each other. Therefore, considering the requirement of EMI standard, the ZCMV scheme can help lower the EMI filter cost and component sizes.

To verify the function of the proposed dual-segment three-phase motor, steady-state experiment has been carried out, as shown in Fig. 7.82. It can be found that the amplitudes of corresponding phase currents keep almost identical, so the ZCMV scheme can also keep the phase currents equally divided in the dual winding structure. This characteristic ensures the same output torque contribution for the two sets of stator windings and makes full use of the capacity of the dual inverter system. In addition, with the phase-shift of PWM signal, the current ripple in the corresponding phase current maintains the homologous interleaved effect shown in the enlarged view, so it can be easily deduced that the torque ripple caused by the ripple current in phase current will not be degraded in this case.

So the ZCMV PWM scheme can cooperate with the specially designed dual winding motor to cancel out the CI in motor drive system and maintain the CMV elimination effect. Moreover, the proposed ZCMV PWM scheme with dual-segment motor can keep corresponding phase current equally distributed in the dual winding,

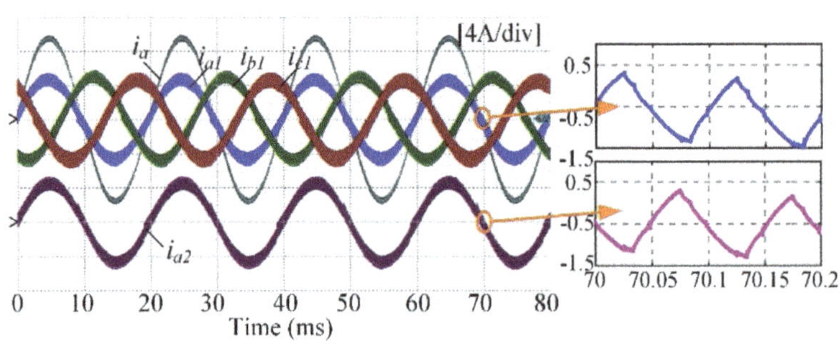

Fig. 7.82 The phase current comparison for the ZCMV scheme with the proposed dual-segment three-phase motor

which guarantee the corresponding phases make the same contribution to motor output torque, thus making full use of the capacity of the dual inverter system.

7.6.2 *Common-Mode Voltage Elimination for Dual Three-Phase PMSM with 30° Angle Displacement [33]*

For dual three-phase machines (DTPM) with 30° angle displacement, it is more attractive than the 0° angle displacement ones due to the cancellation of the sixth torque harmonic. The spatial distribution of the two winding sets is shown in Fig. 7.83, with one set designated as ABC, and the other set lagging 30° electrical angle designated as UVW, and the two neutral points can be connected or isolated. Thus, with the even number of phase-legs, the zero-CMV state can be achieved theoretically if the two VSIs generate three "1" and three "0" states simultaneously.

The 30° DTPM has three subplans based on vector space decomposition (VSD) theory, the fundamental and the $(12\,k \pm 1)$th harmonics are mapped to α-β subplane, while the $(6\,k \pm 1)$th and the $3k$th harmonics can be mapped to z_1-z_2 and o_1-o_2 subplanes, respectively ($k = 1, 2, 3\ldots$). To derivate the ZCMV modulation scheme while simplifying the realization of ZCMV effects, the carrier-based modulation instead of space vector modulation is adopted, and the feasible modulation signal should be first identified.

In terms of the instantaneous CMV elimination demand, the necessary and insufficient condition can be deduced: the average of total CMV which is the summation of six average quasi-duty cycles in every switching cycle must be zero, that is

$$V_{cm\_avg} = \frac{V_{dc}}{12}\left(d_{qa} + d_{qb} + d_{qc} + d_{qu} + d_{qv} + d_{qw}\right) = 0 \tag{7.17}$$

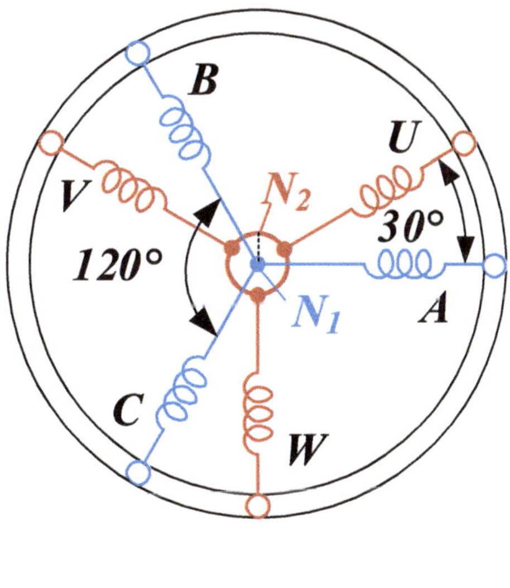

Fig. 7.83 The spatial distribution of the two winding sets for DTPM with 30° angle displacement

where d_{qa}, d_{qb}, d_{qc}, d_{qu}, d_{qv} and d_{qw} are the sampled six-phase quasi-duty cycles from the modulation signals in one switching cycle. Taking the simplest sinusoidal modulation as an example and phase-A as reference, the initial sinusoidal modulation signal for the six phases can be expressed as:

$$\begin{cases} u_A = m\cos\theta \;\; u_B = m\cos(\theta - 2\pi/3) \;\; u_C = m\cos(\theta + 2\pi/3) \\ u_U = m\cos(\theta - \pi/6) \;\; u_V = m\cos(\theta - 5\pi/6) \;\; u_W = m\cos(\theta + \pi/2) \end{cases} \tag{7.18}$$

where m is the modulation index, θ is the real-time reference angle of phase-A, u_A, u_B, u_C, u_U, u_V and u_W are the modulation signals, and the identical equations can be always deduced

$$\begin{cases} u_A + u_B + u_C = 0 \\ u_U + u_V + u_W = 0 \end{cases} \tag{7.19}$$

It is obvious that the conventional sinusoidal modulation for 30° DTPM can naturally satisfy the restrictive condition (7.17). Moreover, considering the situation of harmonic components' injection, Since the injected harmonics in α-β subplane lead to torque ripple of motor, while the injected harmonics in z_1–z_2 subplane generate obvious harmonic currents owing to small impedance of z_1–z_2 subplane, these two kinds of harmonics injection are not considered. In addition, though the zero-sequence components can be injected to the two sets of windings without generating zero-sequence current in the case of isolated neutral points, the influence on the average total-CMV should be discussed. To provide the opposite offset to the peak value of fundamental modulation signals, the offset electrical angles of zero-sequence components should be zero to keep the same phase with the fundamental modulation signals, so the injected zero-sequence components in the two sets of windings can be simplified and expressed as

$$\begin{cases} u_{z\_inj}^{abc} = \sum_{k=1}^{\infty} m_{3k}^{abc}\cos[(3k)(\theta)] \\ u_{z\_inj}^{uvw} = \sum_{k=1}^{\infty} m_{3k}^{uvw}\cos[(3k)(\theta - \pi/6)] \end{cases} \tag{7.20}$$

where $u_{z\_inj}^{abc}$ and $u_{z\_inj}^{uvw}$ are the injected total zero-sequence harmonics for each winding, m_{3k}^{uvw} is the injected harmonic normalization coefficient of UVW phase winding which can be identical or different with m_{3k}^{abc}. In this case, the summation of these two zero-sequence harmonics can be calculated as:

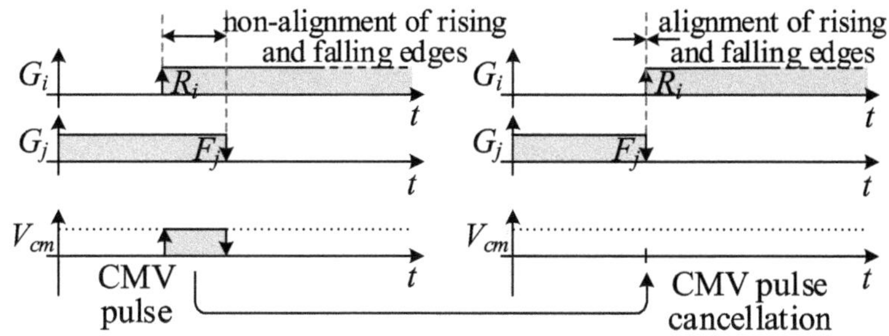

Fig. 7.84 Effect of switching synchronization on CMV pulse cancellation

$$u_{z\_total} = 3\left(u_{z\_inj}^{abc} + u_{z\_inj}^{uvw}\right) = 3\sum_{k=1}^{\infty}\left\{m_{3k}^{abc}\cos[(3k)(\theta)] + m_{3k}^{uvw}\sin[(3k)(\theta)]\right\} \quad (7.21)$$

where u_{z_total} is the summation of the zero-sequence harmonics. It can be seen that the two sets of windings are always injected orthogonal to AC harmonics. In this case, to keep u_{z_total} zero in the whole fundamental period, the only solution is that all harmonic coefficients be kept zero, that is, no zero-sequence components are injected to the fundamental modulation signal.

In addition, to cancel the instantaneous CMV pulses, the simultaneous switching commutations of PWM signals should be arranged to cancel out the CMV pulses, as shown in Fig. 7.84. Thus, in order to realize the CMV elimination in every switching cycle, all the six rising edges should be aligned with the six falling edges, and this requirement can be realized by shifting and setting all the PWM signals in end-to-end arrangement artificially.

To eliminate the CMV while optimizing the DM performance, suitable sequence which retains the two zero voltage vectors V_0 (000) and V_7 (111) is considered in isolated neutral-point condition. In this case, specifically the PWM signals of ABC windings take the central distribution and shifting without overflow, while those of UVW windings take the double-sided distribution. The feasible sub-CMV combinations can be drawn and shown in Fig. 7.85, where V_{cm1} and V_{cm2} are the CMVs for the two sets of windings. V_0, V_7, V_0', V_7' are the zero voltage vectors for V_{cm1} and V_{cm2}. In Fig. 7.85a, d_{abc_max} decides the duty cycles of V_0 and V_7', and it also sets the relative positions of three PWM signals' switching edges, including the maximum duty cycle's PWM signal in ABC windings and two nonmaximum duty cycles' PWM signals in UVW windings. In addition, d_{uvw_max} decides the duty cycles of V_0' and V_7 as well as sets the relative positions of the remaining three PWM signals' switching edges. Similarly, d_{abc_min} decides the duty cycles of V_7 and V_0', and d_{uvw_min} decides the duty cycles of V_0 and V_7', as shown in Fig. 7.85b. Hence, the relative positions of all PWM signals can also be set up.

Considering the maximum duty cycle determined sub-CMV combination pattern shown in Fig. 7.85a, the potential inequalities can be deduced:

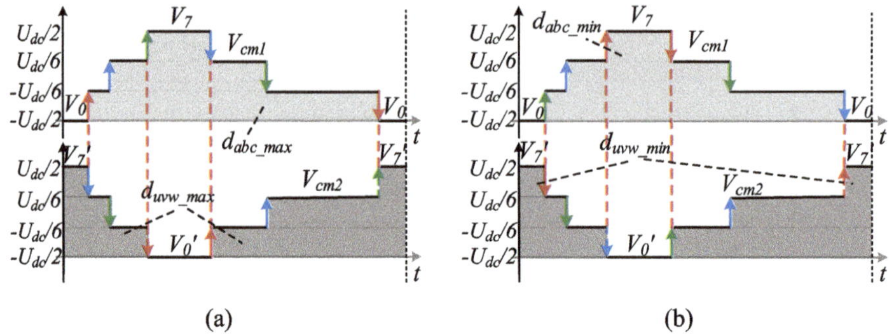

Fig. 7.85 Possible sub-CMV combination patterns: **a** maximum duty cycle determined sub-CMVs combination pattern, **b** minimum duty cycle determined sub-CMVs combination pattern

$$\begin{cases} d_{abc\_\max} + d_{uvw\_\min} > 1 \\ d_{uvw\_\max} + d_{abc\_\min} > 1 \end{cases} \tag{7.22}$$

Similarly, for the minimum duty cycle determined sub-CMV combination pattern shown in Fig. 7.85b, the following inequalities should be satisfied:

$$\begin{cases} d_{abc\_\max} + d_{uvw\_\min} < 1 \\ d_{uvw\_\max} + d_{abc\_\min} < 1 \end{cases} \tag{7.23}$$

With the calculated peak modulation signals, the six regions can be divided by the restrain relationship of peak modulation signal summation in two sets of windings, and the suitable end-to-end cyclic sequence schemes are summarized and illustrated in Table 7.7.

Figure 7.86 shows one of the proposed end-to-end cyclic sequence in Region I, that is, A–V–C–U–B–W–A which takes the central symmetric PWM signal of phase-A as the initial position reference. It can be seen that the total CMV can be cancelled out while retaining the two zero voltage vectors in the two sets of windings. In even regions, the double-sided symmetrical PWM signals are selected as the initial reference to keep the invariance of start and end states for all PWM signals.

Figure 7.87 shows the phase current performance for the proposed ZCMV scheme, where i_a, i_b are the phase currents for ABC windings, and i_u, i_v are the phase currents for UVW windings. In Fig. 7.87a, only the *d-q* axis current control is implemented, and it can be seen that the fundamental current unbalance in the two sets of windings and the obvious fifth and seventh harmonics emerge. With the conventional four-axis current control and harmonic suppression shown in Fig. 7.87b, the fundamental current can be balanced in the two sets of windings, and the fifth and seventh harmonics can be well suppressed. Therefore, the proposed ZCMV scheme can effectively cooperate with the corresponding current controller to keep the phase currents balanced and sinusoidal, which guarantees the normal operation of the DTPM.

Table 7.7 Region division and suitable end-to-end cycle sequence

Region	Peak duty cycles relationship	The decisive peak duty cycle	The duty cycle of zero vectors	End-to-end cyclic sequence
I	$d_{abc\_max} + d_{uvw\_min} > 1$ $d_{uvw\_max} + d_{abc\_min} > 1$	$d_{abc\_max} = d_a$ $d_{uvw\_max} = d_u$	$d_0 = d_7{}' = 1 - d_a$ $d_7 = d_0{}' = 1 - d_u$	A–V–C–U–B–W–A A–W–B–U–C–V–A
II	$d_{abc\_max} + d_{uvw\_min} < 1$ $d_{uvw\_max} + d_{abc\_min} < 1$	$d_{abc\_min} = d_c$ $d_{uvw\_min} = d_w$	$d_0 = d_7{}' = d_w$ $d_7 = d_0{}' = d_c$	W–A–V–C–U–B–W W–B–U–C–V–A–W
III	$d_{abc\_max} + d_{uvw\_min} > 1$ $d_{uvw\_max} + d_{abc\_min} > 1$	$d_{abc\_max} = d_b$ $d_{uvw\_max} = d_v$	$d_0 = d_7{}' = 1 - d_b$ $d_7 = d_0{}' = 1 - d_v$	B–W–A–V–C–U–B B–U–C–V–A–W–B
IV	$d_{abc\_max} + d_{uvw\_min} < 1$ $d_{uvw\_max} + d_{abc\_min} < 1$	$d_{abc\_min} = d_a$ $d_{uvw\_min} = d_u$	$d_0 = d_7{}' = d_u$ $d_7 = d_0{}' = d_a$	U–B–W–A–V–C–U U–C–V–A–W–B–U
V	$d_{abc\_max} + d_{uvw\_min} > 1$ $d_{uvw\_max} + d_{abc\_min} > 1$	$d_{abc\_max} = d_c$ $d_{uvw\_max} = d_w$	$d_0 = d_7{}' = 1 - d_c$ $d_7 = d_0{}' = 1 - d_w$	C–U–B–W–A–V–C C–V–A–W–B–U–C
VI	$d_{abc\_max} + d_{uvw\_min} < 1$ $d_{uvw\_max} + d_{abc\_min} < 1$	$d_{abc\_min} = d_b$ $d_{uvw\_min} = d_v$	$d_0 = d_7{}' = d_v$ $d_7 = d_0{}' = d_b$	V–C–U–B–W–A–V V–A–W–B–U–C–V

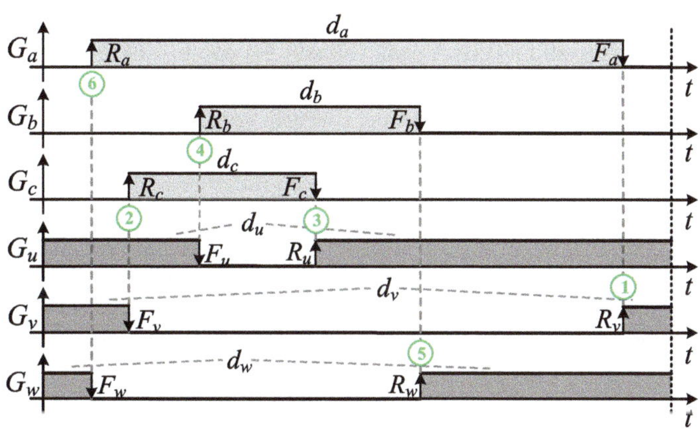

Fig. 7.86 One of the proposed end-to-end cyclic sequence of PWM signals in Region I

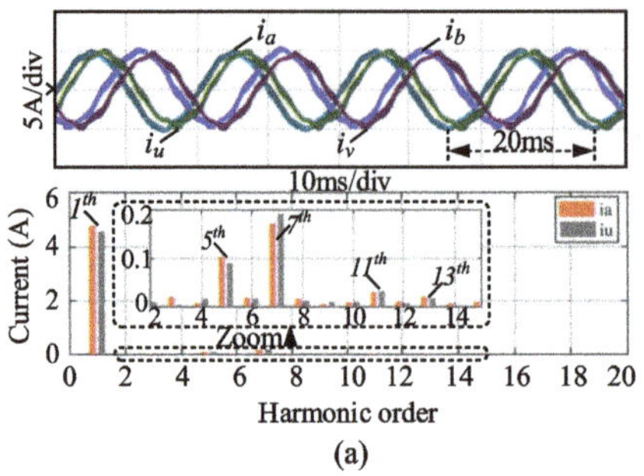

Fig. 7.87 The phase currents for the proposed ZCMV method: **a** with *d-q* axis current control, **b** with four-axis current control and harmonic suppression

To verify the CM reduction effect for the proposed method, the conventional SVPWM scheme is selected to make comparison. Figure 7.88 shows the test results of two sub-CMVs and the motor frame voltage for these two modulation schemes. For conventional scheme without the function of switching cancellation for the two sub-CMVs, the obvious motor frame voltage is generated by the stray capacitance coupled between the stator winding and the motor frame, as shown in Fig. 7.88a. As for the proposed ZCMV scheme shown in Fig. 7.88b, the two sub-CMVs keep the opposite states to cancel out the total CMV, therefore the motor frame potential remains almost unchanged with only small spikes, so the motor frame voltage can be significantly reduced.

To further verify the CM reduction effect for the proposed ZCMV scheme, the CM leakage current and CM EMI are tested and compared, as shown in Fig. 7.89. In Fig. 7.89a, it can be seen that the conventional SVPWM scheme with obvious motor frame voltage leads to about 0.16 A peak value of the output CM leakage current. As for the ZCMV scheme, the output CM leakage current can be effectively suppressed to less than 0.07 A peak value. The remaining CM leakage current is caused by the small frame voltage spikes. Moreover, considering the frequency-domain performance, the DC side CM EMI for different schemes has been measured and compared with the

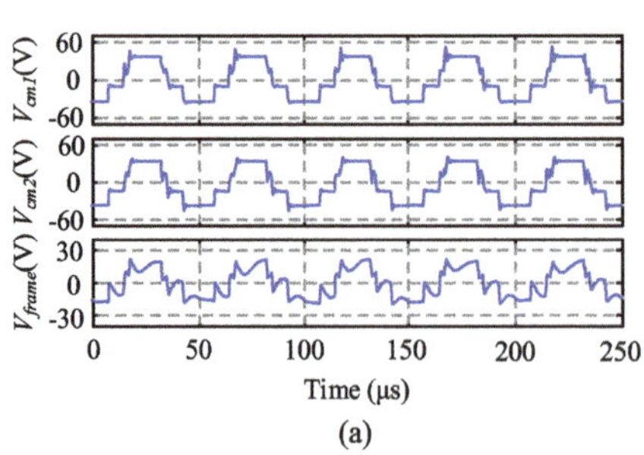

Fig. 7.88 The experimental results of sub-CMVs and motor frame voltage: **a** the conventional SVPWM scheme, **b** the proposed ZCMV scheme

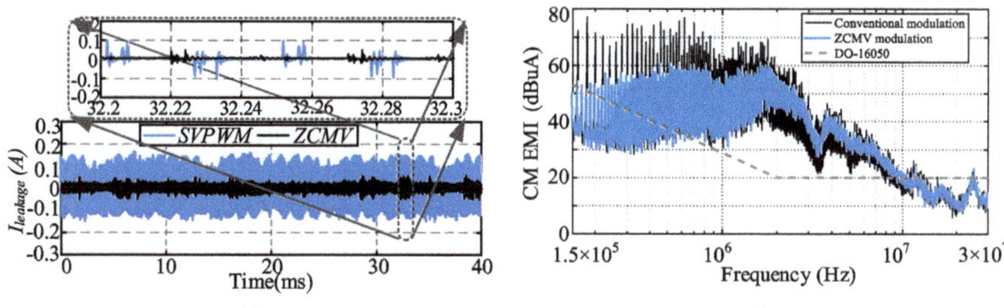

Fig. 7.89 The experimental results for different schemes: **a** the CM leakage current comparison, **b** the CM EMI comparison

EMI standard DO-16050, as shown in Fig. 7.89b. In the low EMI frequency range (150 kHz to 2 MHz) where EMI performance is mainly influenced by the modulation strategy, the conventional SVPWM induces high CM leakage current and results in obvious CM EMI with the peak value of about 78 dB. For the proposed ZCMV scheme, the optimized CM EMI performance can be achieved with about 20 dB average peak EMI reduction compared with SVPWM scheme. Thus, the proposed ZCMV scheme can help to lower the EMI filter weight and reduce the component sizes in the strict EMI requirement condition.

So compared with the conventional SVPWM scheme, the proposed complete carrier-based ZCMV scheme can effectively suppress the total CMV as well as reduce the leakage current and CM EMI which effectively mitigates the CM problem for DTPM with 30° angle displacement.

7.6.3 Common-Mode Voltage Elimination for Dual Three-Phase PMSM with Arbitrary Angle Displacement

Though the above two methods provide the alternative options to eliminate the CMV of different types of DTPMs, these modulation strategies have limited application range which can only be applied to one certain kind of conventional DTPMs with several shifted angles (0°, 30°) between the two sets of windings. More importantly, the above CMV elimination strategies both have the same potential assumption that the stator phase windings and inverter phase-legs have symmetrical parameters. Actually, due to inevitable parameter asymmetries of the motor drive system in reality, the current imbalance could be generated with the conventional CMV elimination methods which will lead to some other unacceptable consequences, such as torque pulsations, efficiency decrease and power derating. Therefore, to maintain the motor performance and suppress the CMV simultaneously, it is important to consider the parameter asymmetry for DTPMs when designing the CMV elimination modulation

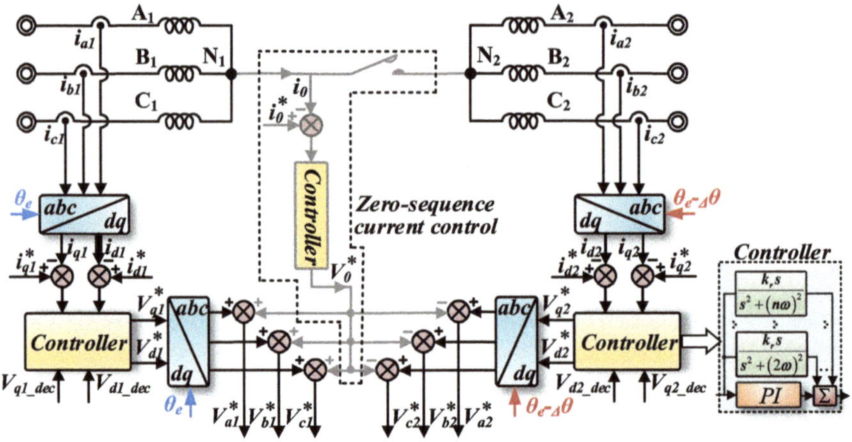

Fig. 7.90 Universal double synchronous frame current control for general DTPM

strategies. Thus, a more flexible and universal CMV elimination PWM strategy for dual-inverter-fed DTPM is provided to deal with the CMV for all kinds of DTPMs both in the cases of symmetrical and asymmetrical parameters.

To derive the suitable PWM scheme to eliminate the CMV for general DTPMs, the feasibility focused on the universal control process, that is, the universal double synchronous frame current control strategy is considered, as shown in Fig. 7.90. With the control method, the phase currents for the two sets of windings can be independently controlled and kept symmetrical. The current controller can be the conventional proportional-integral (PI) regulator, or multiple regulators' combination, for example, the PI controller and extra resonant controllers, which can track the DC reference currents and suppress harmonic AC currents caused by parameter asymmetry or other factors in motor drive system. In addition, when the two sets of windings have the common neutral point, the low-frequency zero-sequence current might be generated, and the opposite zero-sequence voltage could be injected to the two sets of windings to suppress the zero-sequence current if needed.

Regardless of the structure differences of the utilized controllers, assuming the controllers' output reference *d*-, *q*-axis and zero-sequence voltages in the arbitrary switching cycle *k* are $V_{d1}^*(k)$, $V_{d2}^*(k)$, $V_{q1}^*(k)$, $V_{q2}^*(k)$, $V_0^*(k)$ and $-V_0^*(k)$, respectively. Thus, with the corresponding inverse rotational transformations, the reference three-phase voltages in each set of windings can be expressed as

$$\begin{bmatrix} V_{a1}^*(k) \\ V_{b1}^*(k) \\ V_{c1}^*(k) \end{bmatrix} = \begin{bmatrix} \cos(\theta_e) & -\sin(\theta_e) \\ \cos(\theta_e - 2\pi/3) & -\sin(\theta_e - 2\pi/3) \\ \cos(\theta_e + 2\pi/3) & -\sin(\theta_e + 2\pi/3) \end{bmatrix} \begin{bmatrix} V_{d1}^*(k) \\ V_{q1}^*(k) \end{bmatrix} + \begin{bmatrix} V_0^*(k) \\ V_0^*(k) \\ V_0^*(k) \end{bmatrix} \tag{7.24}$$

$$\begin{bmatrix} V_{a2}^*(k) \\ V_{b2}^*(k) \\ V_{c2}^*(k) \end{bmatrix} = \begin{bmatrix} \cos(\theta_e - \Delta\theta) & -\sin(\theta_e - \Delta\theta) \\ \cos(\theta_e - \Delta\theta - 2\pi/3) & -\sin(\theta_e - \Delta\theta - 2\pi/3) \\ \cos(\theta_e - \Delta\theta + 2\pi/3) & -\sin(\theta_e - \Delta\theta + 2\pi/3) \end{bmatrix} \begin{bmatrix} V_{d2}^*(k) \\ V_{q2}^*(k) \end{bmatrix} - \begin{bmatrix} V_0^*(k) \\ V_0^*(k) \\ V_0^*(k) \end{bmatrix} \tag{7.25}$$

where $V_{a1}^*(k)$, $V_{b1}^*(k)$, $V_{c1}^*(k)$, $V_{a2}^*(k)$, $V_{b2}^*(k)$ and $V_{c2}^*(k)$ are the six reference phase voltages, respectively. θ_e and $\theta_e - \Delta\theta$ are the real-time-oriented electrical angles for the two sets of windings.

Figure 7.91 shows the conventional symmetrical CMV pattern for two-level inverter-fed DTPM in one switching cycle, where T_s is the switching period. The CMV has finite positive and negative states which possesses the corresponding pulse-area based on time, that is, the volt-seconds. According to the basic principle of modulation, if the sum of positive and negative volt-seconds has nonzero value, no matter what volt-second mergence method (i.e. the modulation scheme modification) is utilized, there always exists a pulse-area and the CMV cannot be eliminated in this case. Only the zero volt-seconds for CMV can guarantee the possibility of CMV elimination. Thus, in terms of the instantaneous CMV elimination demand, the potential rule should satisfy the following: the volt-seconds corresponding to the CMV must be zero in every switching cycle.

Considering the above CMV elimination potential rule, the output volt-seconds of the two VSIs can be calculated in terms of (7.24) and (7.25).

$$\begin{cases} VT_1(k) = \int_{kT_s}^{(k+1)T_s} [V_{a1}^*(k) + V_{b1}^*(k) + V_{c1}^*(k)]dt = [V_{a1}^*(k) + V_{b1}^*(k) + V_{c1}^*(k)]T_s = 3V_0^*(k)T_s \\ VT_2(k) = \int_{kT_s}^{(k+1)T_s} [V_{a2}^*(k) + V_{b2}^*(k) + V_{c2}^*(k)]dt = [V_{a2}^*(k) + V_{b2}^*(k) + V_{c2}^*(k)]T_s = -3V_0^*(k)T_s \end{cases} \tag{7.26}$$

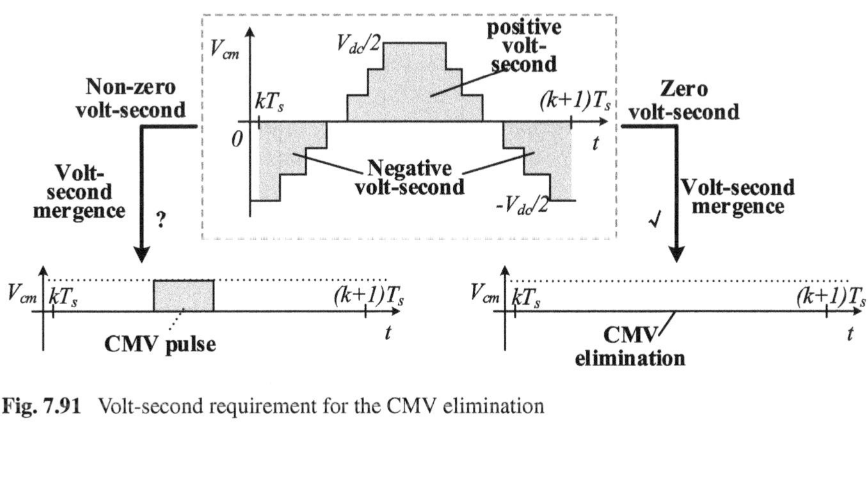

Fig. 7.91 Volt-second requirement for the CMV elimination

Based on (7.26), it can be seen that the real-time d-, q-axis reference voltages and oriented electrical angles have no influence on the volt-seconds for each VSI, and only the possible injected zero-sequence voltages decide the volt-seconds for the two VSIs. Thus, the total volt-seconds of CMV can be calculated as

$$VT(k) = VT_1(k) + VT_2(k) = 0 \tag{7.27}$$

In terms of (7.27), it can be seen that the volt-seconds of CMV always maintain zero in arbitrary switching cycle. So the CMV elimination is feasible for general DTPMs with the universal double synchronous frame current control.

Moreover, considering the possible injected harmonic voltages for the two sets of windings, to roughly judge the weight of injected harmonic voltages in the possible parameter asymmetry condition, the two three-phase total normalization voltage coefficients are defined and expressed as

$$m_i(k) = \frac{2}{V_{dc}}\sqrt{\frac{2}{3}\left[\left(V_{ai}^*(k)\right)^2 + \left(V_{bi}^*(k)\right)^2 + \left(V_{ci}^*(k)\right)^2\right]} \quad (i = 1, 2) \tag{7.28}$$

With no harmonic voltages injected, the two coefficients maintain invariable in a fundamental period and are equal to the modulation indexes in steady state. On the contrary, when some harmonics are injected, these harmonic components will be superimposed into the fundamental phase voltages and generate time-varying coefficients in every switching cycle.

With the control process analysis to ascertain the feasibility of CMV elimination for general DTPMs, further goal is to eliminate the instantaneous CMV pulses, and the restriction of angle displacement and reference voltages between the two sets of windings needs to be lifted. To eliminate the CMV, the synchronous switching cancellation is inevitably utilized, that is, the rising and falling edges corresponding to different PWM signals should be aligned to act simultaneously to cancel out the discrete switching effect on the CMV variation. In this case, the end-to-end arrangement of PWM signals can be implemented to realize the synchronous switching cancellation, as shown in Fig. 7.92. The general process is as follows: first, arbitrarily selecting a PWM signal as the initial reference whose rising edge $p_{r1}(k)$ has $d_{ini}(k)$ ($0 \leq d_{ini}(k) < 1$) interval with the starting position, and then selecting one of the remaining PWM signals and making its rising edge aligned with the falling edge of the previous PWM signal by the shifting operation, and so on; if the overflow is generated, the redundant duty cycle is shifted to the starting position. Thus, with the cyclic operations, the most switching edges can be cancelled out, only the remaining $p_{r1}(k)$ and the falling edge of the final PWM signal $p_{f6}(k)$ should be considered. To derivate the relative position of the remaining switching edges, the duty cycles of PWM signals should be calculated. Assuming that the sorted jth ($j \in [1, 6]$) PWM signal is corresponding to A_1 phase, so the duty cycle can be calculated by the reference phase voltage, that is

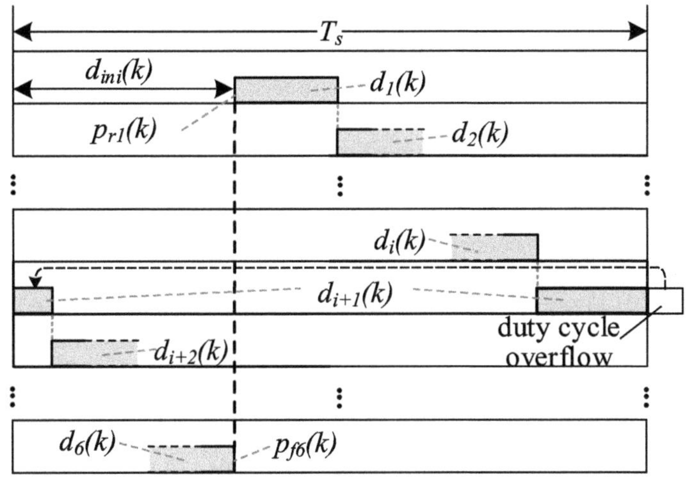

Fig. 7.92 The principle of the general end-to-end arrangement

$$d_j(k) = \left[\frac{V_{a1}^*(k)}{V_{dc}/2} + 1\right] \bigg/ 2 = \frac{V_{a1}^*(k)}{V_{dc}} + \frac{1}{2} \tag{7.29}$$

Other duty cycles of PWM signals follow the same rule. Thus, the sum of duty cycles for all PWM signals be calculated as

$$\begin{aligned} d_{sum}(k) &= \sum_{j=1}^{6} d_j(k) = \frac{V_{a1}^*(k) + V_{b1}^*(k) + V_{c1}^*(k) + V_{a2}^*(k) + V_{b2}^*(k) + V_{c2}^*(k)}{V_{dc}} + 3 \\ &= \frac{VT_1(k) + VT_2(k)}{T_s \cdot V_{dc}} + 3 = \frac{VT(k)}{T_s \cdot V_{dc}} + 3 = 3 \end{aligned} \tag{7.30}$$

So regardless of the different possibilities of duty cycle overflow, the position of $p_{f6}(k)$ can be uniquely determined

$$\begin{aligned} p_{f6}(k) &= d_{ini}(k) + d_{sum}(k) - \text{floor}[d_{ini}(k) + d_{sum}(k)] \\ &= d_{ini}(k) = p_{r1}(k) \end{aligned} \tag{7.31}$$

where the "floor" function is utilized to extract the integer part. In this case, $p_{f6}(k)$ can naturally align with $p_{r1}(k)$, and all the synchronous switching cancellation can be achieved for the proposed method.

Since the general end-to-end arrangement is independent of the real-time reference voltage and the oriented electrical angle for the two sets of windings, so the reference voltages can select different values in the two sets of windings, and the two oriented electrical angles also have no restriction between each other. Thus, the proposed end-to-end arrangement method has no restriction on the modulation index and angle displacement for the two sets of windings.

Though the end-to-end arrangement of six-phase PWM signals can guarantee the function of CMV elimination, the connection sequence of PWM signals can influence the performance of DTPM, so the suitable connection sequence should be derived. Figure 7.93 shows two typical end-to-end connection sequences of PWM signals and the sub-CMVs. In Fig. 7.93a, the three-phase PWM signals in each winding set are arranged with intragroup connection, so the two rising and falling switching edges can be cancelled in the same set of windings which avoid the zero voltage vectors V_0 (000) and V_7 (111) for the two sub-CMVs and can lead to the peak amplitude reduction. Actually, V_0 and V_7 are often retained to achieve the maximum multilevel CMV which can optimize the voltages in nonzero-sequence subplanes and reduce the current ripple. Thus, the reduced sub-CMV pattern which takes the intragroup connection in the same set of windings is not the optimized strategy in the isolated neutral point condition. On the contrary, Fig. 7.93b shows the other possible end-to-end arrangement pattern. The PWM signals in the same set of windings are assigned with interval connection which keeps the distribution patterns identical (central or double-sided distribution). This arrangement can avoid the intragroup connection and provide the possibility of maintaining the zero voltage vectors in each sub-CMVs. So the interval connection pattern is more suitable for the end-to-end arrangement CMV elimination method which has the potential advantage to optimize the current ripple of phase current with the isolated neutral points. The essence of this optimization principle is to transfer the voltage pressure to the blocked zero sequence circuit to optimize the performance of other nonzero sequence loops.

When the DTPM adopts the common neutral point, the current ripple of zero-sequence subplane should be considered, and the current ripple performance is influenced by the motor type and the harmonic weight of all subplanes which has no definite conclusion for the above two connection patterns.

In addition, to maintain the zero vectors for the proposed interval connection pattern, the real-time duty cycle relationship for the PWM signals should be considered, and Table 7.8 sums up an optimized end-to-end arrangement scheme to realize

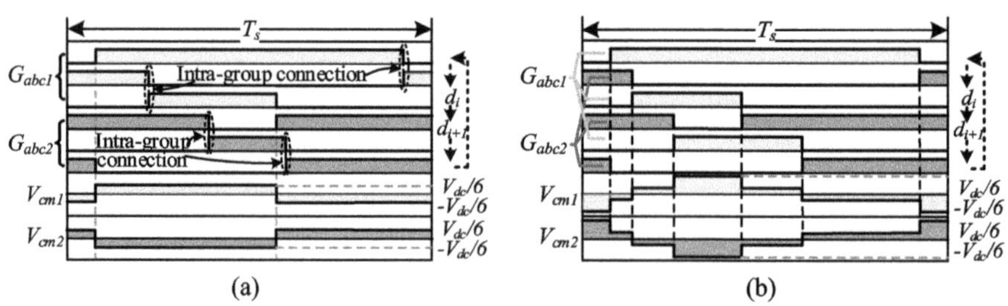

Fig. 7.93 The different end-to-end arrangement patterns: **a** reduced sub-CMV pattern, **b** proposed interval connection pattern

Table 7.8 Proposed end-to-end cycle sequence

Region	Peak duty cycles relationship	The decisive peak duty cycle	End-to-end cyclic sequence
I	$d_{abc1\_max} + d_{abc2\_min} > 1$	$d_{abc1\_max} = d_{a1}$	A1–B2–C1–A2–B1–C2
II	$d_{abc1\_max} + d_{abc2\_min} < 1$	$d_{abc2\_min} = d_{c2}$	C2–A1–B2–C1–A2–B1
III	$d_{abc1\_max} + d_{abc2\_min} > 1$	$d_{abc1\_max} = d_{b1}$	B1–C2–A1–B2–C1–A2
IV	$d_{abc1\_max} + d_{abc2\_min} < 1$	$d_{abc2\_min} = d_{a2}$	A2–B1–C2–A1–B2–C1
V	$d_{abc1\_max} + d_{abc2\_min} > 1$	$d_{abc1\_max} = d_{c1}$	C1–A2–B1–C2–A1–B2
VI	$d_{abc1\_max} + d_{abc2\_min} < 1$	$d_{abc2\_min} = d_{b2}$	B2–C1–A2–B1–C2–A1

the selected PWM pattern in a fundamental period, where the PWM signals in $A_1B_1C_1$ and $A_2B_2C_2$ windings take the central and double-sided distribution.

To evaluate the high-frequency harmonic distortion for the proposed method and compare it with the other schemes, the scanning result in the fundamental subplane is calculated and shown in Fig. 7.94. Considering the undetermined load parameters, the equivalent current THD, which is a universal indicator, can be calculated based on the phase voltage spectrum result and the corresponding harmonic orders,

$$THD = \sqrt{\sum_{n=1,2,3,\cdots}^{\infty} \left(\frac{V_n}{n}\right)^2} \tag{7.32}$$

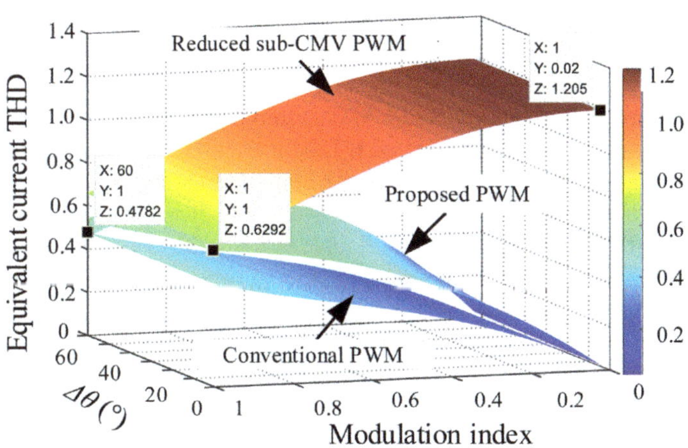

Fig. 7.94 Equivalent current THD comparison in the fundamental subplane

Here V_n is the switching harmonic of phase voltage, and n is the switching harmonic order.

It can be seen that the conventional modulation method has the lowest THD value in the whole modulation index range with different angle displacement values because of the symmetric distribution of PWM signals, but it has no function to eliminate the CMV. For the reduced sub-CMV method with intragroup connection manner, the equivalent current THD is significantly increased, especially in low modulation index range. Considering the proposed method, it can be seen that though the equivalent current THD is relatively bigger than the conventional method with the end-to-end connection operation, the increase of THD is acceptable, and the THD value is obviously lower than the reduced sub-CMV method because of maintaining the zero voltage vectors. This effect can also be extended to other nonzero sequence subplanes. So, the proposed method is more suitable for the DTPM with isolated neutral points.

In order to verify the proposed method and compare it with the conventional schemes, experiments have been done in a prototype experimental platform. Considering that the two DTPM prototypes only provide finite values of angle displacement, to verify the feasibility of arbitrary angle displacement for the proposed strategy, the symmetrical six-phase R-L passive load with common neutral point is first utilized to simulate the DTPMs with an adjustable angle displacement. The fixed 30° angle displacement and identical fundamental modulation index ($m_1 = m_2$) are first implemented, as shown in Fig. 7.95a. It can be seen that the corresponding phase currents are balanced, and the CMV can be almost eliminated in this case. Moreover, a dynamic process in which m_1 and angle displacement increase while m_2 decreases is implemented, as shown in Fig. 7.95b. It can be seen that the fundamental modulation index for the two sets of three-phase load can be arbitrarily combined and the angle displacement can be arbitrarily adjusted such that both have no influence on the CMV elimination effect. Thus, the proposed strategy has the flexible and universal CMV elimination effect which can be utilized in different DTPMs with arbitrary angle displacement and phase voltage combination.

To further verify the effect for the proposed method in actual DTPMs, the two DTPM prototypes are utilized, and the neutral points for the two sets of windings are isolated to validate the proposed sub-CMV pattern. Since the parameter difference is small and ignorable for the fabricated DTPM prototypes, the winding parameters can be regarded as symmetry. In this case, it can be seen that the phase currents in two sets of windings keep symmetrical and have the demand angle displacement for different DTPMs with the universal double synchronous PI control, as shown in Fig. 7.96a–c. In addition, with the enlarged view of two sub-CMVs, it can be found that the two sub-CMVs have four switching states in a switching cycle which means the zero voltage vectors can be retained, and the two sub-CMVs always maintain the opposite states to eliminate the total CMV which is coincident with the theoretical method.

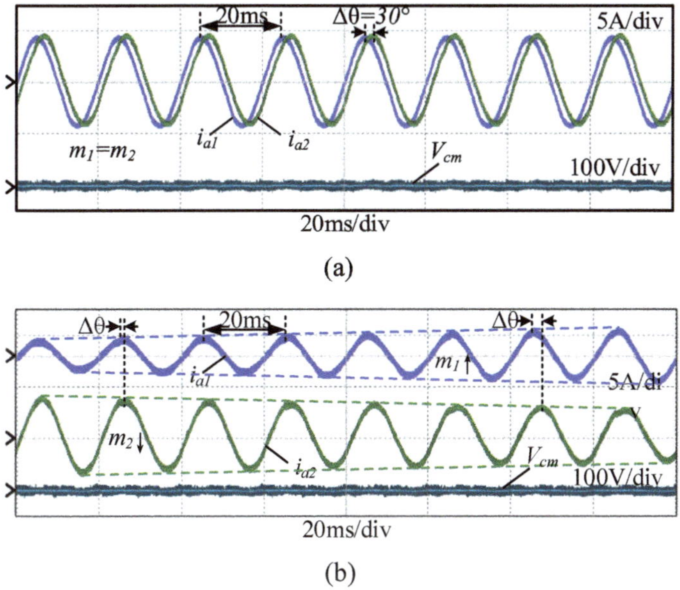

Fig. 7.95 Passive load verification for the proposed PWM scheme: **a** identical modulation index with 30° angle displacement, **b** dynamic process with m_1 and angle displacement increasing while m_2 decreasing

Moreover, to verify the superiority of CM suppression effect for the proposed strategy, the induced motor frame voltage and sub-CMVs are tested. The conventional symmetrical PWM and the reduced sub-CMV PWM schemes are also implemented to make fair comparison for different DTPMs in the same circumstance. Figure 7.97a–c shows the test results of two sub-CMVs, in which the fundamental period reference is given by the motor frame voltage and the phase-leg current. For the DTPM with $\Delta\theta = 0°$ shown in Fig. 7.97a, the conventional symmetrical PWM which does not possess the function of switching cancellation for the two sub-CMVs induces the obvious motor frame voltage in the whole fundamental period. As for the two CMV elimination methods, different from the proposed PWM which takes the interval connection for PWM signals in each set of windings to retain the two zero voltage vectors, the reduced sub-CMV PWM utilizes the intragroup connection and avoids the two zero voltage vectors, so the amplitudes of sub-CMVs are reduced to $\pm V_{dc}/6$. Though the different end-to-end arrangement patterns are utilized, both of them can keep the two sub-CMVs opposite to cancel out the generalized total CMV shown in the enlarged view, and therefore the motor frame potential remains almost unchanged. The remaining small spikes of motor frame voltage are caused by the deadtime and switching cancellation transient process, and so on. Thus, even with some nonideal factors, the motor frame voltage can also be significantly reduced compared with the conventional PWM method. For other different DTPMs shown in Fig. 7.97b, c, similar performance can be achieved which verifies the generalized

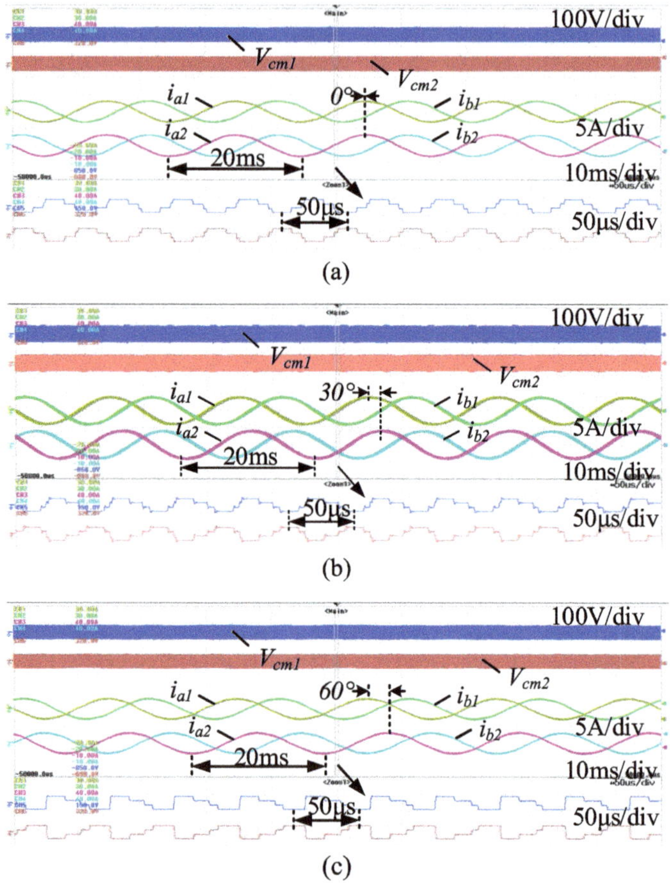

Fig. 7.96 The normal symmetrical parameter condition verification for the proposed scheme: **a** DTPM with $\Delta\theta = 0°$, **b** DTPM with $\Delta\theta = 30°$, **c** DTPM with $\Delta\theta = 60°$

CMV suppression effect for the proposed PWM and reduced sub-CMV PWM strategies. Moreover, it can be seen that the time-domain current ripple of conventional PWM and the proposed PWM are close to each other, while the current ripple of reduced sub-CMV PWM is obviously increased owing to the zero voltage vectors cancellation. So, comprehensively considering the requirement of CMV elimination and current ripple optimization, the proposed method which has better performance is more appropriate.

Moreover, in conventional CMV-elimination PWM methods, the two VSIs are demanded to have identical modulation indexes which only guarantees to generate balance phase voltage. If the parameter difference is non-negligible for the DTPMs in actual application, the contradiction for current balance and CMV elimination emerges for DTPMs. Thus, the parameter asymmetrical cases are also considered

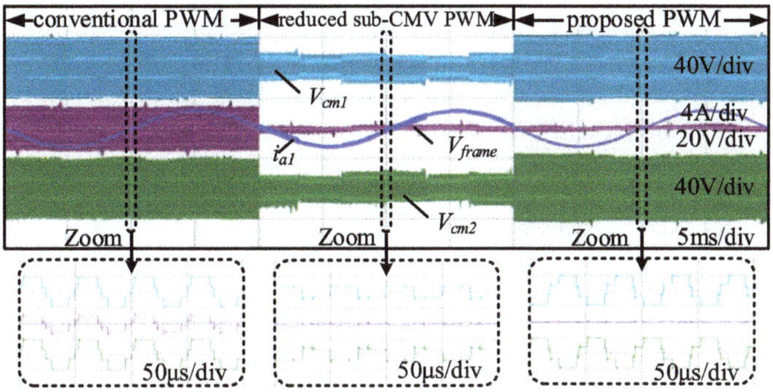

(a)

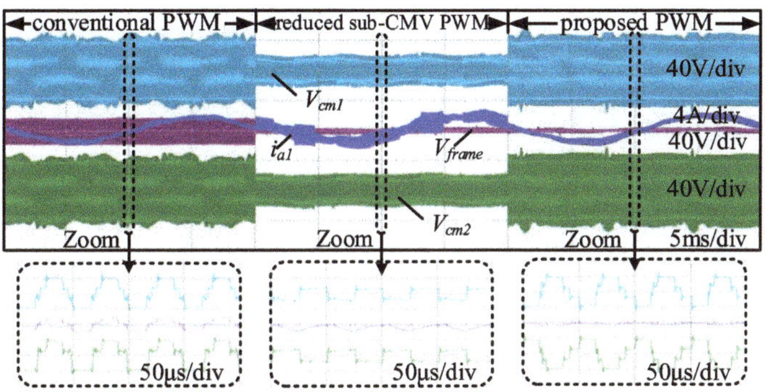

(b)

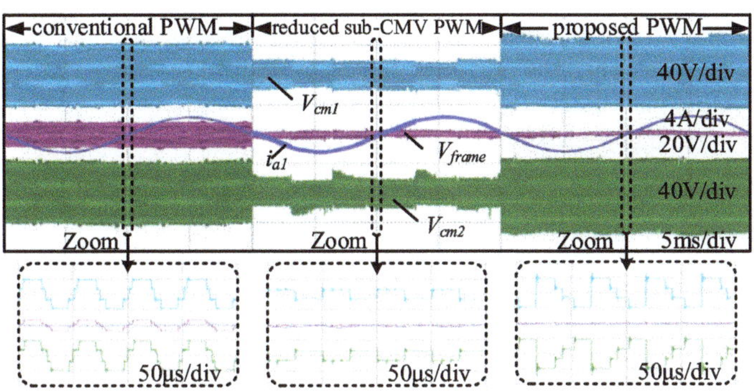

(c)

Fig. 7.97 The sub-CMVs, motor frame voltage, and phase-leg current comparison for three different PWM schemes: **a** DTPM with $\Delta\theta = 0°$, **b** DTPM with $\Delta\theta = 30°$, **c** DTPM with $\Delta\theta = 60°$

for the proposed PWM to verify the effect of current balance and CMV elimination simultaneously. Since the parameter asymmetry is non-negligible for the two motor prototypes, to simulate the possible asymmetrical conditions and prominent the asymmetrical effect, the extra resisters are added in series with phase windings. Figure 7.98 shows the two kinds of constructed parameter asymmetry cases, including the single-phase winding asymmetry and the two sets of windings asymmetry. Other more complicated asymmetrical cases for DTPMs can be deduced from the combination of these two cases.

In the case of parameter asymmetry in the second winding set illustrated in Fig. 7.98a, the phase currents i_{a2}, i_{b2} are unbalanced with the conventional PI control, and obvious negative-sequence currents exist in this winding set, as shown in Fig. 7.99a. With the extra added resonant control to eliminate the negative sequence currents, the three-phase currents can be balanced. In this case, the d- and q-axis reference voltages are time-variant, and the calculated total normalization voltage coefficient m_2 has second-order harmonics which is owing to the negative sequence voltage injection. Even with harmonic voltage injection, there is no influence on the proposed end-to-end connection and sub-CMV construction with the enlarged view, so the total CMV can also maintain the same elimination effect in this condition, which is coincident with theoretical analysis. In the case of two sets of windings parameter asymmetry illustrated in Fig. 7.99b, though there is no harmonic voltage injection in each set of windings, the two total normalization voltage coefficients are not identical, and the difference is increasing with the phase current raise, as shown in Fig. 7.99b. This phenomenon is owing to the voltage drop increase for the extra series resistors. Since the sub-CMV construction is independent of the reference voltage, the total CMV elimination can also be retained even with different reference voltage. Based on the above experimental results, it can be found that the proposed PWM, which provides the function of two three-phase total normalization voltage coefficients decouple, has the adequate freedom of current control which can deal with the possible parameter asymmetrical condition.

So the proposed generalized CMV elimination PWM scheme can lift the restriction on angle displacement between the two sets of windings which can be utilized on universal DTPMs. In addition, the proposed strategy can also decouple the real-time

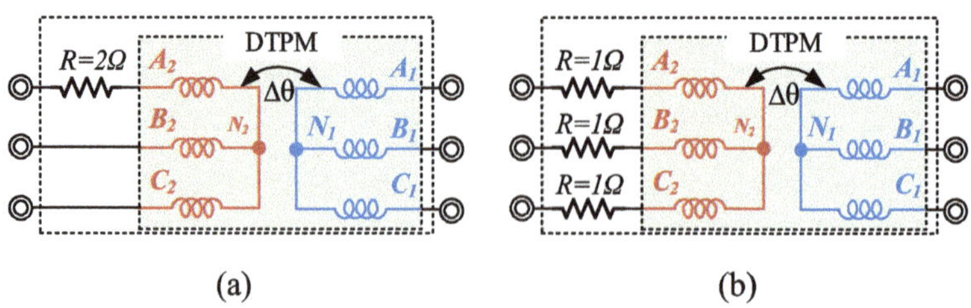

Fig. 7.98 The load configuration for DTPMs: **a** asymmetrical condition in which a resistor in series with single-phase winding, **b** asymmetrical condition in which resistors in series with one set of three-phase windings

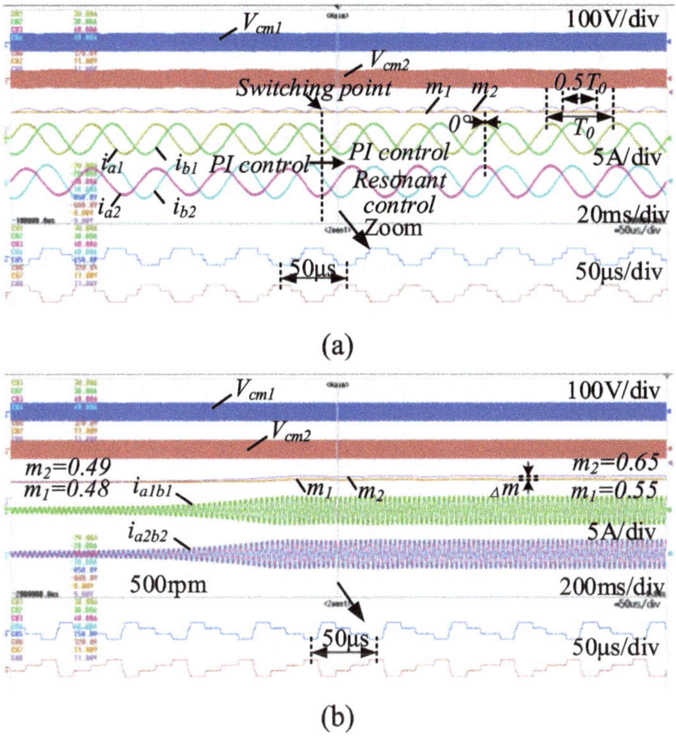

Fig. 7.99 The asymmetrical parameter case verification for the proposed scheme: **a** DTPM with $\Delta\theta = 0°$ for parameter asymmetry in one set of winding, **b** DTPM with $\Delta\theta = 60°$ for parameter asymmetry between two sets of windings

reference voltages for the two sets of windings which provides the function of the phase current balance, both in parameter symmetry and asymmetry conditions, and is more practical than conventional CMV elimination methods. As for the concerned modulation index range, the proposed PWM algorithm can keep it identical as that of the conventional method in common neutral point condition, but is relatively lower than that of the conventional method in separated neutral point condition (reduced from 1.15 to 1 which is similar to most of CMV elimination modulation schemes).

Dual three-phase motor discussed in Sect. 7.6 is a typical kind of multiphase motor. The PWM strategies for CMV elimination come from the idea of paralleled inverters' zero-CM PWM in Sect. 7.4 initially. For more general multiphase motor drives, PWM for CMV reduction will be introduced in Sect. 7.8.

7.7 Dual Inverter-Fed Open-Winding Three-Phase PMSM with Zero-Sequence Current Elimination

The CMV not only induces the CM leakage current and CM EMI in motor drive system but also has the potential to generate the zero-sequence current when the zero-sequence path exists in the motor drive system. A typical example is the open-winding (OW) three-phase motor. The OW three-phase motor has some attracting superiorities compared with conventional three-phase star-connected winding motor, including higher DC bus utilization, power device voltage stress reduction, potential fault-tolerant capability, and so on. However, owing to the open neutral point of stator winding, the zero-sequence path in the motor emerges and the potential zero-sequence current (ZSC) can be generated with single DC power supply, as shown in Fig. 7.100a. The ZSC may cause unwanted power losses and torque ripple which degrades the system performance [34, 35].

Figure 7.100b shows the zero-sequence equivalent circuit model of OW-PMSM. Since the zero-sequence impedance based on L_0 is decided by the stator leakage inductance, the big zero-sequence current (ZSC) can be easily induced by the ZSV disturbance sources. To suppress the ZSC, all zero-sequence voltage (ZSV) disturbance source should be confirmed. Considering the category of ZSV disturbance source, it can be divided into four types [21]. The dual inverter generates two ZSV disturbance sources: u_{0_PWM} and $u_{0\_nonlinearity}$. The u_{0_PWM} is the difference between the two CMVs of VSIs and is decided by the modulation scheme; it has the possibility to be zero if two VSIs generate the identical CMV. Moreover, the inverter nonlinearity caused by the voltage drop of switching device and deadtime effect can induce the other ZSV disturbance source which is also attributed to the inverter system. This ZSV disturbance source has strong correlation with the three-phase currents which can be modeled as the current control voltage source (CCVS) and expressed as $u_{0\_nonlinearity} = f_n(i_a, i_b, i_c)$. In addition, the motor side induces the two kinds of ZSV disturbance sources: $-3\omega_e\Psi_{3f}\sin(3\omega_e t - \theta_1)$, u_{0_id} and u_{0_iq}. The first ZSV disturbance source is the third harmonic $-3\omega_e\Psi_{3f}\sin(3\omega_e t - \theta_1)$ in back-EMF. The other ZSV disturbance sources in motor side are the cross-coupling voltages in

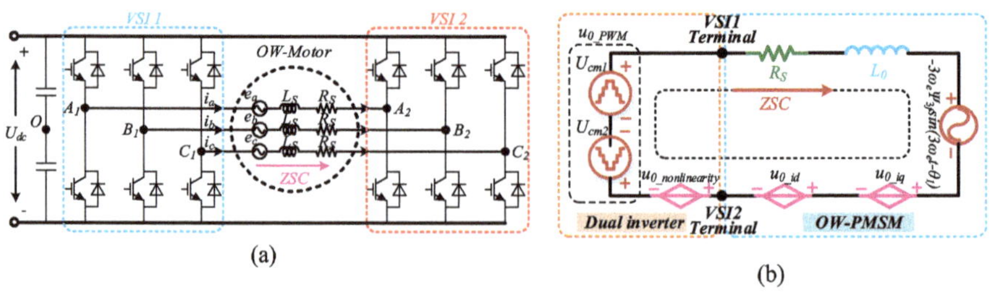

Fig. 7.100 The structures of dual two-level inverter-fed OW motor with common DC-power: **a** the original equivalent circuit model of one phase, **b** the zero-sequence equivalent circuit model of OW-PMSM with common DC-bus

zero-sequence circuit caused by the d- and q-axis currents which also have the CCVS characteristic and can be represented as $u_{0\_id} = f_d(i_d)$ and $u_{0\_iq} = f_q(i_q)$.

In order to suppress the ZSC, the promising controllable ZSV disturbance source should be determined. Generally, all the ZSV disturbance sources except u_{0_PWM} are uncontrollable, so u_{0_PWM} can be recognized as the promising ZSV disturbance source. To eliminate the ZSC, the core mentality is to modulate u_{0_PWM} to zero with appropriate modulation scheme, and then counteract with other ZSV disturbance sources by fine-tuning the u_{0_PWM} based on the ZSC closed-loop control. The modulated ZSV disturbance source caused by the inverter system can be deduced as

$$u_{0\_PWM} = U_{cm1} - U_{cm2} = \frac{1}{3}[(V_{A1O} + V_{B1O} + V_{C1O}) - (V_{A2O} + V_{B2O} + V_{C2O})] \tag{7.33}$$

where U_{cm1} and U_{cm2} are the CMVs of the two VSIs correspondingly, and V_{A1O}, V_{A2O}, V_{B1O}, V_{B2O}, V_{C1O}, V_{C2O} are the output instantaneous pole voltages. Since u_{0_PWM} is the difference between the two CMVs of VSIs and decided by the modulation scheme, it has the possibility to be zero if two VSIs generate the identical CMV.

For dual two-level inverter, each VSI can generate the same eight space voltage vectors shown in Fig. 7.101a, b. The eight space voltage vectors have four CMV values totally, and the peak values are $U_{dc}/2$ and $-U_{dc}/2$ corresponding to the voltage vectors $V_7(111)$ and $V_0(000)$. The voltage vectors $V_1(100)$, $V_3(010)$ and $V_5(001)$ have the same CMV value of $-U_{dc}/6$, and the voltage vectors $V_2(110)$, $V_4(011)$ and $V_6(101)$ have the same CMV value of $U_{dc}/6$. Based on the combination of different voltage vectors for each VSI, the modulated ZSV voltage vector combinations can be calculated, and all the particular ZSV voltage vectors are shown in Fig. 7.101c. These synthesized voltage vectors can be divided into two kinds: the zero voltage vectors which have no effect to synthesis of reference vector, and the effective voltage vectors which can be used to synthesize the reference vector. In this case, ZSV can be realized if suitable arrangement of these special voltage vectors is taken in every switching period.

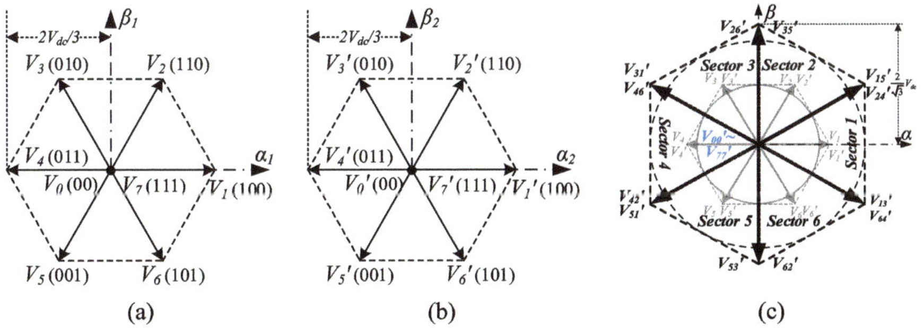

Fig. 7.101 The space voltage vectors of individual VSIs and the particular zero ZSV voltage vectors: **a** VSI1, **b** VSI2, **c** the ZSV voltage vectors for dual inverter

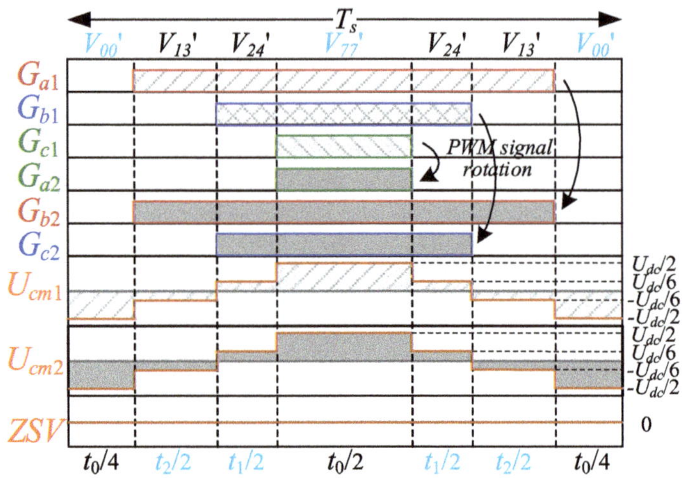

Fig. 7.102 Conventional SVPWM based ZSV elimination modulation scheme

The most utilized ZSV elimination modulation scheme for dual inverter is deduced from conventional SVPWM scheme. The principle is based on the PWM signal rotation which ensures VSI1 and VSI2 taking the same three PWM signals and generates the identical CMVs. Without losing generality, assuming the three-phase duty cycles in VSI1 satisfy the relationship that $d_{a1} > d_{b1} > d_{c1}$. By exchanging the sequence of PWM signals for VSI1, the corresponding PWM signals for VSI2 can be generated shown in Fig. 7.102, where G_{a1}, G_{b1}, G_{c1} and G_{a2}, G_{b2}, G_{c2} are the three-phase PWM signals for VSI1 and VSI2, respectively. In this case, the CMVs of different VSIs can keep identical, and the combined voltage vectors in one switching cycle are $V_{00}'–V_{13}'–V_{24}'–V_{77}'–V_{24}'–V_{13}'–V_{00}'$ which naturally satisfies the requirement of utilizing ZSV voltage vectors, so the ZSV generated by VSI modulation can be eliminated.

Though this modulation method can be utilized, the dual inverter composed of six phase-legs is only driven by three different PWM signals with the manner of PWM signal rotation. So the freedom of the dual inverter is relatively reduced which may degrade the PWM relevant performance. Actually, the unipolar SPWM scheme can control the duty cycles separately for each phase-leg and fully utilize the freedom which has the potential to obtain the double frequency effect. Moreover, the three-phase unipolar SPWM scheme for dual inverter can keep the average CMV of each inverter zero naturally, so it has the possibility to coordinate with the OW motor. The only problem is the six PWM signals taking the symmetric manner and generating the instantaneous ZSV pulses. So the modified method can be considered to eliminate the high-frequency ZSV pulses, that is, phase-shift of voltage pulse for conventional symmetrical unipolar SPWM.

In conventional symmetrical SPWM scheme, the three-phase duty cycles of VSI1 can be expressed as:

$$\begin{cases} d_{a1} = (1 + m\cos\theta)/2 \\ d_{b1} = [1 + m\cos(\theta - 2\pi/3)]/2 \\ d_{c1} = [1 + m\cos(\theta + 2\pi/3)]/2 \end{cases} \tag{7.34}$$

where d_{a1}, d_{b1} and d_{c1} are the three-phase duty cycles, respectively; m is the modulation index within the range of [0, 1]; θ is the instantaneous angle for phase A_1. The above equations can deduce the identical equation

$$d_{a1} + d_{b1} + d_{c1} = 3/2 \tag{7.35}$$

Since the phase voltage is the difference of two phase-leg voltages, the corresponding two phase-leg voltages in one phase should keep the same modulation index and the opposite phase angle to obtain the maximum output phase voltage, so the duty cycles of the corresponding two phase-legs should keep complementary, that is

$$d_{a1} + d_{a2} = 1 \quad d_{b1} + d_{b2} = 1 \quad d_{c1} + d_{c2} = 1 \tag{7.36}$$

Thus, the three-phase duty cycles for VSI2 also satisfies the identical equation

$$d_{a2} + d_{b2} + d_{c2} = 3/2 \tag{7.37}$$

Under the above conditions, without losing generality, assuming d_{a1} is maximal in six phase-legs which can deduce d_{a2} is minimal, and $d_{b1} > d_{c1}$, $d_{b2} < d_{c2}$, so the six PWM signals, CMVs and ZSV for conventional symmetrical SPWM scheme can be drawn and shown in Fig. 7.103a. It can be found that the rising and falling edges of PWM signals are in different positions which generate the high-frequency ZSV pulses which can induce obvious high-frequency ZSC.

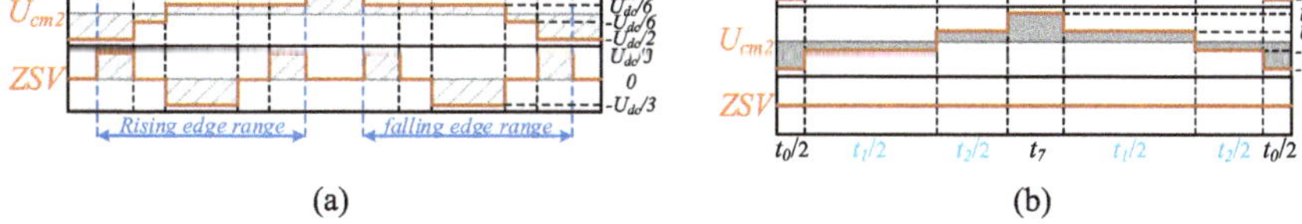

Fig. 7.103 The PWM signals, CMVs and ZSV for different schemes in one switching period: **a** the conventional symmetric SPWM scheme, **b** the novel phase-shift SPWM scheme

Figure 7.103b shows the principle of the proposed phase-shift scheme to eliminate the instantaneous ZSV in one switching period. In this scheme, the PWM signals of phase A which has the maximum or minimum duty cycle keep symmetric while the PWM signals of phases B and C take the phase-shift manner. When the rising edge of C_2 PWM signal aligned with the rising edge of A_1 PWM signal with left side phase-shift, and the falling edge of B_2 PWM signal aligned with the falling edge of A_1 PWM signal with right side phase-shift, the state transition of ZSV caused by A_1 PWM signal can be cancelled out with cooperation of B_2 and C_2 PWM signals. Similarly, the rising edge of B_1 PWM signal and the falling edge of C_1 PWM signal can be aligned with the rising and falling edges of A_2 with proper phase-shift. Moreover, the left rising and falling edge positions can also prove to be aligned, including the falling edges of B_1 and C_2 PWM signals and the rising edges of C_1 and B_2 PWM signal. Thus, the two VSIs can generate the identical CMVs to eliminate the modulated ZSV disturbance source, and this novel method can be called as phase-shift SPWM (PS_SPWM) scheme.

Figure 7.104 shows the normalized phase and pole voltages for phase A (with $U_{dc}/2$ as reference) at the maximal modulation index for the conventional and the novel ZSV elimination schemes. For conventional SVPWM scheme shown in Fig. 7.104a, the pole voltages are nonsinusoidal by injecting the zero-sequence component, and the maximal modulation index can increase to 1.15. With the operation of signal rotation to cancel out the zero-sequence component, though the phase voltage can keep sinusoidal, the peak value of phase voltage only reaches the $\sqrt{3}$ times of pole voltage which is the value of U_{dc}. So the actual maximal modulation index of phase voltage is 2 in this case. As for the novel PS_SPWM scheme shown in Fig. 7.104b, the phase-shift manner has no effect on duty cycle as well as the modulation index range, so there is no zero-sequence component injected to the pole voltage and the maximal modulation index is also kept 1 for the novel SPWM scheme. Thus, the maximal modulation index of phase voltage can also reach the maximal value 2 which has the same value compared with conventional scheme.

Figure 7.105 shows the modulation index scanning result of switching harmonic of normalized phase voltage for the two schemes. For conventional SVPWM scheme shown in Fig. 7.105a, the phase voltage contains abundant odd and even switching harmonics. The dominated odd first switching harmonic is monotonically increased which reaches the peak value 0.7359 at the maximum modulation index, while the

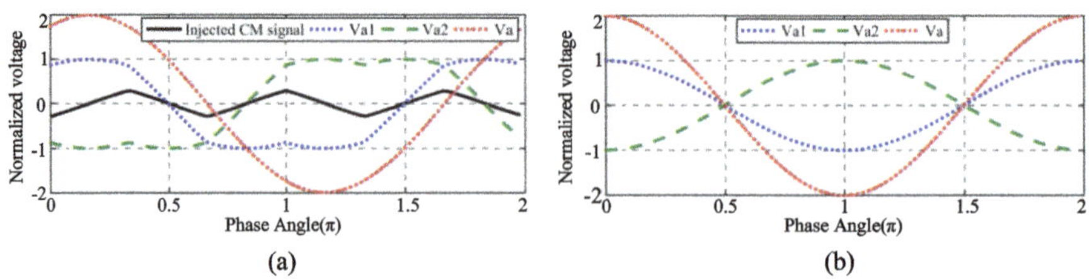

Fig. 7.104 The normalized phase and pole voltages of phase A for two ZSV elimination schemes: **a** the conventional SVPWM with signal rotation scheme, **b** the novel PS_SPWM scheme

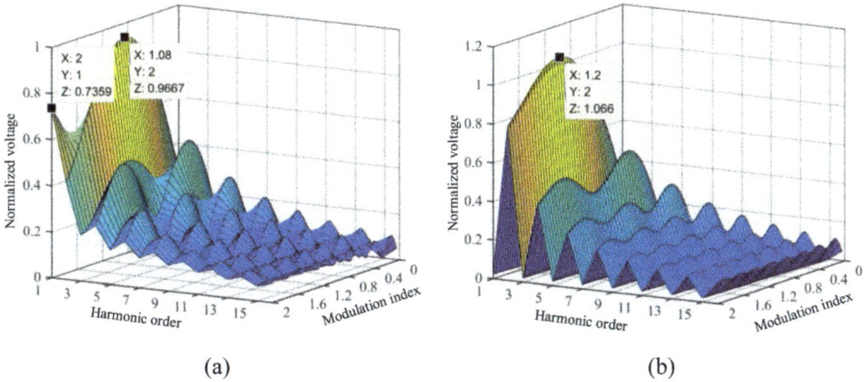

Fig. 7.105 The normalized phase voltage spectrum for the two ZSV elimination schemes: **a** conventional SVPWM with signal rotation scheme, **b** the novel PS_SPWM scheme

dominant even second switching harmonic is nonmonotonic with the peak value of 0.9667 at about half modulation index. As for the novel PS_SPWM scheme shown in Fig. 7.105b, the phase voltage mainly contains the even switching harmonics and almost eliminates the odd switching harmonics. The dominated even second switching harmonic is also nonmonotonic with the peak value of 1.066, which is close to the peak value of second switching harmonic for conventional SVPWM scheme.

Figure 7.106 shows the result of the dominant equivalent switching harmonic currents and equivalent current THD. It can be seen that the equivalent current THD of the two schemes are close to each other when the modulation index is lower than 0.6, so the current THD reduction effect for the PS_SPWM scheme can be

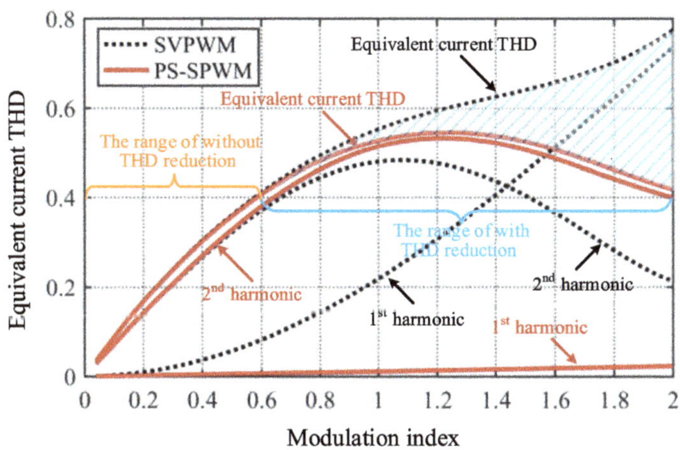

Fig. 7.106 The equivalent current THD and dominant switching harmonic currents comparison

neglected in this range. When the modulation index is further increased, the equivalent current THD of the PS_SPWM scheme is smaller than the conventional SVPWM scheme, obviously in high modulation index range. The equivalent current THD of the SVPWM scheme is monotonically increased which reaches the peak value of 0.7738 at the maximum modulation index, while the equivalent current THD of the PS_SPWM scheme is nonmonotonic and the value is only 0.4184 at the maximum modulation index which means the equivalent current THD can reduce 45.93% compared with the conventional SVPWM scheme theoretically. So the proposed PS_SPWM scheme has better performance than the conventional SVPWM scheme in ripple current reduction.

On the other hand, though the dominated ZSV caused by the modulation scheme can be eliminated in ideal condition, the remaining ZSV disturbance sources should be considered which depends on the ZSC closed-loop control and the zero voltage vector adjustment for the proposed PS_SPWM scheme. The dwell times of zero voltage vectors (000) and (111) are adjustable, which is a freedom to control the average output CMV for each VSI to offset other ZSV disturbance sources. Figure 7.107 shows the principle and effect of zero voltage vector adjustment for the proposed PS_SPWM scheme. For example, the dwell time of zero voltage vector (111) for VSI1 is adjusted to add ΔT while the dwell time of zero voltage vector (000) is adjusted to subtract ΔT shown in Fig. 7.107a, so the three-phase PWM signals change from G_{a1}, G_{b1}, G_{c1} to G_{a1}', G_{b1}', G_{c1}' which leads to the increase of all duty cycles. In this case, the dwell time of active voltage vector keeps unchanged, while the CMV is changed from U_{cm1} to U_{cm1}' which leads to the voltage-second added.

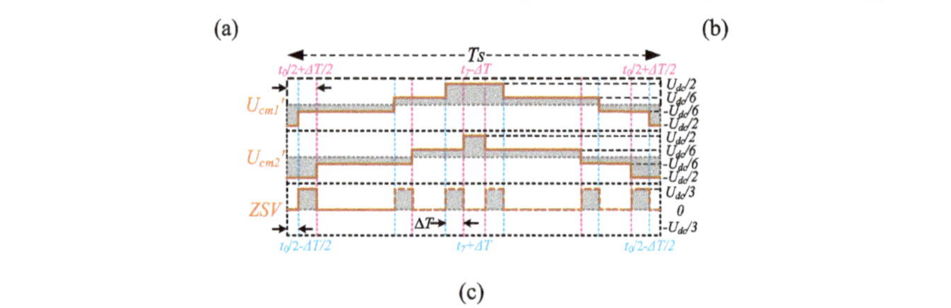

Fig. 7.107 The principle and effect of zero voltage vectors adjustment: **a** the zero-voltage-vector adjustment for VSI1 to increase the volt-second, **b** the zero-voltage-vector adjustment for VSI2 to reduce the volt-second, **c** the effect of zero-voltage-vector adjustment for the output ZSV

Figure 7.107b shows the opposite situation for VSI2 which leads to the total voltage-second reduced. Figure 7.107c shows the effect of zero-voltage-vector adjustment for the output ZSV. With the switching edge adjustment for different phase-legs, the six narrow ZSV pulses can be generated with the same dwell time of ΔT. In this case, with the ZSC controller to generate the reference zero-voltage-vector adjusting time, the ZSC can be effectively suppressed.

Based on the above modulation schemes, the performance of phase current's current ripple is first taken into consideration. The no load tests of three-phase currents and ZSC at low and high rotation speeds are implemented to make clear comparison for the concerned current ripple. Figure 7.108 shows the three-phase currents and ZSC at 150 RPM and without the ZSC closed-loop control for the three different schemes. In Fig. 7.108a, the conventional SPWM scheme is utilized and obvious high-frequency ZSC is generated which is caused by the high-frequency ZSV pulses. This high-frequency ZSC which superimposes on phase current can increase the current ripple of phase current. In addition, though the conventional SPWM scheme does not produce low-frequency ZSV theoretically, the low-frequency ZSC (mainly the triple harmonic current) still exists in phase current which is caused by other ZSV disturbance sources. Figure 7.108b, c shows the performance of SVPWM and PS_SPWM schemes, respectively. It can be found that the high-frequency ZSC is eliminated, and only low-frequency triple ZSC harmonic exists in this case. Moreover, the frequency-domain analysis of phase current is also implemented. Compared with the phase current of the conventional SVPWM scheme which has both the odd and even switching harmonics, it can be seen that the phase currents of SPWM scheme and proposed PS_SPWM scheme only have the double frequency effect. Nevertheless, since the conventional SVPWM scheme have little odd switching harmonics compared with the even switching harmonics in low modulation index condition, the double frequency effect of the proposed PS_SPWM scheme can only reduce little RMS value.

Figure 7.109 shows the three-phase currents and ZSC at 900 RPM for the three different schemes. In Fig. 7.109a, the high-frequency ZSC is further increased for the conventional SPWM scheme which leads to the fundamental current submerged

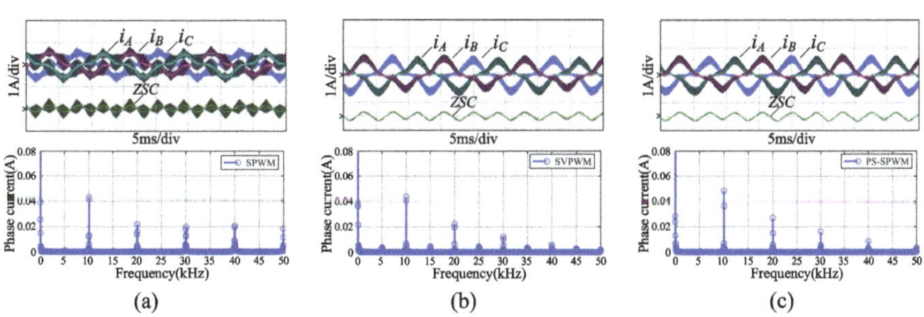

Fig. 7.108 The phase current and ZSC comparison for the three different schemes at 150 RPM: **a** the symmetrical SPWM scheme, **b** the conventional SVPWM scheme, **c** the proposed PS_SPWM scheme

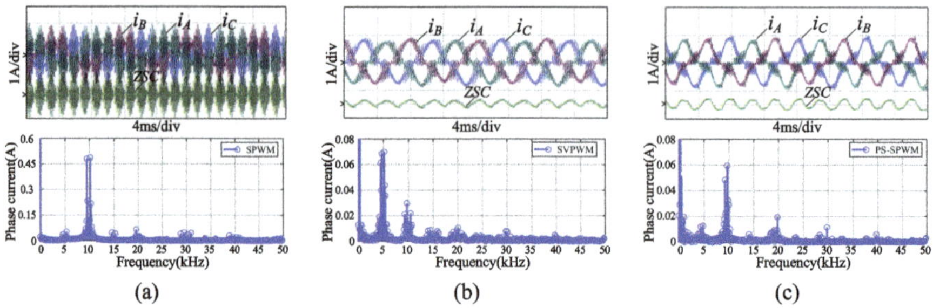

Fig. 7.109 The phase currents and ZSC comparison for the three different schemes at 900 RPM: **a** the conventional SPWM scheme, **b** the conventional SVPWM scheme, **c** the proposed PS_SPWM scheme

by the current ripple. Figure 7.109b, c shows the phase current performance for the SVPWM and PS_SPWM schemes. It can be seen that the current ripple of SVPWM scheme is obviously larger than the proposed PS_SPWM scheme in time domain. This effect can be explained with the FFT analysis in frequency domain. For the SVPWM scheme, the first switching frequency harmonics are dramatically increased and become the dominant part of current ripple in high modulation index. As for the proposed PS_SPWM scheme, even the second switching frequency harmonics are slightly increased, the amplitude is smaller than the first switching frequency harmonics of SVPWM scheme, and it can obviously suppress the first switching frequency current which finally reduces the ripple current compared with the SVPWM scheme.

In addition, to verify the low-frequency ZSC suppression effect for the proposed PS_SPWM scheme, Fig. 7.110 shows the dynamic process in the condition of without ZSC control and with ZSC control. In Fig. 7.110a, it can be seen that though the proposed PS_SPWM scheme can eliminate the modulated ZSV disturbance source, the phase current still takes a rich content of triple harmonics without the ZSC control. This phenomenon is due to other ZSV disturbance sources induced by the ZSC current. After adding the ZSC closed-loop control, the peak value of phase current can be reduced, and the ZSC can be effectively suppressed. With the FFT

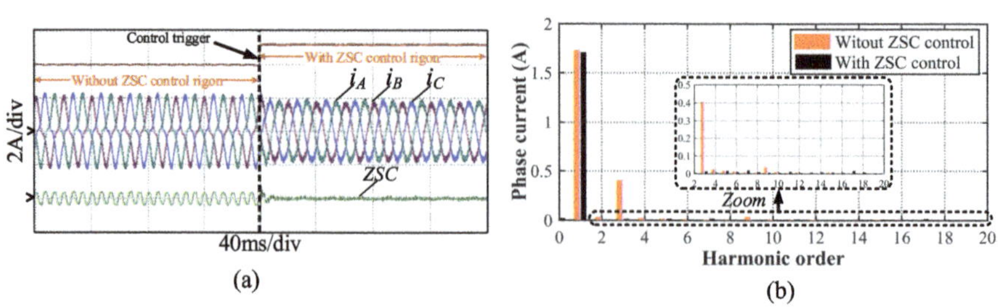

Fig. 7.110 The current waveform for the proposed PS_SPWM method with or without ZSC control: **a** the dynamic process, **b** the frequency-domain analysis

analysis for the phase current shown in Fig. 7.110b, the THD of phase current is 24.1% without the ZSC control, while it can be reduced to 6.09% with the ZSC control. As for other triple harmonics, like the ninth and 15th harmonics, the amplitudes are small enough to be neglected.

Thus, owing to the small impedance of the zero-sequence circuit in OW-PMSM, without the ZSV control, obvious low- and high-frequency ZSC can be found in the motor and brings corresponding problems. The proposed PS_SPWM method which is based on the conventional unipolar SPWM scheme associated with suitable phase-shift can help to cancel the modulated ZSV disturbance source and eliminate the high-frequency ZSC. In addition, it can fully utilize the freedom of dual inverter and maintain the double frequency effect for the phase current and therefore reduce the high-frequency current ripple. Moreover, the proposed PS_SPWM scheme can easily combine with the ZSC regulator and suppress the low-frequency ZSC.

7.8 Common-Mode Voltage Reduction for Multiphase Converters

As discussed in the end of Sect. 7.6, this part will introduce CMV reduction method for general multiphase motor drives [29, 30].

Multiphase drive systems have attracted increasing research interest in recent years [22], because they have several advantages over traditional three-phase drive systems, such as high reliability, high fault-tolerance ability, high torque density and reduced torque ripple. And they are being considered for applications in electric vehicles [23], marine electric propulsion [24] and high-power applications.

CMV is one of the serious problems caused by PWM of the multiphase converters-fed multiphase machines. In this part, CMV is equal to the voltage between the load neutral point and the midpoint of the DC link, shown in Fig. 7.111. In analogy to the three-phase system, the expression of the CMV (v_{com}), for an m-phase VSI is given by

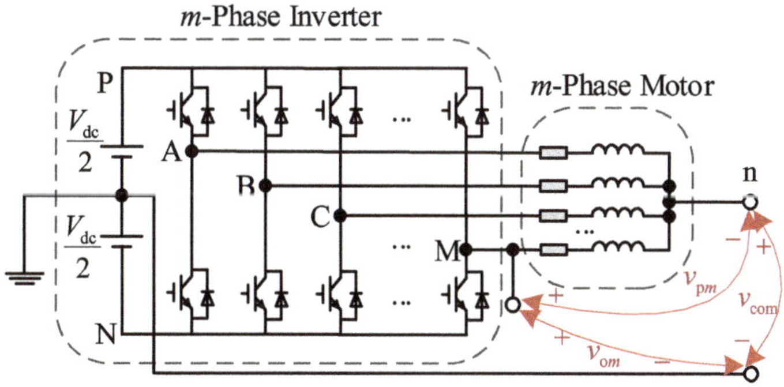

Fig. 7.111 Generalized topology of an m-phase two-level inverter

$$v_{\text{com}} = \frac{1}{m}\sum_{i=1}^{m} v_{\text{o}i} \tag{7.38}$$

For an m-phase two-level inverter, assuming $S_{\text{i}} \in \{0, 1\}$ denotes the switching functions of the ith VSI leg, then $\sum_{i=1}^{m} S_i = 0, 1, 2, \ldots, m$ leads to the $\sum_{i=1}^{m} v_{\text{o}i} = (-m/2)V_{\text{dc}}, (-m/2+1)V_{\text{dc}}, (-m/2+2)V_{\text{dc}}, \cdots, (m/2)V_{\text{dc}}$, which relates to $v_{com} = (-1/2)V_{\text{dc}}, (-1/2+1/m)V_{\text{dc}}, (-1/2+2/m)V_{\text{dc}}, (1/2)V_{\text{dc}}$, respectively. Therefore, for an m-phase two-level single-end inverter, the number of theoretical possible different levels in CMV is $m + 1$.

Most of the existing CMV-reducing PWM methods belong to the modified SVPWM types, which are realized by selecting those voltage space vectors that give out very low or even zero CMV. Modified SVPWM methods for five-phase two-level inverters were proposed in [25], which can reduce the peak-to-peak CMV by 80%. A generic CMV-elimination SVPWM technique was proposed in [26], which can be extended to any multiphase occasion but only suitable for multiphase multilevel occasions. Chen et al. [27] proposed a generalized PWM method with minimum CMV for multiphase two-level VSIs with any odd phase number m, which suppresses the CMV to only two minimum levels, and therefore achieves a CMV reduction rate of 80, 85% for the five-phase, seven-phase two-level inverters, respectively. However, this kind of PWM aims at sinusoidal output voltage on *R-L* loads, without considering the harmonic characteristics of multiphase machines. Besides, only the CMV amplitude is reduced, while the CMV changing frequency remains the same as conventional PWM methods.

In fact, the carrier-based PWM (CBPWM) outweighs SVPWM in the multiphase drive occasions for the following reasons. First, for the m-phase n-level converters, the number of voltage vectors is n^m, which increases exponentially as the phase number increases. Consequently, the selection of voltage vectors and the computation of their action times would be a heavy burden for the industrial controllers. Second, multiphase machines have redundant harmonic subspaces, where low-order harmonics can be eliminated to enhance the machine drive efficiency or injected to improve the torque density. The elimination or injection of these harmonics would be much simple to be determined and implemented under CBPWM methods, through adjusting the modulating wave for each stator phase. In this part of the book, we will introduce two types of the CMV-reduction CBPWM methods for symmetrical multiphase converters, where the phase angle between adjacent phases is $2\pi/m$.

7.8.1 Phase-Shifted Sinusoidal PWM (PS_SPWM)

Carrier-based sinusoidal PWM is a kind of widely used CBPWM techniques, which aims to create a series of switched pulses that have the same volt-second average value as the sinusoidal reference waveform, and these pulses switch at the intersections of sinusoidal reference waveform and triangular carrier waveform. In conventional

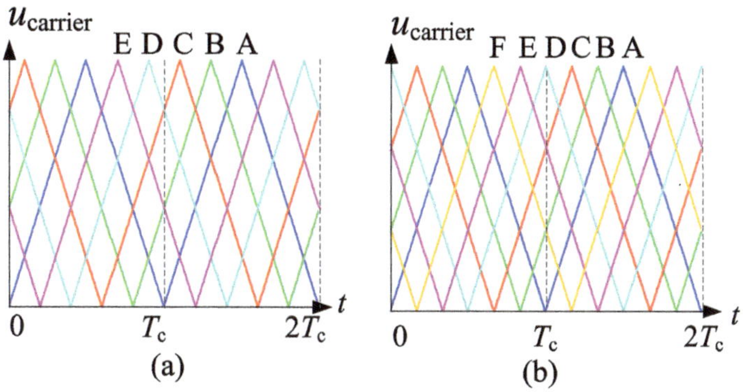

Fig. 7.112 Phase-shifted triangle carriers for: **a** five-phase inverters, **b** six-phase inverters

SPWM, the triangle carrier waves in different phases are the same in phase position. However, the difference in phase positions provides an additional degree of freedom for the improvement of SPWM. In the proposed PS_SPWM for the m-phase two-level inverter, the triangle carriers are the same in amplitude but with phase lag, which means that the carrier in each phase is shifted by $1/m$ of a switching period T_c to the one in the prior phase. The topology of a generalized m-phase two-level inverter is shown in Fig. 7.111. In the proposed PS_SPWM, the phase shift between neighbor carriers is 1/5 T_c for the five-phase inverter and 1/6 T_c for the six-phase inverter, shown conceptually in Fig. 7.112a, b.

Through the intersection plotting method [28], we can explain how the PWM pulses are generated under CBPWM, and further obtain the CMV waveforms [29]. The contours-plotting of the five-phase two-level converter under conventional SPWM can be obtained in Fig. 7.113a. The areas are divided by the contours overlapped with each other, and the new areas can be classified as I, II, III, IV, V, VI. Areas I to VI correspond to $\sum_{i=1}^{5} S_i = 0, 1, 2, 3, 4, 5$, respectively, which finally leads to the voltage amplitudes of $-V_{dc}/2$, $-3V_{dc}/10$, $-V_{dc}/10$, $V_{dc}/10$, $3V_{dc}/10$, $V_{dc}/2$ in the CMV. Then the waveform of the CMV can be obtained in Fig. 7.113b, whose voltage levels correspond to specific areas that the carrier-ratio line passes through. It is clear that the number of voltage levels of CMV is 6, and the v_{com} ranges from $-V_{dc}/2$ to $V_{dc}/2$. Similarly, intersection-plotting of the symmetrical six-phase two-level inverter under conventional SPWM is shown in Fig. 7.114, where the number of voltage levels of CMV is 7, and the v_{com} ranges from $-V_{dc}/2$ to $V_{dc}/2$.

Under the proposed PS_SPWM, the phase shift of neighbor carriers is $1/m$ T_c on the time-axis, which is equal to $2\pi/m$ on the $\omega_c t$ axis. Therefore, in PS_SPWM, the contours of different phases for the m-phase inverter have phase shifts in both y-axis and x-axis. The intersection-plotting for the five-phase two-level inverter under PS_SPWM is shown in Fig. 7.115, and the contours for neighbor phases are shifted by $2\pi/5$ on both y-axis and x-axis in Fig. 7.115a. Compared with Fig. 7.113, there are no areas of I and VI in Fig. 7.115, which means that the switching states corresponding to $\sum_{i=1}^{5} S_i = 0$ or $\sum_{i=1}^{5} S_i = 5$ are avoided. As a result, there are no voltage levels

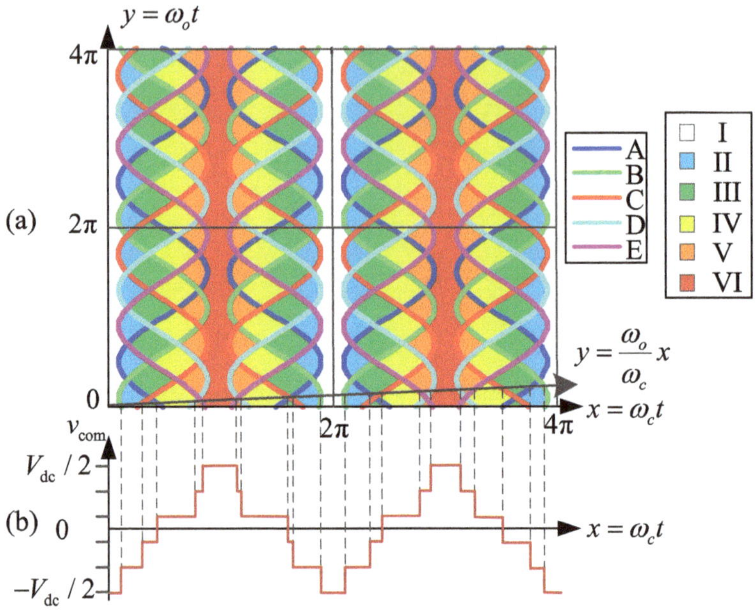

Fig. 7.113 Intersection-plotting of five-phase two-level inverters in conventional SPWM: **a** the contours and the carrier-ratio line, **b** resulting CMV voltage

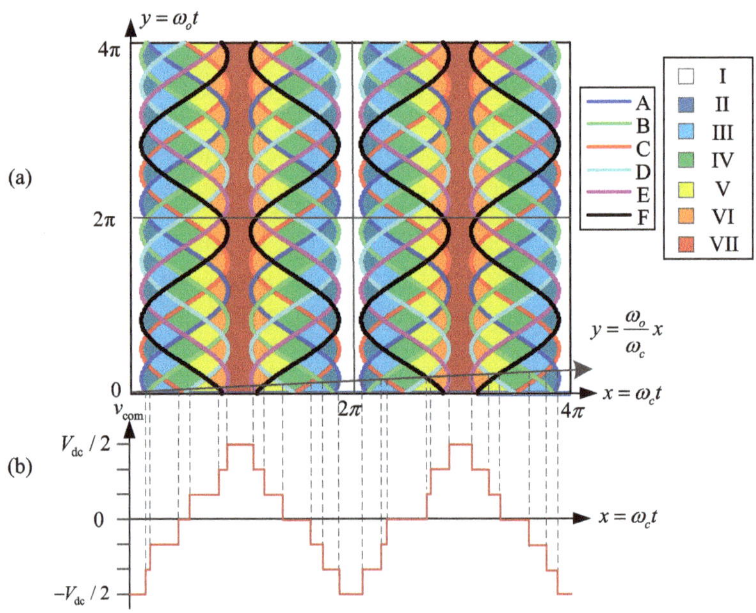

Fig. 7.114 Intersection-plotting of the symmetrical six-phase two-level inverter under conventional SPWM: **a** the contours and the carrier-ratio line, **b** resulting CMV voltage

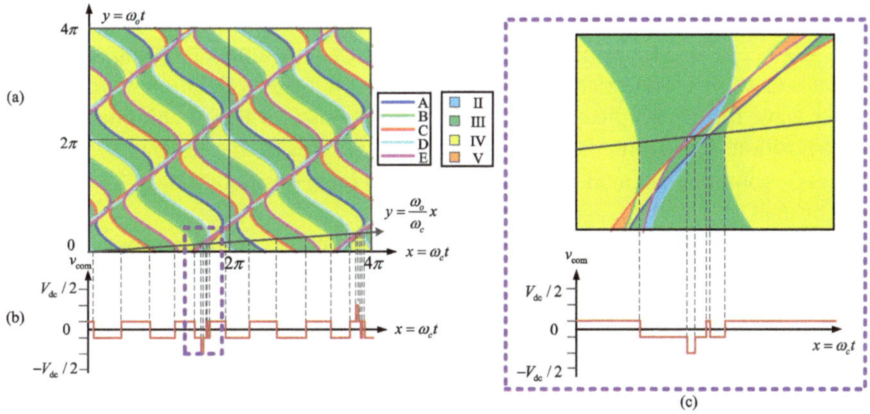

Fig. 7.115 Intersection-plotting of the five-phase two-level inverter in PS_SPWM. **a** The contours and the carrier-ratio line. **b** Resulting CMV voltage. **c** Zoom in on the dash box

of $-V_{dc}/2$ and $V_{dc}/2$ in the CMV, and the amplitude of CMV will be reduced by 40%. Furthermore, types III and IV account for most of areas, so the dwell time of $-3V_{dc}/10$ and $3V_{dc}/10$, which correspond to areas II and V, will be very short. Figure 7.116 shows the CMV for the symmetrical six-phase two-level inverter under PS_SPWM.

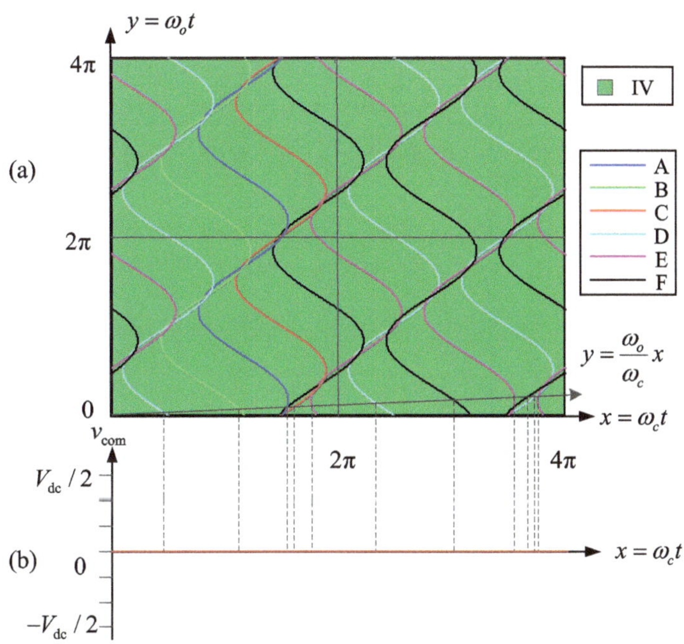

Fig. 7.116 Intersection-plotting of the symmetrical six-phase two-level inverter in PS_SPWM. **a** The contours and the carrier-ratio line. **b** Resulting CMV voltage

It is interesting to find that only area IV exists on the *x*-*y* plane, so the amplitude of the CMV keeps at zero, which means that the CMV is completely eliminated. The reason is that, for symmetrical six-phase inverters, both the modulation wave and the carrier wave for the *i*th phase-leg are opposite to those for the (6-*i*)th phase-leg in phase position. Therefore, the sum of the output voltages of these two phases would always be constant. Then taking all the six phases into consideration, $\sum_{i=1}^{6} S_i$ keeps at a constant of 3, and $v_{\rm com}$ would always be 0.

Simulation in MATLAB/Simulink and experiments on the five-phase and six-phase two-level inverters with *RL* loads were conducted to verify the theoretical analysis. The parameters are as follows: the DC voltage $V_{dc} = 180$ V, the switching frequency (carrier frequency) $f_c = 2$ kHz, the modulating frequency $f_0 = 50$ Hz, the load resistance per phase $R = 10\ \Omega$, the load inductance per phase $L = 50$ mH, and the deadtime of switches $t_{dead} = 3.5\ \mu$s.

Figures 7.117 and 7.118 show the simulation and experimental results of CMV waveforms delivered by two SPWM methods for two types of inverters, with $M = 0.8$. In conventional SPWM, the peak-to-peak values of the phase voltage, $v_{p\text{-}p}$, for the two types of inverters are all 180 V, which is the voltage of DC bus. While in the

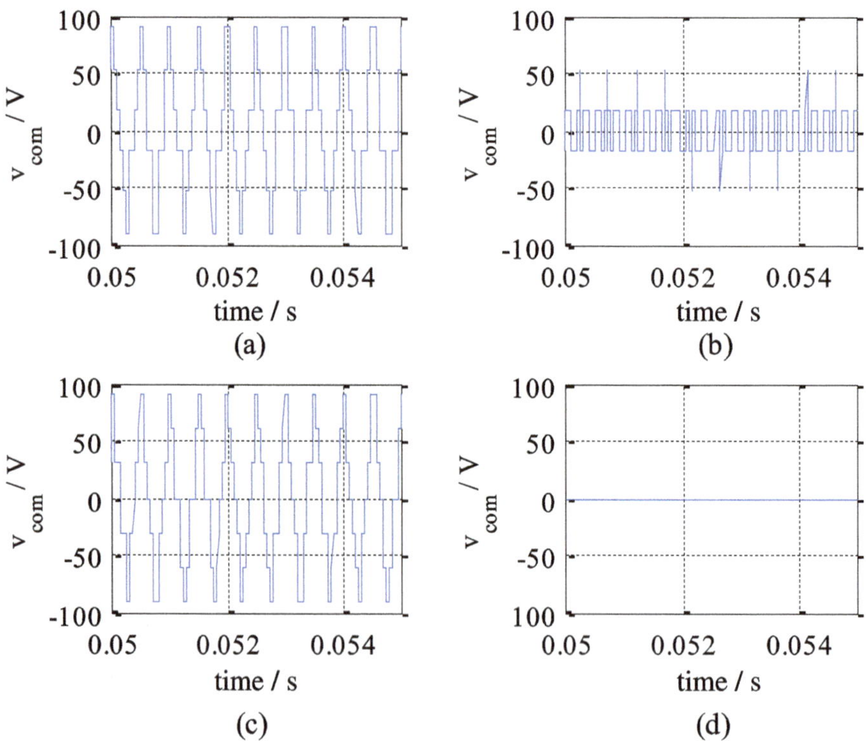

Fig. 7.117 Simulation results of CMV (modulation index $M = 0.8$) for the five-phase inverter under: **a** conventional SPWM, **b** PS_SPWM; and the six-phase inverter under: **c** conventional SPWM, **d** PS_SPWM

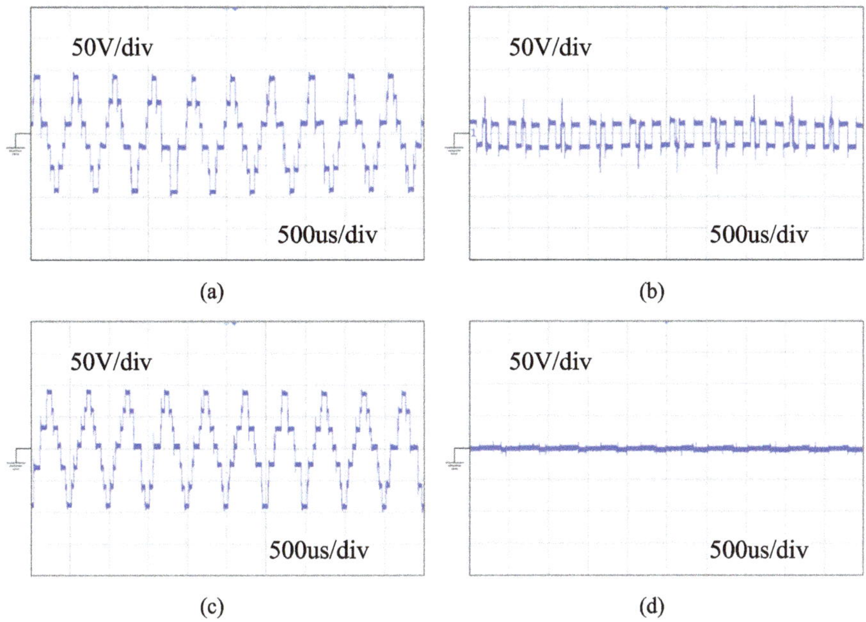

Fig. 7.118 Experimental results of CMV (modulation index $M = 0.8$) for the five-phase inverter under: **a** conventional SPWM, **b** PS_SPWM; and the six-phase inverter under: **c** conventional SPWM, **d** PS_SPWM

PS_SPWM, the $v_{p\text{-}p}$ is reduced to 108 V in the five-phase inverter, 0 V in the six-phase inverter, when $M = 0.8$. Furthermore, the peak-to-peak value of CMV varies as M varies, as shown in Fig. 7.119. Though the PS_SPWM can eliminate the CMV in the even-phase inverter in all the M range, it reduces the maximum peak-to-peak value by 80% at low M but 40% at high M for the five-phase inverter. That demonstrates that the CMV reduction effect of PS_SPWM is affected by the modulation index M in the odd-phase inverter applications .

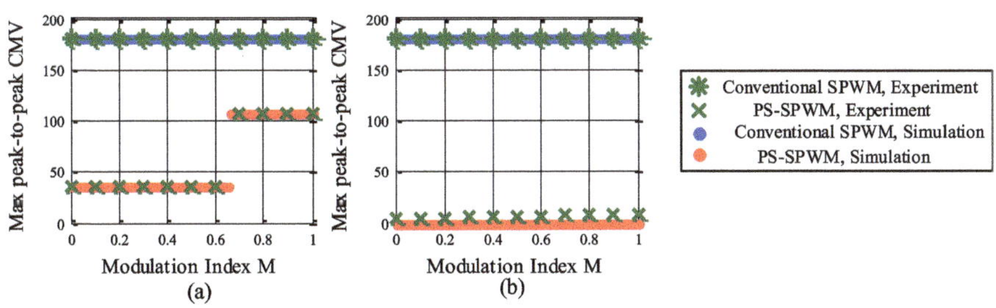

Fig. 7.119 Maximum peak-to-peak CMV value as M varies for: **a** five-phase inverters, **b** six-phase inverters

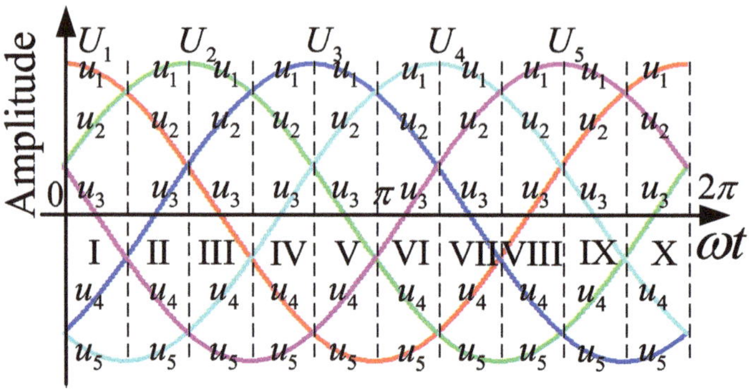

Fig. 7.120 Ten sectors of the five-phase sinusoidal modulation waves

7.8.2 Saw-Tooth Carrier-Based PWM (SCPWM)

The PS_SPWM proposed above can eliminate the CMV in symmetrical inverters with an even phase number. However, for the inverters with odd phase numbers, the CMV reduction effect of PS_SPWM is undermined as the modulation index increases. In this part, we will introduce a sawtooth carrier-based PWM (SCPWM) that can reduce both the amplitude and the frequency of CMV for multiphase inverters with odd phase numbers [30].

For multiphase inverters, the reference values for m phases can be sorted from high to low, and marked as u_1 to u_m. For an m-phase system, each fundamental period can be divided into 2 m sectors according to the relationship between reference values of m phases. For the five-phase case shown in Fig. 7.120, each fundamental period can be divided into 10 sectors.

The saw-tooth carriers are selected to be the carriers of the CBPWM, which are the triangle carriers of the same phase position in the conventional CBPWM, for the following two reasons. First, the saw-tooth carrier for each phase can be left-slanting or right-slanting, which can be utilized to reduce the CMV amplitude by limiting the sum of switching states in a narrow range. Second, the saw-tooth carrier has a straight edge, which enables the coincidence of switching times in different phases, and therefore provides the possibility of reducing the CMV changing frequency. As shown in Fig. 7.120, the references can be divided into two groups according to the odd or even subscript of u_k. In the proposed SCPWM-1, left-slanting saw-tooth carriers marked in blue are applied for the phases in the odd-subscript group, right-slanting sawtooth carriers marked in red are applied in the even-subscript group. In the five-phase case, the switching patterns and respective CMV waveforms are shown in Fig. 7.121, where three left-slanting and two right-slanting carriers (Type I) are employed. Clearly, the CMV changes only six times under SCPWM-1 during one switching cycle. When extended to general m-phase occasions, it can be obtained that the CMV changes $(m + 1)$ times under SCPWM-1, compared with 2 m times under conventional CBPWM employing triangle carriers.

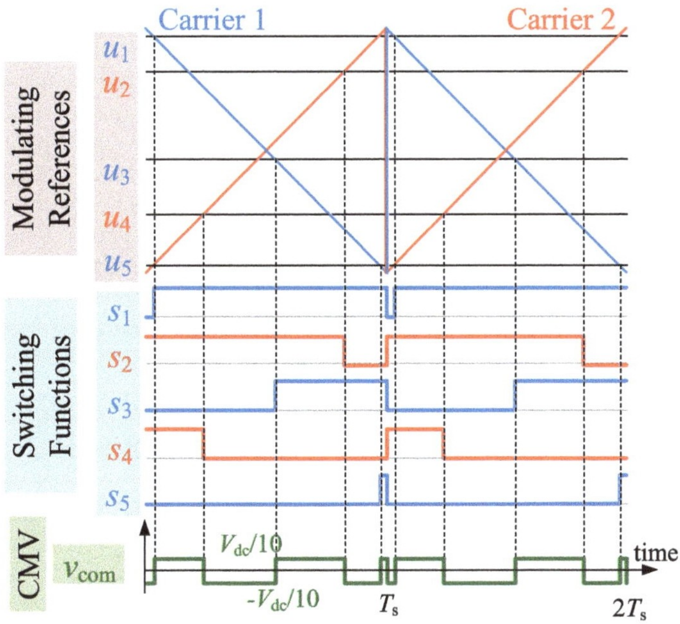

Fig. 7.121 Switching states and CMV of a five-phase inverter under SCPWM-1 of Type I

Because the reference value of sinusoidal modulating wave for each phase is time-varying in reference value, the sorted u_1 to u_m do not correspond to U_1 to U_m in the natural order. Therefore, the carrier for each phase changes eight times during a fundamental period, as shown in Fig. 7.122. The change of carrier for one phase during a fundamental period inevitably deteriorates the output voltage (current) distortion of the CBPWM.

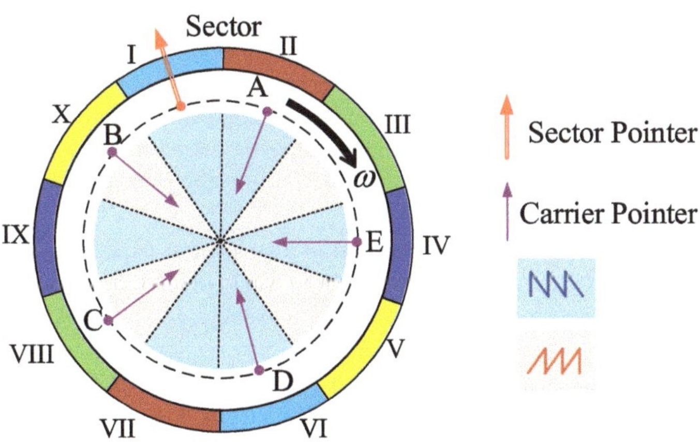

Fig. 7.122 Carriers in one fundamental period for the five-phase system under SCPWM-1

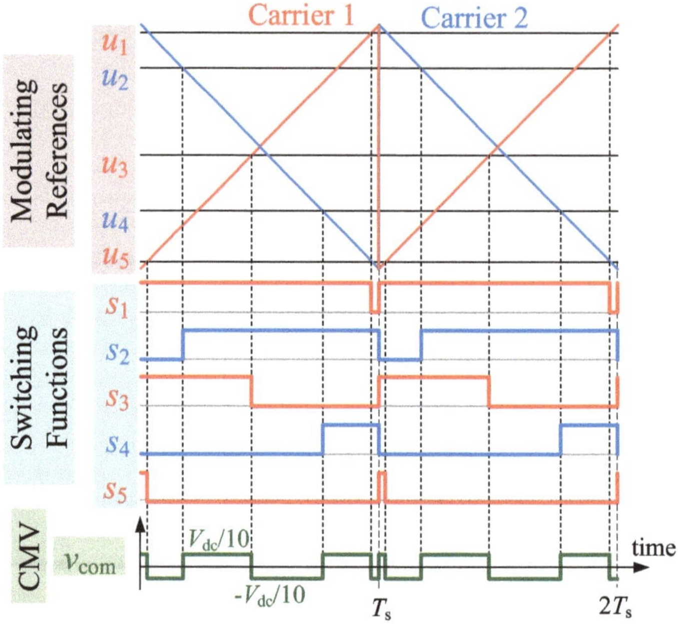

Fig. 7.123 Switching states and CMV of a five-phase inverter under SCPWM-1 of Type II

Actually, two left-slanting and three right-slanting carriers in Fig. 7.123 can result in the same CMV amplitude and frequency as those three left-slanting and two right-slanting carriers in Fig. 7.121. Providing that different types of saw-tooth carriers were adopted for the references with odd and even subscripts, the same CMV performance in peak-to-peak amplitude and changing frequency would be achieved.

In the proposed SCPWM-2, Types I and II of SCPWM-1 are alternately adopted when the sector number in Fig. 7.120 changes, and similar CMV performance can be achieved as that of SCPWM-1. Moreover, the carrier for each phase changes only two times during one fundamental period under SCPWM-2, as shown in Fig. 7.124. In contrast, the number of carrier change times is 8 under SCPWM-1, as shown in Fig. 7.122. Therefore, the proposed SCPWM-2 method manages to reduce the carrier change times by 75%. It can be derived that when extended to the general m-phase case, the carrier for one phase changes only two times under SCPWM-2, compared with 2 m-2 times under SCPWM-1.

Experiments were conducted on the multiphase inverter experimental platform, and the corresponding parameters are shown in Table 7.9. Figure 7.125 shows the waveforms of CMV v_{com}, along with the output voltage v_{o1}, phase voltage v_{p1} and phase current i_{p1} of phase-A. Clearly, the CMV has six levels under conventional CBPWM in Fig. 7.125a, while decreases to two levels under SCPWM methods in Figs. 7.125b, c. In particular, the $v_{com\text{-}pp}$ decreases from 200 V to about 50 V, which corresponds to the 80% reduction rate in the theoretical analysis. Additionally, the

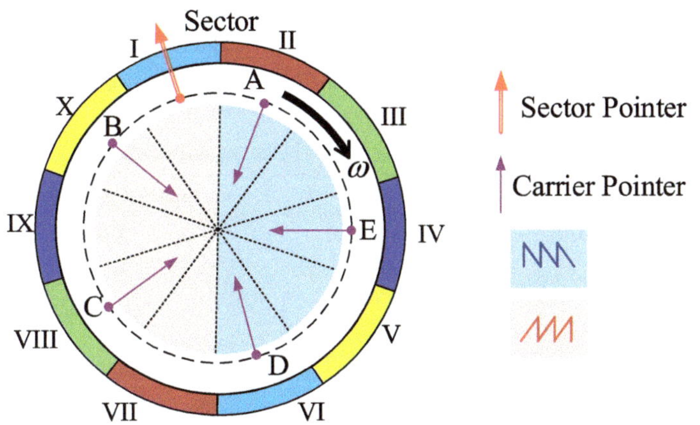

Fig. 7.124 Carriers in one fundamental period for the five-phase system under SCPWM-2

Table 7.9 Parameters for experiments of SCPWM

Five-phase induction machine parameters				Converter parameters	
P_N	4 kW	L_{m1}	203.3 mH	$f_{\text{switching}}$	10 kHz
R_s	1.554 Ω	L_{ls}	14.8 mH	T_{dead}	1 us
R_r	1.582 Ω	L_{lr}	7.6 mH	V_{dc}	200 V
p_{n}	1				

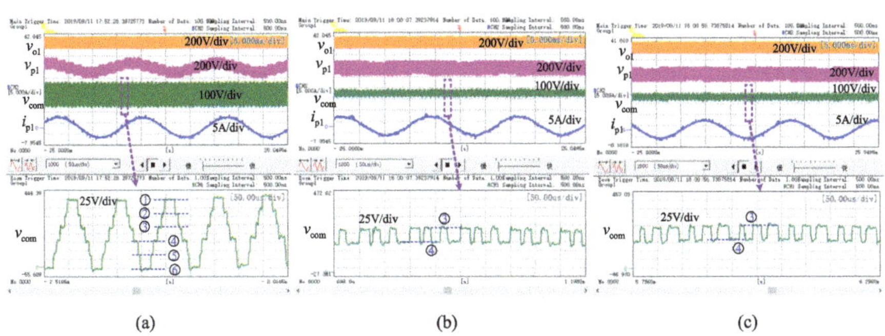

Fig. 7.125 Experiment results of CMV and phase current in five-phase RFOC drive case under: **a** conventional CBPWM, **b** SCPWM-1 and **c** SCPWM-2

CMV changes six times during one switching cycle under the two SCPWM methods in Figs. 7.125b, c, which is 40% less than 10 times in Fig. 7.125a.

The FFT analysis in Fig. 7.126 demonstrates that SCPWM methods lead to higher phase current THD than the conventional CBPWM, and SCPWM-2 achieves the better THD reduction than SCPWM-1. The frequent switching of carriers under SCPWM-1 leads to obvious low-order harmonics, especially the fourth and sixth

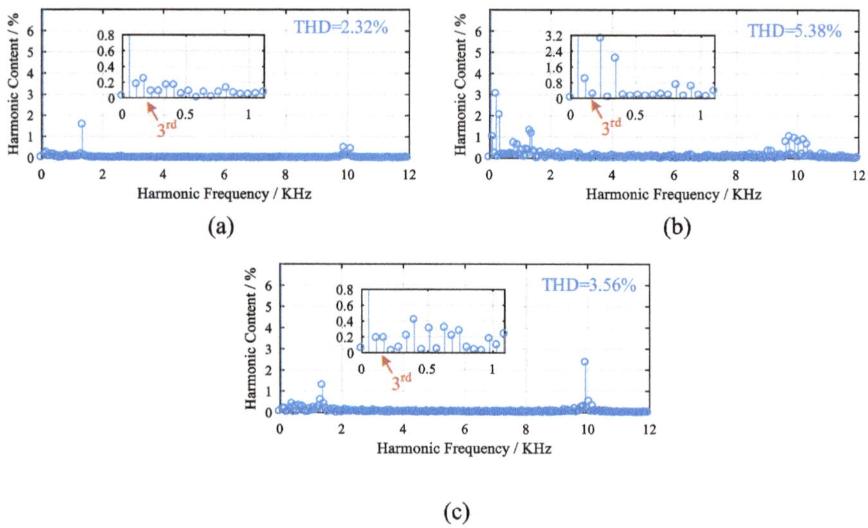

Fig. 7.126 FFT analysis of experiment phase current in five-phase RFOC drive case under: **a** conventional CBPWM, **b** SCPWM-1 and **c** SCPWM-2, with magnified view of harmonics lower than the 20th-order

components, which is about 3.2% and 2.1%, respectively. These low-order harmonics seriously deteriorate the THD of the phase current. With the optimization of carrier switching times, SCPWM-2 manages to suppress the fourth and sixth harmonics to less than 0.5%. Obviously, the decrease of carrier switching times leads to better low-order harmonic performance of SCPWM-2 than SCPWM-1. However, the harmonics of SCPWM-2 are still larger than the conventional CPWM, which results into higher THD of SCPWM-2 than conventional CBPWM.

To verify the extensibility and validity of the proposed PWM methods in other odd-phase occasions, experiments were conducted in seven-phase, nine-phase and eleven-phase cases with RL loads. For each phase of the RL loads, the resistance R is 10 Ω and the inductance L is 50 mH. The parameters of the converter are the same as those in Table 7.9, except that the voltage of DC bus is changed into 100 V.

Testbench CMV results of SCPWM-2 against the conventional CBPWM were shown in Fig. 7.127. Obviously, the CMV has 8 levels in the seven-phase case, 10 levels in the nine-phase case and 12 levels in the eleven-phase case under conventional CBPWM, but is reduced to only two levels in all the multiphase cases under the proposed SCPWM-2 method. Along with the reduction of CMV amplitude, the SCPWM-2 manages to reduce the CMV changing times per switching cycle from 14 to 8 in the seven-phase case, from 18 to 10 in the nine-phase case and from 22 to 12 in the eleven-phase case.

Therefore, the proposed SCPWM methods can be easily extended to any multiphase inverters with odd phase number m, achieving the reduction of CMV amplitude and frequency by $(m - 1)/m$ and $(m - 1)/2\,m$, respectively.

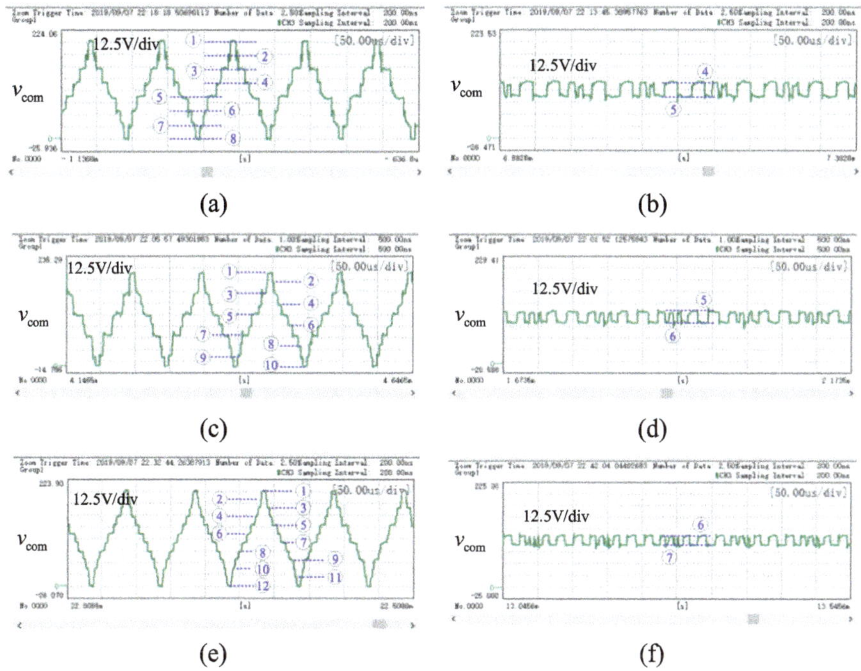

Fig. 7.127 Experimental results of CMV under conventional CBPWM and proposed SCPWM-2 in seven-phase (**a**) and (**b**), nine-phase (**c**) and (**d**), and eleven-phase (**e**) and (**f**) cases

7.9 Zero-CM PWM for Modular Multilevel Converter

Modular multilevel converter (MMC), as a promising power electronics converter interface for medium- and high-voltage applications, has some merits such as lower switching frequency, redundancy, improved harmonic distortion and reduced EMI. Various MMC modulation methods have been discussed in a number of papers, which can be divided into two types: (1) low-frequency modulation such as staircase modulation and selective harmonic elimination (SHE); (2) high-switching frequency modulation based on carrier signals with phase shifting or level shifting. Previous work on MMC modulation methods mainly focuses on the submodule (SM) voltage balance, switching losses and output voltage harmonics, and there is a lack of attention paid to the instantaneous CMV. The high-frequency CMV can generate leakage current and introduce CM EMI when it interacts the stray capacitor. The leakage current is an important concern in motor drive and photovoltaic (PV) system. Currently, [31] proposed a common-mode voltage (CMV) elimination method based on a new modulation scheme for the flying-capacitor MMC (FC-MMC). In this section, an optimal phase disposition PWM (PD_PWM) for CMV elimination is discussed in a normal MMC system.

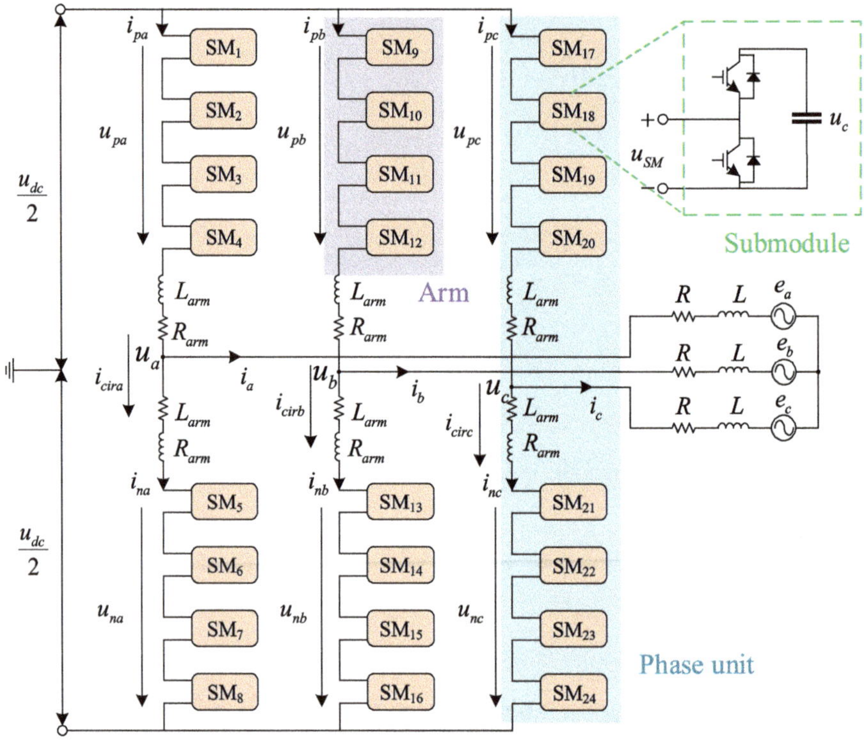

Fig. 7.128 Basic structure of MMC

The circuit diagram of MMC is shown in Fig. 7.128. It is composed of a series of half-bridge submodules (SMs) formed by three-phase legs, and each phase leg consists of an upper and a lower arm. Reactor L_{arm} is necessary for current control and limiting fault current. Without losing generality, each arm including four series-connected SMs is analyzed here.

From Fig. 7.128, the upper and lower voltage equations can be derived for phase-x (x = a, b, c) of the converter:

$$\frac{u_{dc}}{2} - u_{px} - R_{arm}i_{px} - L_{arm}\frac{di_{px}}{dt} - u_x = 0 \tag{7.39}$$

$$-\frac{u_{dc}}{2} + u_{nx} + R_{arm}i_{nx} + L_{arm}\frac{di_{nx}}{dt} - u_x = 0 \tag{7.40}$$

Here, u_{px} and u_{nx} are the sum of the upper and lower arm SM voltages, which can be considered as a controllable voltage source. u_x is the terminal voltage of phase-x. By adding (7.39) and (7.40), the output voltage and current can be derived as:

$$u_x = \frac{u_{nx} - u_{px}}{2} - R_{arm} i_x - L_{arm} \frac{di_x}{dt} \tag{7.41}$$

where

$$i_x = \frac{i_{px} - i_{nx}}{2} \tag{7.42}$$

The CMV of MMC is determined by three-phase terminal voltages and is defined as:

$$u_{com} = \frac{u_a + u_b + u_c}{3} = \frac{(u_{na} + u_{nb} + u_{nc}) - (u_{pa} + u_{pb} + u_{pc})}{6} \tag{7.43}$$

From (7.43), the CMV consists of two parts termed as u_{comp} and u_{comn}, with the definition given by:

$$\begin{cases} u_{comp} = \dfrac{u_{pa} + u_{pb} + u_{pc}}{6} \\ u_{comn} = \dfrac{u_{na} + u_{nb} + u_{nc}}{6} \end{cases} \tag{7.44}$$

In addition, the classical closed-loop control architecture is shown in Fig. 7.129. It can be seen that the circulating current within each phase is controlled by the voltage difference u_{diffx} and the reference voltages of upper and lower arms are determined by:

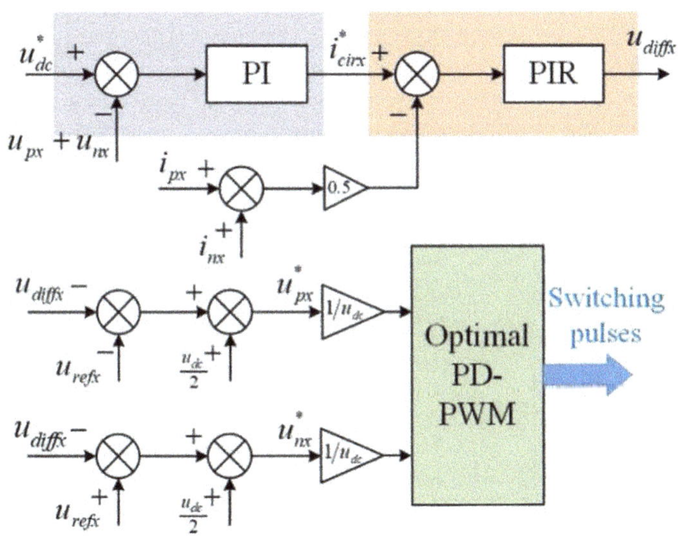

Fig. 7.129 Closed-loop control architecture of MMC with PD_PWM (phase-x)

$$\begin{cases} u^*_{px} = \dfrac{u_{dc}}{2} - u_{refx} - u_{diffx} \\ u^*_{nx} = \dfrac{u_{dc}}{2} + u_{refx} - u_{diffx} \end{cases} \tag{7.45}$$

In (7.45), the reference voltage (u_{refx}) of three-phase is given by:

$$u_{refx} = \begin{cases} m\cos(\omega t)\ (\mathrm{x = a}) \\ m\cos(\omega t - 2\pi/3)\ (\mathrm{x = b}) \\ m\cos(\omega t + 2\pi/3)\ (\mathrm{x = c}) \end{cases} \tag{7.46}$$

Ignoring the nonideal characteristics of MMC, it is supposed that the actual output arm voltage is equal to the reference ($u_{px} = u^*{}_{px}$, $u_{nx} = u^*{}_{nx}$). Obviously, the CMV elimination requires the completed cancellation between u_{comp} and u_{comn} indicated in (7.43) and (7.44). Two principles should be satisfied in each switching cycle [11]: (1) the volt-second areas of u_{comp} and u_{comn} should be equal to each other; (2) the switching pulse positions of upper and lower arms should be consistent. The principle (1) can be investigated by calculating the volt-second areas of u_{comp} and u_{comn}. Combining (7.44–7.46), multiplying (u_{comp}, T_s) and (u_{comn}, T_s) over a switching cycle (T_s) leads to:

$$\begin{cases} A_{comp} = [\dfrac{u_{dc}}{4} - \dfrac{u_{diffa} + u_{diffb} + u_{diffc}}{6}]T_S \\ A_{comn} = [\dfrac{u_{dc}}{4} - \dfrac{u_{diffa} + u_{diffb} + u_{diffc}}{6}]T_S \end{cases} \tag{7.47}$$

Equation (7.47) shows that the volt-seconds of u_{comp} and u_{comn} are always the same, and it is independent of the operation condition. Thus, the principle (1) is satisfied for an MMC. In this case, the reference voltage from the controller would be compared with the saw-tooth waveforms to generate the PWM signals for each SM, as shown in Fig. 7.130. In each switching cycle, only one SM switches twice for each arm, the other SMs keeps inactive. The duty cycles of switching SMs in six arms can be derived as:

$$\begin{cases} d_{px} = [\dfrac{u^*_{px}}{u_{dc}} - floor(\dfrac{u^*_{px}}{4u_{dc}}) \cdot 0.25] \times 4 \\ d_{nx} = [\dfrac{u^*_{nx}}{u_{dc}} - floor(\dfrac{u^*_{nx}}{4u_{dc}}) \cdot 0.25] \times 4 \end{cases} \tag{7.48}$$

In order to satisfy the principle (2), the switching pulses of switching SMs in upper and lower arms are placed end-to-end respectively, as shown in Fig. 7.131. For upper arms, the rising-edge of phase-b is equal to the falling-edge of phase-a, and the rising-edge of phase-c is equal to the falling-edge of phase-b. The switching positions for lower arms follow the same rule. It is noted that the starting point can

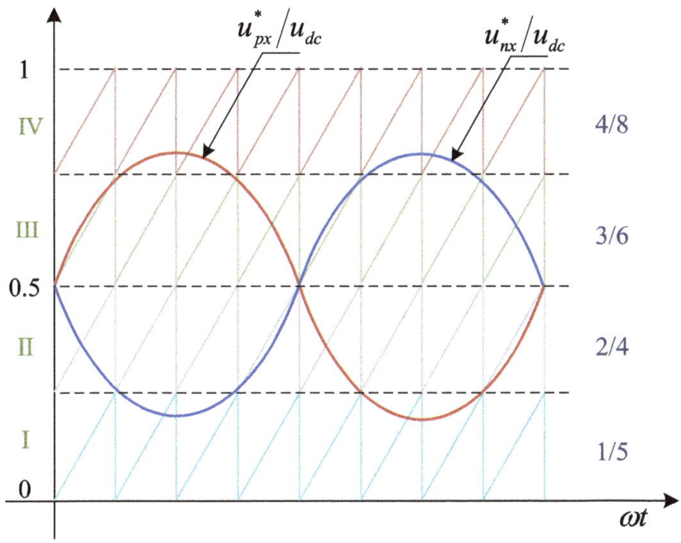

Fig. 7.130 PWM transfer relationships of SMs

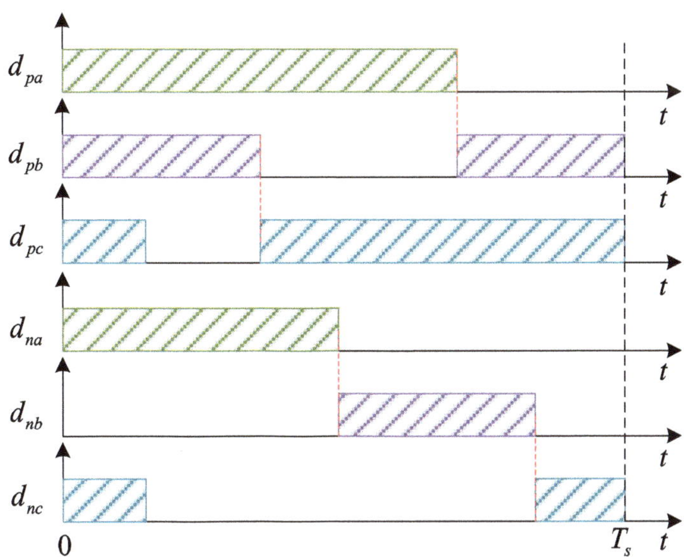

Fig. 7.131 Switching pulse arrangement

be set as any one of the rising-edge of three phases. With this alignment, the CMV elimination is achieved.

The validity of proposed method has been verified through dynamic simulations of an MMC-fed *L-R* load in MATLAB/Simulink, the DC-link voltage is 800 V and the switching frequency is 3 kHz. The output voltage and CMV are given in Fig. 7.132a, b, with modulation index step from 0.3 to 0.8. Compared with the traditional PD_PWM, the CMV is always controlled to be zero under the proposed method.

In addition, the neutral point of load is connected to the ground through a 5 nF series non-inductive capacitor to simulate the CM loop and test the CM noise. It can be seen that the CM noise can be improved a lot with the proposed method, as shown in Fig. 7.133. Considering the THD of output current, it can be seen that the THD of output current for the proposed method would be a little higher with respect to the traditional PD_PWM due to the asymmetrical distribution of switching pulses in a switching cycle, as shown in Fig. 7.134.

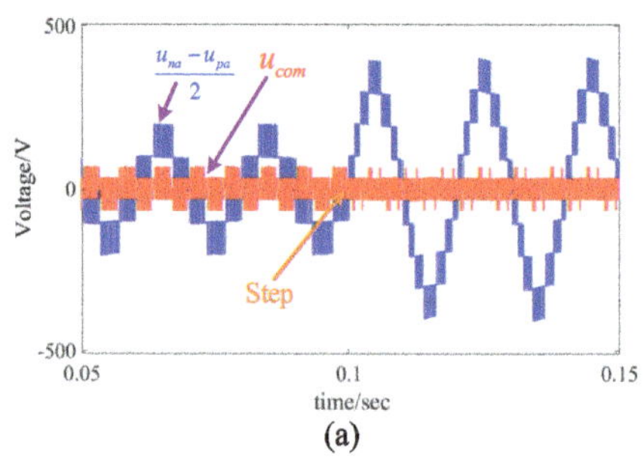

Fig. 7.132 The output voltage and CM voltage comparison for the two modulation schemes: **a** the conventional PD_PWM, **b** the proposed method

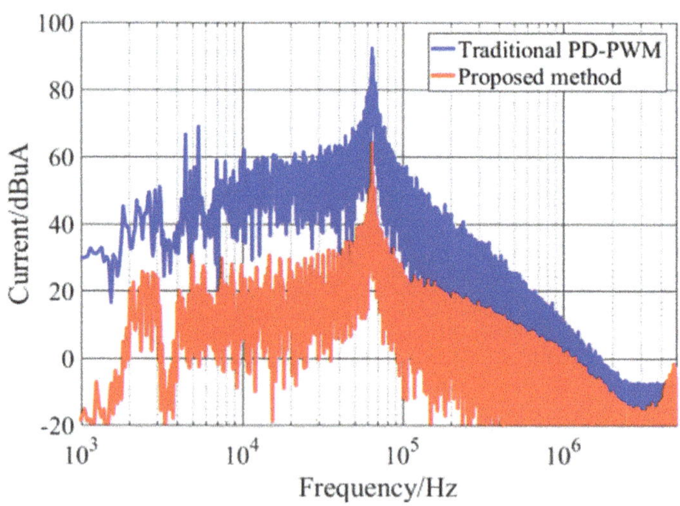

Fig. 7.133 CMC spectrum comparison result

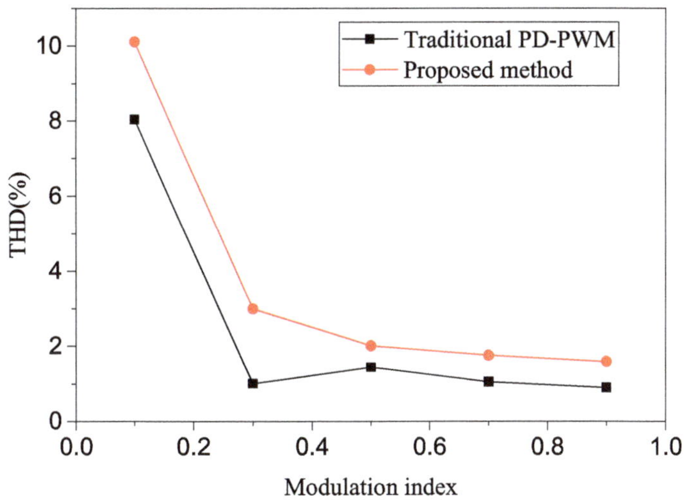

Fig. 7.134 The THD comparison result

7.10 Summary

Common-mode current/EMI is a relatively independent but very serious problem in motor control and power electronics converter systems. This chapter focuses on the research results of using advanced PWM methods to suppress common-mode noise. First, the zero-voltage vector causes the maximum amplitude of CMV in the three-phase converter. By changing the position of PWM signal, the zero-voltage vector can be avoided, so the CMV amplitude can be reduced to one-third of the previous value. This method is equivalent to applying the nonzero vector to achieve the role of zero vector, that is, the AZSPWM and NSPWM methods. However, CM noise is not only caused by the CMV but also closely related to the impedance of conduction path. This chapter studies a comprehensive approach to CM noise suppression combined with conduction path impedance and advanced PWM. The two-level three-phase converter is physically unable to eliminate the CMV. This chapter continues to study the PWM method to realize the CMV suppression and cancellation using complex topologies, such as multilevel and paralleled inverters. In particular, a zero-CM PWM method for paralleled inverters is proposed, which can effectively suppress CM noise. In addition, for the deficiencies of the initial zero-CM PWM method in practical application, the relevant deadtime compensation and circulating current suppression methods are designed to achieve more optimized effects in practical applications. Furthermore, considering the zero-CM PWM method for paralleled inverters combined with the dual-module three-phase motor with identical back-EMF, the similar CM suppression can be realized, and the coupled inductor utilized in paralleled inverters can be cancelled, which can simplify the system structure and increase the power density of the system. Moreover, considering the possibility of CMV elimination for the dual two-level inverter-fed dual three-phase motor with 30° displacement, the special designed carrier-based PWM is developed to deal with the CMV-induced problem. To perfect the CMV elimination effect for the dual two-level inverter-fed general dual three-phase motor, a universal CMV elimination PWM strategy which lifts the restriction on angle displacement and real-time reference voltage between the two sets of windings is developed to provide the function of the phase current balance and CMV elimination both in parameter symmetry and asymmetry conditions. In addition, the CMV of inverter may excite the ZSC in open-winding three-phase motor with single DC power supply, and the special three-phase unipolar SPWM associated with suitable phase-shift scheme can be utilized to perform zero-sequence current suppression. For general multiphase motor drive, PS_SPWM and SCPWM are introduced to reduce the CMV. Finally, the MMC which can be applied in medium- and high-voltage occasions may generate the high amplitude CMV, and an optimal PD_PWM scheme can be utilized to realize the CMV elimination theoretically for a normal three-phase MMC system.

Appendix

Switching sequences for Sector 2–Sector 6, and the corresponding carrier based PWM:

Sector 2

	Inverter 1	Inverter 2
$t_0/4$	111	000
$t_1/2$	110	010
$t_2/2$	100	110
$t_0/2$	000	111
$t_1/2$	010	110
$t_2/2$	110	100
$t_0/4$	111	000

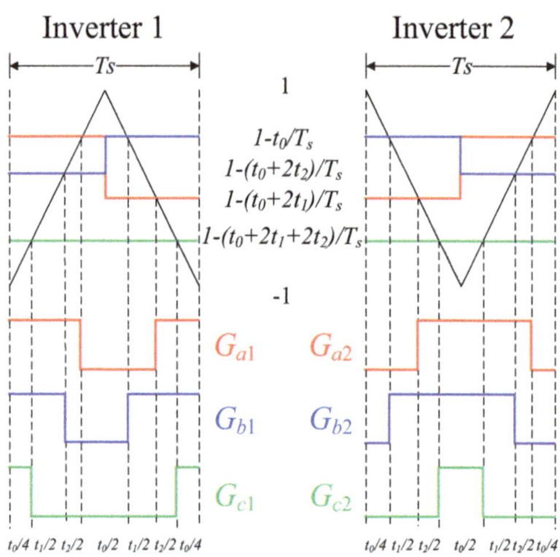

Comparators and pulse generation in Sector 2

Sector 3

	Inverter 1	Inverter 2
$t_0/4$	111	000
$t_1/2$	011	010
$t_2/2$	010	110
$t_0/2$	000	111
$t_1/2$	010	011
$t_2/2$	110	010
$t_0/4$	111	000

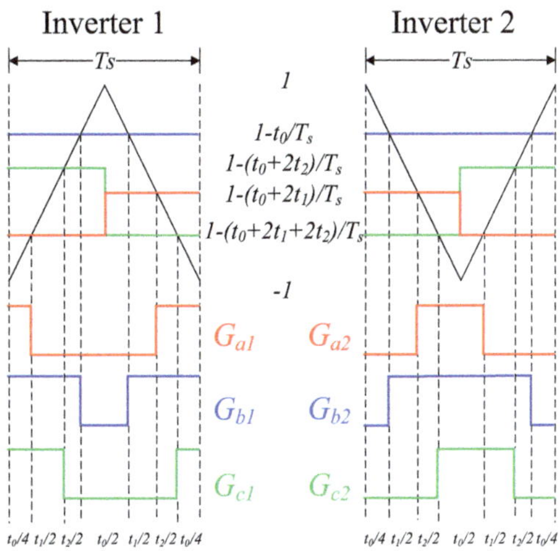

Comparators and pulse generation in Sector 3

Sector 4

	Inverter 1	Inverter 2
$t_0/4$	111	000
$t_1/2$	011	001
$t_2/2$	010	011
$t_0/2$	000	111
$t_1/2$	001	011
$t_2/2$	011	010
$t_0/4$	111	000

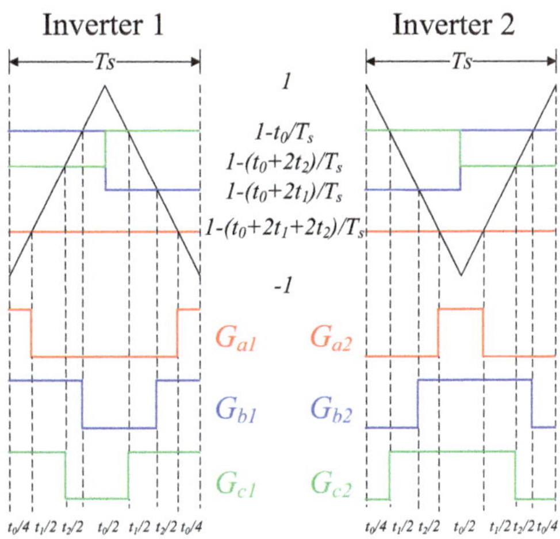

Comparators and pulse generation in Sector 4

Sector 5

	Inverter 1	Inverter 2
$t_0/4$	111	000
$t_1/2$	101	001
$t_2/2$	001	011
$t_0/2$	000	111
$t_1/2$	001	101
$t_2/2$	011	001
$t_0/4$	111	000

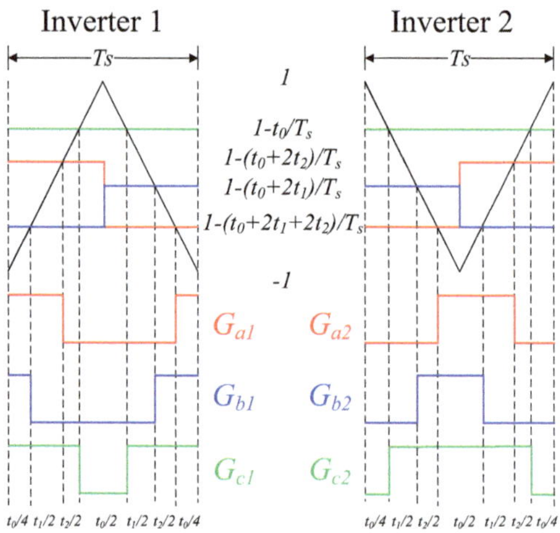

Comparators and pulse generation in Sector 5

Sector 6

	Inverter 1	Inverter 2
$t_0/4$	111	000
$t_1/2$	101	100
$t_2/2$	001	101
$t_0/2$	000	111
$t_1/2$	100	101
$t_2/2$	101	001
$t_0/4$	111	000

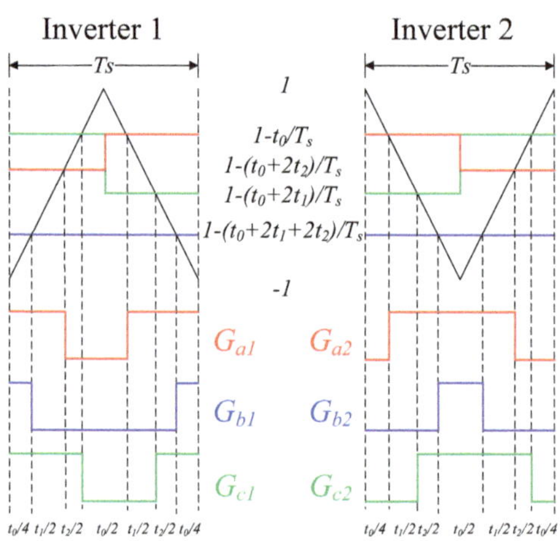

Comparators and pulse generation in Sector 6

References

1. Akagi H, Shimizu T (2008) Attenuation of conducted emi emissions from an inverter-driven motor. IEEE Trans Power Electron 23(1):282–290
2. Muetze A (2009) Scaling issues for common-mode chokes to mitigate ground currents in inverter-based drive systems. IEEE Trans Ind Appl 45(1):286–294
3. Luo F, Wang S, Wang F, Boroyevich D, Gazel N, Kang Y, Baisden AC (2010) Analysis of CM volt-second influence on CM inductor saturation and design for input EMI filters in three-phase DC-fed motor drive systems. IEEE Trans Power Electron 25(7):1905–1914

4. Un A, Hava AM (2009) A near-state PWM method with reduced switching losses and reduced common-mode voltage for three-phase voltage source inverters. IEEE Trans Ind Appl 45(2):782–793
5. Hava M, Cetin NO, Un E (2008) On the contribution of PWM methods to the common mode (leakage) current in conventional three-phase two-level inverters as applied to AC motor drives. In: Proceedings of industry applications conference (IAS), pp 1–8
6. Cetin NO, Hava AM (2010) Interaction between the filter and PWM units in the sine filter configuration utilizing three-phase AC motor drives employing PWM inverters. In: Proceedings of IEEE energy conversion congress and exposition (ECCE), pp 2592–2599
7. Jiang FW, Xue J (2013) PWM impact on CM noise and AC CM choke for variable-speed motor drives. IEEE Trans Ind Appl 49(2):963–972 (2013)
8. Kerekes T, Teodorescu R, Liserre M et al (2009) Evaluation of three-phase transformerless photovoltaic inverter topologies. IEEE Trans Power Electron 24(9):2202–2211
9. Erdman JM, Kerkman RJ, Schlegel DW, Skibinski GL (1996) Effect of PWM inverters on AC motor bearing currents and shaft voltages. IEEE Trans Ind Appl 32(2):250–259
10. Zhang H, von Jouanne A, Dai S, Wallace AK, Wang F (2000) Multilevel inverter modulation schemes to eliminate common-mode voltages. IEEE Trans Ind Appl 36(6):1645–1653
11. von Jauanne A, Zhang H (1999) A dual-bridge inverter approach to eliminating common-mode voltages and bearing and leakage currents. IEEE Trans Power Electron 14(1):43–48
12. Zhang D, Wang F, Burgos R, Lai R, Boroyevich D (2010) Impact of interleaving on AC passive components of paralleled three-phase voltage-source converters. IEEE Trans Ind Appl 46(3):1042–1054
13. Zhang D, Wang F, Burgos R, Boroyevich D (2012) Total flux minimization control for integrated inter-phase inductors in paralleled, interleaved three-phase two-level voltage-source converters with discontinuous space-vector modulation. IEEE Trans Power Electron 27(4):1679–1688
14. Jiang D, Shen Z, Wang F (2018) Common-mode voltage reduction for paralleled inverters. IEEE Trans Power Electron 33(5):3961–3974
15. Jiang D (2015) US20150349662 A1, Pulse-width modulation control of paralleled inverters
16. Jiang D, Shen M (2015) US20150349626 A1, Output filter for paralleled inverter
17. Shen Z, Jiang D, Chen J, Qu R (2018) Circulating current reduction for paralleled inverters with modified zero-CM PWM algorithm. IEEE Trans Ind Appl 54(4):3518–3528
18. Urasaki N, Senjyu T, Kinjo T, Funabashi T, Sekine H (2005) Dead-time compensation strategy for permanent magnet synchronous motor drive taking zero-current clamp and parasitic capacitance effects into account. IEEE Proc Electr Power Appl 152(4):845–853
19. Urasaki N, Senjyu T, Uezato K, Funabashi T (2007) Adaptive dead-time compensation strategy for permanent magnet synchronous motor drive. IEEE Trans Energy Convers 22(2):271–280
20. Shen Z, Jiang D (2019) Dead-time effect compensation method based on current ripple prediction for voltage-source inverters. IEEE Trans Power Electron 34(1):971–983
21. Shen Z, Jiang D, Liu Z, Chen J (2020) A general dead time compensation method based on current ripple prediction and pulse delay measurement for voltage source inverter. In: 2020 IEEE energy conversion congress and exposition (ECCE), Detroit, MI, pp 1–5
22. Levi E, Bojoi R, Profumo F, Toliyat HA, Williamson S (2007) Multiphase induction motor drives—a technology status review. IET Electr Power Appl 1(4):489–516
23. Parsa L, Toliyat HA (2007) Fault-tolerant interior-permanent-magnet machines for hybrid electric vehicle applications. IEEE Trans Veh Tech 56(4):1546–1552
24. Liu Z, Wu J, Hao L (2018) Coordinated and fault-tolerant control of tandem 15-phase induction motors in ship propulsion system. IET Electr Power Appl 12(1):91–97
25. Durán MJ, Prieto J, Barrero F (2013) Space vector PWM with reduced common-mode voltage for five-phase induction motor drives operating in overmodulation zone. IEEE Trans Power Electron 28(8):4030–4040
26. López Ó et al (2016) Space-vector PWM with common-mode voltage elimination for multiphase drives. IEEE Trans Power Electron 31(12):8151–8161
27. Chen K, Hsieh M (2017) Generalized minimum common-mode voltage PWM for two-level multiphase VSIs considering reference order. IEEE Trans Power Electron 32(8):6493–6509

28. Holmes DG, Lipo TA (2003) Pulse width modulation for power converters: principles and practice, vol 18. Wiley, pp 95–153
29. Liu Z, Zheng Z, Sudhoff SD, Gu C, Li Y (2016) Reduction of common-mode voltage in multiphase two-level inverters using SPWM with phase-shifted carriers. IEEE Trans Power Electron 31(9):6631–6645
30. Liu Z, Wang P, Sun W, Shen Z, Jiang D Sawtooth carrier-based PWM methods with common-mode voltage reduction for symmetrical multiphase two-level inverters with odd phase number. IEEE Trans Power Electron
31. Du S, Wu B, Zargari NR (2018) Common-mode voltage elimination for variable-speed motor drive based on flying-capacitor modular multilevel converter. IEEE Trans Power Electron 33(7):5621–5628
32. Shen Z, Jiang D, Zou T, Qu R (2019) Dual-segment three-phase PMSM with dual inverters for leakage current and common-mode EMI reduction. IEEE Trans Power Electron 34(6):5606–5619
33. Shen Z, Jiang D, Zou T, Qu R (2019) Common-mode voltage elimination for dual two-level inverter fed asymmetrical six-phase PMSM. IEEE Trans Power Electron
34. Zhan H, Zhu ZQ, Odavic M (2017) Analysis and suppression of zero sequence circulating current in open winding PMSM drives with common DC bus. IEEE Trans Ind Appl 53(4):3609–3620
35. Shen Z, Jiang D, Zhu L, Zou T, Liu Z, Qu R (2019) A novel zero-sequence current elimination PWM scheme for an open winding PMSM with common DC bus. IEEE Trans Power Electron
36. Jiang D, Chen J, Shen Z (2019) Common mode EMI reduction through PWM methods for three-phase motor controller. CES Trans Electr Mach Sys 3(2):133–142

Chapter 8
Software and Hardware Implementation of Advanced PWM

In the actual system, the advanced PWM needs to be realized by the microprocessor combined with related hardware functions, and also needs to be compiled with reasonable software. This chapter first describes how to implement advanced PWM control algorithms in simulation. After that, a simple digital signal processor (DSP) control system is briefly introduced. Based on these tools, the principle of PWM generation in typical DSP is introduced. Finally, the method of implementing advanced PWM strategy by configuring DSP-related registers is introduced.

8.1 Implementation of Advanced PWM in Simulation

Before the hardware experiment, the control method of the power electronic system should be first verified by simulation. Computer simulation technology can simulate the dynamic and static processes of real systems in the time domain based on the mathematical model of the object. With the development of computer technology, high-performance simulation software can be used to simulate the performance of power electronics systems at different time scales. At present, the main simulation software includes MATLAB/Simulink, Plecs, Pspice, Saber, PSIM and so on. Different software has different application characteristics. Pspice and Saber software can embed more accurate devices and distributed circuit models which are suitable for device characteristics and circuit simulation. Though MATLAB/Simulink software does not have such accurate device and distributed circuit model, its programming and analysis capabilities are strong. The function can be directly embedded in the MATLAB program, and a variety of powerful toolboxes are suitable for simulation analysis of various control algorithms. In particular, the Sim PowerSystem toolbox, which contains various converter, motor and passive device models, can be effectively combined with Simulink-controlled signal modules to implement a power electronics system model. Thus, the MATLAB/Simulink is very suitable for the performance

D. Jiang et al., *Advanced Pulse-Width-Modulation: With Freedom to Optimize Power Electronics Converters*, CPSS Power Electronics Series,
https://doi.org/10.1007/978-981-33-4385-6_8

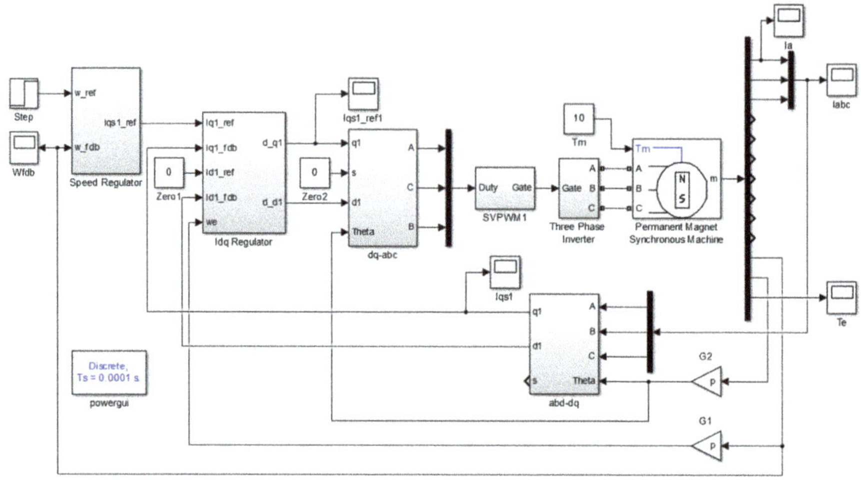

Fig. 8.1 A typical simulation system of motor drive with three-phase power electronics converter

simulation analysis which takes the advanced PWM algorithm as the control strategy [1].

Figure 8.1 shows the Simulink simulation system based on a typical three-phase power electronics converter which is utilized to drive a permanent magnet synchronous motor with vector control. The system includes three parts: outer loop, inner loop and actuator. The outer loop is a speed controller, which generates reference current and feeds it into the inner loop. The inner loop is the current controller with *d-q* decoupling, which generates duty cycles in *d-q* coordinate system and obtains the duty cycles in *a-b-c* coordinate system by coordinate transformation. Then duty cycles in *a-b-c* coordinate system are fed into PWM generating module to generate gate drive pulses, which control the converter to drive the motor. On the other hand, the position, speed and current signals of the motor are fed back to the outer loop and inner loop to complete the closed loop of the system.

In the simulation, the core of the advanced PWM is the PWM module, but the existing PWM generating module in Simulink software cannot realize the controllable changes of the switching period or the pulse position. Therefore, the advanced PWM methods should be developed by the user. Taking three-phase variable switching frequency PWM as an example, Fig. 8.2 shows the internal structure of variable switching frequency PWM module in Simulink environment. First, the duty cycles generated from the current controller are converted to *a-b-c* coordinates. Then they are fed into the first module: variable frequency sampling module. Using the sampling trigger module in Simulink, this module not only realizes pulse trigger sampling and holding but also makes the trigger pulse synchronous with the switching period. In this way, the sampling can be synchronized with the switching action even if the switching cycle changes. The duty cycles after sampling process are fed into the calculation module of MATLAB. The required switching period is calculated

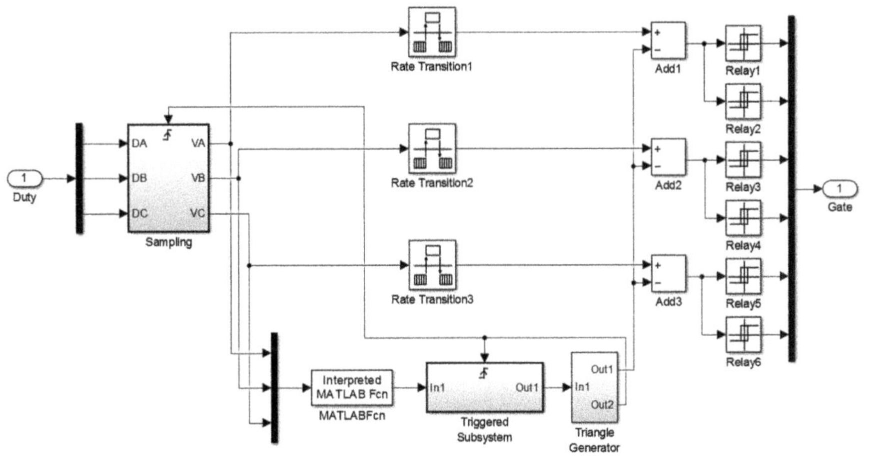

Fig. 8.2 Simulink simulation of variable switching frequency PWM

according to different control requirements and maintained by a trigger sampling module. Then it is fed into the core of variable switching frequency PWM: triangular carrier generator. The triangular wave of variable switching frequency generated by triangular carrier generator can be compared with duty cycles and delivers switching pulse, and also generates square wave sampling trigger signal synchronized with triangular wave, which is sent back to two trigger sampling modules.

The core part of the simulation system of Fig. 8.2 is the triangular wave generator. Its internal structure is shown in Fig. 8.3, which is constructed by Simulink standard modules and can generate triangular wave with variable period. The input is the calculated value of the switching frequency. The input is multiplied by 2 and −2, respectively, and then sent into a switching module (Switch) to determine the slopes of the rising and falling parts of the triangular wave. The slope is integrated by a

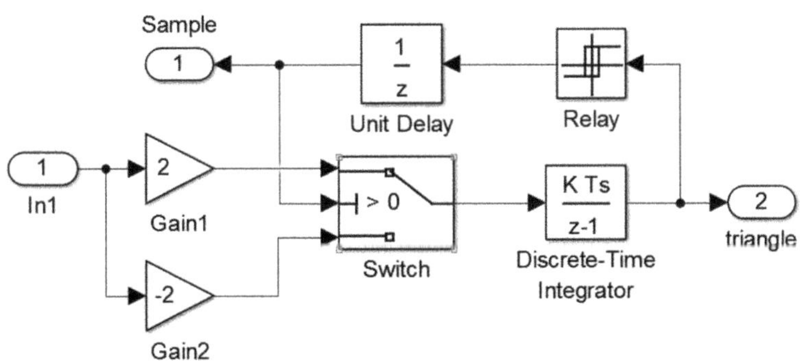

Fig. 8.3 The method of generating variable-cycle triangular wave

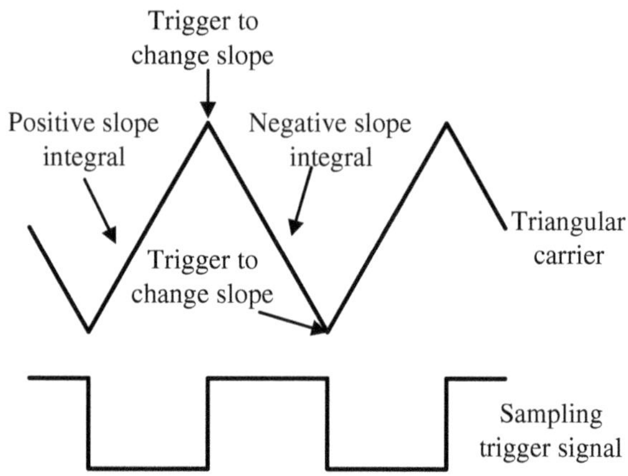

Fig. 8.4 Principle of triangular wave and sampling trigger signal generation in Simulink

discrete integrator and sent to a hysteresis comparison module. Hysteresis comparison module sets the upper and lower bounds of triangular wave. If the integration reaches the upper or lower bound, the output of hysteresis will reverse which triggers the switching module to switch to the other side slope of triangular wave, and the integrator will continue the integration with this new slope. In this way, the corresponding symmetrical triangular wave and sampling trigger signal can be obtained according to the input frequency. In Fig. 8.3, the trigger signal needs a beat delay to avoid the problem of algebraic loop. The generation principle of triangular wave and sampling trigger signal is shown in Fig. 8.4.

Depending on the system and control objectives, different variable switching frequency PWMs can be implemented using the MATLAB calculation block in Fig. 8.2. This approach makes good use of the computational power of the MATLAB language, which can implement the mathematical methods introduced in the previous chapters of this book. Taking the variable-frequency three-level converter as an example, the switching period of the upper and lower carriers is calculated by the three-phase duty cycles which can generate the corresponding upper and lower carriers, and the sampling duty cycles are synchronized with the sampling trigger signal. As shown in Fig. 8.5, the triangular wave periods at different positions are different, and the reference wave is sampled synchronously, ensuring the symmetric regular sampling of duty cycles.

8.2 The Generation Principle of PWM in DSP

Based on the simulation, the advanced PWM algorithm can be further applied in the actual hardware system. Hardware controller is the core component of realizing relevant algorithm in practical system. The basic function of hardware controller is to sample feedback value from actual system and send the executive signal (PWM signal) to the actual system after the corresponding algorithm calculation. Actually,

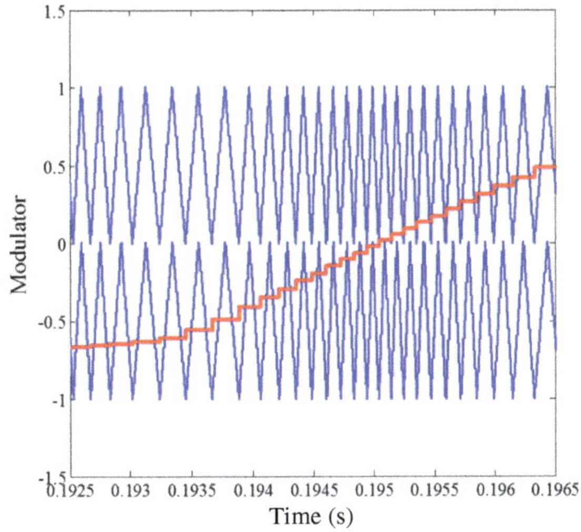

Fig. 8.5 Example: three-level variable switching frequency carrier and sampling reference waves in simulation

software algorithms are implemented in the "bit world", while hardware controllers are the interface between the "bit world" and the real world. Figure 8.6 shows the relationship between hardware controller and control object in power electronics control system. The core part of hardware controller is sampling and output, which corresponds to analog-to-digital converter (ADC) and PWM port.

With the rapid development of microelectronics and computer technology, hardware controllers have experienced a process from low-end to high-end, from slow to fast, from single core to multicore. The most commonly used hardware controller in teaching and scientific research is personal computer (PC), but PC is not applicable in special industrial applications. It is necessary to realize signal processing and control functions through a minimal system composed of core processing elements which is the so-called embedded system. Its core processing elements mainly include three types: central processing unit (CPU), microprocessor (MCU) and digital signal

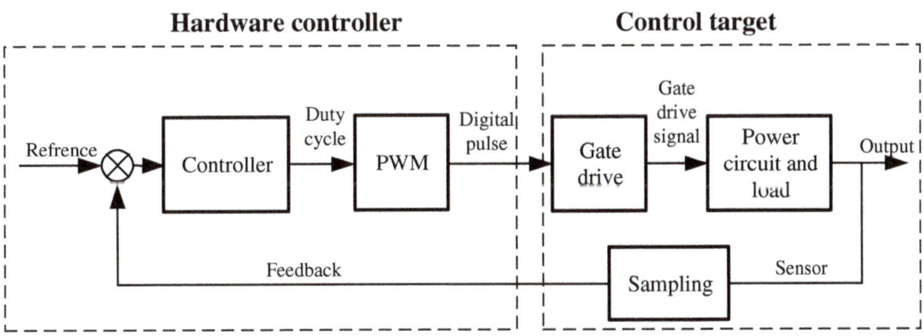

Fig. 8.6 Hardware controller in power electronic control system

processor (DSP). Among them, CPU is mainly used in PC and has the function of multithreading simultaneous processing; MCU is also called single-chip computer, which has strong control function but weak digital signal processing ability; while DSP is a high-performance microprocessor designed for signal processing, which is the main tool for realizing high-performance power electronics and motion control in recent years, and is also the focus of this section.

The DSP chip mainly consists of two parts: the core to perform the calculation operation and the peripheral to realize the external interface. The DSP core completes the basic functions through hardware adder and multiplier, and carries out the calculation through the basic computer architecture and pipeline work mode. However, these functions are not enough to complete closed-loop control. DSP also needs peripheral functions based on ADC and PWM to complete the connection with the control object.

In addition, DSP can be divided into two types according to the different digital representation, that is, fixed point and floating point. Fixed-point DSP is the main type in the early stage which uses integers to implement mathematical operations. The size of integers depends on the length of the word used. The larger the number of digits for the integer, the larger is the expressed range. For example, the number range of 16 bits is −32,768–32,767. Once beyond this range, data overflow will occur, resulting in computational errors. The new floating-point DSP uses floating-point counting method, which greatly enlarges the numerical range and effectively avoids data overflow. Moreover, the counting method of fixed-point DSP requires special process of ordinary digits in programming, while the floating-point DSP can avoid this process and has the same accuracy which is easy to express and program.

Among the major DSP suppliers, Texas Instruments (TI) is a prominent company that has launched multiple series of DSP products for different applications, and the main products include C2000 series, C5000 series and C6000 series. The C2000 series DSPs are designed for power electronics and motor control which feature a wide range of peripheral functions, including multiple PWM and AD channels. The C5000 and C6000 series of DSPs have more powerful digital signal processing capabilities and are used in communications and video signal processing.

In the C2000 series DSPs, TMS320F2407 and TMS320F2812 are early types of fixed-point DSPs, lacking in computing capability but rich in peripheral control functions. In the high-performance DSP control system, the C6000 series DSP with more computing capability is often applied, and can form a dual DSP system with the C2000 series DSP through the dual-port random-access memory (RAM), which can allocate the C6000 series DSP to implement complex algorithms and the C2000 series DSP to implement sampling and PWM control. Some applications use floating-point DSP plus field programmable gate array (FPGA)/ complex programming logic device (CPLD) and allow them to complete PWM and other functions, which can compensate for the lack of control peripheral interface for high-performance floating-point DSP. However, the dual-processor mode structure is complicated, which reduces the efficiency of the operation. In particular, it brings the challenge for the advanced PWM implementation that needs to change the modulation method in real time. In 2007, TI introduced the first widely used floating-point DSP in the C2000 series: the

TMS320F28335. This kind of DSP has the advantages of strong floating-point DSP computing power and convenient programming. A single DSP can independently perform complex power electronics control. This DSP has become one of the most widely used processors in power electronics and motor drive in 2010.

Though the TMS320F28335 is powerful, a single DSP chip cannot perform the control functions, and a minimum system needs to be built in conjunction with the peripheral circuits. External memory, power management circuit, communication interface, motor speed/position sensor interface (CAP/QEP), PWM output signal processing (gate drive isolation), AD input signal processing circuit (sensor and signal conditioning circuit) and other functional modules are integrated on the periphery of the DSP. Combined with the converter and the load motor, the minimum system can achieve closed-loop control of the motor.

The TI company provides evaluation board (EVM board) for its main DSP products. These evaluation boards are minimal DSP systems that integrate basic peripheral circuits (clocks, power supplies, memory, etc.) and can work independently. These EVM boards contain flexible connectors, such as breadboard jacks. Users can directly use these interfaces to make basic connections with the converter. One of the typical EVM boards is the control card which is similar to the computer memory bar, as shown in Fig. 8.7a. It can be inserted into the slot and connected to the breadboard through the interface between the slot and the breadboard. With this control card, the corresponding carrier board can be developed, as shown in Fig. 8.7b. The board is equipped with the functions of digital and analog signal processing, communication circuit and power supply. Based on this system, users can complete the application without developing DSP chip, which simplifies the system development.

Typical control systems need output signal to execute control function and sampling feedback signal to compare. Taking DSP chip TMS320F28335 as an example, it contains many functional submodules, such as serial communication interface (SCI) submodule, serial peripheral interface (SPI) submodule, ADC submodule, enhanced quadrature encoder pulse (eQEP) submodule and ePWM submodule, and so on. The SCI and SPI submodules are mainly used for communication; ADC submodule is responsible for the conversion of analog signal to digital

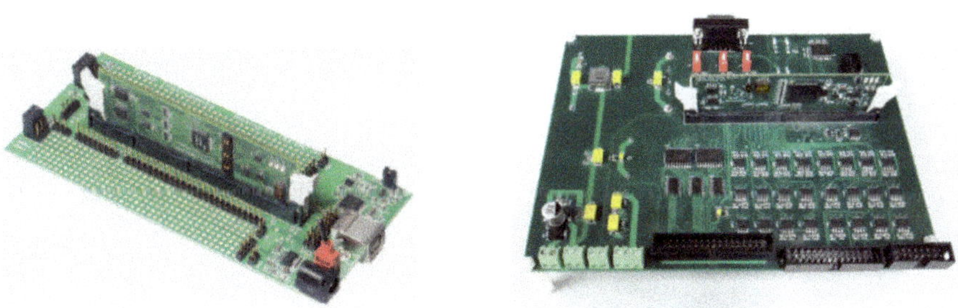

(a) (b)

Fig. 8.7 **a** Control card based on TI TMS320F28335 DSP. **b** Minimum system based on TI TMS320F28335 control card signal which is the commonly used sampling module; eQEP submodule is mainly used for processing the signal of the position encoder in motor control; and ePWM submodule is responsible for controlling signal output.

PWM control is realized mainly through its ePWM module in the DSP chip. Its principle is to simulate the actual carrier with digits, and update the pulse width and position of the PWM in real time by modifying the comparison values in each interruption period. Each ePWM module consists of a number of functional submodules, including time base (TB) submodule, counter compare (CC) submodule, action qualifier (AQ) submodule, dead band (DB) submodule, event trigger (ET) submodule, PWM chopper (PC) submodule, trip zone (TZ) submodule. Each function submodule has its own function. The final complex PWM signal is generally realized by combining the relevant registers of the corresponding submodules.

For the TB submodule, the function is to determine the event sequence of the ePWM module. By setting and configuring the relevant registers, the following functions can be achieved:

(1) Setting the value of the time-base period (TBPRD) register and the time-base counter register (TBCTR) to control the PWM period;
(2) Setting the value of time-base phase (TBPHS) register to modify the count value of TBCTR register when the synchronization event occurs which can change the initial position of the digitized carrier and generate different digitized carrier;
(3) Setting the value of time-base control (TBCTL) register to select the operating mode of TBCTR. There are up-count mode, down-count mode and up-down-count mode, which can define the shape of the digitized carrier, that is, saw-tooth wave or symmetrical triangle wave.

Figure 8.8 shows the value variation of the time base counter in the above three counting modes when TBPRD = 4.

For the CC submodule, the input is the value of TBCTR register, which is continuously compared with the counter-compare A (CMPA) register and the counter-compare B (CMPB) register. When the value of TBCTR register is equal to one of the comparison values, the compare register unit generates a corresponding event which is directly output to the AQ submodule. The principle is similar to the process of modulating wave comparing with carrier during a switching cycle. In the process

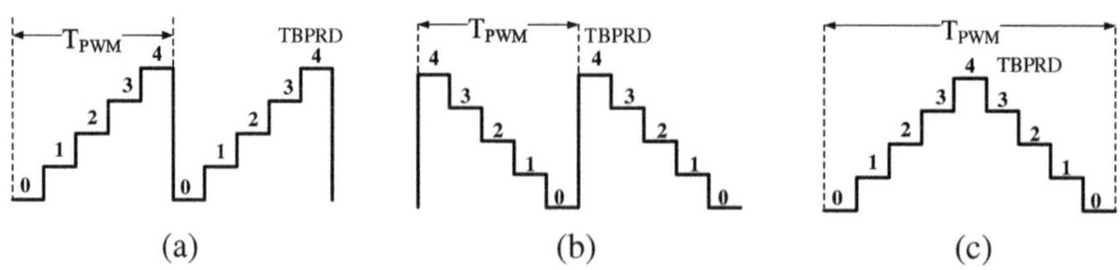

Fig. 8.8 Waveform change of the time base counter in three counting modes: **a** up mode, **b** down mode, **c** up/down mode

of counting comparison, there are four kinds of events that can generate trigger pulse output to the AQ submodule to control the PWM action:

(1) CTR = CMPA: time-base counter equal to the active counter-compare A value;
(2) CTR = CMPB: time-base counter equal to the active counter-compare B value;
(3) CTR = PRD: time-base counter equal to the active period;
(4) CTR = Zero: time-base counter equal to zero.

For the AQ submodule, the input is the trigger pulse of corresponding events for the CC submodule. According to the configuration of the relevant registers, the trigger event can be set to change the states of the two signals (ePWMxA and ePWMxB) in the PWM module, thereby realizing different conversion actions, that is, PWM rising and falling edges. The specific operations that can be implemented by configuring the relevant registers are:

(1) Set high: set output ePWMxA or ePWMxB to a high level;
(2) Clear low: set output ePWMxA or ePWMxB to a low level;
(3) Toggle: if ePWMxA or ePWMxB is currently pulled high, then pull the output low; if ePWMxA or ePWMxB is currently pulled low, then pull the output high;
(4) Do Nothing: keep outputs ePWMxA and ePWMxB at the same level as currently set.

The two PWM signals of ePWMxA and ePWMxB output by the AQ submodule are generally used to drive the upper and lower switching devices of one converter phase-leg. Thus, deadtime must be inserted between the two PWM signals to prevent the phase-leg shoot-through. The DB submodule can insert the deadtime between the two PWM signals of ePWMxA and ePWMxB. The commonly used dead band modes are:

(1) Rising edge delayed (RED);
(2) Falling edge delayed (FED);
(3) Active high complementary (AHC);
(4) Active low complementary (ALC);

The effect of the dead band control on the PWM waveform under different modes is shown in Fig. 8.9.

The ET submodule is an important part of the PWM control. The DSP can generate a function signal through the ET submodule to trigger an interruption. In the ET submodule, different events can be set as the trigger flags of the function signals. The commonly used events are:

Time-base counter equal to zero (TBCTR = 0 × 0000);
Time-base counter equal to period (TBCTR = TBPRD);
Time-base counter equal to the compare A register when the timer is increasing or decreasing (TBCTR = CMPA);
Time-base counter equal to the compare B register when the timer is increasing or decreasing (TBCTR = CMPB).

Fig. 8.9 Dead band waveforms in typical cases

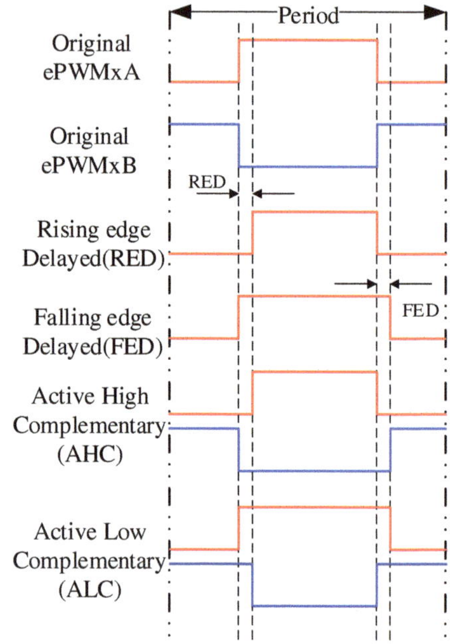

Through the above introduction of each functional submodules, the principle of PWM generation in DSP can be summarized as shown in Fig. 8.10. First, the TB submodule should be configured to set the switching period and carrier mode. At the same time, two kinds of trigger events can be set, that is, the value of the TBCTR register is equal to zero or the value of the TBPRD register. Then the values of TBCTR register and counter-compare register are compared in real time. When the values of two registers are equal, a trigger event is set, and the trigger pulse is input to the AQ submodule, and the effect of controlling PWM state is achieved by setting the relevant switching action. If the PWM signals need to insert the deadtime, the relevant signal can be selected to generate various types of PWM signals with deadtime through the

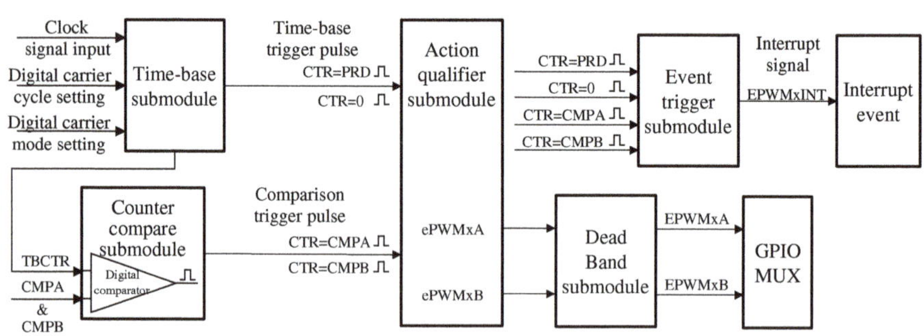

Fig. 8.10 Flowchart of PWM generation

Table 8.1 Comparison of two floating-point DSPs

Performance	TMS320F28335	TMS320F28377
CPU frequency	150 MHz	200 MHz
PWM	18 channels	24 channels
AD	12 bits	16 bits
DA	none	3 channels

DB submodule, and finally output through the GPIO pins. Finally, the trigger signal can be input into the ET submodule to set the interrupt event. Thus, the interrupt signal can be generated, and the interrupt function is executed cyclically to achieve the control effect.

It is also worth mentioning that since TMS320F28335 coming out in 2007, TI has introduced a variety of new products. The introduction of a new DSP-TMS320F28377 is expected to change this situation. Compared with TMS320F28335, the performance of this DSP has been greatly improved in all aspects. As shown in Table 8.1, the main frequency is increased by 1/3 and the calculation speed is increased; the PWM is increased to 24 channels, which can independently drive a three-level three-phase back-to-back converter; the AD accuracy is improved from 12 to 16 bits, so the closed-loop control accuracy is improved; and the increased three-channel DA output can be effectively applied to external output and system identification. TI also develop the corresponding control card for this DSP. So the new type of DSP can be quickly applied in the actual system, including the application of advanced PWM.

8.3 Modified PWM Realization in Hardware

The general PWM methods usually adopt common triangular carrier, which have the characteristics of pulse symmetric distribution and fixed switching frequency. Taking three-phase SVPWM as an example, the waveform is shown in Fig. 8.11.

In fact, the freedoms of PWM are not fully utilized. If the effect of freedom can be derived from the control target performance, the corresponding control law can be designed and the control effect of PWM can be fully exerted.

8.3.1 Modified PWM Realization: Variable Switching Frequency PWM

For general three-phase variable switching frequency PWM, the simplified principle is shown in Fig. 8.12.

From the previous analysis, the frequency control of PWM is mainly to control the carrier period of PWM. In the DSP, the digital carrier can be flexibly adjusted by

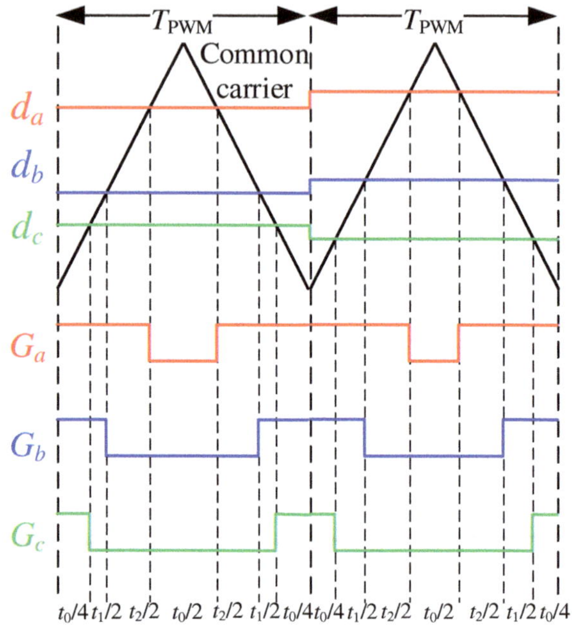

Fig. 8.11 PWM waveforms of symmetric SVPWM scheme with fixed switching frequency

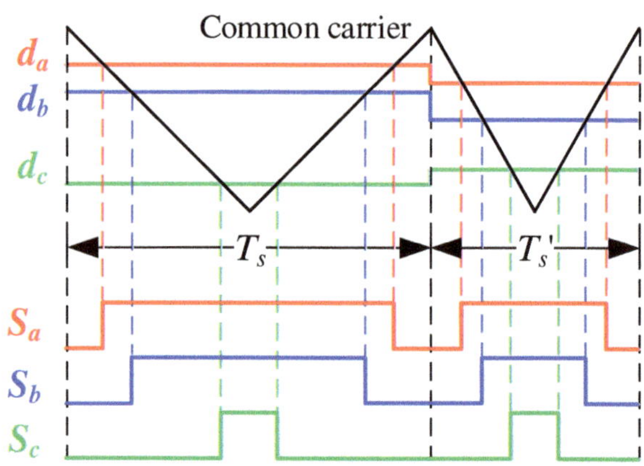

Fig. 8.12 Principle of PWM with variable switching frequency

changing the clock counting period. Therefore, it is simple to implement the variable switching frequency PWM in the DSP. The TBPRD register in the TB submodule is the control register that adjusts the digitized carrier period. If the value of TBPRD register is updated in real time during the execution of interruption service routines, PWM with different switching frequencies can be realized. According to the above principle, the related registers can be set to obtain the three-phase PWM waveforms with the variable switching frequency, as shown in Fig. 8.13.

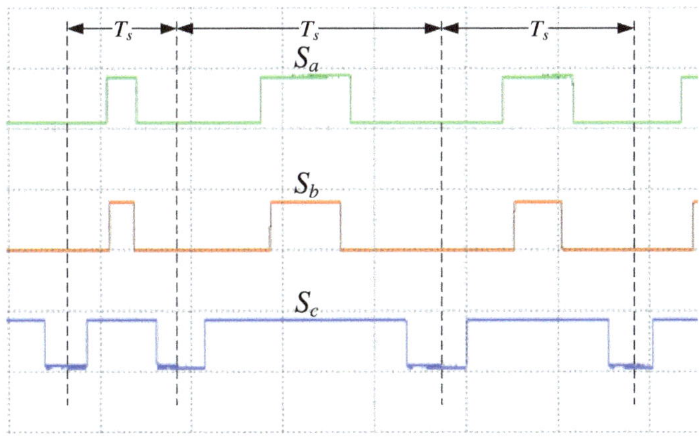

Fig. 8.13 PWM waveforms with variable switching frequency

8.3.2 Modified PWM Realization: Phase-Shift PWM

Phase-shift PWM actually uses different carriers when PWM needs to change the vector combination or vector sending sequence of the converter in one switching cycle. For a three-phase inverter, one phase carrier is generally kept unchanged, and the other two-phases' carriers can be changed to achieve some specific effects, such as common-mode voltage suppression, EMI peak suppression, and so on. The carrier phase-shift PWM is generally divided into two types: PWM with fixed carrier phase-shift and real-time carrier phase-shift. The simplified principle is shown in Fig. 8.14.

For fixed carrier phase-shift PWM, the DSP provides the related configuration register, which is the time-base phase (TBPHS) register in the TB submodule. Taking Fig. 8.14a as an example, the phase *A* carrier is a center-symmetric triangular carrier, and the phase *B* and *C* carriers are fixed phase-shifted carriers, so phase

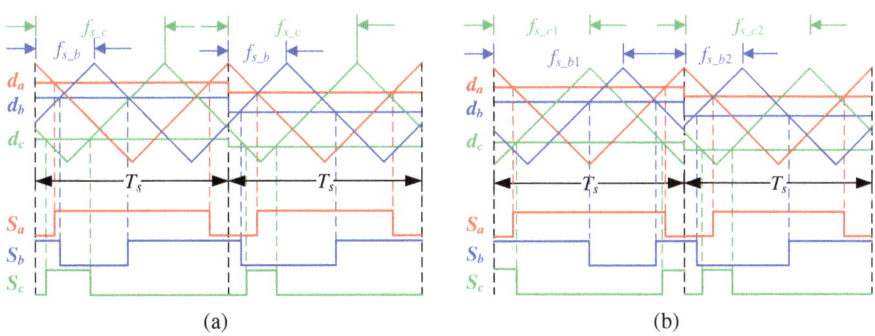

Fig. 8.14 Three-phase inverter carrier phase-shift principle: **a** fixed carrier phase-shift, **b** real-time carrier phase-shift

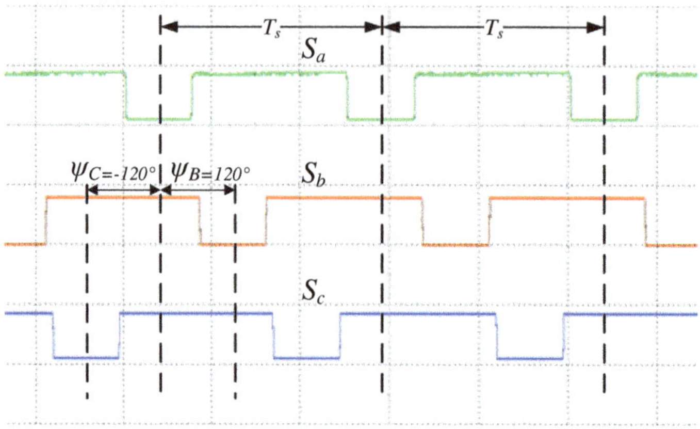

Fig. 8.15 Three-phase phase-shifted PWM waveforms with fixed phase difference of 120°

B and *C* PWMs have corresponding displacement which are no longer with centrally symmetric PWM. Assuming that phase *B* and *C* carriers are phase-shifted by 120° and 240° with reference to the *A*-phase carrier respectively, the related registers can be set to obtain the three-phase phase-shifted PWM with fixed phase difference of 120° as shown in Fig. 8.15.

For phase-shift PWM with real-time variation, the TBPHS register provided by DSP is limited in use, especially in the up-down-count mode. This is owing to the fact that the value relationship between the TBPHS register and counter-compare register needs to be discussed which adds the difficulty to calculate the PWM pulse position and deduce the related action configuration with real-time carrier phase-shift. In fact, the purpose of carrier phase-shift is to change the position of the rising and falling edges of PWM pulse. If the required PWM pulse edge positions can be determined, the carrier phase-shift function can be equivalently realized.

For phase-shift PWM with real-time variation, the up-count mode is recommended. The value of TBCTR register in the up-count mode is monotonically increasing in one switching cycle, which ensures that the value of real-time phase-shift cannot be bigger than the set value of the TBPRD register, thereby avoiding cumbersome logical judgments. At the same time, the PWM asymmetry caused by phase-shift is also easy to implement in the up-count mode, which only needs to modify the corresponding comparison value. Figure 8.16 illustrates the PWM implementation principle of real-time phase-shift with unified saw-tooth carrier in up-count mode.

According to the above method, the random phase-shift function for three-phase PWM can be realized, as shown in Fig. 8.17.

Though the use of up-count mode and up-down-count mode has the same effect of PWM generation, the demands for registers are different. In the up-count mode, the PWM rising and falling edge control need to use two counter-compare registers (CMPA and CMPB) at the same time. Therefore, the DB submodule must be used

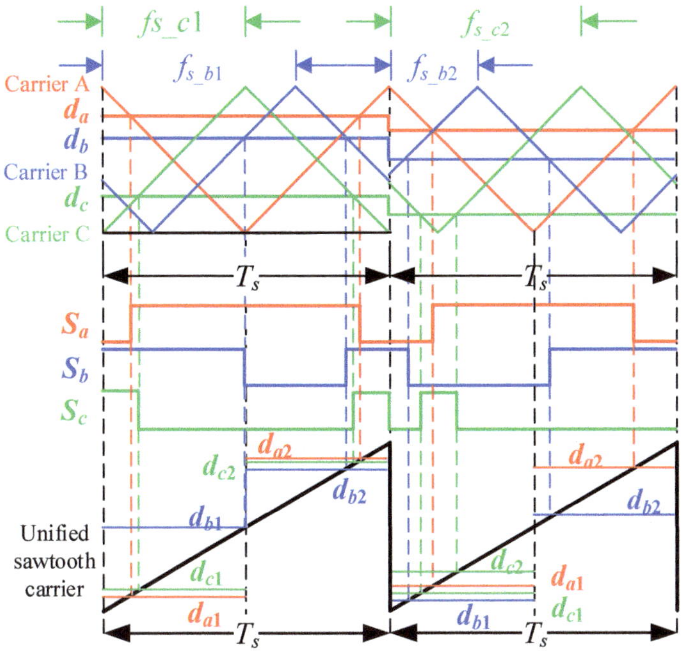

Fig. 8.16 PWM implementation principle of real-time phase-shift with unified saw-tooth carrier in up-count mode

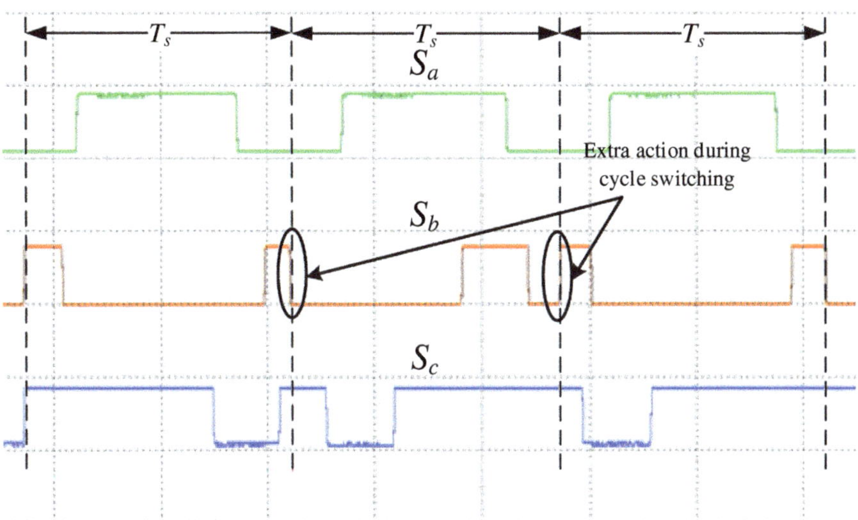

Fig. 8.17 Random phase-shift PWM waveform

to provide the function of inserting deadtime in two PWM signals of one phase-leg. While in the up-down-count mode, the upper and lower symmetrical PWM signals in one phase-leg can be generated by a comparison value, so the deadtime can be inserted with software configuration. Thus, considering the fixed phase-shift case, the configuration of TBPHS register is simple; while in the case of real-time phase-shift, the up-count mode with PWM switching edge calculation is more flexible.

8.3.3 Modified PWM Realization: Asymmetric PWM Realization with Different Comparison Values in Single Switching Cycle [5]

For the asymmetric PWM, though it can be generated in the up-count mode mentioned above, the generation manner with up-down-count mode is easier to understand, especially when the sequence of switching edges are unchanged in each switching cycle. In this case, it is necessary to use two comparison registers CMPA and CMPB to generate the asymmetric PWM. Therefore, to add the deadtime function, the DB submodule in the DSP must be used. Figure 8.18 shows the carrier-based comparison and PWM generation principle for paralleled inverters with zero-CM modulation algorithm in sector one with triangular carrier (up-down-count mode).

It can be seen from Fig. 8.18 that the two inverters use different carriers, while the DSP hardware only has one digital carrier. If the carrier of the Inverter 1 is used as the reference digital carrier, then the PWM generation principle of Inverter 2 needs

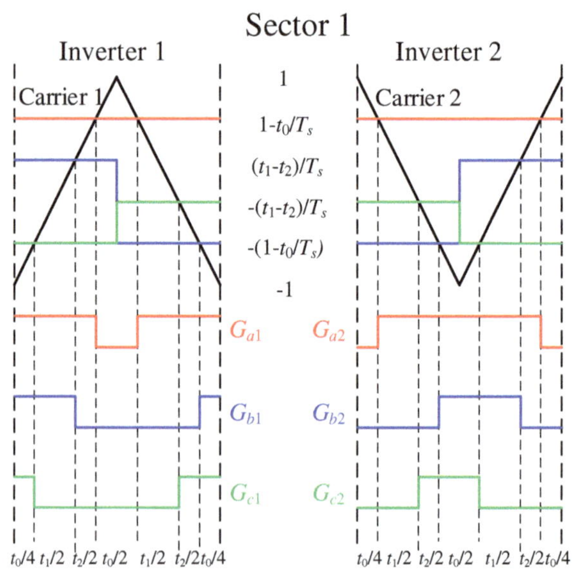

Fig. 8.18 Carrier-based comparison and PWM generation for paralleled inverters with zero-cCM modulation algorithm in sector one

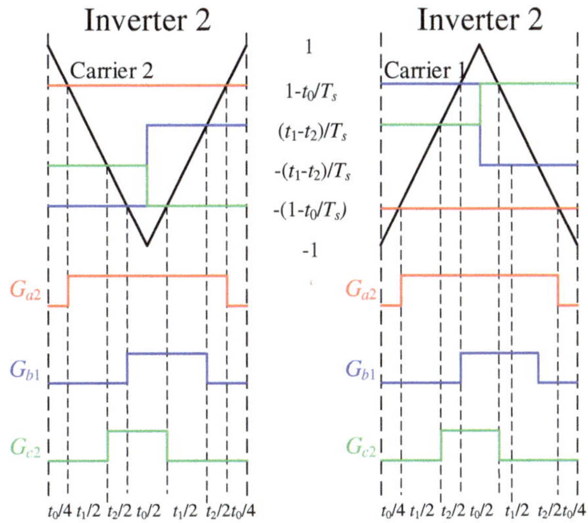

Fig. 8.19 Comparison of the original carrier and unified carrier for inverter 2

to be equivalently transformed. Figure 8.19 shows the comparison of the original carrier and unified carrier for Inverter 2 to generate the identical PWM signals. As can be seen from the Fig. 8.19, in order to obtain the same PWM waveform, the Inverter 2 should take the opposite comparison values while maintaining the same PWM switching actions.

Through the introduction of the above several special PWM implementation methods, it can be seen that the main idea of the modified PWM implementation is to combine the related function registers, such as the TBPRD register, the TBPHS register, etc., with setting counting mode and triggering event actions, which can be flexibly implemented.

8.4 Summary

This chapter focuses on the implementation method of advanced PWM technology in practical systems. First, based on the MATLAB/Simulink simulation software, this chapter describes how to simulate the PWM control method based on model prediction, including the construction of variable-switching frequency triangular wave generator and the realization of variable-frequency synchronous sampling. Later, for some problems of implementing advanced PWM in hardware, this chapter briefly introduces the hardware control system represented by digital signal processor (DSP). The TI-based TMS320F28335 DSP is used as a platform to introduce the register-based PWM generation method. Based on this, the principle of advanced PWMs such as variable switching frequency PWM, carrier phase-shift PWM and zero-CM PWM for paralleled inverters is introduced. According to "Moore's Law", microprocessor chips will also be developed toward miniaturization and high performance, and the space for hardware implementation of advanced PWM will be further expanded, which will provide a better application platform for the various methods described in this book.

References

1. https://www.mathworks.com/
2. Qian S (2001) Introduction to time-frequency and wavelet transforms. Prentice Hall
3. https://www.ti.com/
4. Texas instrument, TMS320X2833X, 2832X enhanced pulse width modulator (ePWM) module—reference guide (2008)
5. Jiang D, Shen Z, Wang F (2017) Common-mode voltage reduction for paralleled inverters. IEEE Trans Power Electron 33(5):3961–3974

GPSR Compliance
The European Union's (EU) General Product Safety Regulation (GPSR) is a set of rules that requires consumer products to be safe and our obligations to ensure this.

If you have any concerns about our products, you can contact us on

ProductSafety@springernature.com

In case Publisher is established outside the EU, the EU authorized representative is:

Springer Nature Customer Service Center GmbH
Europaplatz 3
69115 Heidelberg, Germany

www.ingramcontent.com/pod-product-compliance
Ingram Content Group UK Ltd.
Pitfield, Milton Keynes, MK11 3LW, UK
UKHW021315130926
15127UKWH00002B/8

* 9 7 8 9 8 1 3 3 4 3 8 4 9 *